# Diffractive Optics and Optical Microsystems

# Diffractive Optics and Optical Microsystems

Edited by

## S. Martellucci
*The University of Rome "Tor Vergata"*
*Rome, Italy*

and

## A. N. Chester
*Hughes Research Laboratories, Inc.*
*Malibu, California*

Plenum Press • New York and London

Library of Congress Cataloging-in-Publication Data

```
Diffractive optics and optical microsystems / edited by S. Martellucci
  and A.N. Chester.
      p.   cm.
    Proceedings of the 20th Course of the International School of
  Quantum Electronics on Diffractive Optics and Optical Microsystems,
  held Nov. 14-24, 1996, Erice, Italy.
    Includes bibliographical references and index.
    ISBN 0-306-45770-9
    1. Electrooptical devices--Congresses.  2. Integrated optics-
  -Congresses.  3. Electromechanical devices--Congresses.  4. Optical
  fiber detectors--Congresses.  5. Diffraction--Congresses.
    I. Martellucci, S.  II. Chester, A. N.  III. 20th Course of the
  International School of Quantum Electronics on Diffractive Optics
  and Optical Microsystems (1996 : Erice, Italy)
  TA1750.D53  1997
  621.36--dc21                                               97-40564
                                                                  CIP
```

Proceedings of the 20th Course of the International School of Quantum Electronics on Diffractive Optics and Optical Microsystems, held November 14–24, 1996, in Erice, Sicily, Italy

ISBN 0-306-45770-9

© 1997 Plenum Press, New York
A Division of Plenum Publishing Corporation
233 Spring Street, New York, N.Y. 10013

http://www.plenum.com

10 9 8 7 6 5 4 3 2 1

All rights reserved

No part of this book may be reproduced, stored in a retrieval system, or transmitted in any form or by any means, electronic, mechanical, photocopying, microfilming, recording, or otherwise, without written permission from the Publisher

Printed in the United States of America

# PREFACE

There is a consistent trend towards miniaturization of devices and systems in many fields of engineering, in order to achieve significant reductions in size, weight, power consumption and cost. This trend is especially evident in optics and optoelectronics, where recent years have seen rapid growth in such new or renewed areas as microoptics, integrated optics, integrated optoelectronics, and diffractive optics.

In November 1996, an international group of scientists convened in Erice, Sicily, for a meeting on the subject of "Diffractive Optics and Optical Microsystems." This Conference was the 20th Course of the International School of Quantum Electronics, under the auspices of the "Ettore Majorana Center for Scientific Culture" and was directed by Prof. Franco Gori of the Third University of Rome, Italy, and Prof. Giancarlo Righini of the "Nello Carrara" Institute of Research on Electromagnetic Waves (IROE-CNR) in Florence, Italy. This book presents the Proceedings of this Conference, providing a fundamental introduction to the topic as well as reports on recent research results.

The aim of the Conference was to bring together some of the world's acknowledged scientists who have as a common link the use of optoelectronics instrumentation, techniques and procedures related to the fields of diffractive optics and optical microsystems. Most of the lecturers attended all the lectures and devoted their spare hours to stimulating discussions. We would like to thank them all for their admirable contributions. The Conference also took advantage of a very active audience; most of the participants were active researchers in the field and contributed with discussions and seminars. Some of these seminars are also included in these Proceedings.

The Conference was an important opportunity to discuss the latest developments and emerging perspectives on the use of optoelectronic techniques for diffractive optics and optical microsystems.

The Chapters in these Proceedings are not ordered exactly according to the chronology of the Conference but they give a fairly complete accounting of the Conference lectures with the exception of the informal panel discussions. The contributions presented at the Conference are written as extended, review-like papers to provide a broad and representative coverage of the fields of diffractive optics and optical microsystems. We did not modify the original manuscripts in editing this book, except to assist in uniformity of style; but we did group them according to the following Sections:
• "Theory and Design" begins with tutorial treatments of diffractive optics theory and design by Profs. Gori and Herzig, then seven additional chapters treat refractive (graded index) structures, holographic optical elements, nonlinear effects, and spherical microparticles;
• "Materials and Processes" provides a detailed discussion of the materials and fabrication technologies of diffractive optical elements in eight chapters;
• "Components and MEMS" is a series of seven chapters treating optical components,

including binary optics, as well as micro-electro-mechanical systems (MEMS);
• "Fiber Sensors" presents three chapters which provide an overview of the broad variety of fiber-based sensors and their applications;
• "Measurements" offers chapters addressing two important areas for microoptics: micromechanical measurements, and the use of Cerenkov radiation to characterize optical waveguides; and, finally,
• "Microsystems and Applications" offers seven chapters which survey a variety of the many practical applications of diffractive optics and microoptics, including chemical detection, optical measurement, spectral analysis, manipulation of microparticles, velocimetry, and medicine.

These Proceedings update and augment the material contained in the previous ISQE volumes, "Integrated Optics: Physics and Applications," S. Martellucci and A.N. Chester, Eds., NATO ASI Series B, Vol. 91 (Plenum, 1983), and "Advances in Integrated Optics," S. Martellucci, A.N. Chester and M. Bertolotti, Eds. (Plenum, 1994). For some closely related technology, to the topical Section devoted to "Fiber Sensors," the reader may also wish to consult the ISQE volume, "Optical Fiber Sensors," A.N. Chester, S. Martellucci and A.M. Scheggi, Eds., NATO ASI Series E, Vol. 132 (Nijhof, 1987).

We are grateful to Profs. Gori and Righini for their able organization and direction of the Course, to our editor at Plenum Press London, Joanna Lawrence, for outstanding professional support. We also greatly appreciate the expert help from our assistants Carol Harris and Margaret Hayashi, and the support of Eugenio Chiarati for much of the computer processing work. This International School was held under the auspices of the "Ettore Majorana" Center for Scientific Culture, Erice, Italy. We acknowledge with gratitude the cooperation of the Quantum Elecronics and Plasma Physics Research Group of the Italian Research Council (GNEQP – CNR) and support from the Italian Ministry of Education, the Italian Ministry of University and Scientific Research, and the Sicilian Regional Government.

Sergio Martellucci
Professor of Physics
University of Rome "Tor Vergata"
Rome (Italy)

Arthur N. Chester
Chairman and President
Hughes Research Laboratories, Inc.
Malibu, California (USA)

CONTENTS

**I. Theory and Design**

Diffractive Optics: An Introduction
F. Gori ............................................................................................. 3

Design of Refractive and Diffractive Micro-Optics
H.P. Herzig ....................................................................................... 23

GRIN Planar Structures for Focusing, Collimation and Beam Size Control
C. Gómez-Reino, C. Bao, and M. V. Pérez ........................................... 35

Numerical Analysis of Surface Relief Gratings
R. Orta, S. Bastonero, and R. Tascone ................................................ 47

The New Possibility to Produce Holographic Optical Elements
Using the Lowest Cost Laser Diodes
M.I. Nemenov, J.S. Utochkina, and S.V. Utochkin ................................ 57

Diffraction with Second-Harmonic Generation
for the Formation of Self-Guided or "Solitary" Waves
G. Assanto ........................................................................................ 65

Light Scattering by a Dielectric Cylinder Near a Flat Substrate
R. Borghi, M. Santarsiero, F. Frezza, and G. Schettini ........................ 75

Steady States in Spherical Microparticles and Their Instabilities
L.A. Kotomtseva and G. P. Lednyeva ................................................. 83

Modelling of a Microlaser Based on a Spherical Microparticle
L.A. Kotomtseva and G. P. Lednyeva ................................................. 91

**II. Materials and Processes**

Microfabrication Technologies for Integrated Optical Devices
G.C. Righini and M.A. Forastiere ...................................................... 103

Fabrication of Diffractive Optics: Surface Reliefs and Artificial Dielectrics
C. Arnone, C. Giaconia, and G. Lullo ................................................ 119

Low Cost High Quality Fabrication Methods and CAD
for Diffractive Optics and Computer Holograms Compatible
with Micro-Electronics and Micro-Mechanics Fabrication
*S. H. Lee and W. Däschner* ............................................................... 133

Design and Fabrication Aspects of Continuous-Relief Diffractive Optical Elements
*T. Hessler and M. Rossi* ................................................................... 139

Fabrication of Diffractive Optical Elements by Electron Beam Lithography
*E. Di Fabrizio, L. Grella, M. Baciocchi, and M. Gentili* ......................... 149

Pulsed Laser Deposition: Perspectives as a Micro-Optics Fabrication Technique
*S. Martellucci, M. Richetta, A. Tebano, and A. Spena* .......................... 161

Fabrication of Thin-Film Microlens Arrays by Mask-Shaded Vacuum Deposition
*R. Grunwald, S. Woggon, and R. Ehlert* .............................................. 169

New Photoluminescent Materials Based on LiF:NaF Microstructures
*G. Baldacchini, E. De Nicola, R.M. Montereali, M. Cremona,
M. Passacantando, and F. Somma* ...................................................... 179

## III. Components and MEMS

Advances in Optical Microsystems Combining Microtechnologies
and Batch Processing Fabrication
*S. Valette* ..................................................................................... 189

Actuation Mechanisms for Micromechanics
*E. M. Yeatman* ............................................................................... 209

Design of Computer Generated Binary Holograms
for Free Space Optical Interconnections
*I. Montrosset, D. Cojoc, and F. Sartori* .............................................. 223

Holographic Diffractive Components for Beam Coupling
*M. Miler* ...................................................................................... 239

A 16 level $CO_2$ Laser Beam Shaper : Design and Fabrication
*P. Antuofermo, A. Cacucci, A. M. Losacco, and O. De Pascale* ................ 251

Microoptics for Chromatic Control of Extended Polychromatic Sources:
Design Methodologies and Technologies to Implement Them Over Large Areas
*P. Perlo, C. Biglianti, V. Lambertini, P.M. Repetto, and S. Sinesi* .......... 259

An Unconventional Optical Element
for Splitting and Focusing High Power Laser Beams
*V. Russo, G. De Angelis, and A. Scaglione* ......................................... 279

## IV. Fiber Sensors

Fiber Bragg Gratings as Temperature and Strain Sensors
*R. Falciai, R. Fontana, A. Schena, and A.M. Scheggi* ........................... 293

Radially Gradient-Index Lenses: Applications to Fiber Optic Sensors
*A.G. Mignani, A. Mencaglia, M. Brenci, and A.M. Scheggi* .................................. 311

Integrated Optical Instrumentation for Fiber Grating Sensors
*M. Varasi* ................................................................................................. 327

## V. Measurements

Micromechanics: New Challenges for Optical Measurements
*K. Patorski and M. Kujawinska* ................................................................ 341

Waveguides in $LiNbO_3$ for Optical Sensors: Characterisation by Cerenkov Effect
*R. Ramponi* ............................................................................................ 353

## VI. Microsystems and Applications

Microoptical Components and Systems Fabricated by the LIGA Process
*C. Müller* ............................................................................................... 365

Sensors and Microsystems: Electronic Nose
*C. Di Natale, A. D'Amico, F. Davide, and G. Saggio* ........................................ 371

Advances in the Development of Optical Microsystems at DEA- OptoLab:
Integrated Optical Measuring Devices and Diffractive Optical Components
for Industrial Applications
*F. Docchio and U. Minoni* ...................................................................... 381

Micro-optics for Micro Total Analysis Systems
*A.E. Bruno, B. Krattiger, S. Barnard, M. Ehrat, R. Völkel,
Ph. Nussbaum, H. P. Herzig, and R. Dändliker* .......................................... 387

Optical Tweezers: Laser Manipulation of Microparticles
*G.C. Righini* .......................................................................................... 391

Laser Time-of-Flight Velocimetry: Proposals for Miniaturisation
*H. Imam, B. Rose, S. G. Hanson, and L. Lading* ......................................... 399

A Robotic Microsystem for Colon Visualization and Sampling
*L. Lencioni, P. Dario, M.C. Carrozza, B. Magnani, and S. D'Attanasio* ..................... 411

Index ........................................................................................................ 417

**THEORY AND DESIGN**

# DIFFRACTIVE OPTICS: AN INTRODUCTION

F. Gori
"E. Amaldi" Physics Department
The Third University of Rome
Via della Vasca Navale 84, 00146 Rome, Italy

## 1. FOREWORD

Today diffractive optics is a flourishing field of research. Conference reports (see, e. g. Ref. 1) and special issues of optical journals [2,3] keep on appearing, and authoritative assessments of the present state of the art and future prospects are available.[4,5] There are several reasons for this state of affairs. Diffractive optical elements (or DOE's, for short) a) allow us to manipulate light fields in ways that would be unthinkable with traditional refractive and reflective optics; b) can be cheap and lightweight; c) stimulate in a challenging way the researchers in the design process.

A DOE changes a wavefield by means of diffraction. Basically, DOE's of current interest are designed for use with coherent light. In spite of widely used phrases such as "three-dimensional images" and the like, the information conveyed by a coherent wavefront is inherently two-dimensional. Strange as it may sound at first, this statement is supported by the well-known fact that if the values of a field impinging on a plane are specified all across the plane then the field can be evaluated throughout the half space where the wave propagates. This is in fact the key for producing any prescribed wavefront by diffraction and lies at the core of the holographic method of wavefront reconstruction. In this sense, it can be said that the first diffractive optical elements are half a century old, since Gabor realised his first holograms at the end of the forties. This could suggest that we are simply dealing with holograms, possibly generated by computer. Actually, there is more. From the practical point of view, one wants to reduce energy losses. Consequently, ordinary holograms based on absorptive photographic emulsions are of modest interest. Instead, one would prefer that DOE's only act on the phase of the incident field. This practical requirement has important consequences. On the one hand, it leads to simple replication techniques that make feasible low cost mass production of DOE's. On the other hand, the need for reaching a prescribed goal of acting on the phase only makes design problems of diffractive optics far from trivial.

As an introduction to a course on diffractive optics, the present Chapter is addressed to the newcomer. As such it will make use of very simple concepts. Most of our discussion will rely on one-dimensional gratings and similar devices. The aim is to present some basic issues in the analysis and design of DOE's. The reader will find more advanced treatments of both

## 2. PRELIMINARIES

Before discussing diffractive optics let us recall a few basic facts from diffraction theory.[6,7] We shall start with diffraction produced by plane screens. Suppose a certain monochromatic wavefield, a plane wave, say, is incident on a thin diffracting structure such as a slide transparency with known optical properties. We want to find the field in the half space, z > 0 beyond the screen. Disregarding the thickness of the screen, we can divide the diffraction problem into two sub problems: a) Interaction problem. This means finding the field at $z = 0^+$, knowing it at $z = 0^-$. b) Propagation problem. This means finding the field across a typical plane z = const > 0 from the knowledge of it at $z = 0^+$. The difficulty lies in problem a), whereas problem b) can always be solved rigorously.

In many cases of practical interest the following simplified approach gives good results. First, we ignore the vectorial character of the field by describing the wave through a single complex scalar function V(**r**) where **r** is a position vector, leaving aside the time dependence, which is accounted for by the usual function exp(-i ω t). In a homogeneous region the scalar field V satisfies the Helmholtz equation

$$\nabla^2 V + k^2 V = 0 \tag{1}$$

where $k = 2\pi / \lambda$ is the wave number (λ being the wavelength).

Second, we assume that the screen can be characterised by a transmission function defined as

$$\tau(x, y) = \frac{V(x, y, 0^+)}{V(x, y, 0^-)} . \tag{2}$$

It is further assumed that τ(x, y) can be estimated "a priori" by knowledge of the structure of the screen. For example, if the screen is an opaque mask with some holes, τ is taken to be unity inside the holes and zero otherwise. Subtleties such as reirradiation from the boundaries of the holes are disregarded. If the transmission function is known the interaction problem is solved at once by means of Eq. (2).

Such an idea is so appealing that one may well ask why it could not be made rigorous. One might argue that the ratio of Eq. (2) could be evaluated rigorously, or perhaps measured, once and for all. We could then legitimately use the function τ to characterise the screen. Unfortunately, there is no such thing as a transmission function of the screen independently from the incident field. As rigorous solutions show, the value of the ratio (2) at a certain point (x, y) actually depends on the incident field. In other words, there are infinitely many transmission functions for a given screen and we are back to the interaction problem again. In spite of this criticism, the use of the transmission function as a unique and independent characteristic of the screen leads to correct predictions if the details of the diffracting structure have linear dimensions that are rather large with respect to the wavelength. We shall adopt this picture for most of our discussion. We shall see later some cases where one is forced to abandon it.

As for the propagation problem, there are no fundamental difficulties. Standard tools such as plane wave expansion or the (equivalent) Rayleigh-Sommerfeld formula can be used for a rigorous solution.[6,7]

## 3. BEAM DIVIDERS

DOE's that divide an input beam into two or more copies are among the simplest and most useful examples in diffractive optics. We shall begin our discussion about such beam dividers (or beam splitters) with the elementary problem of dividing a beam into two parts with equal power by using a diffracting structure. We could of course solve the problem by using an ordinary transmitting-reflecting beam splitter (a half-silvered mirror), but we are interested in the diffractive case for clarifying certain basic points. A first schematization of the problem is as follows. A plane wave impinges, say orthogonally, on a plane screen at z = 0. At the exit we want two plane waves of equal amplitude symmetrically directed with respect to the z-axis. What is the transmission function of the screen to be used? The answer is easily found because the two plane waves produce a sinusoidal pattern at z = $0^+$. Hence the incident field having a uniform distribution V = A (a constant) has to be transformed into a sinusoid. Accordingly, the required transmission function is of the form

$$\tau(x) = \sin\left(\frac{2\pi}{P}x + \alpha\right), \qquad (3)$$

where $2\alpha$ is the phase difference between the interfering waves at x = 0, P is the period of the structure and the wave vectors of the output plane waves are assumed to lie on the xz plane. In the following we shall set $\alpha$ = 0 by using a suitable shift of the origin of the x-axis. In conclusion, the diffractive element is a sinusoidal grating affecting both the amplitude and the phase of the incident wave (note that the phase is $\pi$ where $\tau$ < 0). The directions of the output waves are found by the grating formula

$$\sin\vartheta_{\pm 1} = \pm\frac{\lambda}{P}, \qquad (4)$$

where $\vartheta_{\pm 1}$ are the angles between the z-axis and the directions of the diffracted waves (orders +1 and -1 of diffraction).

This is very simple indeed but has a drawback. Since the spatial average of $\tau^2$ is 0.5, only 50% of the incident power is found at the output. The remaining power is absorbed. For most practical uses this is unacceptable because: a) one does not want to waste too much power; and b) absorption causes detrimental heating of the element. On the other hand, if the problem is exactly in the form specified before there is nothing we can do. In particular, we cannot solve our problem by a pure phase screen. Eq. (4) is the only solution. Here we realise that even a simple synthesis problem such as the present one can be unsolvable by phase objects. This is because the starting requirements are too tight, so to speak. Luckily they can generally be relaxed. Let us see how this works for the case at hand. Generally, the incident field is a beam (most often a laser beam), not a plane wave. Therefore, at a certain distance from the plane z = 0, the diffracted beams become spatially separated. Consequently we can accept that more than two diffracted orders are produced by the screen because the ones different from ±1 can be eliminated through stops. Taking this into account we can try to transform the transmission function of Eq. (3) into one acting only on the phase of the incident field. To do this, we keep only the phase of the function (3) and define a new transmission function

$$\tau(x) = \text{sgn}\left[\sin(\beta x)\right], \qquad (5)$$

where $\beta = 2\pi/P$ and sgn denotes the signum function. The transparent element defined by Eq. (5) gives infinitely many orders (including evanescent orders). Indeed, it can be written as the Fourier series

$$\tau(x) = \sum_{m=-\infty}^{\infty} \tau_m e^{2\pi i m \frac{x}{P}}, \qquad (6)$$

where the Fourier coefficients, which are easily shown to be

$$\tau_m = \begin{cases} \dfrac{2}{\pi i m}, & m \text{ odd}, \\ 0, & m \text{ even and } m = 0, \end{cases} \qquad (7)$$

are proportional to the amplitudes of the diffracted waves. Now, what about the fraction of the incident power that is transferred to the first order beams?

On defining the efficiency $\eta$ as the ratio between the power of the desired orders and the overall power we easily obtain

$$\eta = 2|\tau_1|^2 = \frac{8}{\pi^2} \cong 81\%, \qquad (8)$$

where the Parceval theorem

$$\frac{1}{P} \int_{-P/2}^{P/2} |\tau(x)|^2 dx = \sum_{m=-\infty}^{\infty} |\tau_m|^2 = 1, \qquad (9)$$

has been used. As seen from Eq. (8) a considerable increase of efficiency is obtained with respect to the 50% efficiency pertinent to the grating of Eq. (4). In addition there is no absorption (hence no heating of the optical element) and the physical structure of the grating can be extremely simple. Since only two values (0 and $\pi$) are required for the phase function, the grating can be realised with a dielectric slab whose thickness varies periodically (and abruptly) between two suitable values. Paradoxically, we found a more efficient solution than the one afforded by Eq. (3) by increasing the number of diffracted beams. We thus learn that unwanted diffraction orders can actually play a beneficial role in diffractive optics.

Our next example is the design of a divider producing three beams of equal power. Again, we can start by considering the field that would be formed at $z = 0^+$ by the three waves alone (say the orders 0 and ±1 of a grating). Here another significant point comes out. The phase relationship among the required waves is usually of no concern and we can choose it at will. Changing the phase difference between orders 1 and -1 merely shifts their interference pattern along the x-axis (see Eq. (3)). Therefore, suppose the first order waves are arranged to produce a $\sin(\beta x)$ distribution. Further, let $\alpha$-$\pi/2$ be the phase difference between orders +1 and 0. The resulting field distribution is of the form

$$V(x,0^+) = A\left(1 + 2e^{i\alpha} \sin \beta x\right). \qquad (10)$$

Since we are looking for a phase device, we can exploit our freedom in choosing □ so as to minimise the intensity variations of the field (10) as a function of x. It is easily seen that this is obtained by choosing $\alpha = \pm\pi/2$. The intensity profiles corresponding to $\alpha = 0$ and $\alpha = \pi/2$ are shown in Fig. 1, where we let $A = 1$.

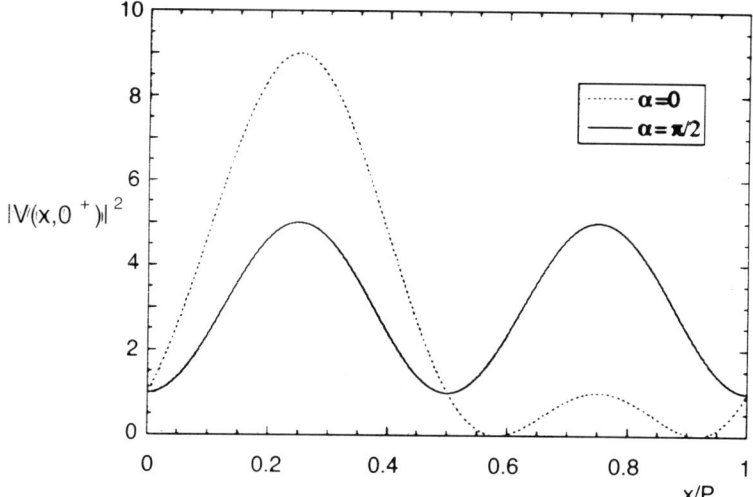

Fig. 1 Squared modulus of the function $1 + 2\exp(i\alpha)\sin(2\pi x/P)$ for two values of $\alpha$.

As a first step, we take a transmission function proportional to the field distribution (10) with $\alpha = \pi/2$

$$\tau(x) = \frac{1}{\sqrt{5}}\left[1 + 2i\sin 2\pi\left(\frac{x}{P}\right)\right]. \tag{11}$$

This describes an (amplitude and phase) object that produces only three beams, with equal power, when illuminated by an input beam. The squared modulus has a mean value 3/5, so that the efficiency is 60%. Keeping only the phase leads to the new transmission function

$$\tau(x) = \exp\left\{i\tan^{-1}\left[2\sin\left(2\pi\frac{x}{P}\right)\right]\right\}. \tag{12}$$

This change, however, alters the Fourier coefficients and we cannot expect $\tau_0$, $\tau_1$ and $\tau_{-1}$ to have still the same modulus. A way out is to replace the factor 2 in front of the sin function by a parameter a to be determined through the condition $\tau_0 = \tau_1$. In other words, we take a phase distribution of the form

$$\phi(x) = \tan^{-1}\left[a\sin\left(2\pi\frac{x}{P}\right)\right]. \tag{13}$$

On computing the Fourier coefficients $\tau_0$, $\tau_1$ we find

$$\tau_0 = \frac{2}{\pi}K(-a^2), \qquad \tau_1 = \frac{2}{\pi a}\left[E(-a^2) - K(-a^2)\right], \tag{14}$$

where K and E denote the complete elliptic integrals of the first and second kind respectively[8]. As for $\tau_{-1}$, it turns out to equal $-\tau_1$. We next search numerically for the value of a that makes $\tau_0 = \tau_1$. Such value is a = 2.6572 and gives $\tau_0 = 0.55544$. The

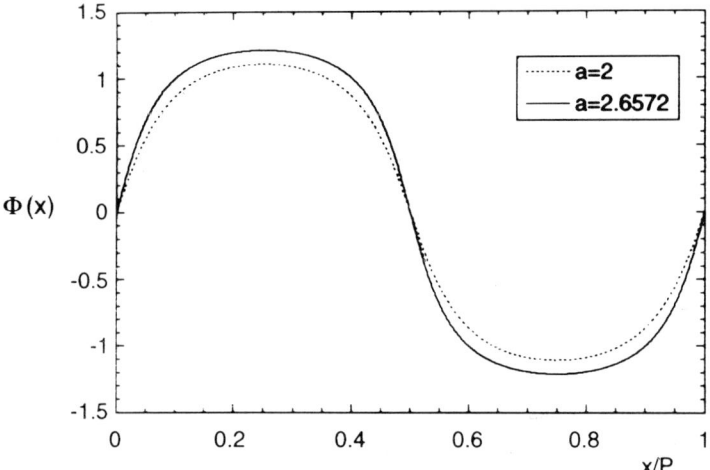

Fig. 2 The function $\tan^{-1}[a\sin(2\pi x/P)]$ for two values of $a$.

corresponding efficiency, namely $3\tau_0^2$, is 92.5%. The function $\phi(x)$ is shown in Fig. 2 for the value $a = 2$ pertaining to Eq. (12) as well as for the value $a = 2.6572$.

From a practical point of view the realisation of DOE's with continuous phase profiles is more demanding than the case of discontinuous phase changes. We then ask what results can be obtained on replacing the law (13) with the simpler law

$$\phi(x) = \begin{cases} \phi_M, & 0 < x \leq P/2, \\ -\phi_M, & -P/2 \leq x < P, \end{cases} \tag{15}$$

where the value of $\phi_M$ will be dictated by the condition that zero and first orders carry the same power. On computing the Fourier coefficients of the function $\exp[i\phi(x)]$ we easily obtain

$$\tau_0 = \cos(\phi_M), \qquad \tau_1 = \frac{2}{\pi}\sin(\phi_M), \qquad \tau_{-1} = -\tau_1. \tag{16}$$

The condition $\tau_0 = \tau_1$ then gives

$$\phi_M = \tan^{-1}\left(\frac{\pi}{2}\right) \cong 1. \tag{17}$$

The corresponding efficiency is

$$\eta = 3\tau_0^2 = \frac{12}{4 + \pi^2} \cong 86.5\%. \tag{18}$$

While lower than the 92.5% efficiency of the continuous case, this figure is quite remarkable when we take into account the simplicity of the present divider. It should be noted in fact that the present DOE has the same structure as the divide by two device described by Eq. (5). The only difference occurs in the phase values used in the two cases.

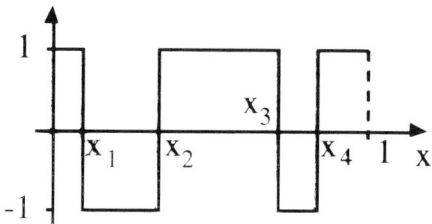

Fig. 3 Transmission function of a Dammann grating.

The problem of dividing a beam into an arbitrary number N of copies with equal power becomes more complex when N exceeds three.[9-11] A popular solution is to use Dammann gratings.[12] These are binary phase gratings in which the transmission function jumps from 1 to -1 and vice versa at, say, 2N points $x_1, x_2, ..., x_{2N}$, within a period (see Fig.3). Using suitable units we can let P = 1 and write the transmission function in the form

$$\tau(x) = 1 - 2\sum_{s=1}^{2N} (-1)^s \text{rect}\left(\frac{x}{2x_s}\right), \qquad (0 \leq x \leq 1). \tag{19}$$

The corresponding Fourier coefficients are found to be

$$\tau_0 = 1 - 2\sum_{s=1}^{2N} (-1)^s x_s, \quad \tau_m = \frac{1}{\pi i m}\sum_{s=1}^{2N} (-1)^s e^{-2\pi i m x_s}, \qquad (m = \pm 1, \pm 2, ...). \tag{20}$$

We can try to determine the values of the transition abscissas in such a way that a number of Fourier coefficients have equal modulus. For example, we can require $|\tau_0| = |\tau_{\pm 1}| = ... = |\tau_{\pm(2N-1)}|$. Let us consider the simplest case N = 1. Then, letting $x_1 = 0$, we have from Eq. (20)

$$\tau_0 = 1 - 2x_2, \quad \tau_1 = -\frac{2e^{-\pi i x_2}}{\pi}\sin(\pi x_2) = \tau^*_{-1}. \tag{21}$$

The condition $|\tau_0| = |\tau_{\pm 1}|$ leads to

$$1 - 2x_2 = \pm \frac{2}{\pi}\sin(\pi x_2). \tag{22}$$

This simple trascendental equation has two solutions: $x_2 = 0.2647$ and $x_2 = 0.7353$. Being complementary they actually correspond to the same grating. The associated efficiency is

$$\eta = 3|\tau_0|^2 = 3(1 - 2x_2)^2 \cong 66.4\%. \tag{23}$$

This is not a very high efficiency (compare with Eq. (15)) but better results are obtained for higher values of N. On the other hand, when N increases the system of trascendental equations to be solved becomesmore and more extensive. More than one solution exists and one would like to choose the one corresponding to the maximum efficiency. A common approach is to treat the problem using optimisation theory, searching for minimisation of the

Fig. 4 Phase grating with a sawtooth profile.

mean square difference among the $|\tau_m|$ and simultaneous maximisation of $\eta$. For example, the method of simulated annealing can be used.[13] As a sample of results we present a solution for the case $N = 4$:

$x_1 = 0$, $x_2 = 0.0939$, $x_3 = 0.1334$, $x_4 = 0.2364$, $x_5 = 0.3207$, $x_6 = 0.4143$, $x_7 = 0.5967$, $x_8 = 0.6885$.

The corresponding efficiency is $\eta = 83.2\%$.

A significant question can be posed. Could we find any continuous profile phase grating such that there is only a finite number of diffracted beams, possibly with the same power? In such a case the efficiency would be equal to one. Unfortunately, this is impossible. Phase gratings are rather peculiar in this respect. They have either one or an infinite number of non-zero Fourier coefficients (see Appendix). Consequently a unitary efficiency can be obtained only for a phase grating possessing a transmission function of the form $\exp(2\pi i M x)$, with arbitrary integer $M$, in each period. The cross section of such a grating has the form of a sawtooth and can be thought of as the juxtaposition of (infinitely many) tiny prisms (see Fig. 4). It is easily seen that this grating behaves as a beam deflector when a plane wave impinges on it.

The deflector case gives us the opportunity of discussing another point of interest. Since continuous phase profiles can be difficult to control experimentally, these profiles can be replaced by stepwise approximations. This of course affects the efficiency and we want to get an idea of how large is the effect of this phase quantization process. A discrete counterpart of the sawtooth deflector of Fig. 4 is depicted in Fig. 5. If $N$ is the number of phase steps, the transmission function is

$$\tau(x) = e^{is\delta\varphi}, \qquad \left(\frac{s-1}{N} < x < \frac{s}{N}, s = 1, 2, \ldots, N\right), \tag{24}$$

where $\delta\varphi = 2\pi M/N$. On evaluating the Fourier coefficients we find

$$\tau_m = e^{i\left[\frac{M(N+1)}{N} - m\right]\pi} \frac{\sin(\pi m/N)}{\pi m} \frac{\sin[\pi(M-m)]}{\pi(M-m)/N}, \qquad (m = 0, \pm 1, \ldots). \tag{25}$$

In particular, the coefficient of order $M$, which equals one for the continuous deflector, is

$$\tau_M = e^{i\pi\frac{M}{N}} \frac{\sin(\pi M/N)}{\pi M/N}. \tag{26}$$

Fig. 5 A discrete version of the sawtooth grating.

Suppose M = 1. Then, we have $|\tau_1| = .9$ for N = 4, $|\tau_1| = .974$ for N = 8 and so on. We see that with a limited number of steps a good approximation of the continuous case is obtained. Similar results hold for more general profiles.

## 4. DIFFRACTIVE ELEMENTS IN THE PARAXIAL REGIME

In the previous Section we relied on Fourier analysis of the diffracted field. The beams generated by the dividers separate from one another if we recede enough from the grating plane. In several applications, e.g. changing the transverse shape of a beam, we are interested in the field distribution in the neighbourhood of the diffractive element. In addition, the paraxial approximation can often be used. This is the regime in which the propagation of the field is studied with the help of the Fresnel transform, a mathematical tool for which several theorems hold.[14,15]

Here, we shall limit ourselves to a couple of simple points. The first one refers to the possibility of transferring to near field devices results from the previous section. There is a general procedure that can be used towards this purpose. Suppose that a transmission function of the form of Eq. (6) is given. Now replace x/P by $(r/P)^2$, where r denotes the distance from the origin in a plane. This leads to a transmission function, say t(r), with the series expansion

$$t(r) = \sum_{m=-\infty}^{\infty} \tau_m e^{2\pi i m \left(\frac{r}{P}\right)^2}. \tag{27}$$

This describes a circular grating. Of course t(r) is not a periodic function of r. Instead it takes the same values at $r = P\sqrt{s}$, (s = 0, 1,...). When a plane wave impinges orthogonally on such transparent element, the emerging field is proportional to the function t. Each term of the sum can be read as the paraxial approximation of a spherical wave by letting

$$\frac{2m}{P^2} = \frac{1}{\lambda R_m}, \quad (m = 0, \pm 1,...), \tag{28}$$

where $R_m$ is the corresponding radius of curvature. Accordingly, the circular grating gives rise to a set of diverging (m > 0) and converging (m < 0) waves whose amplitudes $\tau_m$ are the same that were possessed by the plane waves diffracted by the linear grating. In a sense, the circular grating behaves as a set of lenses with the (positive and negative) focal lengths given by Eq. (28). If we want to obtain only one spherical wave we have to start from the sawtooth grating of Fig. 4. The transformation is illustrated in Figs. 6 and 7. If these continuous profiles are replaced by stepwise approximations the resulting efficiencies can be evaluated using Eq. (26).

The second point refers again to linear gratings. While in the general case the evaluation of the near field can be cumbersome, things are easier and very interesting when the paraxial approximation can be used. The most important phenomenon observed in this case is the Talbot effect,[15,16] i. e., the production of images of the gratings at distances equal to integer multiples of the so called Talbot distance $z_T = 2P^2/\lambda$.

Let us briefly recall how this phenomenon can be explained. A typical plane wave whose wavevector has components $k_x$ and $k_y$ along the transverse axes has a z-component $k_z$ given by

$$k_z = \sqrt{k^2 - k_x^2 - k_y^2}, \tag{29}$$

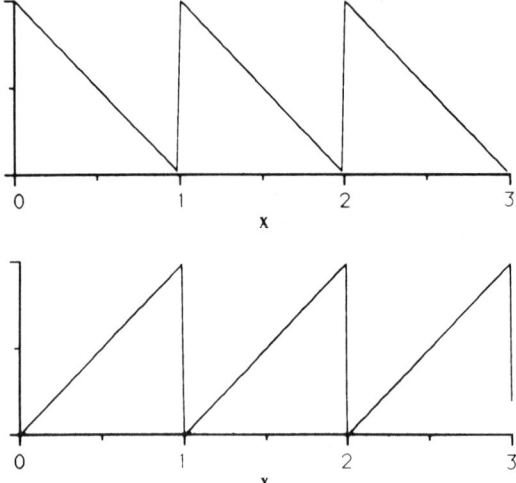

Fig. 6 a), and b) Sawtooth profiles with different orientations.

because the magnitude of the wavevector **k** must equal $2\pi/\lambda$. If both $k_x$ and $k_y$ are much smaller than k then the paraxial approximation

$$k_z = k\sqrt{1 - \frac{k_x^2 + k_y^2}{k^2}} \cong k - \frac{k_x^2 + k_y^2}{2k}, \qquad (30)$$

can be used. Let us now consider the field emerging from the grating under orthogonal plane wave illumination. We can write it in the form

$$V(x,0^+) = A \sum_{m=-\infty}^{\infty} \tau_m e^{ik_{mx}x}. \qquad (31)$$

where we let

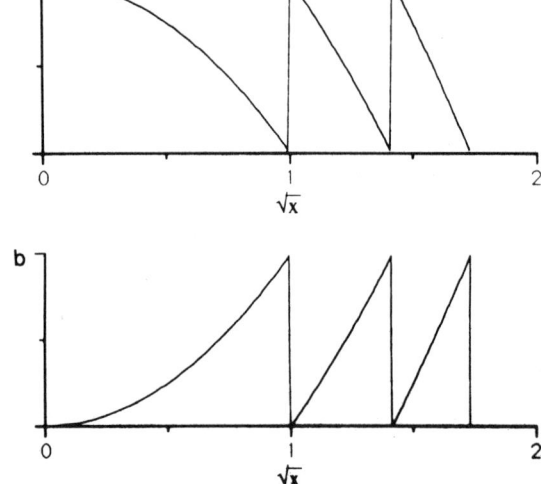

Fig. 7 a), and b) Applying the transformation $(x/P \rightarrow (r/P)^2$ to the profiles of Fig.6.

$$k_{mx} = \frac{2\pi m}{P}, \qquad (m = 0, \pm 1, \pm 2, \ldots). \qquad (32)$$

We shall assume that the coefficients $\tau_m$ are significantly different from zero within a range $(-m_{max}, m_{max})$ such that

$$k_{m_{max}} = \frac{2\pi m_{max}}{P} \ll k = \frac{2\pi}{\lambda}. \qquad (33)$$

The field that propagates at a distance z from the grating can be evaluated by multiplying each term in the sum (31) by the corresponding propagation factor $\exp(ik_{mz}z)$, where $k_{mz}$ is to be computed using Eq. (29), taking into account that $k_{my} = 0$. Thanks to Eq. (33) the paraxial expression for $k_{mz}$ can be used and the propagated field becomes

$$V(x,z) = Ae^{ikz} \sum_{m=-\infty}^{\infty} \tau_m e^{i\left(k_{mx} x - \frac{k_{mz}^2}{2k} z\right)}, \qquad (34)$$

or, by using Eq. (32),

$$V(x,z) = Ae^{ikz} \sum_{m=-\infty}^{\infty} \tau_m e^{2\pi i m \frac{x}{P}} e^{-2\pi i m^2 \frac{\lambda}{2P^2} z}. \qquad (35)$$

It is now seen at once that for the distances $z_s$ satisfying the condition

$$\frac{\lambda}{2P^2} z_s = s, \qquad (s = 1, 2, \ldots), \qquad (36)$$

all the exponentials containing $m^2$ in Eq. (35) equal one. Therefore the field (35) is identical to the field $V(x, 0^+)$ of Eq. (31) for all distances of the set

$$z_s = s \frac{2P^2}{\lambda}, \qquad (s = 1, 2, \ldots), \qquad (37)$$

the first one (s = 1) being known as the Talbot distance $z_T = P^2/(2\lambda)$. This explains the basic Talbot effect.

Many other phenomena occur at distances that are fractions of the Talbot distance. Let us work out the case $z = z_T/4$. The exponentials containing $m^2$ in Eq. (35) now become

$$e^{-\frac{\pi}{2} i m^2} = (-i)^{m^2}. \qquad (38)$$

For even m (m = 2s, with integer s) and odd m (m = 2s+1) these quantities take the respective values

$$(m = 2s): \quad (-i)^{4s^2} \equiv 1; \quad (m = 2s+1): \quad (-i)^{4s^2+4s+1} \equiv -i. \qquad (39)$$

On inserting from Eq. (39) into Eq. (35) we see that, except for the common factor $\exp(ikz)$ in front of the sum, the effect of propagation is equivalent to multiplying by $-i$ all the Fourier

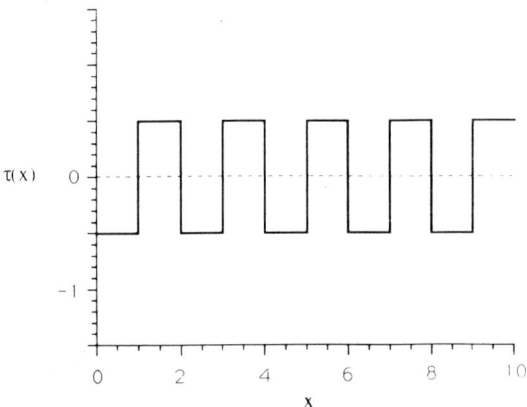

Fig. 8 Transmission function of Ronchi grating.

coefficients with odd indexes, leaving unaffected those of even indexes. We can realise what this can produce in the transverse field distribution by considering the transmission functions of Figs. 8 and 9. Figure 8 describes a Ronchi grating. It can be thought of as a set of parallel slits separated by opaque strips with the same width as the slits. The function of Fig. 9 differs from the previous one only for the mean value, which is zero instead of 1/2. The transmission function of the Ronchi grating can be written

$$\tau(x) = \frac{1}{2} - \frac{1}{2}\text{sgn}\left[\sin(\beta x)\right] , \qquad (40)$$

where $\beta = 2\pi / P$ (P equals 2 in Figs. 8 and 9). Suppose the Ronchi grating is illuminated by an orthogonal plane wave. The emerging field is proportional to $\tau(x)$. Since the second term on the r. h. s. has only Fourier components with odd indexes (see Eq. (7)) while the first one gives the zero order wave, we conclude from Eqs. (35) and (39) that the propagated field has the form

$$V(x, z_T/4) = \frac{A}{2} e^{ik\frac{z_T}{4}} \left\{1 + i\,\text{sgn}\left[\sin(\beta x)\right]\right\} . \qquad (41)$$

It is easily seen that the field (41) has uniform intensity along the x-axis whereas its phase jumps from $-\pi/4$ to $\pi/4$ at each half period. In conclusion, the propagation process alone has transformed a pure amplitude modulation into a pure phase modulation. The roles of planes $z = 0$ and $z = z_T/4$ could be interchanged and we could tart from the grating of Fig. 9 at $z = 0$. Then the light received at $z = z_T/4$ would be distributed into an array of spots separated by dark regions. This is the principle of array illuminators based on the Talbot effect.[17,18]

Talbot effect is even richer than this. For any distance equal to a rational multiple of $2P^2/\lambda$ the transverse field distribution is obtained by summing a finite number of suitably shifted and properly phased replicas of the field at $z = 0$. This gives rise to rather bizarre changes of the field distribution when z is varied. At the core of this behaviour there is the law [15] for computing the Fresnel transform of the comb function, i. e. an array of equally spaced Dirac delta functions. A novel approach has been proposed by Berry and Klein.[19] They showed that the phenomenon can interpreted in fractal terms. Thus, we touch here the topic of fractal optics,[20] a subject where DOE's should play a significant role.

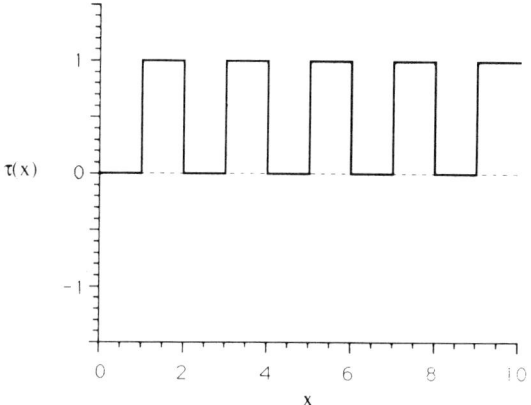

Fig. 9 Same as Fig. 8 except that the mean value has been set to zero.

## 5. THE NEED FOR ELECTROMAGNETIC APPROACHES

Up to now we have used a simple approach based on a scalar field description and on the transmission function concept. In certain cases, however, such an approach is not adequate. This typically occurs when the details of the diffracting structure have linear dimensions that are not much larger than the wavelength. It is generally said that in such cases one has to make recourse to the full vectorial electromagnetic theory. It could be misleading to stress the term "vectorial", since in many significant instances even the electromagnetic problem can be cast in scalar form. Let us quote as celebrated examples diffraction from a half plane [6] or from a cylinder.[21] The important point instead is that the interaction problem can no longer be treated by simple means such as the transmission function. To understand why, we shall work out an elementary example. Let us consider two identical gratings, each of them possessing the transmission function

$$\tau(x) = \frac{1 + \cos\beta x}{2} \ . \tag{42}$$

If the two gratings are superposed (without mutual lateral shift) the overall transmission function becomes

$$\tau^2(x) = \frac{1}{4}\left[\frac{3}{2} + 2\cos\beta x + \frac{1}{2}\cos 2\beta x\right]. \tag{43}$$

Let a plane wave of amplitude A impinge orthogonally on the gratings. The emerging field then consists of five diffraction orders, 0, ±1, ±2, with amplitudes

$$A_0 = \frac{3}{8}A, \quad A_{\pm 1} = \frac{A}{4}, \quad A_{\pm 2} = \frac{A}{16} \ . \tag{44}$$

Suppose now that the gratings are at different values of z, say z = 0 and z = d. The field emerging from the first grating is

$$V(x,0^+) = \frac{A}{2}(1 + \cos\beta x) \ . \tag{45}$$

On propagating from z = 0 to z = d it changes into

$$V(x,d^-) = \frac{A}{2}e^{ik_zd}\left(e^{i\Delta k_z d} + \cos\beta x\right), \qquad (46)$$

$$k_z = \sqrt{k^2 - \beta^2}, \ \Delta k_z = k - k_z = k\left[1 - \sqrt{1 - \left(\frac{\lambda}{P}\right)^2}\right]. \qquad (47)$$

Passing through the second grating the field becomes

$$V(x,d^+) = \frac{A}{4}e^{ik_zd}\left[\frac{1}{2} + e^{i\Delta k_z d} + \left(1 + e^{i\Delta k_z d}\right)\cos\beta x + \frac{1}{2}\cos 2\beta x\right], \qquad (48)$$

which is generally different from the field produced by the transparency function of Eq.(43). As an example, let $\Delta k_z d = \pi$. Then Eq.(48) gives

$$V(x,d^+) = \frac{A}{8}e^{ik_zd}\left(-1 + \cos 2\beta x\right), \qquad (49)$$

so that only three diffraction orders emerge from the second grating, namely, 0 and ±2, with respective amplitudes

$$A_0 = -\frac{A}{8}e^{ik_zd}, \qquad A_{\pm 2} = \frac{A}{16}e^{ik_zd}. \qquad (50)$$

There is a drastic difference between the two cases. Not only has the real amplitude of the zero order wave changed, but the second order waves have disappeared altogether. The overall absorption of the incident power has also changed. This slightly surprising result can be understood by noting that during propagation the transverse power distribution changes. It could be easily seen that under the hypothesis $\Delta k_z d = \pi$ the power peaks are concentrated in the regions where the transmission function (42) has lower values.

To understand the role of this elementary example in discussing the need for electromagnetic treatments we note that the distance d over which the field changes considerably can be rather small if the grating period is not very large compared with the wavelength. Suppose $P = 5\lambda/3$. Using Eq. (47) we find that the distance d satisfying the condition $\Delta k_z d = \pi$ is $d = 5\lambda/2$ and this means less than 2 µm in the visible range. Now, e. g., holographic emulsions have typical thicknesses of about 10 µm. We see then that when the grating period is not very large with respect to $\lambda$ we are no longer allowed to describe the transparency as a flat one because the field changes its transverse shape during propagation within the diffractive structure itself. We also understand from our example that the predictions of a model based on the transmission function concept can be very different from those obtained by a proper theory in which the effects of propagation in the diffracting volume are taken into account. It can be noted that it is for this reason that volume holograms behave so differently from thin holograms.

## 6. SUBWAVELENGTH STRUCTURES

The remarks of the previous section apply "a fortiori" when the diffracting structure has details smaller than the wavelength. Here, new types of phenomena may appear. To give

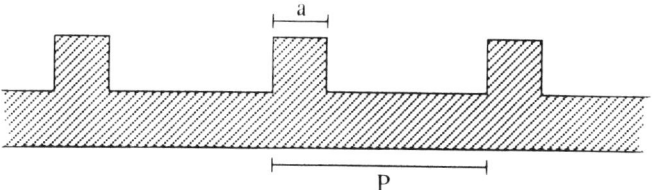

Fig. 10 Dielectric grating with a subwavelength structure.

an idea of these, suppose that the surface of a dielectric specimen has a large number of hills and valleys with linear dimensions and mutual distances smaller than the wavelength. Then, roughly speaking, the incident light performs a sort of spatial average between the index of refraction of the air that occupies the valleys and that of the material constituting the hills.

In order to estimate the effects of this type of operation let us refer (see Fig. 10) to a linear binary phase grating whose height is larger in an interval of length a (the hill) and lower in the remaining part of the period P (the valley). A linearly polarised plane wave impinges orthogonally on the grating having its electric field parallel to the grooves. We assume P $\ll \lambda$ so that the electric field is approximately uniform across each period. Because of the boundary conditions, the magnitude E of such field has to remain unaltered on passing from the air to the dielectric. As a consequence, the magnitude of the displacement vector has different values $D_1$ and $D_2$ in the specimen and the air respectively. We assume that the magnetic permeability is the same as in the vacuum for both the air and the specimen. The refractive index of the specimen is denoted by n while that of the air is approximated by one. Therefore we have

$$D_1 = \varepsilon_0 n^2 E, \qquad D_2 = \varepsilon_0 E, \tag{51}$$

where $\varepsilon_0$ denotes the permittivity of the vacuum. The spatial average of D over a period is then

$$D = \frac{aD_1 + (P-a)D_2}{P} = \varepsilon_0 \left[ \frac{a}{P} n^2 + \left(1 - \frac{a}{P}\right) \right] E, \tag{52}$$

or equivalently the index of refraction "seen" by the wave is

$$n_o = \sqrt{\frac{a}{P} n^2 + \left(1 - \frac{a}{P}\right)}. \tag{53}$$

We can thus vary the equivalent index of refraction from one to n by adjusting the value of a/P. In this way we can synthesise a dielectric coating with prescribed refractive index by actually acting on the surface relief of a homogeneous dielectric. Then, by a proper choice of the height of the hills we can, for example, produce an antireflection layer. It is noteworthy that antireflection subwavelength structures are found in nature in the eye of certain night-flying moths. Since light reflected by the moth's eye can be detected by predators it is thought that the antireflection structure increases the survival chances of the moth.

There is a further point to be noted. If the state of polarisation of the wave is rotated by $\pi/2$ (electric field orthogonal to the grooves) the boundary condition to be used is that D be the same in both media. This implies that the electric field has different values in the two media. More explicitly we have

$$E_1 = \frac{D}{\varepsilon_0 n^2} \; ; \quad E_2 = \frac{D}{\varepsilon_0} \; . \tag{54}$$

The average value of E is

$$E = \frac{aE_1 + (P-a)E_2}{P} = \frac{D}{\varepsilon_0}\left[\frac{a}{P}\frac{1}{n^2} + \left(1 - \frac{a}{P}\right)\right], \tag{55}$$

so that the equivalent index of refraction becomes

$$n_e = \frac{1}{\sqrt{\dfrac{a}{Pn^2} + \left(1 - \dfrac{a}{P}\right)}} \; . \tag{56}$$

This phenomenon is known as form birefringence.[6] Our structure behaves as a birefringent material having its optical axis orthogonal to the grooves. (The suffixes o and e stand for ordinary and extraordinary in Eqs. (52) and (56)).

Above, we considered a periodic object. Therefore the equivalent index of refraction is the same all across the surface. Extensions are easily envisaged. For example, we can change the spacing and the width of the hills and valleys from one surface region to another. In this way we can control the equivalent local index of refraction. This in turn affords a mean for impressing any phase structure to our object.

## 7. DIFFRACTIVE ELEMENTS AND PARTIALLY COHERENT LIGHT

Up to this point we have assumed a monochromatic field. This type of idealised field has strictly predictable time behaviour and consequently is fully coherent. We can ask what changes need to be made when diffractive elements are used with partially coherent fields. In certain cases the answer can be relatively simple. In the general case, however, one enters a potentially broad field of study in which little has been done up to now. Here, we shall limit ourselves to a few remarks relating to the simplest cases and to the role that diffractive optics can play in the synthesis of partially coherent fields.

Let us recall [6, 22] that the most important effects in stationary partially coherent fields are adequately described through the use of the mutual coherence function $\Gamma(P_1, P_2, \tau)$ defined as

$$\Gamma(P_1, P_2, \tau) = \langle V^*(P_1, t+\tau) \, V(P_2, t) \rangle \tag{57}$$

where $P_1$ and $P_2$ are two points in the wavefield, V is the analytic signal associated to the field and $\tau$ denotes a temporal delay. The angular brackets stand for a time average.

In particular, for $P_1 = P_2$ and $\tau = 0$ Eq. (57) gives the time averaged optical intensity. It is generally said that for $\tau = 0$ the function $\Gamma(P_1, P_2, 0)$ accounts for the spatial coherence properties of the field, whereas for $P_1 = P_2$, $\Gamma(P_1, P_1, \tau)$ describes the temporal coherence properties.

The celebrated theorem of Wiener-Kintchin [6, 22] asserts that $\Gamma(P_1, P_1, \tau)$ is the Fourier transform of the power spectrum, or spectral density, of the field. Roughly speaking, this means that a high temporal coherence is associated with fields which have a limited spectral content (e.g. quasi-monochromatic fields). Another fundamental theorem, namely that of van Cittert-Zernike, relates the spatial coherence properties of the field generated by a spatially uncorrelated source to the intensity distribution across the source itself. In current research on partially coherent light the distinction between space and time coherence is made sharper by introducing the cross-spectral density $W(P_1, P_2, \nu)$, which is related to the mutual coherence function $\Gamma$ by the equation

$$W(P_1, P_2, \nu) = \int_{-\infty}^{\infty} \Gamma(P_1, P_2, \tau) e^{2\pi i \nu \tau} d\tau . \tag{58}$$

At any temporal frequency $\nu$ the cross-spectral density accounts for the spatial correlations of the field between $P_1$ and $P_2$. In particular, for $P_1 = P_2$, the cross-spectral density coincides with the power spectrum or spectral density, generally denoted by $G(P,\nu)$. The function $G(P,\nu)$ is also called the optical intensity (at frequency $\nu$) when the immediate concern is the spatial distribution of power. A normalised version of W, known as the spectral degree of coherence $\mu(P_1, P_2, \nu)$, is obtained through the following definition:

$$\mu(P_1, P_2, \nu) = \frac{W(P_1, P_2, \nu)}{\sqrt{G(P_1, \nu) G(P_2, \nu)}} , \tag{59}$$

in that the modulus of $\mu$ lies between zero and one. A field is then said to be spatially coherent (at frequency $\nu$) if $|\mu| = 1$ for any pair $P_1$, $P_2$.

Let us briefly discuss the simple cases. First, suppose that a DOE has to work with a partially coherent field endowed with full spatial coherence (at any frequency). As an example, we can think of a polychromatic plane wave. The problem then reduces to inquiring about the dependence on $\nu$ of the response of the DOE. After that, a superposition of the optical intensities pertaining to the various frequencies is made. Another simple situation is encountered when the source is spatially incoherent. An elementary procedure consists in summing up the optical intensity produced by the various source points. It is therefore sufficient to inquire about the behaviour of the DOE when the source point is changed.

Let us proceed to a much more difficult task, namely, recording and reconstructing a general wavefield. With coherent light the problem is solved by holography. There is not, however, an equivalent of holography for partially coherent fields. It is not difficult to explain why.

To this end, we note that information in a partially coherent field is carried by the correlation function. Indeed, let us consider the following problem. A partially coherent field, described at a fixed frequency by its cross-spectral density W, is incident on the plane z=0. Dropping the explicit dependence on $\nu$, let $W_0(\rho_1, \rho_2)$ be the cross-spectral density across that plane, $\rho_1$ and $\rho_2$ denoting position vectors. There exists a propagation formula[6, 22] that reads

$$W_z(r_1, r_2) = \iint W_0(\rho_1, \rho_2) K^*(r_1, \rho_1) K(r_2, \rho_2) d^2\rho_1 d^2\rho_2 , \tag{60}$$

where $W_z(r_1, r_2)$ is the cross-spectral density between two typical points $r_1$, $r_2$ at the plane z = const and $K(r,\rho)$ is a kernel, known as the propagator,[22] describing the propagation process from a point-like source at $\rho$ to point $r$. In particular, for $r_1 = r_2$ we obtain the optical intensity at the selected frequency across the plane z = const. Hence the evaluation of such optical intensity requires the knowledge of $W_0$ for all possible pairs $\rho_1$ and $\rho_2$.

Therefore, in a well-defined sense, the information carried by the wavefield is four-dimensional (at each frequency). This means that using partially coherent light we should be able to record and recreate not the field, which is a random quantity, but its correlation function. This is really a formidable task because of the enormous amount of information that should be managed. So far very little has been done. To see where diffractive optics might come in handy, we shall limit ourselves to the subproblem of synthesising a prescribed partially coherent wavefield at a fixed frequency (this is equivalent to the reconstruction problem). Let us suppose we start with a fully coherent field. How can we modify it in such a way that it acquires a prescribed form of the correlation function? An answer is furnished by the Wolf's modal theory of coherence.[22, 23] Without going into details, the theory shows that any partially coherent field can be thought of as the superposition of certain mutually uncorrelated, coherent fields known as the modes associated to the cross-spectral density. Such fields could be generated at different times for short intervals, and then a temporal average would be performed.[24] This would impress onto the field the required correlation function. One may speculate that in such a process the modes could be generated by suitable DOE's. On the other hand, different modes should be produced at different times. This could be obtained by using a sequence of suitable DOE's in a movie-like system. A more interesting and flexible solution would be afforded by DOE's whose properties could be changed by the application of electrical signals. DOE's of this type, i. e., reconfigurable DOE's are still in their infancy. Something can be obtained by using liquid crystal devices but much remains to be done. It is clear, however, that the possibility of reconfiguration would greatly increase the potentialities of DOE's. This applies of course not only to the problem of field synthesis but, more generally, to any type of optical information handling.

## REFERENCES

1. S. H. Lee, Diffractive and Miniaturized Optics, *SPIE Press*, Vol CR49 (1994).
2. *Applied Optics*. Special issue on Diffractive Optics, (May 1995).
3. *Journal of Modern Optics*. Special issue on Diffractive Optics, (July 1996).
4. J. Turunen and F. Wyrowski, Diffractive optics: from promise to fruition, in *Trends in Optics*, A. Consortini, ed., Academic Press, London, (1996).
5. A. A. Friesem and Y. Amitai, Planar diffractive elements for compact optics, in *Trends in Optics*, A. Consortini, ed., Academic Press, London, (1996).
6. M. Born and E. Wolf, *Principles of Optics*, Pergamon Press, Oxford (1980), 6th ed.
7. J. W. Goodman, *Introduction to Fourier Optics*, McGraw-Hill, New York, (1968).
8. M. Abramowitz and I. A. Stegun, *Handbook of mathematical functions*, Dover, New York, (1965).
9. P. Ehbets, H. P. Herzig, D. Prongué and T. M. Gale, High-efficiency continuous surface-relief gratings for two-dimensional array generation, *Opt. Lett.* 17:908, (1992).
10. U. Krackhardt, J. N. Mait and N. Streibl, Upper bound on the diffraction efficiency of phase-only fanout elements, *Appl. Opt*, 31:27, (1992).
11. F. Wyrowski, Design theory of diffractive elements in the paraxial domain, *J. Opt. Soc. Am.* A 10:1553 (1993).
12. H. Dammann and K. Görtler, High-efficiency in-line multiple imaging by means of multiple phase holograms, *Opt. Commun.* 3:312 (1971).
13. W. H. Press, B. R. Flannery, S. A. Teukolsky and W. T. Vetterling, *Numerical Recipes*, Cambridge University Press, Cambridge, (1995).
14. F. Gori, Fresnel transform and sampling theorem, *Opt. Comm.*, 39:293, (1981).
15. F. Gori, Why is the Fresnel Transform so little known?, in Current Trends in Optics, J. C. Dainty, ed., Academic Press, London, (1994).
16. K. Patorski, The Self-Imaging Phenomenon and its Applications, in Progress in Optics, Vol XXVII, E. Wolf, ed., North-Holland, Amsterdam (1989).

17. A. W. Lohmann and J. A. Thomas, Making an array illuminator based on the Talbot effect, *Appl. Opt*, 29:4337, (1990).
18. J. R. Leger and G. J. Swanson, Efficient array illuminator using binary-optics plates at fractional Talbot planes, *Opt. Lett*. 18:1, (1990).
19. M. V. Berry and S. Klein, Integer, Fractional and fractal Talbot effects, *J. Mod Optics*, 43:2139, (1996).
20. J. Uozumi and T. Asakura, Fractal Optics, in *Current Trends in Optics*, J. C. Dainty, ed., Academic Press, London, (1994).
21. W. K. H. Panofsky and M. Phillips, *Classical Electricity and Magnetism*, Addison-Wesley, Reading, (1969).
22. L. Mandel and E. Wolf, *Optical Coherence and Quantum Optics*, Cambridge University Press, Cambridge (1995).
23. E. Wolf, New theory of partial coherence in the space-frequency domain. Part I: Spectra and cross-spectra of steady state sources, *J. Opt. Soc. Am.*, 72:343, (1982).
24. P. De Santis, F. Gori, G. Guattari and C. Palma, Synthesis of partially coherent fields, *J. Opt. Soc. Am.* A, 3:1258, (1986).

## APPENDIX

Here, we want to show that a periodic phase function has either one or an infinite number of Fourier coefficients. We shall argue by contradiction. Suppose that $\exp[i\Phi(x)]$ has a finite number of Fourier coefficients different from zero and let N be the maximum integer such that either $c_N$ or $c_{-N}$ (or both) is different from zero. The Fourier expansion of $\exp[i\Phi(x)]$ then reads

$$e^{i\Phi(x)} = \sum_{m=-N}^{N} \tau_m e^{2\pi i m \frac{x}{P}} . \qquad (1)$$

Computing to the squared modulus of $\exp[i\Phi(x)]$ we obtain

$$1 = \sum_{s=-2N}^{2N} c_s e^{2\pi i s \frac{x}{P}}, \qquad (2)$$

where the coefficients $c_s$ can be found by multiplying the sum on the r.h.s. of Eq.(1) by its complex conjugate. With some index manipulation one finds

$$c_s = \sum_{h=0}^{2N-s} \tau_{-N+h+s} \tau^*_{-N+h} , \qquad (s = 0,1,...,2N) ,$$
$$c_{-s} = c^*_s, \qquad (s = 1,...,2N) . \qquad (3)$$

By virtue of Eq.(2) all of the coefficients $c_s$ must vanish except $c_0$. Let us write the equations obtained by imposing this condition and by using Eq. (3):

$$c_{2N} = \tau^*_{-N} \tau_N = 0 \qquad (4)$$

$$c_{2N-1} = \tau^*_{-N} \tau_{N-1} + \tau^*_{-(N-1)} \tau_N = 0 , \qquad (5)$$

$$c_{2N-2} = \tau^*_{-N} \tau_{N-2} + \tau^*_{-(N-1)} \tau_{N-1} + \tau^*_{-(N-2)} \tau_N = 0 , \qquad (6)$$

....

$$c_1 = \tau^*_{-N} \tau_{-(N-1)} + \ldots\ldots\ldots\ldots\ldots + \tau^*_{N-1} \tau_N = 0 . \qquad (7)$$

According to our hypothesis, at least one of the coefficients $\tau_{-N}$ and $\tau_N$ differs from zero. Suppose $\tau_N \neq 0$. Then it follows from Eq.(4) that $\tau_{-N} = 0$. Consequently, Eqs.(5) and (6) become

$$\tau^*_{-(N-1)} \tau_N = 0 , \tag{8}$$

$$\tau^*_{-(N-1)} \tau_{N-1} + \tau^*_{-(N-2)} \tau_N = 0 . \tag{9}$$

We see from eq.(8) that $\tau_{-(N-1)} = 0$, so that eq.(9) implies $\tau_{-(N-2)} = 0$. Proceeding in this way we aveable to prove through Eq.(7) that $\tau_{N-1} = 0$. Therefore, only $\tau_N$ is different from zero. A similar procedure can be followed if the assumption $\tau_{-N} \neq 0$ is made. In conclusion, if the Fourier coefficients form a finite set, no more than one of them can be different from zero.

# DESIGN OF REFRACTIVE AND DIFFRACTIVE MICRO-OPTICS

H. P. Herzig

Institute of Microtechnology University of Neuchâtel
Rue A.-L. Bregued 2, 2000 Neuchâtel, Switzerland

## 1. INTRODUCTION

Passive optical components are used in optical systems to collect, distribute, or modify optical radiation. Widely known examples include refractive components, such as lenses, mirrors and prisms, which have been applied successfully in optical instruments. Following the trend of miniaturization, novel technologies have been developed to shrink the size of the elements (see Fig. 1). Refractive microlenses with diameters of 10 µm – 1 mm can be fabricated with high quality.

In parallel, diffractive optics has emerged from holography.[1-3] Typical diffractive optical elements (DOEs) have multilevel microreliefs («binary optics») or continuous microreliefs, with features ranging from submicron to millimeter dimensions and relief amplitudes of a few microns. Novel structures can be realized complementing and exceeding the possibilities of traditional lenses, prisms and mirrors. Almost any structure shape, including asymmetric aspherics, can be manufactured, which provides many degrees of freedom for the design.

Refractive elements consisting of macroscopic surface-relief structures are designed using the laws of geometrical optics, treating light by the refraction and reflection of geometrical rays at optical interfaces. The eikonal equation as the basis of geometrical optics can be derived from Maxwell's equations in the limit where the wavelength tends to zero. Therefore, no wavelength dependent properties, apart from those due to material dispersion, are intended. In contrast, diffractive optical elements (DOEs) are planar elements consisting of zones, which retard the incident wave by a modulation of the refractive index or by a modulation of the surface profile. The light emitted from the different zones interferes and forms the desired wavefront. Since these phenomena are strongly dependent on the wavelength of light, DOEs are restricted to monochromatic applications. In order to combine the advantages of refractive optics (low dispersion) and diffractive optics (arbitrary shape), considerable work has been invested in the development of hybrid elements (Chapt. 10 in Ref. 4). In the following, we introduce the design fundamentals of diffractive optical elements (DOEs) and we compare their optical properties with those of refractive elements.

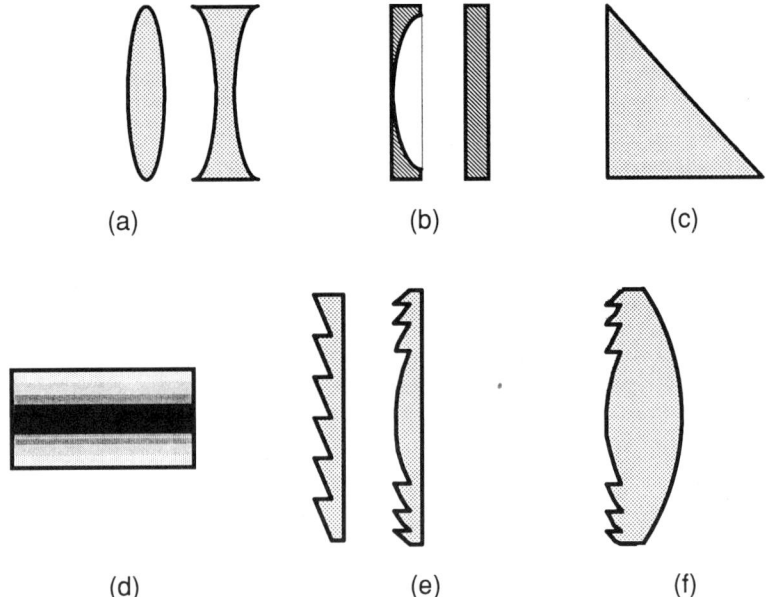

Fig. 1 Optical elements: (a) refractive lenses, (b) mirrors, (c) prisms, (d) GRIN lenses, (e) diffractive optical elements (DOEs), (f) hybrid elements.

## 2. DESIGN OF DIFFRACTIVE OPTICAL ELEMENTS

In scalar diffraction theory, a diffractive optical element with phase profile $\Psi(x,y)$ is modeled as a thin phase screen with a complex amplitude transmittance of

$$t(x,y) = \exp[i\Psi(x,y)] . \tag{1}$$

The DOE retards the incident wavefront and propagation of the new wavefront is modeled by the appropriate scalar formulation (e.g., angular spectrum, Fresnel diffraction, Fraunhofer diffraction, see Ref. 5). Note, that there is a difference between the phase profile $\Psi(x,y)$ of a DOE and the phase $\Phi(x,y)$, which is generated in the first (or another) diffraction order. Fig. 2 shows diffractive lenses with different phase profiles $\Psi(x,y)$, which generate all the same phase function $\Phi(x,y)$ in the first order. The elements perform the same wavefront conversion, but with different diffraction efficiency. The propagation of the first diffraction order can be modeled by replacing $\Psi(x,y)$ with $\Phi(x,y)$ in Eq. (1).

## 3. PHASE FUNCTION: DESCRIPTION AND REALIZATION

### 3.1. Phase function

A thin phase element that is illuminated by an incident wave $F_{in}(x,y)$ generates an output wave $F_{out}(x,y)$. The wavefront conversion is described by

$$\Phi_{out}(x,y) = \Phi_{in}(x,y) + \Phi(x,y) . \tag{2}$$

From Eq. (2) we can easily find the phase function $\Phi(x,y)$ of the phase element for a given

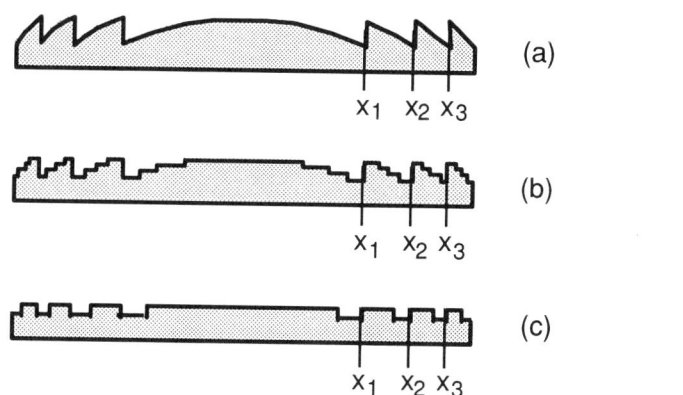

Fig. 2 Diffractive lens with (a) continuous profile, (b) multilevel profile, and (c) binary profile.

pair of waves:

$$\Phi(x,y) = \Phi_{out}(x,y) - \Phi_{in}(x,y). \qquad (3)$$

For the diffractive lens shown in Fig. 3, which has to connect an object point $(x_1,y_1,z_1)$ with an image point $(x_2,y_2,z_2)$, the phases $\Phi_{out}$ and $\Phi_{in}$ are of the form

$$\Phi_i(x,y) = \frac{2\pi}{\lambda_0}\sqrt{(x-x_i)^2 + (y-y_i)^2 + (z_i)^2},$$

where $\lambda_0$ is the design wavelength and $i = 1, 2$.

In general, the optical task is more complex, e.g., if an extended object has to be imaged. In that case, the DOE phase function $\Phi(x,y)$ is typically described by a polynomial:

$$\Phi(x,y) = \frac{2\pi}{\lambda_0} \sum_m \sum_n a_{mn} x^m y^n. \qquad (4)$$

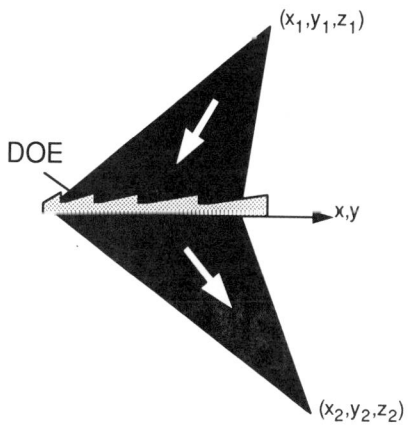

Fig. 3 Diffractive lens which connects an object point $(x_1,y_1,z_1)$ with an image point $(x_2,y_2,z_2)$.

The DOE is then optimized by optimizing the polynomial coefficients $a_{mn}$.

## 3.2. The implementation of the phase function as diffractive element

The phase function $\Phi(x,y)$ can be implemented as a refractive or as a diffractive element. A refractive element generates the phase distribution $\Phi$ by varying the optical path length through a phase plate. In the case of a diffractive element, the phase function is mainly generated by the position and the grating period of a local grating. The shape of the grating period determines the efficiency of the element, which is the amount of light that goes into a particular diffraction order.

In order to realize a diffractive element, the phase function $\Phi$ is wrapped to an interval between 0 and an integer multiple of $2\pi$. In the ollowing, without loss of generality, we restrict the discussion to the case of maximum modulation depth equal to $2\pi$. The phase profile $\Psi$ of the DOE is then given by

$$\Psi(x,y) = [\Phi(x,y) + \varphi_0] \bmod 2\pi, \qquad (5)$$

where $\varphi_0$ is a constant phase offset. The surface-relief profile $h(x,y)$ for a thin DOE in transmission is related to the phase profile $\Psi(x,y)$ by

$$h(x,y) = \frac{\lambda_0}{n(\lambda_0)-1} \frac{\Psi(x,y)}{2\pi}, \qquad (6)$$

where n is the refractive index of the grating material and $\lambda_0$ is the design wavelength. In the case of large diffraction angles, Eqs. (5) and (6) have to be calculated with a more accurate method.[6]

## 3.3. Binarization and diffraction efficiency

Neglecting Fresnel losses, scalar theory predicts an ideal diffraction efficiency of 100 % for DOEs with continuous surface-relief profile. The technology for fabricating these elements has made significant progress, however, experimental results are still 5 % - 15 % lower than the theoretical prediction. The standard method to fabricate DOEs in rigid material, such as glass and quartz, is based on multiple mask projection and subsequent etching. These elements have staircase like phase profiles as shown in Fig. 2(b) and (c). For further reading on the fabrication methods, we refer to the literature (see Refs.7-9). The diffraction efficiency of staircase gratings depends on the number of phase levels M. For linear gratings the first-order diffraction efficiency $\eta$ is given [1] by

$$\eta = \left( \frac{\sin(\pi/M)}{\pi/M} \right)^2. \qquad (7)$$

For 8 phase levels the diffraction efficiency is already 95 %. The diffraction efficiency for binary gratings (2 phase levels) is only 40.5 %. However, they are interesting for applications such as fan-out elements or diffusers, where several diffraction orders are used.

Eq. (7) can also be applied to estimate the efficiency of diffractive lenses. The lens structure is considered as local gratings with varying periods. In the central part of the lens, the grating periods are large and therefore the quantization of the phase profile into eight phase levels is simple and not critical. The grating periods become smaller towards the rim.

The number of phase levels is then limited by the resolution of the lithographic fabrication process. The diffraction efficiency decreases rapidly for small grating periods if only four or two phase levels are feasible.

## 4. OPTIMUM DESIGN

Fig. 4 shows two basically different design problems: (a) imaging and (b) beam-shaping. In the case of imaging [10] a set of continuous waves emitted by the object has to be converted into another continuous set of output waves forming the image. The ideal input and output waves are known, but the optical element that images best all input waves is not known yet. The term «imaging» includes here not only classical image formation, but also the design of Fourier lenses, laser scanners, or other design problems, where we want to minimize geometrical aberrations. In the case of beam shaping [Fig. 4(b)], one input wave illuminates the DOE which should generate the desired intensity distribution in another plane. Now, the output wave is unknown. Typical applications are the conversion of a Gaussian beam into a uniform beam with rectangular shape, or fan-out elements for array generation. Different design methods have been investigated.[11] We present here some of the most common methods.

### 4.1. Ray-tracing

Ray-tracing is the standard method to design optical elements.[12] In ray-tracing through lens systems the path of the light is determined with the help of elementary geometry by

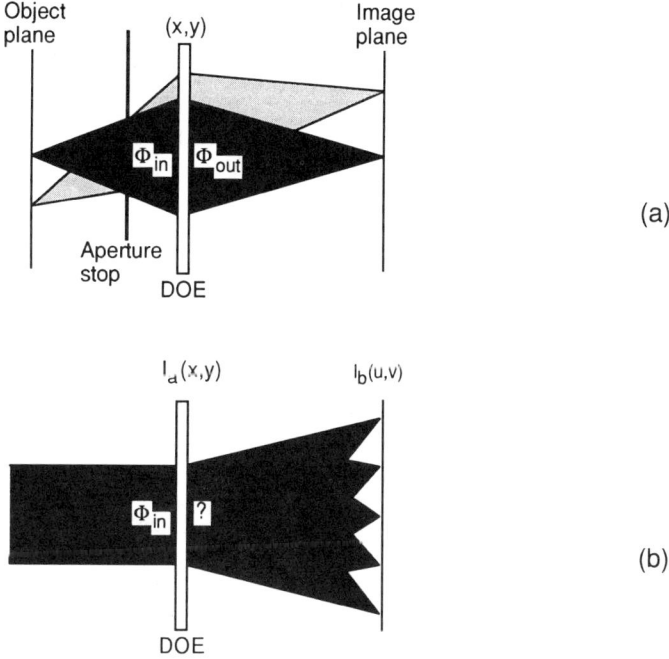

Fig. 4 (a) Imaging, (b) beam-shaping.

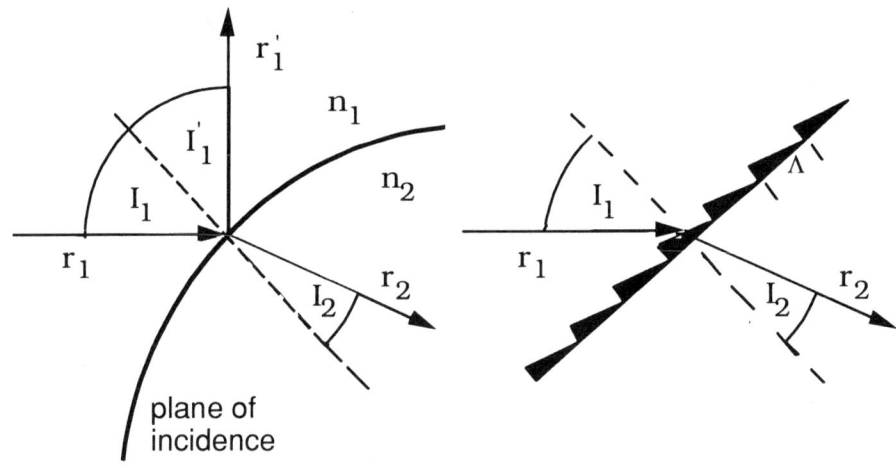

| Refraction | Reflection | Grating diffraction |
|---|---|---|
| $n_2 \sin I_2 = n_1 \sin I_1$ | $I_1 = -I_1'$ | $\sin I_2 = \sin I_1 + m\lambda/\Lambda$ |

$n_i$ : refractive index  $\quad\quad$ m : diffraction order
$r_i$ : ray vector $\quad\quad\quad\quad\quad$ $\lambda$ : wavelength

Fig. 5 Ray-tracing.

successive application of the law of refraction (or reflection). In diffractive optics, the law of refraction has to be replaced by the grating diffraction equation (Fig. 5).

In a ray-tracing program, the diffractive phase element is described by the first order phase function $\Phi(x,y)$. The reconstruction process is essentially governed by the condition of phase matching in the DOE plane $(x,y)$, which is given by Eq. (2). The phase matching condition yields relations for the normal projection $k_{x,i}$, $k_{y,i}$ of the wavevectors $\mathbf{k}_i$, onto the $(x,y)$ plane. The vectors $(k_{x,i}, k_{y,i})$ and the phase functions $\Phi_i$ are related by $k_{x,i} = \partial \Phi_i / \partial x$ and $k_{y,i} = \partial \Phi_i / \partial y$. By derivation of Eq. (2), we obtain

$$K_{x,out} = k_{x,in} + m \frac{\partial \Phi}{\partial x}, \quad\quad\quad (8a)$$

$$K_{y,out} = k_{y,in} + m \frac{\partial \Phi}{\partial y}, \quad\quad\quad (8b)$$

where m is the diffraction order. In general, the elements are designed for the first diffraction order (m = 1). The gradient $(\partial\Phi/\partial x, \partial\Phi/\partial y)$ in Eq. (8) describes the local grating vector of the diffractive structure. The length of the wavevectors at the reconstructing wavelength is given by $|\mathbf{k}| = |\mathbf{k}_{in}| = 2\pi/\lambda_{in}$. For a transmission element, the component $k_{z,out}$ of the outgoing wave normal to the x,y-plane is determined by

$$k_{z,out} = \text{sign}(k_{z,in}) \sqrt{(2\pi/\lambda_{in})^2 (k_{x,out})^2 (k_{y,out})^2}, \quad\quad\quad (9)$$

where $\text{sign}(k_{z,in})$ denotes the sign of $k_{z,in}$ For reflective elements, $\text{sign}(k_{z,in})$ has to be replaced by $-\text{sign}(k_{z,\text{in}})$.

Eqs. (8) and (9) describe grating diffraction, and allow the tracing of a bundle of finite rays through a diffractive component. The results are, e.g., presented as spot diagrams, which are the points of intersection of the calculated rays with the image plane.

## 4.2. Optimum design for beam shaping

Up to now, we have only considered the optimization of the first order phase function $\Phi$, because we were mainly interested in an aberration-free imaging. In principle, a beam shaping element can also be implemented as a first order DOE, which generates the desired wave in the first diffraction order. However, in many cases several orders are involved in the beam shaping process (fan-out elements, diffusers), therefore we consider here the direct optimization of the DOE phase profile $\Psi(x,y)$.

In general, the beam shaping problem has no analytical solution. Many different numerical methods have been investigated. Here, we concentrate on the two most successful methods: the iterative Fourier transform (IFT) algorithm and simulated annealing. For further reading, we recommend the Refs. 13-16

For the discussion, we consider a wave with an intensity distribution $I_a(u,v) = a^2(x,y)$ and a uniform phase. It is desired to add a DOE with phase $\Psi(x,y)$ in the x,y-plane such that a given intensity distribution $I_b(u,v) = b^2(u,v)$ results in another plane, which we call here the image plane [Fig. 4(b)]. The problem is now to find the phase profile $\Psi(x,y)$ of the DOE. The phase distribution $\phi(u,v)$ in the image plane is usually a free parameter. In the following, we assume that the complex amplitudes a(x,y) and b(u,v) in the DOE plane and in the image plane, respectively, are related by a simple Fourier transform. However, the algorithms are not restricted to this case.

The iterative Fourier transform algorithm was developed [1] by Gerchberg and Saxton in 1972 (see Ref. 17). This algorithm has since been modified and improved by a number of authors (see, e.g., Refs. 18, 19). The basic principle is shown in Fig. 6. The algorithm is started in the image plane with the desired amplitude distribution b(u,v) and a guess for the phase distribution $\phi(u,v)$. Then, the field in the DOE plane (x,y) and in the image plane (u,v) are iteratively calculated by using the fast Fourier transform (FFT) and projected in each domain onto a set of specific constraints until it converges. In the DOE plane, the amplitude modulation is clipped to the amplitude a(x,y) of the illumination beam, in order to calculate a phase-only element. In addition, fabrication constraints have to be satisfied. For multilevel surface-relief elements, the fabrication constraints include the discretization of the phase function to N phase levels. In the case of fabrication by direct writing, the constraints include the compensation of point spread function (PSF) effects of the writing spot.

The IFT algorithm is very efficient and can handle large amounts of data. On the other hand, the algorithm is sensitive to the starting parameters. It is difficult to avoid the process converting to a local rather than global minimum.

Global optimization methods do not depend on the starting point. They are very effective in avoiding local minima. The main problem with these methods is that the computing time may become extremely large. An example of global optimization is the simulated annealing algorithm.[20] The DOE is characterized by a parameter set q. The set q stands, e.g., for the pixels of a binary structure, or the coefficients of a polynomial. Then, a cost function has to be defined which describes the difference between the desired pattern [$I_b(u,v)$ in Fig. 4(b)] and the calculated pattern as a function of the parameter set q. During the process the parameter set is altered, which yields a change of $\Delta E$ in the cost function. If E is reduced ($\Delta E \leq 0$), the change is accepted. If E increases ($\Delta E > 0$), the change is

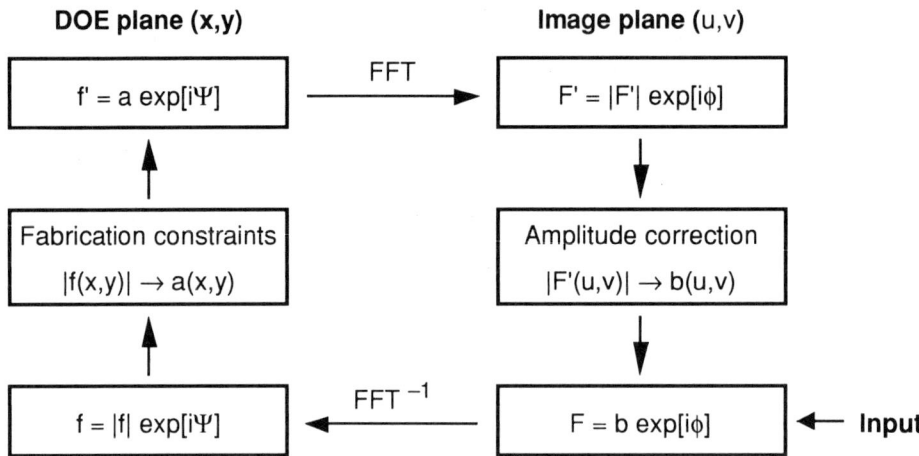

Fig. 6. Schematic representation of the iterative Fourier algorithm.

accepted with probability $\exp(-\Delta E/T)$. T is a control parameter analog to the temperature in thermodynamics. After a number of cycles, the temperature T is lowered gradually and the process continues. As a result, the probability of a change being accepted, with $\Delta E > 0$, is reduced. The temperature associated with the annealing decision process is continuously reduced until the cost function reaches a minimum value and shows no further signs of decreasing. At this point, the simulated annealing algorithm is terminated. The success of simulated annealing depends strongly on the annealing schedule and on the strategy of the move from one parameter set to the other.

## 5. REFRACTIVE PROPERTIES OF DIFFRACTIVE ELEMENTS

A diffractive element consists of zones which diffract the incoming light. The light diffracted from the different zones interferes and forms the desired wavefront. We can calculate this wavefront, e.g., by ray-tracing according to Eqs. (8) and (9). If only a few zones of the element are illuminated, the element does not behave like a pure diffractive element. Consequently, the results are not accurately modeled.[21, 22] Critical elements are, e.g., diffractive field lenses, which are close to an intermediate image plane.

Fig. 7 shows the refractive and diffractive properties of a blazed grating, assuming a shrinkage error of 10 %. Q is a normalized value for the variation of the deflection angle due to shrinkage error. Q equals 1 means that the deflection angle changes according to the laws of refraction at a prism with thickness change of 10 %. Q equals zero means that the deflection angle stays constant. The element behaves like a refractive element if only one zone is illuminated, and it behaves diffractively, if many zones are illuminated.

## 6. PARAXIAL PROPERTIES OF DIFFRACTIVE AND REFRACTIVE LENSES

Fig. 8 shows examples of refractive and diffractive lenses. A refractive lens is described by a refractive index $n(\lambda)$ and two curvatures $c_1$ and $c_2$. A diffractive lens on a planar substrate is described by a phase function. The phase function of a rotationally-symmetric diffractive lens with an arbitrary profile can be expressed in the form

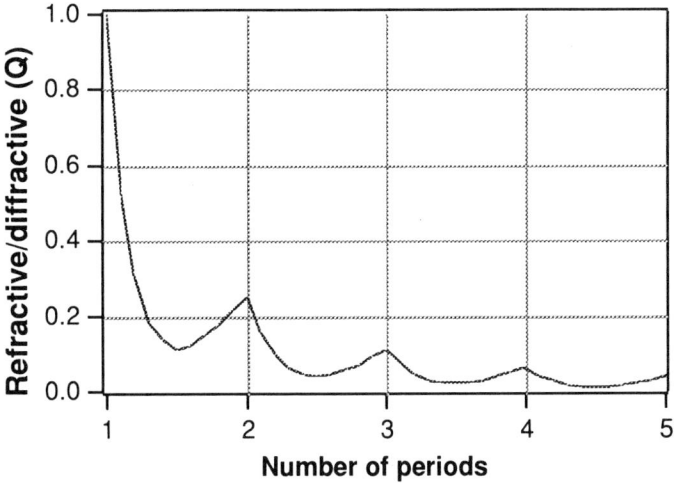

Fig. 7 Refractive and diffractive properties of a blazed grating, assuming a shrinkage error of 10 %. Q is a normalized value for the variation of the deflection angle due to shrinkage error. The element behaves refractively if only one zone is illuminated and it behaves diffractively if many zones are illuminated.

$$\Phi(r) = 2\pi \, (a_2 r^2 + a_4 r^4 + ...), \tag{10}$$

where r is the radial coordinate in the plane of the diffractive lens. The optical power of the diffractive lens in the m-th diffraction order is then given by

$$\phi = 1/f_0 = -2a_2 \lambda_0 m, \tag{11}$$

where $\lambda_0$ is the design wavelength and $f_0$ is the design focal length.

We compare now the focal length $f_r$ of a refractive lens with the focal length $f_d$ of a diffractive lens, which are given by

$$f_r(\lambda) = \frac{1}{n(\lambda)-1} \frac{1}{c_1 - c_2} \quad \text{and} \quad f_d(\lambda) = f_0 \frac{\lambda_0}{\lambda}. \tag{12}$$

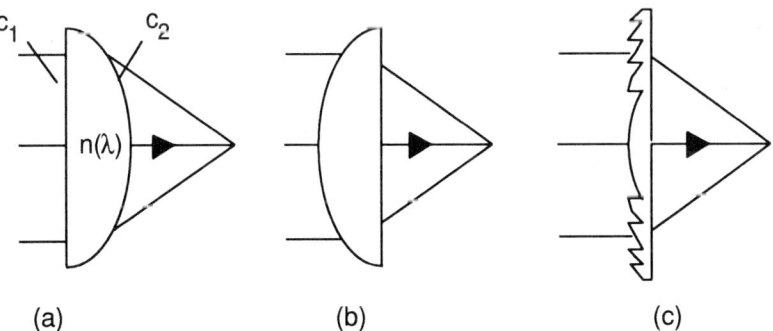

Fig. 8 Refractive and diffractive lenses.

Table I. Comparison (refractive lens - diffractive lens)

| | **Refractive lens** (plano-convex) | **Diffractive lens** |
|---|---|---|
| **Focal length** | $f_{a,b} = \dfrac{r}{n(\lambda) - 1}$ | $f_c = \dfrac{\lambda_0}{\lambda} f_0$ |
| **Airy disk radius** | $r(\lambda) = 1.22 \dfrac{\lambda f}{D}$ | $r(\lambda) = 1.22 \dfrac{\lambda_0 f_0}{D} \approx \text{const}$ |
| **Spherical aberration** $S_I = (NA)^4 f C_i$ | $C_a = \dfrac{n^2}{(n-1)^2} \quad C_b = \dfrac{n^3 - 2n^2 + 2}{n(n-1)^2}$<br>can be eliminated using aspheric lens shape | $C_c = 1 - \left(\dfrac{\lambda_0}{\lambda}\right)^2$<br>arbitrary phase functions are standard |

[Indices a,b,c refer to Fig. 8: Focal length f, Numerical aperture NA ≈ ρ/f, Refractive index n, Lagrange invariant H]

For a refractive lens the variation of the focal length with respect to the wavelength is small and depends on the factor $n(\lambda) - 1$. The dispersion of such a lens is described by the Abbe number $v_r$, defined as

$$v_r = \frac{n(\lambda_1) - 1}{n(\lambda_2) - n(\lambda_3)}. \tag{13}$$

For a diffractive lens the focal length is linearly proportional to the wavelength $\lambda$. Therefore, the Abbe number becomes

$$v_d = \frac{\lambda_1}{\lambda_2 - \lambda_3}, \tag{14}$$

where $\lambda_3 > \lambda_1 > \lambda_2$. Since the refractive index of glass decreases with increasing wavelength, $v_r$ is always positive, whereas $v_d$ becomes negative. The Abbe number is normally given for the standard wavelengths $\lambda_1 = 587.6$ nm, $\lambda_2 = 486.1$ nm, and $\lambda_3 = 656.3$ nm. For the various optical glasses, the Abbe number $v_r$ ranges from about 80 to 20, whereas $v_d$ becomes $-3.45$, independent of the material. Note, that a strong dispersion corresponds to a small Abbe number.

The strong negative dispersion of diffractive elements is used in hybrid (refractive/diffractive) elements in order to compensate the chromatic aberration of the refractive component. This enables relatively thin and light-weight achromatic systems made of only one material.

## 7. ABERRATIONS

A summary of the main differences between the aberrations [23] of refractive and diffractive lenses is shown in Table I. The refractive lenses are plano-convex lenses as shown in Fig. 8. For these lenses, the spherical aberration depends on the direction of light propagation through the lens ($C_a \approx 4 C_b$). Diffractive lenses have no spherical aberration for a single wavelength $\lambda_0$. For other wavelengths, spherical aberration is present, but still relatively small. The dominant error is the change in the focal length, which varies proportionally to the wavelength λ. An interesting consequence of this wavelength dependence is that the Airy disk radius generated by diffractive lenses does not vary with λ. For a fixed aperture, the spherical aberration scales with the focal length.

## REFERENCES

1. G.J. Swanson, and W.B. Veldkamp, Diffractive optical elements for use in infrared systems, *Opt. Engin.*, 28, 605-608 (1989).
2. H.P. Herzig, and R. Dändliker, Holographic optical elements for use with semiconductor lasers, in Goodman, J.W. (Ed.) *International Trends in Optics*, pp. 57-75, New York: Academic Press (1991).
3. H.P. Herzig, and R. Dändliker, Diffractive components: Holographic optical elements in *Perspectives for Parallel Interconnects*, Ph. Lalanne, P. Chavel, (Eds.), pp. 43-69, Berlin: Springer (1993).
4. H. P. Herzig, (Ed.), *Micro-Optics: Elements, Systems, and Applications*, London: Taylor & Francis (1997).
5. J.W.Goodman, *Introduction to Fourier Optics*, San Francisco: McGraw-Hill (1968).
6. M. Rossi, C.G. Blough, D.H. Raguin, E.K. Popov, and D Maystre, Diffraction efficiency of high-NA continuous-relief diffractive lenses, in *Diffractive Optics and Micro-Optics*, 1996 OSA Technical Digest Series, Vol. 5, pp. 233-236, Washington, DC: Optical Society of America 1996.
7. H.P. Herzig, M.T. Gale, H.W. Lehmann, and R. Morf, Diffractive components: Computer generated elements, in *Perspectives for Parallel Interconnects*, Ph. Lalanne,., P. Chavel, (Eds.), pp. 71-107, Berlin: Springer (1993).
8. H.P. Herzig, D. Prongué, and R. Dändliker, Design and fabrication of highly efficient fan-out elements, *Jpn. J. Appl. Phys.*, 29, L 1307-L 1309 (1990).
9. P. Ehbets, M.Rossi, and H. P. Herzig, Continuous-relief fan-out elements with optimized fabrication tolerances, *Opt. Engin.* 34, 3456-3464 (1995).
10. D.A Buralli,. and G.M Morris, Design of a wide field diffractive landscape lens, *Appl. Opt.*, 28, 3950-3959 (1989).
11. D. Malacara, and Z. Malacara, *Handbook of lens design*, New York: Marcel Dekker (1994).
12. W.J. Smith, *Modern Optical Engineering*, New York: McGraw-Hill (1990).
13. F. Wyrowski, Design theory of diffractive elements in the paraxial domain, *J. Opt. Soc. Am. A.*, 10, 1553-1561 (1993).
14. D.E.G. Johnson, A.D. Kathman, D.H. Hochmuth, A.L. Cook, D.R. Brown, and B. Delanay, Advantages of genetic algorithm optimization methods in diffractive optic design, in *Diffractive and Miniaturized Optics*, S.-H. Lee, Ed., Vol. CR49 of Critical Reviews, pp. 54-74, Bellingham, Wash.: SPIE (1993).
15. J.N. Mait, Understanding diffractive optic design in the scalar design, *J. Opt. Soc. Am. A*, 12, 2145-2158 (1995).
16. L.A. Romero, and F.M. Dickey, Lossless laser beam shaping, *J. Opt. Soc. Am. A*, 13, 751-760 (1995).
17. R.W. Gerchberg, and W.O. Saxton, A practical algorithm for the determination of phase from image and diffraction plane pictures, *Optik*, 35, 237-246 (1972).
18. J.R. Fienup, Iterative method applied to image reconstruction and to computer-generated holograms, *Opt. Engin.*, 19, 297-305 (1980).
19. F. Wyrowski, and O. Bryngdahl, Iterative Fourier-transform algorithm applied to computer holography, *J. Opt. Soc. Am. A.*, 5, 1058-1065 (1988).
20. S. Kirkpatrick, C.D. Gelatt, and M.P.Vecchi, Optimization by simulated annealing, *Science*, 220, 671-680 (1983).
21. S. Sinzinger, and M. Testorf, Transition between diffractive and refractive micro-optical components, *Appl. Opt.*, 34, 5970-5976 (1995).
22. M. Rossi, R.E. Kunz,.and H.P. Herzig, Refractive and diffractive properties of planar micro-optical elements, *Appl. Opt.*, 34, 5996-6007 (1995).
23. W.T. Welford, *Aberrations of Optical Systems*, Bristol: Adam Hilger (1986).

# GRIN PLANAR STRUCTURES FOR FOCUSING COLLIMATION AND BEAM SIZE CONTROL

C. Gómez-Reino, C. Bao, and M. V. Pérez

Faculty of Physics, and Optometry and Optics School
University of Santiago de Compostela
15706 Santiago de Compostela, Galicia Spain

## 1. INTRODUCTION

Currently the terms GRadient INdex or GRaded INdex (GRIN) are often used to describe an inhomogeneous medium, that is, a medium in which the refractive index varies from point to point within it.[1-2] In a GRIN medium, the optical rays follow curved trajectories, instead of straight lines as in a homogeneous medium. By an appropriate choice of the refractive index profile, a GRIN medium can have the same effect on light rays as conventional optical components, such as a prism or a lens.

Light propagation in GRIN media with a wide variety of gradient profiles has been analysed with emphasis on the waveguiding properties of importance for integrated optics, microoptics, and optical sensing.[3] In designing of GRIN structures for guiding purposes, one goal is to select an index profile that provides exact focusing and collimation. The simplest GRIN medium with this capability is the parabolic one (selfoc medium), where the equation governing ray propagation can be solved analytically in the paraxial approximation.[4] There is another GRIN medium with focusing and collimation properties. It is characterized by a hyperbolic secant (hereafter referred to as HS) profile which presents the following advantages in comparison with the parabolic one.[2, 5-7] First, it is a more general profile, because the parabolic function can be considered as the first-order approximation of the Taylor expansion series of the hyperbolic secant function. Second, the ray equation in this medium can be solved analytically without carrying out any type of approximation. Third, this medium is free of aberrations for meridional rays (rays propagating in planes that they include the optical axis).

The aim of this lecture is to study and discuss light propagation through a planar waveguide, whose refractive index profile is described by a 1D-hyperbolic secant function, and its application in integrated optics and microoptics. Focusing and collimation conditions are analysed and the design of devices to focus (focuser), to collimate (collimator), and to contract or expand a light beam (beam-size controller) are presented.

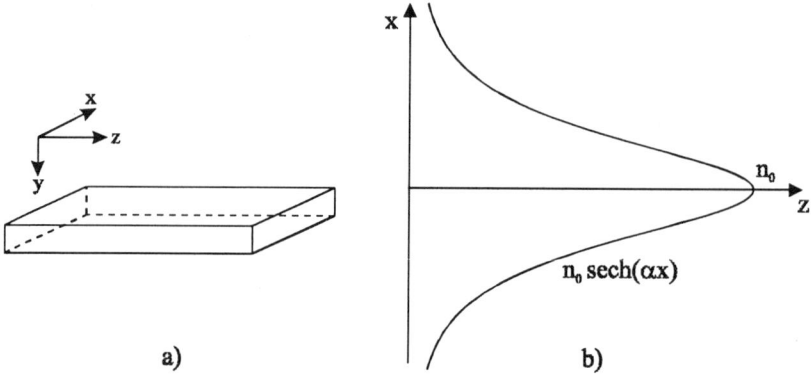

Fig. 1 a) Planar waveguide and b) Hyperbolic secant refractive index profile.

## 2. MATRIX 2. RAY EQUATIONS: FORMALISM

We consider light transmission through a GRIN medium whose unidimensional transverse refractive index profile is given by

$$n(x) = n_0 \operatorname{sech}(\alpha x) \tag{1}$$

where $\alpha$ is a constant, called the profile or gradient parameter, and $n_0$ is the index on the z-axis (Fig. 1). For this profile of the refractive index $\partial n/\partial z = 0$ and the third optical direction cosine of any ray at each point within the medium, that is, $l_0 = n \, (dz/ds)$, is a constant.

While the propagation of rays in selfoc media is studied in real space (x), in the HS media it is best handled in what we term "hyperbolic space" (u) with the following transformation [8]

$$u = \sinh(\alpha x) \tag{2}$$

The above transformation allows the use of the (2×2) matrices which have been developed to handle ray-tracing equations for describing light propagation through any optical system within the framework of the paraxial approximation (ABCD law). For example, the ray position u and ray slope $\dot{u}$ after propagating the light a distance z in the HS waveguide in the new coordinate system u-z, are related to the initial ray position $u_0$ and ray slope $\dot{u}_0$ by the following matrix representation

$$\begin{pmatrix} u(z) \\ \dot{u}(z) \end{pmatrix} = \begin{pmatrix} H_2(z) & H_1(z) \\ \dot{H}_2(z) & \dot{H}_1(z) \end{pmatrix} \begin{pmatrix} u_0 \\ \dot{u}_0 \end{pmatrix} \tag{3}$$

where dots denote derivates with respect to z and $H_1(z)$, $H_2(z)$, are the axial and field rays given by Luneburg.[1] They are written as

$$H_1(z) = \frac{\sin(\alpha z)}{\alpha} \Rightarrow \dot{H}_1(z) = \cos(\alpha z) = H_2(z) \tag{4a}$$

$$H_2(z) = \cos(\alpha z) \Rightarrow \dot{H}_2(z) = -\alpha \sin(\alpha z) = -\alpha^2 H_1(z) \tag{4b}$$

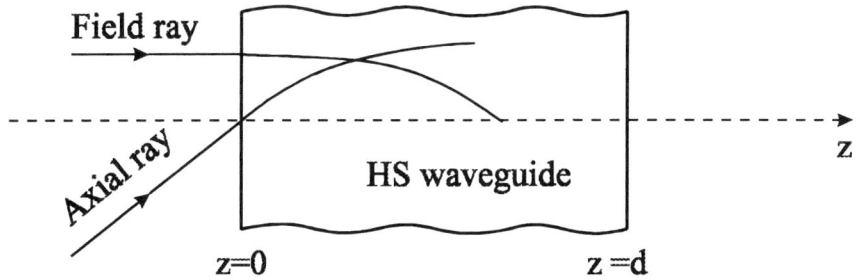

Fig. 2 Axial and Field rays in a HS waveguide.

The initial conditions for these rays are given by

$$H_1(0)=0 \; ; \; \dot{H}_1(0)=1 \tag{5a}$$

$$H_2(0)=1 \; ; \; \dot{H}_2(0)=0 \tag{5b}$$

Thus, the axial ray $H_1$ is a ray which originates at the point of the axis on the input with ray slope of $\pi/4$ and the field ray $H_2$ is a ray which leaves the input at unit height parallel to the axis (Fig. 2).

Taking into account Eqs.(2) and (3), we can determine the ray position x and the ray slope $\dot{x}$ at $z > 0$ in the cartesian system x-z, that is

$$x(z) = \frac{1}{\alpha}\sinh^{-1}[u(z)] = \frac{1}{\alpha}\sinh^{-1}[u_0 H_2(z) + \dot{u}_0 H_1(z)] \tag{6a}$$

$$\dot{x}(z) = \frac{\dot{u}(z)}{\alpha \cosh[\alpha x(z)]} = \frac{u_0 \dot{H}_2(z) + \dot{u}_0 \dot{H}_1(z)}{\alpha \cosh\{\sinh^{-1}[u_0 H_2(z) + \dot{u}_0 H_1(z)]\}} \tag{6b}$$

## 3. FOCUSING AND COLLIMATION CONDITIONS

From Eqs.(3-4) and (6) it follows that any ray leaving the input of the waveguide describes a sinusoidal path through it. This result shows light transmission capabilities and optical transformation in HS waveguides, such as focusing and collimation. On the other hand, there is a method for determining the image and transform conditions in GRIN media based on the evaluation of the zeros of functions $H_1$ and $H_2$. This method can be applied for obtaining the focusing and collimation conditions for a HS waveguide when it is illuminated by a point source at the input.

For a point source in the input of a HS waveguide, the focusing condition is given by

$$H_1(z_m) = \dot{H}_2(z_m) = 0 \tag{7}$$

where $z_m$ are the planes satisfying

$$z_m = \frac{m\pi}{\alpha} \qquad \text{(m, natural)} \tag{8}$$

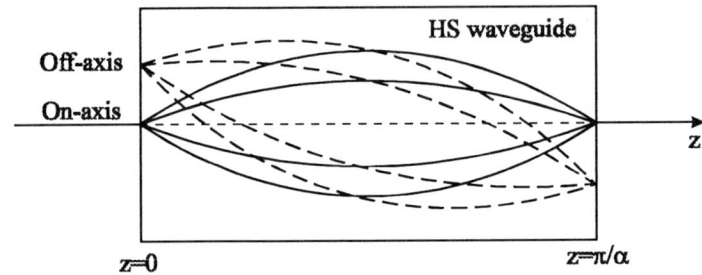

Fig. 3 Light focuser Perfect focusing for an on-axis and an off-axis point source by a HS waveguide.

For $z_m$ the field ray $H_2$ given by Eq.(4b) becomes

$$H_2(z_m) = (-1)^m = \dot{H}_1(z_m) \tag{9}$$

Thus, inserting Eqs.(7) and (9) into (6), the ray position and ray slope at $z_m$ are

$$x(z_m) = (-1)^m x_0 \tag{10a}$$

$$\dot{x}(z_m) = (-1)^m \dot{x}_0 \tag{10b}$$

The ray position at the output remains constant and it is independent of the ray slope at the input. In other words, planar HS waveguides have perfect equalization at $z_m$ for all rays. Therefore, cutting the waveguide at $z_m$ we can obtain stigmatic focusing for a source located at the input and the waveguide may be applied as a light focuser (Fig. 3).

On the other hand, the collimation condition for a HS waveguide is given by

$$H_2(z_p) = \dot{H}_1(z_p) = 0 \tag{11}$$

where $z_p$ are the planes satisfying

$$z_p = \frac{(2p+1)\pi}{2} \quad \text{(p, natural)} \tag{12}$$

For $z_p$ the axial ray reduces to

$$H_1(z_p) = \frac{(-1)^p}{\alpha} = -\frac{\dot{H}_2(z_p)}{\alpha^2} \tag{13}$$

and the ray position and ray slope are

$$x(z_p) = \frac{(-1)^p}{\alpha} \sinh^{-1}[\dot{x}_0 \cosh(\alpha x_0)] \tag{14a}$$

$$\dot{x}(z_p) = (-1)^{p+1} \frac{\sinh(\alpha x_0)}{\cosh\{\sinh^{-1}[\dot{x}_0 \cosh(\alpha x_0)]\}} \tag{14b}$$

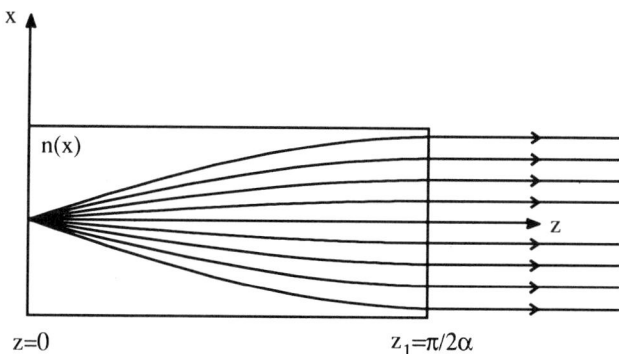

Fig. 4 Light collimator: Perfect collimation for an on-axis point source by a planar HS waveguide.

From Eq.(14b) it follows that we obtain perfect collimation only if $x_0 = 0$, that is, at the output all the rays will be parallel to the z-axis ($x(z_p) = 0$) when the point source is located on-axis at the input of the waveguide (Fig. 4).

On the other hand, from the principle of reversibility of rays, a collimated beam propagating along the axis at $z = 0$ will be focused at $z_p$, just as an on-axis point source at $z = 0$ will produce a parallel beam of rays at $z_p$. So $H_2(z_p) = 0$ is also the focusing condition when a collimated beam propagating along the z-axis impinges on the input of the waveguide.

## 4. TILTED PLANE ILLUMINATION: LIGHT DEFLECTOR AND LIGHT SHIFTER

Now we will analyse light propagation through a planar HS waveguide when it is illuminated by a tilted plane beam.

We assume that a tilted beam of parallel rays impinges on the input of a planar HS waveguide making an angle $\beta$ with the z-axis. In Fig. 5 we show the ray tracing for this tilted

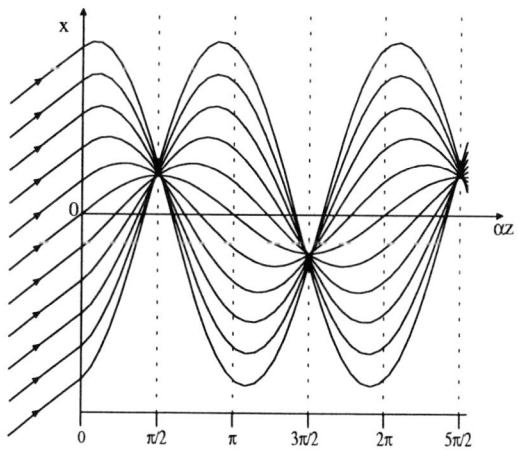

Fig. 5 Propagation of a tilted plane beam through a planar HS waveguide.

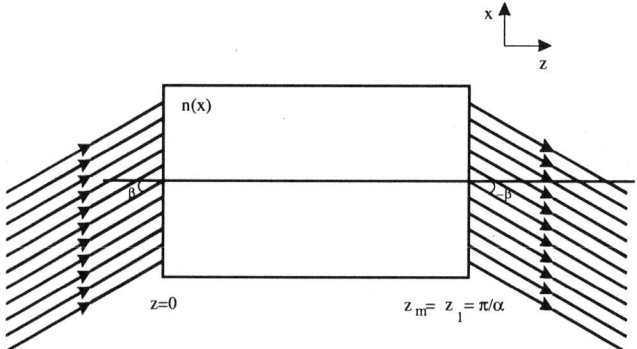

Fig. 6 Light deflector by a planar HS waveguide.

beam through the waveguide. The ray tracing is periodic and we can see that the trajectories of the rays in two consecutive periods are symmetrical with respect to the z-axis.

On the other hand, we can analyse the behavior of the light at planes $z_m$ (focusing condition for a point source on the input) for tilted illumination. In this case, the initial conditions for any ray are given by

$$x_i(z=0) = x_{0i} \tag{15a}$$

$$\dot{x}_i(z=0) = \dot{x}_{0i} = \tan\gamma_i \tag{15b}$$

$\gamma_i$ being the refraction angle given by Snell's law, which can be written as

$$\sin\beta = l_0\,\dot{x}_{0i} \tag{16}$$

provided that the waveguide is surrounded by air.

At these planes, the real position and slope of the rays are expressed as

$$x_i(z_m) = (-1)^m\, x_{0i} \tag{17a}$$

$$\dot{x}_i(z_m) = (-1)^m\, \dot{x}_{0i} = (-1)^m\,\frac{\sin\beta}{l_0} \tag{17b}$$

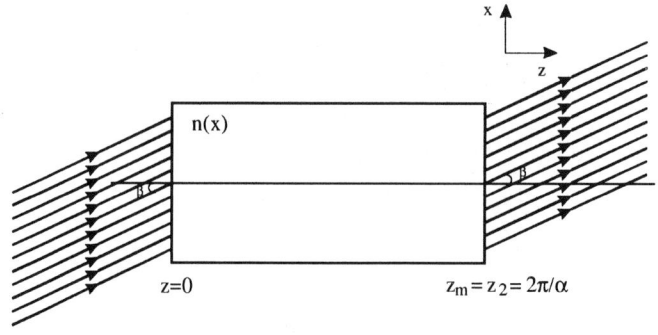

Fig. 7 Light shifter by a planar HS waveguide.

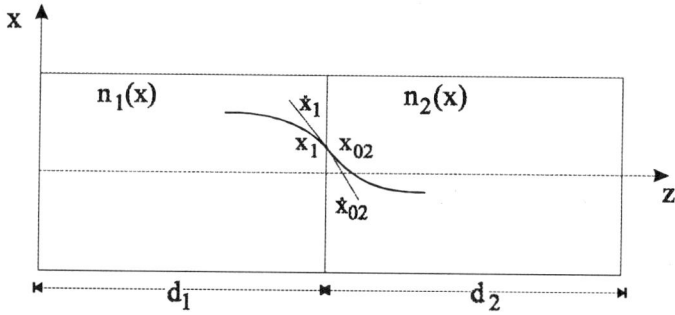

Fig. 8 Geometry, coordinate system, and coupling conditions for HS planar waveguides.

From Eqs.(17) it follows that, at planes $z_m$, the position and slope of the rays depends on the m-order, that is, on the waveguide length and that, at each plane $z_m$, the slope of the rays is a constant.

Thus, if we cut a planar HS waveguide at planes $z_m$, with m an odd integer ($m = 2q + 1$), the exit beam will be a collimated beam tilted at an angle $-\beta$ (Fig. 6). In other words, cutting the waveguide at these planes we can obtain a light deflector of angle $2\beta$. Likewise, if we cut the waveguide at planes $z_m$ with m an even integer ($m = 2q$), the exit beam will have the same propagation direction as the input one. Therefore, the planar HS waveguide only moves the input beam as shown in Fig. 7 and works as a light shifter.

## 5. BEAM SIZE CONTROLLER BY BUTT-JOINING COUPLING OF TWO HS PLANAR WAVEGUIDES

We consider the butt-joining coupling of two HS planar waveguides of lengths $d_1$ and $d_2$ whose refractive index profiles $n_1(x)$ and $n_2(x)$, respectively, are given by functions such as Eq.(1). Fig. 8 shows the geometry of the problem and the coordinate system.[9] Assuming that the z-axis refractive indices of both waveguides are different ($n_{01} \neq n_{02}$) and the same for the profile parameters ($\alpha_1 \neq \alpha_2$), the ray equations for the two waveguides are

$$x_i(z) = \frac{\sinh^{-1}(u_i)}{\alpha_i} \tag{18a}$$

$$\dot{x}_i(z) = \frac{\dot{u}_i}{\alpha_i \cosh(\alpha_i x_i)} \quad , i = 1, 2 \tag{18b}$$

where Eqs.(6) have been used.

The input coordinates of any ray in the second waveguide, in hyperbolic space, expressed as a function of the output coordinates in the first waveguide, can be written as

$$u_{02}(d_1) = \sin\left\{\frac{\alpha_2}{\alpha_1} \sinh^{-1}[u_1(d_1)]\right\} \tag{19a}$$

$$\dot{u}_{02}(d_1) = \frac{l_{01} \alpha_2 \cosh[\alpha_2 x_1(d_1)]}{l_{02} \alpha_1 \cosh[\alpha_1 x_1(d_1)]} \dot{u}_1(d_1) \tag{19b}$$

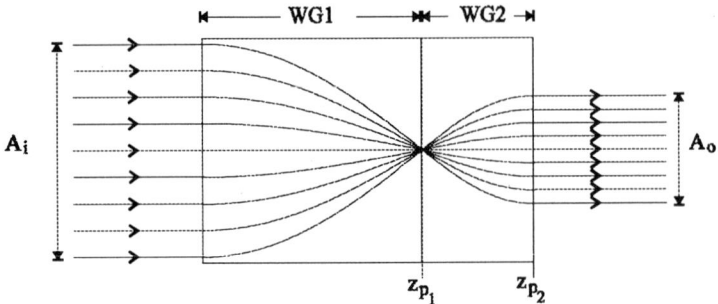

Fig. 9 Beam-size contraction by butt-joining of HS optical waveguides. Ray tracing calculations have been made for $n_{01}=n_{02}$ and $\alpha_2 = 2\,\alpha_1$.

where Snell's law for the interface between both HS waveguides

$$l_{01}\,\dot{x}_{01} = l_{02}\,\dot{x}_{02} \tag{20}$$

has been used, $l_{01}$ and $l_{02}$ being the third optical direction cosines in both waveguides.

From Eqs.(19) and (3) it follows that the ray equations for the second waveguide are given by

$$u_2(z) = \dot{u}_{02}(d_1)H_1^{(2)}(z) + u_{02}(d_1)H_2^{(2)}(z) \tag{21a}$$

$$u_2(z) = \dot{u}_{02}(d_1)\dot{H}_1^{(2)}(z) + u_{02}(d_1)\dot{H}_2^{(2)}(z), \qquad d_1 \le z \le d_1 + d_2 \tag{21b}$$

where $H_1^{(2)}$ and $H_2^{(2)}$ are the axial and field rays in the second waveguide, respectively.

Assuming that the lengths of the waveguides are $d_1 = (2p_1 + 1)\,\pi/2\alpha_1$ and $d_2 = (2p_2 + 1)\,\pi/2\alpha_2$, respectively, ($p_1$ and $p_2$ are integers), and that the coupled system is illuminated by a collimated beam of rays propagating parallel to the z-axis, Eqs.(21) at the output reduce to

$$u_2(d_1 + d_2) = (-1)^{p_1+p_2+1}\,\frac{l_{01}}{l_{02}}\,u_{01} \tag{22a}$$

$$\dot{u}_2(d_1 + d_2) = 0 \tag{22b}$$

where the initial conditions

$$u_1(z=0) = u_{01}\ ;\ \dot{u}_1(z=0) = \dot{u}_{01} = 0 \tag{23}$$

and Eqs.(18) and (19) have been used.

From Eqs.(22) and (23) it follows that at the output we have a collimated beam propagating in the same direction as the input beam (Fig. 9). However, the apertures of the input and output beams are not equal and the relationship between them can be obtained from the ray equation of the input marginal ray, denoted by $u_{0M1}$. This relationship is given by

$$\frac{A_o}{A_i} = \frac{\alpha_1}{\alpha_2}\,\frac{\sinh^{-1}\left(\frac{l_{01}}{l_{02}}u_{0M1}\right)}{\sinh^{-1}(u_{0M1})} = \frac{1}{\alpha_2\,x_{0M1}}\sinh^{-1}\left(\frac{l_{01}}{l_{02}}u_{0M1}\right) \tag{24}$$

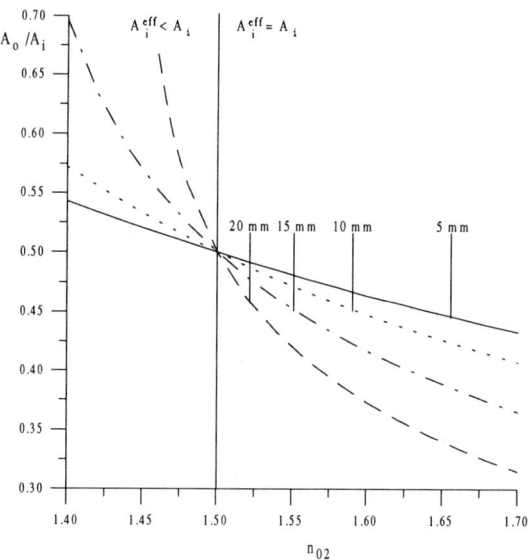

Fig. 10 Variation of the input and output beam aperture relationship versus $n_{o2}$ for different heights of the input marginal ray. Calculations have been made for $n_{01} = 1.5$, $\alpha_1 = 0.1$ mm$^{-1}$ and $\alpha_2 = 2\alpha_1$.

where $A_i$ and $A_o$ are the input and output beam apertures, respectively, and the following equation has been used

$$u_{0M1} = \sinh(\alpha_1 x_{0M1}),\qquad(25)$$

$x_{0M1}$ being the real position of the marginal ray at the input of the coupled system.
Note that when $n_{01} = n_{02}$ ($l_{01} = l_{02}$), eq.(24) becomes

$$\frac{A_o}{A_i} = \frac{\alpha_1}{\alpha_2}\qquad(26)$$

On the other hand, the relationship between beam apertures must be defined in two ways:
- If $n_{01} > n_{02}$ there is a threshold value for the input ray height given by

$$x_{0M1}^{max} = \frac{1}{\alpha_1}\cosh^{-1}\left[\frac{n_{01}}{(n_{01}^2 - n_{02}^2)^{1/2}}\right]\qquad(27)$$

where the total reflection condition has been used

$$n_i^2(x_{0M1}^{max}) = n_{01}^2 - n_{02}^2\qquad(28)$$

Beyond this value all the rays are totally reflected at the interface of waveguides and we must define an input effective aperture as

$$A_i^{eff} = 2x_{0M1}^{max} \qquad (A_i^{eff} < A_{wg})\qquad(29)$$

$A_{wg}$ being the waveguide aperture.

Then, at the input there is a ray limitation that is due to total reflection on the interface of waveguides. In this case Eq.(24) can be rewritten as

$$\frac{A_o}{A_i^{eff}} = \frac{1}{\alpha_2 \, x_{OM1}^{max}} \sinh^{-1}\left(\frac{l_{01}}{l_{02}} u_{OM1}^{max}\right) \tag{30}$$

where Eq.(25) has been used.
- If $n_{01} < n_{02}$ there is no possibility of total reflection at the interface of waveguides. In this case

$$A_i^{eff} = A_i = A_{wg} \tag{31}$$

and Eq.(24) gives the relationship between beam apertures.

Fig. 10 shows the variation of the aperture relationship versus $n_{02}$ for different height values of the input marginal ray. The two cases described above are clearly represented. Note that the cut-off point of curves represents the critical angle at the interface, and that we can contract the beam size by choosing the appropriate values of the waveguide parameters. From the principle of reversibility of rays, we can also expand the size of a beam and this coupled system works like a beam size controller device. Likewise, the same aperture relationship can be achieved in different ways. For instance, if $n_{01} = n_{02} = 1.5$ and $\alpha_2 = 2\alpha_1$, $A_0 = A_i/2$. One can also design this beam controller device by coupling two HS planar waveguides whose parameters are $n_{01} = 1.5$, $n_{02} = 1.4$, and $\alpha_2 = 2.125\alpha_1$; however, in this case it is necessary to take into account the input effective aperture, in particular, the value of $\alpha_2$ has been calculated for $A_i^{eff} = 10$ mm.

## 6. CONCLUSIONS

Using geometrical optics we have obtained the conditions of focusing and collimation for a planar waveguide with HS refractive index profile. Light propagation through this GRIN medium illuminated by a tilted plane beam has been studied, as well as the design of a light deflector and a light shifter. We have designed an optical device for transforming a beam size in such a way that expansion or contraction can be controlled by butt-joining of two HS planar waveguides. The beam size controller, the focuser and the collimator that have been presented are three important options for designers of integrated optics to take into account.

ACKNOWLEDGMENTS. This work was supported by the CICYT, Ministerio de Transporte, Turismo y Comunicaciones under contract TIC no 0846/95.

## REFERENCES

1. R. K. Luneburg, *Mathematical Theory of Optics*, U. California Press, Berkeley, Ca. (1964).
2. E. W. Marchand, *Gradient Index Optics*, Academic Press, New York (1978).
3. K. Iga, Y. Kokubun, and M. Oikawa, *Fundamentals of Microoptics*, Academic Press, New York (1984).
4. C. Gómez-Reino, GRIN optics and its application in optical connections, *Int. J. Optoelectron.* 7: 607 (1992).
5. J. Liñares, and C. Gómez-Reino, The optical propagator in a graded-index medium with hyperbolic secant refractive index profile, *Appl. Opt.* 33: 3427 (1994).

6. A. Fletcher, T. Murphy, and A. Young, Solutions of two optical problems, *Proc. R. Soc. London Ser. A.* 233: 216 (1954).
7. D. Bertilone, A. Ankiewicz, and C. Pask, Wave propagation in a graded-index taper, *Appl. Opt.* 26: 2213 (1987).
8. C. Bao, and C. Gómez-Reino, Collimation and focusing properties of a planar waveguide with hyperbolic secant refractive index profile: light deflector and collimator design, *Opt. Quantum. Electron.* 27: 897 (1995).
9. C. Bao, M. V. Pérez, and C. Gómez-Reino, Coupling of planar waveguides with hyperbolic secant refractive-index profile by butt-joining: beam size controller design, *Opt. Lett.* 21: 1078 (1996).

# NUMERICAL ANALYSIS OF SURFACE RELIEF GRATINGS

R. Orta, S. Bastonero, and R. Tascone

Electronics Department and CESPA (C.N.R.)
Polytechnic of Turin,
Corso Duca degli Abruzzi 24, 10129, Torino

## 1. INTRODUCTION

Diffraction gratings are components of great interest for their many applications, for instance in integrated optics, holography and spectroscopy.[1] Among the methods introduced for the analysis of diffraction gratings one may recall the integral method,[2] the coupled wave method,[1,3] and the differential method.[4] In this paper we will discuss a modal method for the analysis of lamellar surface relief diffraction gratings.

Due to the step-like form of the surface of such a grating (Fig.1), the approach is to view the structure as a series of junctions between two periodic arrays of slab waveguides with the same period and different slab thickness and to compute the Generalized Scattering Matrix (GSM) of each junction. With the term GSM we indicate the scattering matrix relative to both propagating and cutoff slab modes. Then the characterization of the grating is obtained by combining the various GSMs.

In carrying out this program we will use a *modal transmission line approach* which is based on a reformulation of field theory in microwave network terms.[5]

The structure of the Chapter is the following. Section 2 is devoted to a brief summary of the normal mode theory, with the essential purpose of defining the notation. Section 3 presents the solution of the basic scattering problem in the form of the GSM of the junction. Section 4 discusses how to compute the efficiency of the complete grating and some numerical results are presented.

## 2. REVIEW OF MODE THEORY FOR DIELECTRIC WAVEGUIDES

To establish the notation, let us review the basic facts of normal mode theory. Consider a dielectric waveguide with axis parallel to $\hat{z}$. As well known,[5] the transverse components of the electric and magnetic field are the independent state variables, whereas the longitudinal variables $E_z$, $H_z$ are dependent and can be eliminated. By carrying out this

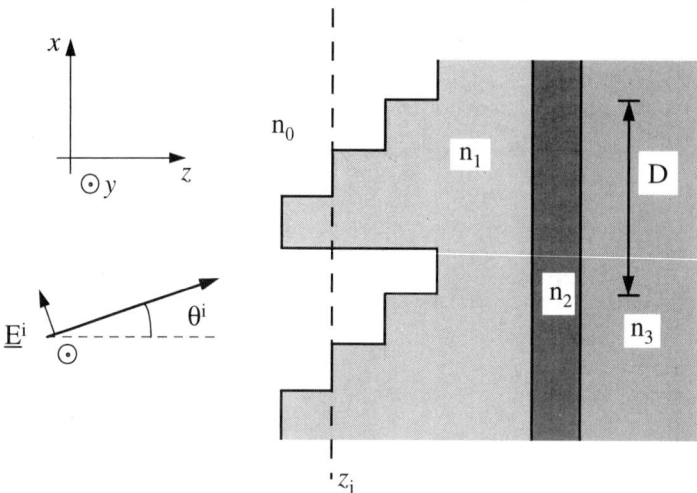

Fig. 1 Geometry of a lamellar surface relief diffraction grating

procedure, which requires the decomposition of all vectors in transverse (subscript t) and longitudinal components, the Maxwell equations become:

$$\begin{cases} -\dfrac{\partial}{\partial_z} \underline{E}_t = j\omega \left[ \mu \underline{\underline{1}} + \dfrac{1}{\omega^2} \nabla_t \dfrac{1}{\epsilon} \nabla_t \right] \cdot (\underline{H}_t \times \hat{z}) + \underline{M} \times \hat{z} \\ -\dfrac{\partial}{\partial_z} \underline{H}_t = j\omega \left[ \epsilon \underline{\underline{1}} + \dfrac{1}{\omega^2} \nabla_t \dfrac{1}{\mu} \nabla_t \right] \cdot (\hat{z} \times \underline{E}_t) + \hat{z} \times \underline{J} \end{cases} \quad (1)$$

where $\underline{J}$ and $\underline{M}$ are transverse electric and magnetic current densities. The temporal factor $\exp(j\omega t)$ is implied and suppressed. Dyadics[6] are underlined twice and $\underline{\underline{1}}$ denotes the identity dyadic in the transverse plane. Eq.(1), completed with the appropriate boundary conditions, are called the Marcuvitz - Schwinger equations and form the modern mathematical basis for the formulation of all types of waveguide problems.[5]

The transverse electric and magnetic fields can be expressed as linear combinations of modes, i.e. source free solutions of Eq. (1):

$$\underline{E}_t(\underline{\rho},z) = \sum_i V_i(z) \underline{e}_i(\underline{\rho}) \qquad \underline{H}_t(\underline{\rho},z) = \sum_i I_i(z) \underline{h}_i(\underline{\rho}) \quad (2)$$

In general, the normal modes of dielectric waveguides are hybrid and the mode vector eigenfunctions $\underline{e}_i, \underline{h}_i$ possess nonvanishing longitudinal components $e_{zi}$, $h_{zi}$. The expansion coefficients $V_i$, $I_i$ are the modal voltage and current and satisfy the transmission line equations

$$\begin{cases} -\dfrac{dV_i}{dz} = jk_{zi}Z_i I_i + \upsilon_i \\ -\dfrac{dI_i}{dz} = jk_{zi}Y_i V_i + i_i \end{cases} \quad (3)$$

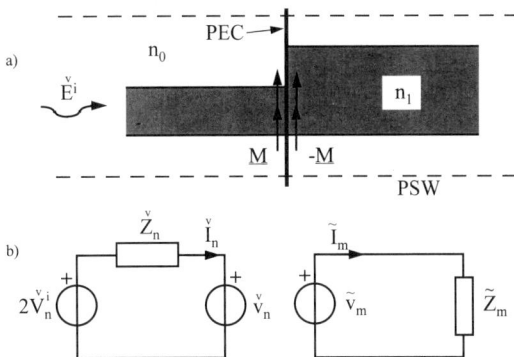

Fig. 2 Basic junction problem. a) Application of the equivalence theorem. PEC: Perfect electric conductor. PSW: Phase shift wall. b) Modal equivalent circuit.

where $k_{zi}$ is the mode propagation constant and $Z_i = 1/Y_i$ is the modal impedance. The generator terms $v_i$, $i_i$ represent the electromagnetic sources in circuit terms. These equations are obtained by projection of Eq. (1) on the mode eigenfunctions, exploiting their orthonormality properties. It can be proved in fact that

$$<\underline{e}_i, \underline{h}^+ \times \hat{z}> = \int \underline{e}_i(\underline{\rho}) \cdot \underline{h}_j^+(\underline{\rho}) \times \hat{z} \, d\underline{\rho} = \delta_{ij} \quad (4)$$

where the superscript "+" denotes the eigenfunction of the adjoint problem and $<,>$ denotes a symmetric inner product.

The expressions for the voltage and current generators are:

$$v_i = <\underline{M}, \underline{h}_i^+> \qquad i_i = <\underline{J}, \underline{e}_i^+> \quad (5)$$

## 3. JUNCTION BETWEEN DIELECTRIC WAVEGUIDES

As anticipated above, the main problem in the analysis of the grating of Fig.1 is the characterization of the junction, at $z = z_j$, between two regions consisting of periodic arrays of dielectric slabs, with the same periodicity (lattice step D) but with different "duty cycles". The incident field is assumed to be that of a plane wave with wavevector in the xz plane, forming the angle $0^i$ with the z axis. The incident plane wave causes a phase shift over the period, and the structure can be modelled as an inhomogeneously filled waveguide with phase shift walls (see Fig.2a).

This scattering problem can be formulated in terms of an integral equation, which is obtained by the Equivalence Theorem.[7] The two regions are decoupled by introducing a perfectly conducting electric or magnetic plate, on which suitable magnetic or electric currents guarantee that the new problem is equivalent to the original one.

The boundary conditions require the continuity of the tangential electric and magnetic fields at the junction. In the case where an electric conductor is introduced, the continuity of the electric field is ensured by assuming that the magnetic current distributions are opposite to each other on the two sides of the junction. Enforcing the continuity of the magnetic field leads to an integral equation that is termed the H-field integral equation (HFIE). The unknown is the magnetic current distribution $\underline{M}(\underline{\rho}) = \underline{E}_t(\underline{\rho}, z_j) \times (-\hat{z})$, which is essentially the electric field at the junction. The kernel of the equation is the Green function of the structure, which is easily derived as an eigenfunction expansion in terms of the modes of the two regions. Fig.2b shows the equivalent modal circuit of the junction. The superscripts ˘ and ˜ are used to denote quantities relative to the left ($z < z_j$) and right region, respectively. The unknown voltage generators $\breve{v}_n$ and $\tilde{v}_n$ are given by Eq. (5):

$$\breve{v}_n = <\underline{M}, \underline{\breve{h}}_n^+ > \qquad \tilde{v}_m = <\underline{M}, \underline{\tilde{h}}_m^+ > \qquad (6)$$

In general, $\underline{h}_i^+$ is the magnetic field eigenfunction corresponding to the $-\theta^i$ incidence angle. For lossless dielectrics, changing the sign of the incidence angle has the same effect as taking the complex conjugate of the mode function.

The voltage generator $2\breve{V}_n^i$ accounts for the incident field and is computed via

$$\breve{V}_n^i = <\underline{\breve{E}}_t^i, \underline{\breve{h}}_n^+ \times \hat{z}> \qquad (7)$$

The magnetic field on the two sides of the junction is computed easily from the modal circuit of Fig.2:

$$\underline{\breve{H}}_t = \sum_n \breve{I}_n \underline{\breve{h}}_n = \sum_n (2\breve{V}_n^i - \breve{v}_n)\breve{Y}_n \underline{\breve{h}}_n \qquad (8)$$

$$\underline{\tilde{H}}_t = \sum_m \tilde{I}_m \underline{\tilde{h}}_m = \sum_m \tilde{v}_m \tilde{Y}_m \underline{\tilde{h}}_m \qquad (9)$$

The boundary condition to be enforced requires that

$$\underline{\breve{H}}_t(\underline{\rho}) - \underline{\tilde{H}}_t(\underline{\rho}) = 0 \qquad \forall \underline{\rho} \quad . \qquad (10)$$

This equation, by substitution of Eqs. (6-9), can be rewritten as the following *magnetic field integral equation* (HFIE):

$$\int \underline{\underline{G}}^H(\underline{\rho},\underline{\rho}') \cdot \underline{M}(\underline{\rho}') d\underline{\rho}' = 2\underline{\tilde{H}}_t^i(\underline{\rho}) \qquad \forall \underline{\rho} \qquad (11)$$

where the kernel (magnetic Green function) is given by the eigenfunction expansion:

$$\underline{\underline{G}}^H(\underline{\rho},\underline{\rho}') = \sum_n \breve{Y}_n \underline{\breve{h}}_n(\underline{\rho}) \underline{\breve{h}}_n^+(\underline{\rho}') + \sum_m \tilde{Y}_m \underline{\tilde{h}}_m(\underline{\rho}) \underline{\tilde{h}}_m^+(\underline{\rho}') \qquad (12)$$

As remarked above, the equivalence theorem can also be applied in a different way, by inserting a magnetic plane at $z = z_j$. The unknown, in this case, is the equivalent electric current distribution $\underline{J}(\underline{\rho})$ introduced on the two sides of the junction ( with opposite signs, so that the continuity of the magnetic field is guaranteed). Enforcing the continuity of the tangential electric field yields an *electric field integral equation* (EFIE), the dual of Eq. (11):

$$\int \underline{\underline{G}}^E(\underline{\rho},\underline{\rho}') \cdot \underline{J}(\underline{\rho}') d\underline{\rho}' = 2\underline{\breve{E}}_t^i(\underline{\rho}) \qquad \forall \underline{\rho} \quad , \qquad (13)$$

where the electric Green function is computed through the eigenfunction expansion:

$$\underline{\underline{G}}^E(\underline{\rho},\underline{\rho}') = \sum_n \breve{Z}_n \underline{\breve{e}}_n(\underline{\rho})\underline{\breve{e}}_n^+(\underline{\rho}') + \sum_m \tilde{Z}_m \underline{\tilde{e}}_m(\underline{\rho})\underline{\tilde{e}}_m^+(\underline{\rho}') \quad, \tag{14}$$

which is derived from a modal circuit dual of that of Fig.2. Note that Eqs. (11) and (13) are just two examples; other integral equations can be obtained by applying the equivalence theorem in different forms.

## 3.1. The Method of Moments

The integral Eqs. (11) and (13) can be solved numerically by the *method of moments* [8, 9] *(also known as the method of weighted residuals)*. Let us illustrate the method with reference to a linear functional equation which, in abstract terms, can be written as

$$L\underline{f} = \underline{g} \tag{15}$$

where $\underline{f}$ (unknown) and $\underline{g}$ (known) are Hilbert space vectors and $L$ is a linear operator. Two (infinite in principle) sets of functions are introduced, the *basis functions*, represented by the vectors $\{\underline{f}_n\}$ and the *test functions*, represented by $\{\underline{w}_n\}$. The former are used to expand the unknown $\underline{f}$:

$$\underline{f} = \sum_n x_n \underline{f}_n \tag{16}$$

where $x_n$ are unknown coefficients. Substituting into Eq. (15) we get

$$\sum_n x_n L\underline{f}_n = \underline{g}_m \tag{17}$$

which is converted to a linear system of algebraic equations by taking projections on $\underline{w}_m$:

$$\sum_n <L\underline{f}_n, \underline{w}_m> x_n = <\underline{g}, \underline{w}_m> \quad \forall m \tag{18}$$

or

$$[L_{mn}][x_n] = [g_m] \tag{19}$$

from which $[x_n]$ is obtained by truncation to size N x N and inversion of the matrix $[L_{mn}]$.

## 3.2. Solution of the Magnetic Field Integral Equation

Let us apply the method of moments to the solution of the HFIE Eq. (11). Let $\{\underline{f}_k(x)\}$ be a suitable set of basis functions: a small number of them must be capable of accurately approximating the unknown. Moreover, let $\{\underline{w}_\ell(x)\}$ be a set of suitable test functions, i.e., such that a small number of them can span the range of the linear operator. It is convenient to introduce a matrix formalism:

$$\breve{F}_{nk} = <\underline{f}_k, \underline{\breve{h}}_n^+> \qquad \breve{W}_{n\ell} = <\underline{w}_\ell, \underline{\breve{h}}_n> \tag{20}$$

$$\tilde{F}_{nk} = <\underline{f}_k, \underline{\tilde{h}}_n^+> \qquad \tilde{W}_{n\ell} = <\underline{w}_\ell, \underline{\tilde{h}}_n> \quad. \tag{21}$$

Then the linear system Eq. (19) takes the form:

$$\left[\underline{\underline{\breve{W}}}^T \cdot \underline{\underline{\breve{Y}}} \cdot \underline{\underline{\breve{F}}} + \underline{\underline{\widetilde{W}}}^T \cdot \underline{\underline{\widetilde{Y}}} \cdot \underline{\underline{\widetilde{F}}}\right] \cdot \underline{x} = 2 \underline{\underline{\breve{W}}}^T \cdot \underline{\underline{\breve{Y}}} \cdot \underline{\breve{V}}^i \qquad (22)$$

where $\underline{\underline{\breve{Y}}}$ and $\underline{\underline{\widetilde{Y}}}$ are the diagonal matrices of the modal admittances. Let us denote by $\underline{\underline{Q}}$ the inverse of the coefficient matrix:

$$\underline{\underline{Q}} = \left[\underline{\underline{\breve{W}}}^T \cdot \underline{\underline{\breve{Y}}} \cdot \underline{\underline{\breve{F}}} + \underline{\underline{\widetilde{W}}}^T \cdot \underline{\underline{\widetilde{Y}}} \cdot \underline{\underline{\widetilde{F}}}\right]^{-1}. \qquad (23)$$

Then the magnetic current at $z = z_j$ is known through the coefficients of the basis functions:

$$\underline{x} = 2 \underline{\underline{Q}} \cdot \underline{\underline{\breve{W}}}^T \cdot \underline{\underline{\breve{Y}}} \cdot \underline{\breve{V}}^i . \qquad (24)$$

The most convenient way to characterize the junction between the two arrays of dielectric waveguides is to use the Generalized Scattering Matrix (GSM), i.e. the one referred to both propagating and cut-off modes. This can be computed easily on the basis of the circuits of Fig.2. In fact, the scattered voltages are given by

$$\breve{V}_n^s = \breve{v}_n - \breve{V}_n^i \qquad \widetilde{V}_n^s = \widetilde{v}_n \qquad (25)$$

and noting that

$$\breve{v}_n = \sum_k \breve{F}_{nk} x_k \qquad \widetilde{v}_n = \sum_k \widetilde{F}_{nk} x_k \qquad (26)$$

we obtain

$$\underline{\breve{V}}^s = 2 \underline{\underline{\breve{F}}} \cdot \underline{\underline{Q}} \cdot \underline{\underline{\breve{W}}}^T \cdot \underline{\underline{\breve{Y}}} \cdot \underline{\breve{V}}^i - \underline{\breve{V}}^i \qquad (27)$$

$$\underline{\widetilde{V}}^s = 2 \underline{\underline{\widetilde{F}}} \cdot \underline{\underline{Q}} \cdot \underline{\underline{\breve{W}}}^T \underline{\underline{\breve{Y}}} \cdot \underline{\breve{V}}^i . \qquad (28)$$

If we now proceed in the same way as before by assuming a right hand side incidence $\underline{\widetilde{V}}^i$, we derive the GSM of the junction:

$$\begin{bmatrix} \underline{\breve{V}}^s \\ \underline{\widetilde{V}}^s \end{bmatrix} = \begin{bmatrix} 2 \underline{\underline{\breve{F}}} \cdot \underline{\underline{Q}} \cdot \underline{\underline{\breve{W}}}^T \cdot \underline{\underline{\breve{Y}}} - \underline{\underline{1}} & 2 \underline{\underline{\breve{F}}} \cdot \underline{\underline{Q}} \cdot \underline{\underline{\widetilde{W}}}^T \cdot \underline{\underline{\widetilde{Y}}} \\ 2 \underline{\underline{\widetilde{F}}} \cdot \underline{\underline{Q}} \cdot \underline{\underline{\breve{W}}}^T \cdot \underline{\underline{\breve{Y}}} & 2 \underline{\underline{\widetilde{F}}} \cdot \underline{\underline{Q}} \cdot \underline{\underline{\widetilde{W}}}^T \cdot \underline{\underline{\widetilde{Y}}} - \underline{\underline{1}} \end{bmatrix} \cdot \begin{bmatrix} \underline{\breve{V}}^i \\ \underline{\widetilde{V}}^i \end{bmatrix}. \qquad (29)$$

### 3.3. Solution of the electric field integral equation

The procedure for the solution of the electric field integral equation Eq. (13) is very similar and will not be reported in detail. In this case the basis functions are chosen to expand the electric current distribution. The GSM of the same junction, according to the EFIE formulation can be shown to be

$$\begin{bmatrix} \underline{\breve{V}}^s \\ \underline{\widetilde{V}}^s \end{bmatrix} = \begin{bmatrix} \underline{\underline{1}} - 2 \underline{\underline{\breve{Z}}} \cdot \underline{\underline{\breve{F}}} \cdot \underline{\underline{Q}} \cdot \underline{\underline{\breve{W}}}^T & 2 \underline{\underline{\breve{Z}}} \cdot \underline{\underline{\breve{F}}} \cdot \underline{\underline{Q}} \cdot \underline{\underline{\widetilde{W}}}^T \\ 2 \underline{\underline{\widetilde{Z}}} \cdot \underline{\underline{\widetilde{F}}} \cdot \underline{\underline{Q}} \cdot \underline{\underline{\breve{W}}}^T & \underline{\underline{1}} - 2 \underline{\underline{\widetilde{Z}}} \cdot \underline{\underline{\widetilde{F}}} \cdot \underline{\underline{Q}} \cdot \underline{\underline{\widetilde{W}}}^T \end{bmatrix} \cdot \begin{bmatrix} \underline{\breve{V}}^i \\ \underline{\widetilde{V}}^i \end{bmatrix}. \qquad (30)$$

where the matrix $\underline{\underline{Q}}$ is defined by

$$\underline{\underline{Q}} = \left\{ \underline{\underline{\breve{W}}}^T \cdot \underline{\underline{\breve{Z}}} \cdot \underline{\underline{\breve{F}}} + \underline{\underline{\tilde{W}}}^T \cdot \underline{\underline{\tilde{Z}}} \cdot \underline{\underline{\tilde{F}}} \right\}^{-1} \tag{31}$$

and the matrices $\underline{\underline{F}}$ and $\underline{\underline{W}}$ are defined by equations similar to Eq. (20) with the magnetic field eigenfunctions $\underline{h}_i$ substituted with the electric field ones $\underline{e}_i$.

### 3.4. Selection of Basis and Test Functions

It is well known that in the presence of a dielectric wedge with a dielectric constant $\epsilon_1$, embedded in a medium with dielectric constant $\epsilon_2$, some field components may diverge approaching the edge. In particular,[10] the magnetic field and the component of the electric field along the edge are finite, whereas the electric field components normal to the edge are $O(\rho^t)$ where $\rho$ is the distance from the edge and the critical exponent t is given by

$$t = \frac{2}{\pi} \cos^{-1}\left( \frac{1}{2} \frac{\epsilon_1 - \epsilon_2}{\epsilon_1 + \epsilon_2} \right) - 1 \tag{32}$$

when the wedge angle is 90° or 270°. For these reasons, it is convenient to distinguish the cases of TE and TM polarization. In the first case, both the electric field ($E_y$) and the magnetic field are finite and continuous at the edge and sets of Legendre polynomials, defined in correspondence of each dielectric region, may be used as basis and test functions. Another possibility is to use a Fourier basis.

In the case of TM polarization, the relevant field components are $E_x$ (divergent) and $H_y$ (finite). If the HFIE formulation is used, it is convenient to select basis functions with the divergent behaviofour Eq. (32). To this end, let us introduce in each dielectric region a set of Gegenbauer polynomials $\{C_n^\alpha(x)\}$ multiplied by their weight function $(1 - x^2)^{\alpha - 0.5}$, where $x \in [-1, 1]$ is a normalized variable. By a suitable selection of the parameter a, these functions are characterized by a divergence rate equal to that required by Eq. (32). In this way, a few expansion functions can accurately approximate the electric field at $z = z_j$. As for the test functions, they are used to enforce the continuity of the magnetic field, which is finite at the edge. Hence, Legendre polynomials, as in the TE case, can be used. It should be remarked that with these choices of expansion and test functions, all the matrix elements can be computed in closed form, without numerical integration.

A very common method for the analysis of waveguide discontinuities is the Mode Matching technique, whereby the electric and magnetic fields on each side of the junction are expressed in modal terms and the unknown modal coefficients are determined by enforcing the continuity of the tangential fields. It can be shown[11] that this technique is a particular case of the method of moments, where the mode eigenfunctions of the two regions are used as basis and test functions. According to which set of modes is taken, different implementations of the method are possible.

For the numerical evaluation of the junction GSM given by Eqs. (29) or (30) it is clear that we must use a finite number of basis and test functions, as well as a finite number of modes in the two regions. As is well known, the two numbers need not be equal but are related because of the relative convergence phenomenon. The number of expansion functions, and hence the size of the matrix to be inverted, can be small thanks to a careful choice of functions, as discussed above. The modes of the two regions adjacent to the

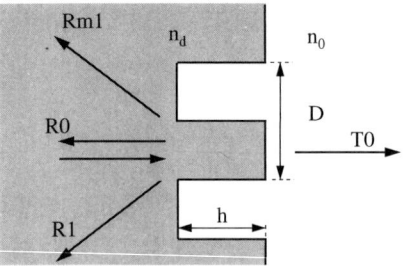

Fig. 3 Geometry of a GIRO grating. no = 1, nd = 3.49, D = 930pm, h = 465pm. Normal incidence.

junction play roles that it is convenient to keep distinct. On one hand, the modes are the ports with respect to which the junction GSM is defined, and their number depends on the accuracy with which the interaction with the adjacent discontinuities is to be described. On the other hand, the modes are used to represent the Green function of the problem (see Eqs. (12, 14)), and hence the accuracy of the solution of the junction problem grows with the number of modes employed to evaluate the matrices Q defined in Eqs. (23) and (31).

It can be noted that the HFIE formulation is convenient in the case of TM polarization and the EFIE in the case of TE polarization. In fact, the convergence of the series is accelerated by the presence of the modal admittances in the former case and of the impedances in the latter, both decreasing functions of the modal order. Specific series acceleration techniques can be useful to reduce the computer time.

## 4. GRATING ANALYSIS

According to the strategy indicated in the Introduction, the grating is decomposed in a series of junctions, each of which is characterized by its GSM. The GSM of the complete structure is computed by a cascade combination of the various GSMs. To this end, it is extremely convenient to subdivide the modes on the two sides of a junction in two classes. *Accessible modes* are those (above and below cut-off) that effectively contribute to the coupling with the adjacent discontinuity located a distances apart, because their attenuation

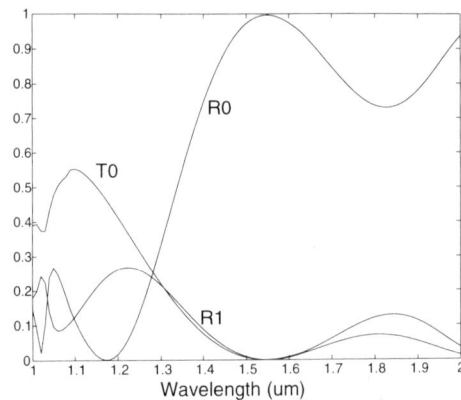

Fig. 4 TM response of the GIRO grating of Fig.3. Normal incidence.

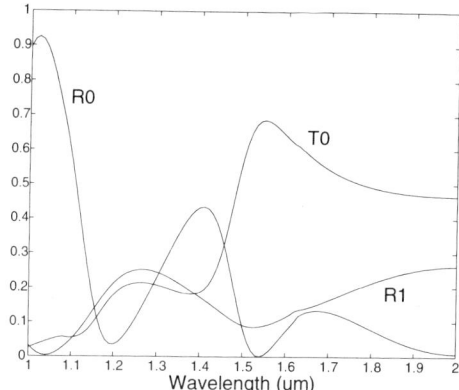

Fig. 5 TE response of the GIRO grating of Fig.3. Normal incidence.

over s is less than a specified threshold. *Localized modes* are cut-off modes that are so attenuated that they "do not see" the neighbouring discontinuity and contribute just to reactive energy storing. Clearly, in the combination process, only the ports corresponding to the accessible modes need to be connected and this results often in the inversion of very small matrices.

It is to be remarked, furthermore, that this procedure is absolutely stable and accurate from a numerical point of view. The well known problems encountered in the analysis of deep gratings were due to the fact that a formulation based on the multiplication of the transmission matrices of the junctions was used. In presence of modes with high attenuation, even if overflow does not occur, linear dependence between the columns of the transmission matrices is introduced because of the finite arithmetic, which prevents one from obtaining the grating efficiencies.

The theory presented above has been applied, as an example, to the analysis of a GIRO[12] grating, shown in Fig.3, optimized for TM polarization at $\lambda = 1.55$ μm and normal incidence. Fig.4 and Fig.5 show the response of this structure for the TM and TE polarizations. R0,R1 and Rm1 are the reflection efficiencies for the orders 0, 1 and –1; T0 is the transmission efficiency for the order 0, the only order in propagation. These curves have been obtained by the method of moments using 7 expansion functions and 21 modes in the dielectric waveguides. The accessible modes, for the computation of the interaction between the two junctions, defined with reference to an attenuation threshold of 30 dB, are 7 in number.

The use of the mode matching technique, with 7 or a higher number of modes, produces essentially the same results. The behaviour is very different for an incidence angle $0^i = 30°$, where the method of moments performs better, as shown in Fig.6. In this case, convergence is already attained with 7 expansion functions, whereas the mode matching approach is still far from convergence.

## 5. CONCLUSION

In this chapter we have presented a modal method for the analysis of lamellar surface relief dielectric gratings. The scattering problem is formulated in terms of an integral equation, and the use of circuit techniques proves to be very convenient in its derivation.

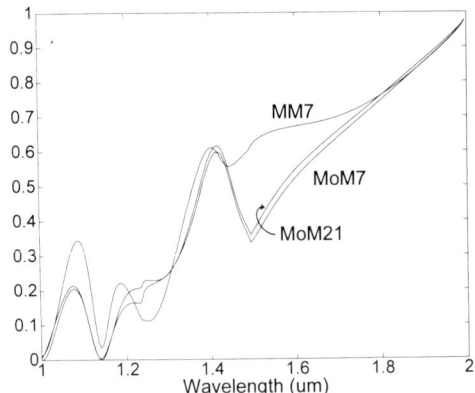

Fig. 6 Zero order reflection efficiency of the GIRO grating of Fig.3 for TM polarization and oblique incidence $\theta^i = 30^0$ MoM: Method of moments. MM: Mode Matching.

The method of moments is used in the numerical solution and it is possible to select the basis and expansion functions so as to obtain rapid convergence.

AKNOWLEDGEMENT. This work has been carried out with the partial support of CNR, Progetto Speciale InP OEIC.

## REFERENCES

1. T. K. Gaylord and M. G. Moharam, Analysis and applications of optical diffraction by gratings, *Proc. IEEE*, vol. 73, pp. 894-937, (1985).
2. R. Petit, ed., *"Electromagnetic Theory of Gratings"*, Springer-Verlag, Berlin, (1980).
3. N. Chateau and J. Hugonin, Algorithm for the rigorous coupled-wave analysis of grating diffraction, J. *Opt. Soc. Am. A*, vol. 11, pp. 1321-1331, (1994).
4. F. Montiel and M. Neviere, Differential theory of gratings: extension to deep gratings of arbitrary profile and permittivity through the Rmatrix propagation algorithm, *J. Opt. Soc. Am. A*, vol. 11, pp. 3241-3250, (1994).
5. L. B. Felsen and N. Marcuvitz, *"Radiation and Scattering of Waves"*, Prentice-Hall, Englewood Cliffs, NJ, (1973).
6. J. Van Bladel, *"Electromagnetic Fields"*, Hemisphere Publishing Co., Washington, DC, (1985).
7. R. F. Harrington, *"Time Harmonic Electromagnetic Fields"*, McGraw-Hill, New York, NY, (1961)
8. R. F. Harrington, *"Field computation by moment method"*, The Macmillan Company, New York, NY, (1964).
9. I. C. Gohberg and I. A. Feld'man, *"Convolution equations and projection methods for their solution"*. American Mathematical Society, Providence, Rhode Island, (1974).
10. J. Bach Andersen and V. V. Solodukhov, Field behavior near a dielectric wedge, *IEEE 73rans. Antennas Propagat.*, vol. AP-26, pp. 598-602, (1978).
11. H. Auda and R. F. Harrington, A moment solution for waveguide junction problems, *IEEE Trans. Micro?l)ave Theory Tech.*, vol. MTT-31, pp. 515-520, (1983).
12. B. Dhoet, S. Goeman, B. Demeulenaere, B. Baekelandt, D. De Zutter and R. Baets, GIRO gratings: a novel concept for mirrors with high reflectivity and polarization selectivity, in *"Integrated Photonics Research"*, Boston, (1996).

# THE NEW POSSIBILITY TO PRODUCE HOLOGRAPHIC OPTICAL ELEMENTS USING THE LOWEST COST LASER DIODES

M. I. Nemenov[1], J. S. Utochkina[2], and S. V. Utochkin[2]

[1] Department of Solid State Electronics
St.Petersburg State Technical University
195251 St.Petersburg, Russia

[2] Department of Physics
St.Petersburg Institute of Fine Mechanics and Optics
197101 St.Petersburg, Russia

## 1. INTRODUCTION

The history of semiconductor heterolasers and microoptics for them started in 1968 with continuous wave double heterojunction semiconductor lasers operating at room temperature. The lasers were developed by Prof. Zh.I. Alferov and his research team at A.F. Ioffe Physico-Technical Institute.[1]

Effective coupling of semiconductor lasers to the other elements of integrated and fibre optics, and the use of laser diodes for pumping solid state lasers, are important for laser diode applications. Single mode semiconductor lasesr without astigmatism, with near diffraction limited modes and asymmetry factor (Q) equal to 1 would be ideal for all applications. However, even laser diodes with a single narrow stripe and the only the lowest single transverse mode, are operating on two, three or more longitudinal modes.[2]

A single stripe laser diode operating in zero transverse mode close to the diffraction limit can be easily focused to a point. It is more difficult to provide high efficiency coupling of the radiation of a multimode laser into a waveguide or active crystal. The technique of beam division from a laser array with emitting area about 10 mm allows us to achive a Q factor around 1 and to obtain total efficiency up to 90 %.[3] However, adjustment of such multielement coupling devices is rather complicated.

Special devices with improved far field and near field patterns, such as a device consisting of a master laser and tapered amplifier, allow one to obtain zero hyper mode operation and simpler adjustment.[4] However, these are very expensive and rather complicated devices, as are also DFB and DBR laser diodes.

One of the solutions for the problem of effective coupling would be for example a beam-correcting holographic doublet recorded at 488 nm, which could successfully correct the astigmatism of a multimode laser diode at 820 nm.[5] However only such holographic optical elements (HOE) that could be recorded by laser diode itself, that is, self aligned HOEs, allow coupling close to ideal, and recording such holograms with a multimode laser diode is a problem.

## 2. THE CRITERIA FOR APPLICABILITY OF MULTIMODE SEMICONDUCTOR LASERS IN HOLOGRAPHY

The ideal source for recording is a single longitudinal mode, single transverse mode laser diode. Using this laser is not difficult, because the width of lasing spectrum is in the range of 20-500 MHz and the worst case - 500 MHz - gives us a coherence length of about 60 cm. However, to the best of our knowledge, using standard commercial laser diodes, especially high power single mode lasers or high power multimode lasers, was never considered except in Ref. 6,7.

Of course, if the longitudinal modes are equidistant and their widths are narrow, it is possible to record the hologram with an efficiency depending only on the width of longitudinal mode, by fine tuning of the photoplate position. This is the case of mode locked operation of the laser diode (LD). This type of operation can be achived with the LD with a saturable absorber and an operating current to threshold current ratio of 3 to 10. This is the standard amplitude - modulation mode locking case, which corresponds to the phases of all modes being fixed and approximately equal.

Mode locking operation may be understood more broadly as a regime where the modes are equidistant with fixed phases but the difference between the phases of two adjacent modes is not small. Mode locking can be obtained with no saturable absorber, using a fast optical nonlinearity in the active layer of a semiconductor laser. Possibly this type of mode locking can be realised for laser diodes with both narrow and broad stripes.

## 3. EXPERIMENTS

We investigated the possibility of recording holograms using a multimode laser diode. We used (AlGa)InP quantum-well LDs with 5-8 μm wide stripe and with a maximum intensity of lasing around 690 nm (Fig. 1). The operating current to threshold current ratio was in the range 1.2-1.6. The spectral width at half power was 40 - 60 nm.

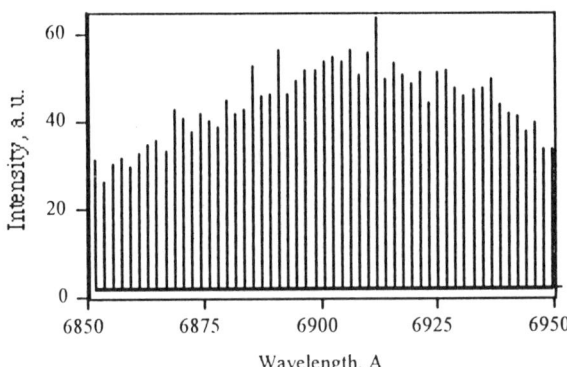

Fig. 1 Typical Spectrum of LDs.

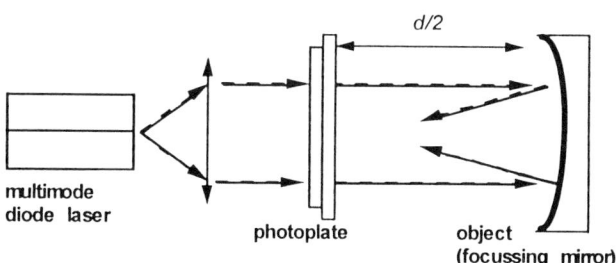

Fig. 2 Experimental scheme.

The holograms were recorded using the standard Denisyuk-type scheme[8] (Fig. 2). A focusing mirror was chosen as the object for recording. Silver halide photoplates PFG-03 were used. The optical path difference was 2 to 25 cm and the time of exposure varied from 0.5 to 15 minutes. The optical output power was kept constant by means of a monitoring photodiode current feedback. For temperature stabilisation a large heatsink for the laser diode was used. No special fine tuning was carried out to adjust the position of the photoplate to coincide with the maximum degree of coherence.

In all cases, after the standard developing procedure, holographic images of the focusing mirror were registered on photoplates.

The resulting values of the efficiency were 8 to 25 %. The maximum specified efficiency of PFG-holograms for He-Ne laser radiation at 633 nm is 40%. The efficiency value depended on time of exposure and power density. This result was somewhct unexpected for us. In fact, we supposed that our laser diode was operating in the mode locked regime.

## 4. DISCUSSION

Only two dynamic regimes of laser operation can be implemented for our laser. One of these is mode locking in a broadsense, and the other is free-running emission.

To understand the phenomenon of recording holograms when the position of photoplate does not coincide with the maximum of the coherence function it is necessary to analyse the coherence properties of laser diodes, compared to the mode locking and free running emission cases.[7] We will use the simple analysis from Ref. 7.

It is well known[9] that the complex degree of temporal coherence for a light source with an intensity emission spectrum $S(w)$ at a distance $ct$ from the source may be expressed as

$$\mu(t) = \int_\omega S_\omega(\omega) e^{i\Delta\omega t} d\omega \qquad (1)$$

where the spectrum is normalised so that the numerical value of the integral is = 1 and $\Delta\omega = \omega - \omega_o$ is the optical frequency relatively to some reference frequency $\omega_o$. We assume that the multimode laser emission spectrum consists of $N$ modes with relative intensities $S_k$, $k = 1,..,N$, and relative frequencies $\Delta\omega_k$. For the case of locked modes, we simply take

$$\Delta\omega_k = k\Omega, \qquad \Omega \approx c/(2n_g l) \qquad (2)$$

being the intermodal frequency interval with $l$ the cavity length and $n_g$ the group refractive index. For the case of free-running emission, we assume that group - velocity dispersion makes the modes non-equidistant:

$$\Delta\omega_k = k\Omega(1 + \vartheta k) \quad (3)$$

where

$$\vartheta = \frac{1}{2n_g} \cdot \frac{dn_g}{d\omega} \cdot \Omega = \frac{1}{2n_g} \cdot \frac{dn_g}{d\lambda} \cdot \Delta\lambda \quad (4)$$

is the measure of group velocity dispersion (GVD), $\Delta\lambda \approx \lambda^2/(2n_g l)$ being the intermodal interval in units of wavelength. Without much loss of generality, we may assume that each mode has the same spectral width and an identical line-shape $L(\Delta\omega)$ which we take to be Lorentzian. Then we may re-write Eq. (1) as

$$\mu(t) \approx L(t) \cdot \sum_{k=-N/2}^{N/2} S_k e^{i\Delta\omega_k t} \quad (5)$$

where $L(t)$ is an overall decay due to the finite line width of an individual mode and is a Fourier transformation of the line-shape $L(\Delta\omega)$, e.g. for homogeneously-broadened Lorentzian lines the decay is exponential.[9]

The Eq. (5) was used to calculate the degree of coherence of an LD with parameters approximately corresponding to the experimental case, for two cases: locked modes and free-running emission. A total of 60 modes separated by an intermodal interval of 0.2 nm and with a Lorentzian energy distribution with a FWHM of $\Delta\Omega = 60$ ps$^{-1}$ are taken into account, close to the experimental situation shown in Fig. 1; the line shape of the individual modes is also assumed to be Lorentzian, with a line width of 70 MHz. For small distances (0-1 cm), the degree of coherence exhibits a series of sharp spikes at the round-trip distance of the laser; at larger distances, the difference between locked and free-running emission is significant, as illustrated by Fig.s 3-4.

In bulk GaAlAs lasers, values of GVD about 7-10 $\mu m^{-1}$ have been measured.[10] The general trend is that with shorter wavelength, the dispersion values become higher; however, in multiple QW lasers they are smaller than in bulk lasers.[11] We therefore use a value $\vartheta = 10^{-4}$, which corresponds to $dn_g/d\lambda = 4$ $\mu m^{-1}$, as an estimate.

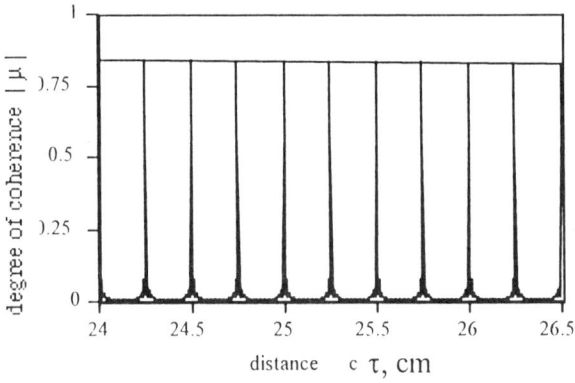

Fig 3 Degree of coherence of light emitted by a LD with locked (equidistant, $\vartheta=0$) modes. Parameters as in the text.

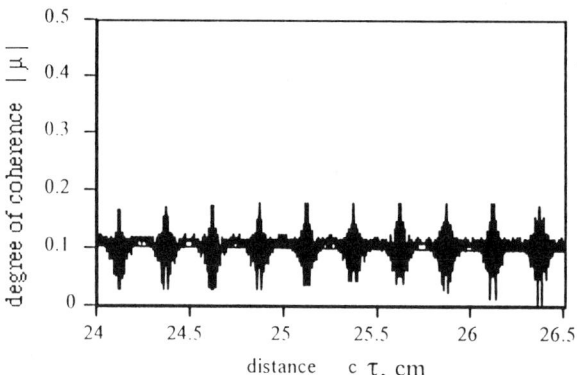

Fig. 4 Degree of coherence of light emitted by LD with free-running multimode emission; $\vartheta = 10^{-4}$.

As can be seen from Fig.s 3 and 4 the GVD smoothes down the sharp peaks seen in the degree of coherence calculated for a laser with equidistant modes. This leads to an increase in the value of the degree of coherence in the greater part of the laser free range and to a decrease of the maximum degree of coherence. The decreased values of the sharp spikes lead to their smoothing; such broadening of spikes due to a temporally smooth degree of coherence can appear in the free running range for free-running multimode emission. Of course, this value of temporal degree of coherence function depends on value of group velocity dispersion. Therefore, it is possible to record a hologram with an efficiency of 10-25% with a multimode laser, which is what was found experimentally.

In conclusion, the main criterion of applicability of a multimode laser diode to Holography is its operating regime (self-pulsation, free-running emission, mode locked operation). Only two regimes of operation allow one to record holograms: free running generation and mode locked operation.

A laser diode under a free running regime of operation allows one to record a hologram without tuning the photoplate position to coincide with the maximum of the coherence function. Using this regime it is possible to record a hologram with efficiency of at least 8-15 %, in a photosensitive layer 5-25 μm thick.

Recording a hologram by a laser diode operating in the mode locked regime is possible in a photosensitive plate with thickness equal to twice the length of the Fabry-Perot cavity of the laser diode, or by fine tuning the position of the photoplate to coincide with the maximum of the coherence function, with efficiency limited only by the linewidth of an individual longitudinal mode. Therefore, it has been shown that multimode lasers can be used for recording self-aligned phase holographic elements.

## 5. PROSPECTS FOR PRODUCING HOLOGRAPHIC OPTIC ELEMENTS BY LASER DIODES

The possibility of using self-aligned holographic optical elements (SAHOE) is not limited only to the purpose of effective coupling. SAHOEs may also be used for obtaining single transverse mode operation. Such operation is possible for multimode edge emitting laser diodes, where the different transverse modes can interact in the active layer, as well as for VCSEL arrays, where all channels of generation are independent.

An external cavity consisting of a monolithic integrated SAHOE with a Fabry-Perot cavity or the cavity of the VCSEL may provide a means for single trasverse mode operation.

It is necessary to solve the velocity equation[12] of laser diodes, where the coefficient of interaction between different channels of generation is taken into account. In practice, the solution of this problem depends on the two following tasks. The first is the determination of conditions under which the laser array with independent channels of generation will have enough interchannel coupling[13]. The second is looking for holographic media which will allow recording the HOE and be technically compatible with III-V semiconductor compounds.

In fact, it is possible to get zero hyper transverse mode for theVCSEL or laser array if the distance between the output mirror of the LD and an additional external cavity mirror isadjustable. This allows obtaining enough coupling between adjacent channels of generation. If it is possible to place photosensitive media between these mirrors we will be able to keep the same mode regime when the output mirror is removed.

In Ref. 14 this problems was addressed for the situation when different channels of generation were coupled in a semiconductor, but to our knowledge that was the most recent publication on this topic.

A new holographic medium can help solve this problem. In fact, some recent publications[15, 16] discuss the possibility of using the semiconductor materials AlGaAs and $CdF_2$ containing bistable defects as reversible photorefractive media. AlGaAs structures with DX centers can operate at temperatures not more than 120 K, and $CdF_2$ layers-at temperatures near 250 K. However, it is difficult to make them compatible with $A_3B_5$ layers. Therefore, the problem of creating new photorefractive elements that may ensure the large change of refractive index of about $10^{-4}$-$10^{-3}$ typical for layers with DX centers, the potential to operate at room temperature, and technological compatibility with a substrate of $A_3B_5$ is still of interest.

Such photorefractive materials can be produced if the thermal activation energy of the DX center is increased from the 0.4eV typical for III-V and II-VI structures to values of 0.6-0.7 eV. An increase of the thermal activation energy may also lead to an increase in defect localisation. In investigations of GaAs and AlGaAs epitaxial layers irradiated with 1MeV electrons, the formation of bistable defects of an E1-like type which had properties similar to those of the DX center was observed.[17]

The proposed technology, for producing photorefractive media by means of irradiation makes it possible to integrate optoelectronic and integral optical elements with arbitrary optical elements synthesised or recorded by holographic techniques; such as optical stationary and dynamic interconnections, memory, etc. The applicability of a self-aligned HOE (SAHOE) recorded by the laser diode itself, both for edge emitting high power laser diodes and for VCSE laser arrays, is determined by the operating temperature and technological compatibility of the holographic medium with the growth processes of semiconductor lasers. Such a holographic medium might be developed by irradiation of III-V compounds by electrons. However, the principles of operation of SAHOEs can be demonstrate with photopolymer films.

ACKNOWLEDGEMENTS. The authors are grateful to Yr.N. Denisyuk and R.A. Suris for useful discussions and St.Petersburg Laser Medical Centre, Pavlov Medical University for support.

## REFERENCES

1. Zh.I. Alferov, and R.F. Kazarinov, Double Heterostracture Laser, *Author's Certificate No. 181 737 [in Russian]*, appl. 3 March, 1965; publ. 15 April (1975) [*Byull. Izobret*. No. 14, 147 (1975); Zh.I.Alferov et al, *Fiz. Tekh. Poluprovod. (Sov. Phys. Semicon.)* 2:1545 (1968).
2. *SDL 1994 Laser Product Catalog.*

3. M. Khaleev, A. Mak, A. Mikhailov, G. Novokov, O. Orlov, V.Ustugov, et al, Combining and Beam Shaping Technique for High Brightness Circular Symmetric Focusing of Several Laser Diode Bars, *Technical Digest EUROPE CLEO'96, Hamburg* (1996).
4. D. Mehuys, D. Welch, and L. Goldberg, 2.0 W CW Diffraction-Limited Tapered Amplifier with Diode Injection, *Electronics Letters* 28:1944 (1992).
5. A. Aharoni, J.W. Goodman, and Y. Amitai, Beam - Correcting Holographic Doublet for Focusing Multimode Laser Diodes, *Optics Letters*, 18:179 (1993).
6. M.I. Nemenov, and E.A. Avrutin, Applicability of multimode diode lasers in holography, *Technical Digest 12th National Conference on Quantum Electronics, Southampton, UK* (1995).
7. M.I.Nemenov, and E.A.Avrutin, Peculiarities of application of multimode semiconductor lasers for recording deep holograms, *Proc. of Second International Conference on Optical Information Processing, St.Petersburg* , (1996).
8. Yr.N. Denisyuk, Photographic Reconstruction of the Optical Properties of an Object in Its Own Scattering Radiation Field, *Sov. Phys.-Dokl.* 144:1275 (1962).
9. R.J. Collier, C.B. Burckardt, and L.H. Lin, *Optical Holography*, Academic Press (1971).
10. J.P. Vanderziel, and R.A. Logan, Dispersion of the Group-Velocity Refractive -Index in GaAs Double Heterostructure Lasers, *IEEE J. Quantum Electronics*, 19:164 (1983).
11. K. L. Hall, G. Lenz, and E. P. Ippen, Femtosecond Time Domain Measurements of Group Velocity Dispersion in Diode Lasers at 1.5 µm, *J. of Lightwave Technology*, 10:616 (1992).
12. W.A. Hamel, M.P.V. Exter, and J.P. Woerdman, Coherence Properties of a Semiconductor Laser, *IEEE J. Quantum Electronics*, 26:1459 (1992).
13. E. Firsova, M. Nemenov, and V. Halfin, unpublished.
14. M. Orenstein, E. Kapon, N.G. Stoffel, J.P. Harbison, and L.T. Florez, Two-Dimentional Phase-Locked Arrays of Vertical-Cavity Surface Emitting Lasers by Mirror Reflectivity Modulation, *Appl. Phys. Lett.* 58:804 (1991).
15. R.A. Linke , T. Thio, J.D. Chadi, and G.E. Devlin, Diffraction from Optically Written Persistent Plasma Gratings in Doped Compound Semiconductors. *Appl. Phys. Lett.* 65:16 (1994).
16. R.A. Linke, R.L. MacDonald, G.E. Devin, T. Thio, D.J. Chadi, and M. Mizuta, High Diffraction Efficiency from Sub-Micron Period Plasma Gratings, *preprint NEC Research Institute* (1996)
17. M.M. Sobolev, I.V. Kochnev, M.I. Papentsev, and V.S. Kalinovsky, Metastable defects in AlGaAs, *Proc. of Material Science Forum* (1995).

# DIFFRACTION WITH SECOND-HARMONIC GENERATION FOR THE FORMATION OF SELF-GUIDED OR "SOLITARY" WAVES

G. Assanto

Third University of Rome, Department of Electronic Engineering
Via della Vasca Navale 84, 00146 Rome (Italy)

## 1. INTRODUCTION

Diffraction occurring during free propagation of a beam finite in its transverse plane tends to widen the beam lateral extension and, whenever its characteristic length exceeds the propagation distance, will weaken its intensity and ability to spatially resolve a given area or spot. Propagation in optical waveguides is generally considered as the most effective approach for counterbalancing diffraction over finite propagation distances, utilizing a geometrical resonance occurring in configurations with media of different optical parameters.[1] An entirely different approach to this problem relies on employing a nonlinear material response to oppose the beam lateral spreading through effects such as self-focussing or inverse bleaching or photorefraction etc. A nonlinear solution in order to counterbalance diffraction neither alters the bulk nature of the medium nor introduces interfaces or graded regions. Conversely, it typically needs high intensities and low attenuations in material systems which are additionally required to exhibit fast responses for applications to all-optical signal processing.[2] Using the optical Kerr effect in self-focussing materials with an intensity (I) dependent refractive index $n = n(I) = n_0 + n_2 I$, experimental demonstrations of self-guided solutions or spatial solitons have been achieved in carbon disulfide,[3] glass[4] and AlGaAs[5] planar waveguides, with geometrical confinement in one transverse dimension y and self-induced guidance in the orthogonal one x. These solitons are solutions of a nonlinear Schrödinger equation (NLSE) of the form:

$$\frac{\partial E}{\partial z} - \frac{i}{2k}\frac{\partial^2 E}{\partial x^2} - ik\frac{n_2}{n_0}|E|^2 E = 0 \qquad (1)$$

where $k = \omega n_0 / c$ is the propagation constant, and the second and third terms in Eq.(1) account for diffraction and self-focussing, respectively. These are eigenwaves of the nonlinear material system, i.e. they propagate with no alterations over unlimited distances in an ideal lossless medium. While they are stable waves in the one dimensional (1D) case, they are unstable in bulk (2D), due to the quadratic scaling of both peak intensity through diffraction ($I(L) \approx L^{-2}$) and self-focussing ($I(L) \approx L^2$) through the Kerr index. [6] Additional fifth-order nonlinear terms in the expansion of a perturbing polarization field (i.e. $n(I) = n_0 + n_2 I + n_4 I^2$) or saturating effects limiting $\Delta n = \Delta n(I)$ are therefore necessary to have stable spatial solitons of a cubic Kerr-like nonlinearity, [7] as demonstrated in vapors [8] and in an organic PTS crystal. [9]

A radically different way to attain self-guided waves in bulk is based on second-order nonlinearities. While a quadratic or three-wave effect is never able to induce a real refractive index change, the parametric interaction can give raise to gain and to the consequent gain-guiding effect which can overcome diffraction. A large enough parametric gain, i.e. a short enough nonlinear interaction length, will prevail over diffraction and cause a distortion of the beam phasefront through up- and down-conversion processes occurring between photons of different frequencies. As a result, fields at the three frequencies involved will exchange energy in a dynamic fashion, forming a multifrequency spatially confined transverse distribution also known as *simulton*.[10-13] One of the relevant advantages of a 2nd-order versus a 3rd-order process stems from the size of nonlinearities in existing materials and the degree of development of the pertinent technology. Conversely, a 2nd-order interaction is efficient only if wavevector match or synchronism is, at least approximately, guaranteed. The latter implies either the use of specific (i.e. *ordinary* or *extraordinary*) field polarizations with respect to the axes of a nonlinear non-centrosymmetric crystal, or operation at an appropriate *phase-matching* temperature when propagating along principal directions.

The main peculiarity of a three-wave mixing mechanism in forming a spatially confined diffractionless beam is, however, its ability to propagate even in 2D, due to the different scaling of the nonlinear effects with intensity.[14] Moreover, such multifrequency solitary waves, which are not rigorous solutions of an NLSE and therefore not properly *solitons*, rely on a coherent process and exhibit more degrees of freedom than their Kerr counterparts. This characteristic can be exploited in the implementation of control or switching functions, as discussed later.

## 2. SOLITARY WAVES IN SPACE VIA SECOND HARMONIC GENERATION

### 2.1. Model and Fundamentals

The simplest three-wave process is Second-Harmonic Generation (SHG) in the presence of two spatially co-polarized fundamental frequency (FF) components (i.e. Type I SHG). For this simple case, the equations governing the evolution of the electric field amplitudes in a planar waveguide are:

$$\frac{\partial E^{\omega}}{\partial z} - \frac{i}{2k^{\omega}} \frac{\partial^2 E^{\omega}}{\partial x^2} - i\Gamma \ E^{2\omega}(E^{\omega})^* e^{-i\Delta k z} = 0$$

$$\frac{\partial E^{2\omega}}{\partial z} - \frac{i}{2k^{2\omega}} \frac{\partial^2 E^{2\omega}}{\partial x^2} - i\Gamma E^{\omega} E^{\omega} e^{+i\Delta k z} = 0$$

(2)

with $\Gamma$ the nonlinear coefficient and $\Delta k = k^{2\omega} - 2k^{\omega}$ the wavevector mismatch. Although eigensolutions of (2) contain field components at both FF and SH, stationary self-guided waves also form when launching only the FF field at the input. These solutions, of more practical interest, will exhibit oscillations in amplitude along the propagation direction but will remain transversely confined. An observation of self-guided waves due to Type I SHG in a waveguide was recently reported in a Ti:indiffused Lithium Niobate slab guide, kept at 335 °C for phase-matching.[15] They launched a Gaussian input beam at 1.32μm and let it propagate in a 47 mm long crystal. At peak power densities above 20 W/μm clean self-guided waves could be produced, in good agreement with predictions based on Eqs. (2).

In the more general Type II SHG in 2D, the two FF components are orthogonally polarized in space and, when birefringent phase-matching is used, one of them will be an extraordinary ray and exhibit walk-off, i.e. will carry power along a direction slightly noncollinear to its wavevector. In the hypothesis of an SH field also affected by walk-off, using the subscripts "e" and "o" for the extraordinary and ordinary rays, respectively, the SHG equations read:

$$\frac{\partial E_e^\omega}{\partial z} - \rho^\omega \frac{\partial E_e^\omega}{\partial x} - i\frac{\Delta_\perp^2 E_e^\omega}{2k_e^\omega} - i\Gamma E_e^{2\omega}(E_o^\omega)^* e^{-i\Delta k z} = 0$$

$$\frac{\partial E_o^\omega}{\partial z} - i\frac{\Delta_\perp^2 E_o^\omega}{2k_o^\omega} - i\Gamma E_e^{2\omega}(E_e^\omega)^* e^{-i\Delta k z} = 0 \qquad (3)$$

$$\frac{\partial E_e^{2\omega}}{\partial z} - \rho^{2\omega} \frac{\partial E_e^{2\omega}}{\partial x} - i\frac{\Delta_\perp^2 E_e^{2\omega}}{2k_e^{2\omega}} - i2\Gamma E_e^\omega E_o^\omega e^{+i\Delta k z} = 0$$

with $\rho^\omega$ and $\rho^{2\omega}$ the walk-off angles in the x-z plane and $\Delta k = k_e^{2\omega} - k_e^\omega - k_o^\omega$. When the FF components are equally excited at the input, i.e. an FF wave is launched with its electric field at 45° with respect to the "e" and "o" polarizations, the process is similar to the Type I case, except for the presence of walk-off. Walk-off will, generally speaking, reduce the nonlinear effectiveness, because it displaces the various field distributions in space upon propagation. When the nonlinear length becomes shorter than both the diffraction and the walk-off distances, however, parametric gain will force the three fields to interact and exchange photons in the transverse region where their amplitudes are higher. If this process is effective, a spatial solitary wave will form and propagate in an "e" direction in between the walking-off FF and SH components and depending also on the phasefront distortion due to the cascaded interaction. This has been indeed verified in a KTP crystal using ps pulses from a Nd:YAG laser operating at 1.064 μ m. The 20mm-waist input beam would linearly diffract and weaken at the output face of a 1cm long sample, whereas it would form a 2D stable bi-frequency undiffracted solitary wave when the input peak intensity (FF) exceeded 5 GW/cm$^2$ for the phase-matched case.[16]

## 2.2. Solitary Wave Steering

An entire range of possibilities is opened up by considering unequal inputs in the FF polarizations, i.e. when taking advantage of the two distinct frequency-degenerate fields (FF) launched into the crystals. When the "e" ("o") component is larger than the "o" ("e"), one would expect the parametric gain to be stronger along the corresponding extraordinary (ordinary) direction of propagation, thereby resulting in the formation and propagation of an SSW (spatial solitary wave) which will evolve at an angle with respect to the unbiased case. The BPM (beam propagation method) numerical prediction is shown in Fig.1 vs propagation with reference to KTP and an input Gaussian beam of waist 20 μm at 1.064 μm. Clearly, angular steering is expected based on polarization imbalance at the input. When the launch conditions are somewhat intermediate between the two extreme cases, a double-hump solitary wave forms. Notice, however, that due to walk-off in the SH component, this case does not correspond to the perfectly balanced input employed above. A diagram of an all-optical angular beam steering

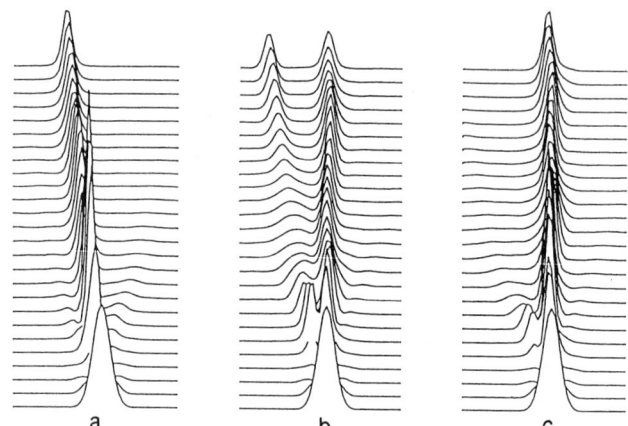

Fig. 1 Beam Propagation Method (split-step) calculations of the propagation of a fundamental frequency Gaussian input along a KTP crystal. Intensity profiles for a) $I_o = 20\ J_0$, $I_e = 20.5\ J_0$; b) $I_o = 20\ J_0$, $I_e = 19.5\ J_0$; c) $I_o = 20\ J_0$, $I_e = 18.5\ J_0$ and $J_0 = 1\ GW/cm^2$. We chose $\Delta k = 0.1\ \pi\ cm^{-1}$, an input waist $w_0 = 20$ mm and a length of 1 cm, corresponding to five diffraction lengths $L_D$.

device with the indication of calculated ordinary and extraordinary FF inputs and the corresponding lateral displacement in the ouput plane is shown in Fig. 2. In this figure, notice the intermediate two-hump situation shown in Fig. 1b. Experimental demonstrations of SSW steering were performed in a 1cm KTP crystal, varying the orientation of a half-wave plate in front of a linearly polarized beam at 1.064 μm, according to the schematic set-up in Fig. 3.[17]. Typical results obtained by rotating the half-wave plate are shown in Fig. 4 and are in excellent agreement with the predictions. Because the lateral displacement is limited by the crystal length and the natural walk-off due to the intrinsic birefringence, it is interesting to explore possibilities for the enhancement of such a steering effect. In particular, non-normal incidence can introduce a bias in the of such a steering effect. In particular, non-normal incidence can introduce a bias in the resulting direction of propagation, counteracting or emphasizing the effect of walk-off.

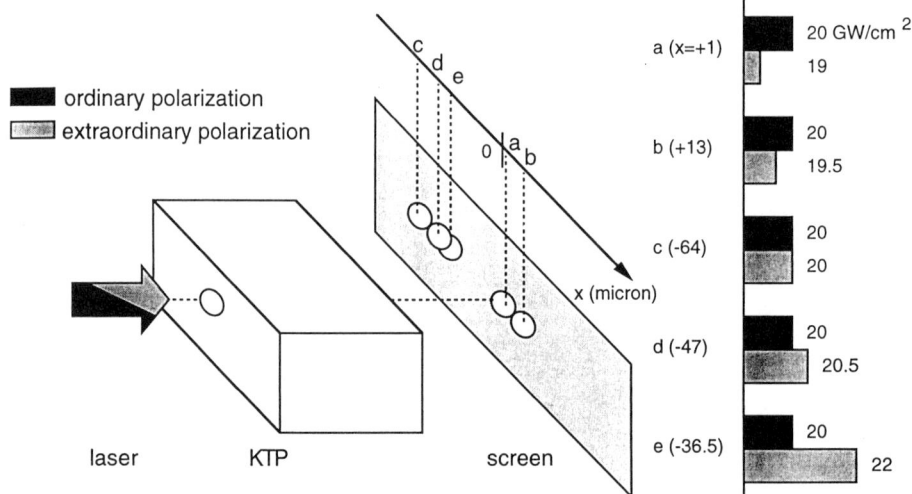

Fig. 2 Pictorial view of an imbalance-controlled all-optical beam steering device. With reference to the parameters of Fig. 1, various combinations of input $I_e$ and $I_o$ cause different lateral displacements along x at the output of the crystal. Displacements (in brackets) are in μm.

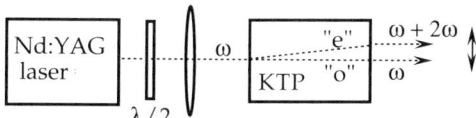

Fig. 3 Experimental set-up. The waveplate (λ/2) permits to alter the relative weight of ordinary and extraordinary components at the input for a given linearly polarized laser beam at the fundamental frequency.

An example is presented in Fig. 5, where the lateral displacement is augmented by inputting the fundamental beam with a wavefront tilt $\theta = \rho^\omega$. Clearly, the steering range is substantially enlarged and no longer limited by the inherent natural walk-off.[18] Finally, since the y-coordinate is not affected by walk-off, the bidimensional nature of these waves can be exploited in the whole output plane by introducing suitable wavefront tilts. Tilts in the y-z plane will move the solitary beam up and down, whereas combined walk-off (natural + tilt) and imbalance will steer it in the x-z plane. An illustration of tilt-mapping in x and y is given in Fig. 6.

## 2.3. Solitary Wave All-optical Switching

The lateral displacement through the angular steering effect in the vectorial interaction described above can be translated into efficient switching action by the use of an output aperture, in order to allow complete transmission only at the proper intensity imbalance. Such an all-optical switch is shown in Fig. 7, in conjunction with the curve relative to the lateral shift as discussed in the previous paragraph. Fig. 8 shows the experimental results obtained in KTP with a 30 μm aperture. Without any optimization, a contrast better than 10 :1 is achieved through an intensity modulation of less than 20 %. The use of pulses and the inevitable temporal walk-off prevented a sharper response from being experimentally demonstrated.

## 2.4. Solitary Wave Interactions

Kerr solitons are often considered true solitons because they tend to behave as particles, characterized by a well-defined phase, unmodified upon propagation, and capable of elastic

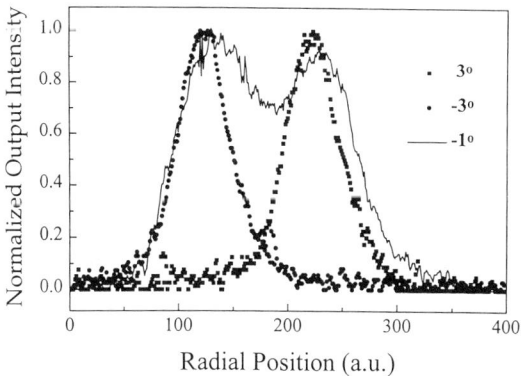

Fig. 4 Measured output beam profiles, for $\Delta k L = 3\pi$, $w_0 = 12$ μm and a total ($I_e + I_o$) peak intensity of 80 GW/cm$^2$. The legend indicates the half-wave plate rotation (in degrees) around π/4.

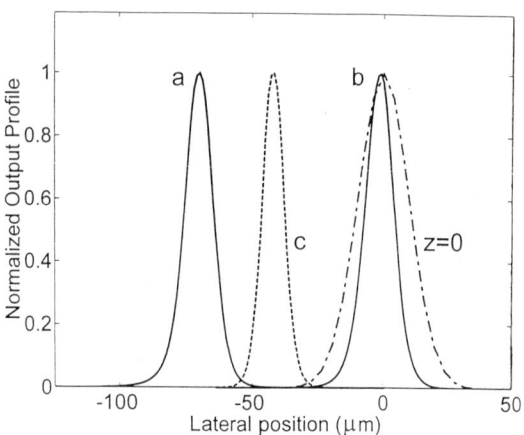

Fig. 5 Output cross sections along x for three different launch conditions ($J_0 = 1$ GW/cm$^2$). Solid lines: "e" field with an input tilt equal to the natural walk-off for a) $I_o = 20$ $J_0$, $I_e = 17.5$ $J_0$; b) $I_o = 20$ $J_0$, $I_e = 22.5$ $J_0$. Dashed line c) refers to the no-tilt input case, whereas the dot-dashed line is the input (z = 0) Gaussian profile.

collisions.[19-20] Interactions between soliton-like waves are also the basis for a number of all-optical logic operations and switching schemes.[21] The interaction between quadratic simulton-like waves in space has been addressed as well, with emphasis on Type I SHG processes controlled by the relative phase of the two incoming beams.[22-24] Here, in line with the previous development, we rather address the problem of interactions with Type II SHG processes and in two transverse dimensions, i.e. utilizing the self-guided beams already discussed.[25] The interaction between two solitary waves which overlap in some region of space after forming can be controlled by several parameters, i.e. the collision angle, their individual intensity or power, their relative phase, their individual intensity imbalances in the vectorial case. In general, the phases will determine the interference effects occurring between co-polarized components of the fields in the overlap region, and the intensities will affect the nonlinear

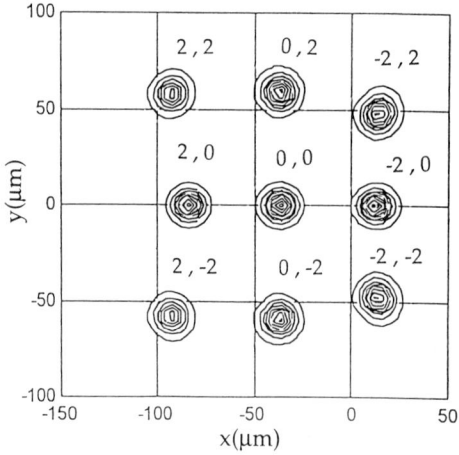

Fig. 6 Countour plots of the output profiles in the x-y plane, for various pairs of input tilts (in units of $\rho^{(o)}$) in the "e" component, and $I_o = 20$ $J_0$, $I_e = 22$ $J_0$.

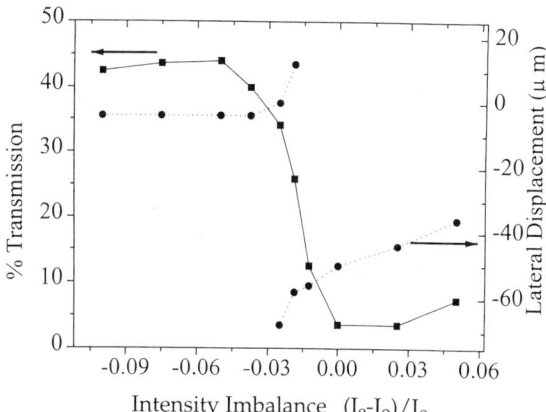

Fig. 7 Calculated output displacement versus imbalance and corresponding all-optical switching through a 30 μm aperture. Here $I_o = 20$ GW/cm$^2$ was kept fixed, and the other parameters are as above.

length and the strength of the parametric process. It is particularly instructive to examine two simple cases: a) interactions between two identical solitary waves with an overall phase difference between the launched beams; b) interactions of in-phase solitary waves for varied intensity imbalances.

**2.4.1. Phase controlled collisions.** The interaction between identical Type II SHG spatial solitary waves out-of-phase with respect to each other can be studied by changing either of the phase differences between co-polarized components in the input beams at the fundamental frequency. Since the phase difference existing in each input beam between "e" and "o" components is irrelevant to the formation of the solitary wave, the "relative" phase variation will affect the interaction through interference in the overlap region. A simpler case is that of identical beams (A and B) with individual "e" and "o" components in-phase, but with an overall relative phase between them. Fig. 9 provides some examples of such interactions, with fixed intensity, mismatch, imbalance and collision angle, for various input relative phases $\Delta\phi = \phi_A - \phi_B$, with $\phi_{A,B} = [\phi_{A,B}]_e = [\phi_{A,B}]_o$. Clearly, various outcomes are obtained, from repulsion to fusion and coalescence to mixing with unequal and/or steered output profiles.

**2.4.2. Imbalance controlled collisions.** Of greater interest is the ability to control the interaction of two colliding solitary waves by changing their intensities. Furthermore, since vectorial Type II

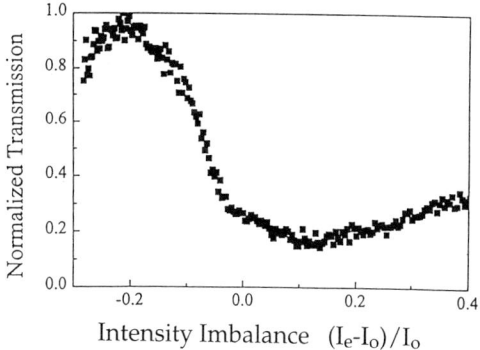

Fig. 8 Measured normalized transmission through an aperture, for a total intensity of 80 GW/cm$^2$ and a waist of 12 μm at the fundamental frequency input.

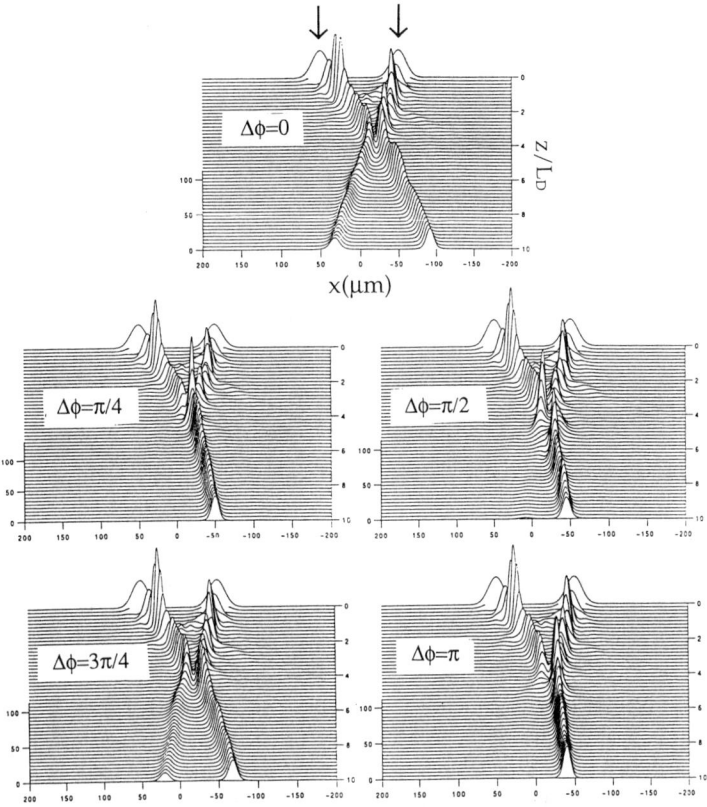

Fig. 9 Phase controlled collisions: 3D plots vs x (μm) and z (in diffraction lengths $L_D$), varying the overall relative phase at the inputs. The Gaussian inputs are launched at an angle $\theta = 4\rho^{\omega}$, with $I_o = 20 J_0$ and $I_e = 18.5 J_0$ in the Left beam, and $I_o = 18.5 J_0$ and $I_e = 20 J_0$ in the Right beam.

SHG solitary waves are dependent on the imbalance between input polarizations at the fundamental frequency, it seems quite desirable to investigate situations in which, keeping overall intensities and phases fixed, we can shape the interaction through an input imbalance.[25] Examples of such cases are presented in Fig.10. Crossing, repulsion, coalescence are obtained, without the need for impractical phase conditioning. Since natural walk-off introduces an asymmetry in the problem, imbalance controlled interactions can also take place when the input beams are launched parallel to each other, i.e. with zero collision angle. By acting on the imbalance, in fact, one can steer the forming solitary waves making them collide or diverge, provided their initial distance and induced steering are sufficient to cause a collision to occur within the available propagation length.[25] Typical examples are shown in Fig.11, where different outcomes are observed upon the injection of vectorial beams with equal or opposite imbalances.

## 3. CONCLUSIONS

A rich and stimulating scenario has been opened up by the investigation of self guided waves through parametric processes such as second-harmonic generation. These novel solitary waves are stable in fully dimensional space and have already been observed experimentally in available nonlinear crystals and waveguides. The intrinsic flexibility of a three-wave process can be fully exploited even in

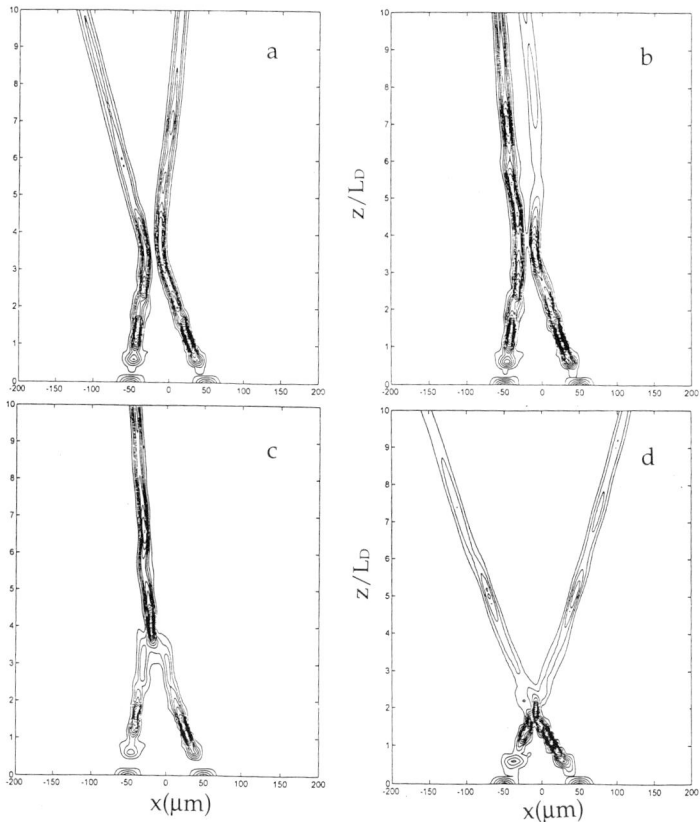

Fig. 10 Imbalance controlled collisions: countour plots for a zero relative phase, varying the imbalance $\delta$ of two symmetrically launched beams (100 mm apart along x in z = 0) and holding the total intensity at 40 GW/cm$^2$. a) $\delta = 15\%$, $\theta = 4\rho^\omega$; b) $\delta = 10\%$, $\theta = 4\rho^\omega$; c) $\delta = 0$, $\theta = 4\rho^\omega$; d) $\delta = 0$, $\theta = 8\rho^\omega$. The propagation distance is in diffraction lengths, $\Delta k = 10\pi$ m$^{-1}$.

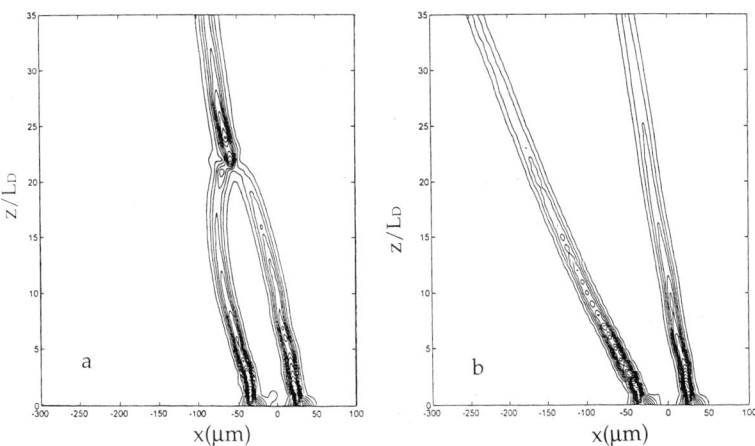

Fig. 11 As in Fig.10, but for two Gaussian beams (Left and Right) launched parallel along z ($\theta = 0$), 60 μm apart in x and with imbalances: a) $\delta_R = \delta_L = -0.5$; b) $\delta_L = -\delta_R = 0.5$.

the frequency degenerate case by adopting Type II (vectorial) SHG schemes, whereby the imbalance between orthogonal polarization components at the fundamental frequency can be used as a control parameter for all-optical switching, angular steering and collisional interactions. These phenomena, once material systems with the appropriate nonlinear strength become available, will open the way to a new generation of all-optically reconfigurable multidimensional interconnects. Based on recent progress in organic crystals and polymeric systems for quadratic effects, this goal appears neither too ambitious nor too far ahead in the future.

## REFERENCES

1. D. Marcuse, *Theory of Dielectric Optical Waveguides,* Academic Press, London, Chap. I (1974)
2. G.I. Stegeman, E.M. Wright, All-optical waveguide switching, *Opt. & Quantum Electron.* 22:95 (1990)
3. A. Barthelemy, S. Maneuf, C. Froehly, Propagation soliton et auto-confinement de faisceaux laser par non linearité optique de Kerr, *Opt. Comm.* 55:201 (1985)
4. J.S. Aitchison, Y. Silberberg, A.M. Weiner, D.E. Leaird, M.K. Oliver, JL. Jackel, E.M. Vogel, P.W.E. Smith, Spatial optical solitons in planar glass waveguides, *J. Opt. Soc. Am. B* 8:1290 (1991)
5. J.S. Aitchison, K. Al-hemyari, C.N. Ironside, R.S. Grant, W. Sibbett, Observation of spatial solitons in AlGaAs waveguides, *Electron. Lett.* 28:1879 (1992)
6. P.L. Kelley, Self-focusing of optical beams, *Phys. Rev. Lett.* 15:1005 (1965)
7. Y. Chen, Self-trapped light in saturable nonlinear media, *Opt. Lett.* 16:4 (1991)
8. J.E. Bjorkholm, A. Ashkin, CW self-focusing of light in Na vapor, *Phys. Rev. Lett.* 32:129 (1974)
9. B.L. Lawrence, W.E. Torruellas, G.I. Stegeman, Solitary waves and ring-formation in polydiacetylene para-toluene sulfonate, in *Nonlinear Guided Waves and Their Applications*, 1996 Optical Society of America Tech. Dig. Series, Washington DC, 15:272 (1996)
10. Yu.N. Karamzin and A.P. Sukhorukov, Mutual focusing of high-power light beams in media with quadratic nonlinearity, *Sov. Phys. JETP* 41:414 (1976)
11. M.J. Werner and P.D. Drummond, Strongly coupled nonlinear parametric solitary waves, *Opt. Lett.* 19:613(1994)
12. L. Torner, C.R. Menyuk, G.I. Stegeman, Bright solitons with second-order nonlinearities, *J. Opt. Soc. Am. B* 12:389(1995)
13. K. Hayata, M. Koshiba, Multidimensional solitons in quadratic nonlinear media, *Phys. Rev. Lett.* 71:275(1993)
14. L. Torner, E. M. Wright, Soliton excitation and mutual locking of light beams in bulk quadratic nonlinear crystals, *J. Opt. Soc. Am. B* 13:864 (1996)
15. R. Schiek, Y. Baek, G.I. Stegeman, One-dimensional spatial solitary waves due to cascaded second-order nonlinearities in planar waveguides, *Phys. Rev. E* 53:1138 (1996)
16. W.E. Torruellas, Z. Wang, D. J. Hagan, E. W. VanStryland, G.I. Stegeman, L. Torner, C. R. Menyuk, Observation of two-dimensional spatial solitons in a quadratic medium, *Phys. Rev. Lett.* 74:5036 (1995)
17. W.E. Torruellas, G. Assanto, B.L. Lawrence, R.A. Fuerst, G.I. Stegeman, All-optical switching by spatial walk-off compensation and solitary-wave locking, *Appl. Phys. Lett.* 68:1449 (1996)
18. G. Leo, G. Assanto, W. E. Torruellas, Beam pointing control with spatial solitary waves in quadratic nonlinear media, *Opt. Commun.* 134:223 (1997)
19. M. Shalaby and A. Barthelemy, Experimental spatial soliton trapping and switching, *Opt. Lett.* 16:1472 (1991)
20. J.S. Aitchison, A.M. Weiner, Y. Silberberg, D.E. Leaird, M.K. Oliver, J.L. Jackel, P.E. Smith, Experimental observation of spatial soliton interactions, *Opt. Lett.* 16:15 (1991)
21. T. Shi and S. Chi, Nonlinear photonic switching by using the spatial soliton collision, *Opt. Lett.* 15:1123 (1990)
22. M.J. Werner, P.D. Drummond, Simulton solutions for the parametric amplifier, *J. Opt. Soc. Am. B* 10:2390 (1993).
23. D.-M. Baboiu, G.I. Stegeman, and L. Torner, Interaction of one-dimensional bright solitary waves in quadratic media, *Opt. Lett.* 20:2282 (1995).
24. C. Etrich, U. Peschel, F. Lederer, B. Malomed, Collision of solitary waves in media with a second-order nonlinearity, *Phys. Rev. A* 52:R3444 (1995)
25. G. Leo, G. Assanto, W.E. Torruellas, Intensity controlled interactions between vectorial spatial solitary waves in quadratic nonlinear media, *Opt. Lett.* 22:7 (1997)

# LIGHT SCATTERING BY A DIELECTRIC CYLINDER NEAR A FLAT SUBSTRATE

R. Borghi[*], M. Santarsiero[*], F. Frezza[§], and G. Schettini[§]

[*] "E. Amaldi" Physics Department
The Third University of Rome
Via della Vasca Navale 84, 00146 Rome, Italy
[§] Department of Electronics Engineering
University of Rome "La Sapienza"
V. Eudossiana 18, 00184 Rome Italy

## 1. INTRODUCTION

The study of light scattering by dielectric cylinders, with or without a reflecting surface, has received attention in recent years, due to the great number of possible applications in several fields of physics and engineering. For example, we could cite applications in optical fiber characterization, microwave heating, defect detection in semiconductors, and near-field optics.[1-9]

Recently the problem of scattering of a plane wave by a perfectly conducting circular infinite cylinder placed in front of a plane discontinuity for the electromagnetic constants has been treated.[10-12] In particular, in Ref. [12] a method was proposed that, starting from the customary expansion of the scattered field, exploits the plane-wave representation of cylindrical waves [13], and provides a rigorous solution for this problem, at least on a theoretical basis. Here we show how this method can be extended to treating case of a dielectric cylinder characterized by a refractive index $n_c$, which can also assume complex values, corresponding to lossy materials.

With the present method, since the presence of the interface is taken into account only through its reflection coefficient, very different types of surfaces can be considered. Moreover, both the near and the far-zone diffracted fields can be determined, for both polarization states (TE and TM, with respect to the cylinder axis). In a practical implementation of the algorithm, even if some numerical approximations must be made due to the necessity of truncating the relevant series, a very rapid convergence and a remarkable stability with respect to the input data are obtained. These characteristics make the technique very attractive for a wide range of applications.

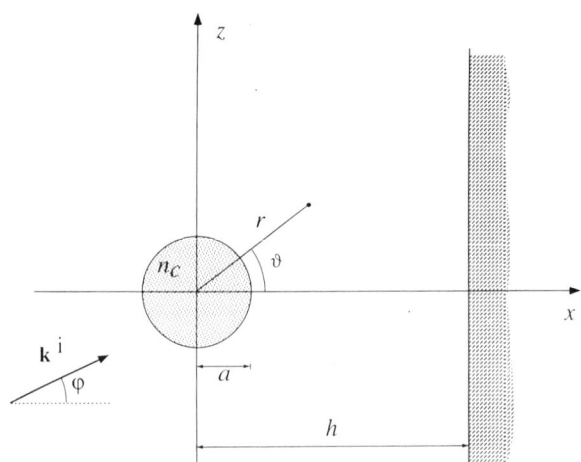

Fig.1 Geometrical layout for the scattering problem.

Fig. 1 shows the geometrical layout of our problem: a monochromatic plane wave with wavelength $\lambda$ impinges on a dielectric circular cylinder of radius $a$ and refractive index $n_c$. The cylinder axis is placed at a distance h from a general reflecting flat surface. This diffractive structure is assumed to be infinite along the y-direction, which is parallel to the cylinder axis, so that the problem is reduced to a two-dimensional form. Moreover, we utilize the following dimensionless variables:

$$\xi = k_0 x \qquad \zeta = k_0 z \qquad \chi = k_0 h \qquad \rho = k_0 r, \tag{1}$$

where $k_0 = 2\pi/\lambda$ is the wave number in vacuum. In the following, $\mathbf{n} = \mathbf{k}/k_0$ denotes the unit vector associated with a typical plane wave whose wave vector is $\mathbf{k}$, while $n_\parallel$ and $n_\perp$ are the components of $\mathbf{n}$ parallel and perpendicular to the reflecting surface, respectively. The surface is characterized by a complex reflection coefficient $\Gamma$, which is a function of $n_\parallel$. Moreover, $\mathbf{k}^i$ is the wave-vector of the incident field and $\varphi$ denotes the incidence angle with respect to the z-axis. The polarization of the fields involved is assumed to be either TM or TE (electric or magnetic field directed along the axis of the cylinder). In both cases the amplitude of the field parallel to the cylinder axis is represented by a scalar function $V(\xi, \zeta)$.

To solve the scattering problem we consider the field $V(\xi, \zeta)$, due to the interaction between the incident plane wave and the diffractive structure, as the superposition of five contributions: • $V_i$: field of the incident plane wave; • $V_r$: field due to the reflection of $V_i$ by the plane surface; • $V_c$: field present inside the cylinder; • $V_d$: field diffracted by the cylinder; and, • $V_{dr}$: field due to the reflection of $V_d$ by the plane surface.

The boundary conditions on the cylinder surface can be imposed by expressing each of these contributions in the polar reference frame centered on the cylinder axis. The expressions of $V_i$, $V_r$, $V_d$, and $V_{dr}$ are discussed in Ref. [12], and have the form

$$V_i(\xi, \zeta) = V_0 \sum_{m=-\infty}^{+\infty} i^m \exp(-im\varphi) J_m(\rho) \exp(im\vartheta), \tag{2a}$$

$$V_r(\xi, \zeta) = V_0 \Gamma(n_\parallel^i) \exp(i n_\perp^i 2\chi) \sum_{m=-\infty}^{+\infty} i^m J_m(\rho) \exp[im(\vartheta + \varphi - \pi)], \tag{2b}$$

$$V_d(\xi, \zeta) = V_0 \sum_{m=-\infty}^{+\infty} i^m \exp(-im\varphi) \, c_m \, CW_m(\xi, \zeta) , \qquad (2c)$$

$$V_{dr}(\xi, \zeta) = V_0 \sum_{m=-\infty}^{+\infty} i^m \exp(-im\varphi) \, c_m \, RW_m(2\chi - \xi, \zeta) = \qquad (2d)$$

$$= V_0 \sum_{m=-\infty}^{+\infty} i^m \exp(-im\varphi) \, c_m \sum_{\ell=-\infty}^{+\infty} i^\ell J_\ell(\rho) \exp(i\ell\vartheta) \, RW_{\ell+m}(2\chi, 0)$$

In these equations $V_0$ is the amplitude of the incident field, $(\rho, \vartheta)$ are coordinates of the polar reference frame centered on the cylinder axis, $J_m$ is the Bessel function of the first kind of m-th order, and the coefficients $c_m$ represent the unknown quantities of the problem. Moreover, the functions CW and RW are defined as [12]

$$CW_m(\xi, \zeta) = H_m(\rho) \exp(im\vartheta) , \qquad (3a)$$

$$RW_m(\xi, \zeta) = \frac{1}{2\pi} \int_{-\infty}^{+\infty} \Gamma(n_\parallel) \, F_m(\xi, n_\parallel) \exp(in_\parallel \zeta) \, dn_\parallel \qquad (3b)$$

where $H_m(\rho)$ is the Hankel function of the first kind and order m, while the function $F_m(\xi, n_\parallel)$, representing the angular spectrum of the cylindrical function $CW_m(\xi, \zeta)$ is [12,13]

$$F_m(\xi, n_\parallel) = \frac{2\exp(in_\perp \xi)}{n_\perp} \exp(-im\cos^{-1} n_\parallel) \qquad (4)$$

The field transmitted into the cylinder can be written as [14]

$$V_c(\xi, \zeta) = V_0 \sum_{m=-\infty}^{+\infty} i_m \exp(-i_m \varphi) \, d_m \, J_m(n_c \rho) \exp(i_m \vartheta) , \qquad (5)$$

with unknown coefficients $d_m$.

Now, from Eqs. (2a) ... (2d) and (5), it is straightforward to impose the appropriate boundary conditions on the surface of the cylinder, in order to determine the values of the coefficients $c_m$ and $d_m$.

Let us analyse this procedure in some detail for both polarization states. If we denote by $\tilde{\nabla}$ the gradient operator with respect to the dimensionless coordinates $(\xi, \zeta)$, Maxwell's equations read [12]

$$\mathbf{E}(\xi, \zeta) = \frac{iZ_0}{\upsilon^2} \tilde{\nabla} \times \mathbf{H}(\xi, \zeta) \qquad (6)$$

$$\mathbf{H}(\xi, \zeta) = -iY_0 \tilde{\nabla} \times \mathbf{E}(\xi, \zeta), \qquad (7)$$

$Z_0$ and $Y_0$ being the characteristic impedance and admittance, respectively, of the medium outside the cylinder, while $\upsilon$ is defined as

$$\upsilon = \begin{cases} 1 & \text{outside the cylinder,} \\ n_c & \text{inside the cylinder .} \end{cases} \qquad (8)$$

The expression of the curl operator in cylindrical coordinates is

$$\tilde{\nabla} \times [V(\xi, \zeta) \hat{y}_0] = \frac{1}{\rho} \partial_\vartheta V(\xi, \zeta) \hat{r}_0 - \partial_\rho V(\xi, \zeta) \hat{\vartheta}_0, \qquad (9)$$

where $\hat{y}_0$, $\hat{r}_0$, and $\hat{\vartheta}_0$ are the unit vectors associated to the cylindrical reference frame. In the case of TM polarization we have $E(\xi, \zeta) = V(\xi, \zeta) \hat{y}_0$. Then the boundary conditions, arising from the continuity of the tangential components of electric and magnetic fields, may be written as

$$\begin{aligned}\left[V_i + V_r + V_d + V_{dr}\right]_{\rho=ka} &= \left[V_c\right]_{\rho=ka}, \\ \left[\partial_\rho V_i + \partial_\rho V_r + \partial_\rho V_d + \partial_\rho V_{dr}\right]_{\rho=ka} &= \left[\partial_\rho V_c\right]_{\rho=ka}.\end{aligned} \qquad (10)$$

For TE polarization the roles of $V(\xi, \zeta)$ and $\partial_\rho V(\xi, \zeta)$ are interchanged, so that the boundary conditions become

$$\begin{aligned}\left[V_i + V_r + V_d + V_{dr}\right]_{\rho=ka} &= \left[V_c\right]_{\rho=ka}, \\ \left[\partial_\rho V_i + \partial_\rho V_r + \partial_\rho V_d + \partial_\rho V_{dr}\right]_{\rho=ka} &= \frac{1}{n_c^2}\left[\partial_\rho V_c\right]_{\rho=ka}.\end{aligned} \qquad (11)$$

By using Eqs. (2a) ... (2d), (5), (10), and (11), after some algebra we obtain the following linear system for the unknown coefficients $c_m$ and $d_m$:

$$\sum_{\ell=-\infty}^{+\infty} A_{m\ell}^{(1)} c_\ell - E_m^{(1)} d_m = b_m^{(1)}, \qquad (12a)$$

$$\sum_{\ell=-\infty}^{+\infty} A_{m\ell}^{(2)} c_\ell - E_m^{(2)} d_m = b_m^{(2)}, \qquad (12b)$$

the superscripts (1) and (2) referring to the boundary condition on the field and on its normal derivative, respectively. Here,

$$E_m^{(1)} = \exp(-im\varphi) \frac{J_m(n_c ka)}{H_m(ka)}, \qquad (13a)$$

$$E_m^{(2)} = p \exp(-im\varphi) \frac{J_m'(n_c ka)}{H_m'(ka)}, \qquad (13b)$$

where the prime denotes the derivative and p is defined as

$$p = \begin{cases} n_c & \text{for TM polarization}, \\ n_c^{-1} & \text{for TE polarization}. \end{cases} \qquad (14)$$

Furthermore, the coefficients of the system of Eqs. (12a) and (12b) are

$$A_{m\ell}^{(1,2)} = \exp(-i\ell\varphi)\left[\delta_{m\ell} + i^{\ell-m}T_m^{(1,2)}(ka)RW_{\ell+m}(2\chi,0)\right], \quad (15a)$$

$$b_m^{(1,2)} = -T_m^{(1,2)}(ka)\left\{\exp(-im\varphi) + \Gamma\left(n_\parallel^i\right)\exp\left(in_\perp^i 2\chi\right)\exp\left[-im(\pi-\varphi)\right]\right\}, \quad (15b)$$

where the symbol $\delta_{m\ell}$ denotes the Kronecker delta, while the function $T_m^{(j)}$, containing the information about the boundary conditions, is defined as follows:

$$T_m^{(j)}(x) = \begin{cases} \dfrac{J_m(x)}{H_m(x)} & \text{for } j=1, \\[2mm] \dfrac{J_m'(x)}{H_m'(x)} & \text{for } j=2. \end{cases} \quad (16)$$

One way to solve the system of Eqs. (12a) and (12b) is to eliminate the coefficients $d_m$, thus obtaining the following linear system for the sole $c_\ell$ coefficients:

$$\sum_{\ell=-\infty}^{+\infty} D_{m\ell} c_\ell = L_m, \quad (17)$$

where

$$D_{m\ell} = E_m^{(2)} A_{m\ell}^{(1)} - E_m^{(1)} A_{m\ell}^{(2)}, \quad (18a)$$

$$L_m = E_m^{(2)} b_m^{(1)} - E_m^{(1)} b_m^{(2)}. \quad (18b)$$

Once the $c_\ell$ coefficients are known, it is possible to evaluate the internal field $V_c$ by means of the $d_m$ coefficients in a straightforward way. Indeed, by subtracting term by term Eqs. (12a) and (12b), we obtain

$$d_m = -\frac{1}{E_m^{(1)} - E_m^{(2)}}\left\{b_m^{(1)} - b_m^{(2)} - \sum_{\ell=-\infty}^{+\infty}\left[A_{m\ell}^{(1)} - A_{m\ell}^{(2)}\right]c_\ell\right\}, \quad (19)$$

and, by recalling Eqs. (14) ... (16), after some algebra we get the following equality:

$$d_m = \frac{J_m(ka)H_m'(ka) - J_m'(ka)H_m(ka)}{J_m(n_c ka)H_m'(ka) - pJ_m'(n_c ka)H_m(ka)}\left\{1 + \Gamma\left(n_\parallel^i\right)\exp\left(in_\perp^i 2\chi\right)\exp\left[im(2\varphi-\pi)\right] + \right.$$

$$\left. + \sum_{\ell=-\infty}^{+\infty} RW_{\ell+m}(2\chi,0) i^{\ell-m} \exp\left[i(m-\ell)\varphi\right]c_\ell\right\}, \quad l, m = 0, \pm 1, \pm 2, \ldots \quad (20)$$

By the knowledge of the coefficients $c_m$ and $d_m$ ($m = 0, \pm 1, \pm 2, \ldots$) the scattering problem is exactly solved. Of course, since the procedure outlined above involves infinite series, some truncation criterion needs to be adopted. As we shall see, one can use the same criteria that proved to be adequate for other problems involving scattering from cylinders.[12, 15÷17]

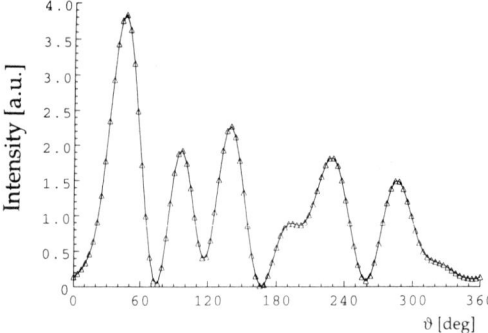

Fig. 2 Comparison between intensity values (in arbitrary units) at the cylinder sutface from the internal (full curve) and external (triangles) expansions. The plane surface is the interface between vacuum and a homogeneous dielectric medium (ns = 3.8), $\lambda$ = 632.8 nm, $\varphi$ = 30°, $n_c$ = 1.46, k$a$ = 3.95, N = 12.

Fig. 3 Far-field diffraction intensity as a function of the scattering angle $\theta$ for the case of Fig. 2.

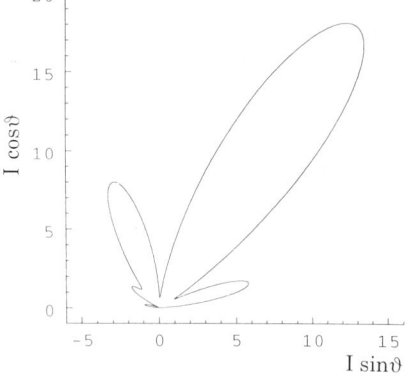

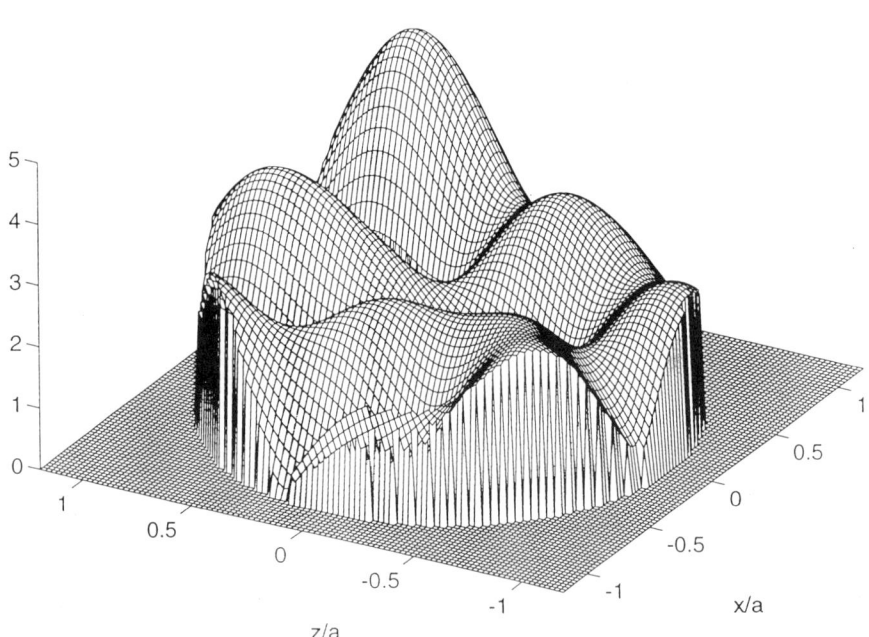

Fig. 4 Intensity distribution inside the dielectric cylinder for the case of Figs. 2 and 3. The field outside the cylinder has been set to zero for clarity.

In the following we will show that, in order to obtain an accurate description of the electromagnetic field, the required value of the index of truncation of the series can be reasonably small.

The case under test was studied in Ref. [6] and concerns the scattering problem of a monochromatic plane wave of wavelength $\lambda$ = 632.8 nm by a dielectric circular cylinder of $SiO_2$ ($n_c$ = 1.46), whose radius is $a$ = 0.35 µm, placed onto a flat substrate of sylicon (refractive index $n_s$ = 3.8). The incidence angle of the impinging wave is $\varphi$ = 30°, and the reflection coefficient of the surface is computed by means of the well-known Fresnel formulas.[14] The polarization is assumed to be TM.

It is well known that when a series related to the scattering from circular cylinders has to be truncated, the truncation index, say $N$, can be related to the cylinder radius a by the rule $N \sim 3ka$.[12, 15÷17] We verified that this choice also leads to good results in the present problem, by controlling the convergence of the expansion coefficients for increasing values of $N$.

To show the good match between the internal ($V_c$) and external ($V_i + V_r + V_d + V_{d\,r}$) field on the cylinder surface, the squared modulus of the electric field (here and in the following loosely referred to as the intensity) is reported in Fig. 2 for $\rho \to ka^-$ (internal expansion) as a full line and for $\rho \to ka^+$ (external expansion) by means of triangles, for values of $\vartheta$ ranging from 0 to 360 degrees. Fig. 3 shows the far-zone diffracted intensity I as a function of the scattering angle $\vartheta$ (see Fig. 1) for the case under test, in agreement with Fig. 2 of Ref. [6]. Figure 4 shows a 2D-plot of the intensity distribution inside the dielectric cylinder, which may be an important information for various applications.[1-5, 7-9]

Finally, it is worth stressing that the case of a lossy cylinder could be treated by means of the same procedures by suitably choosing the complex refractive index of the cylinder. Moreover, the method can be extended to cases in which the incident field differs from a single plane wave,[18] or for treating scattering by a set of parallel cylinders.[19]

## REFERENCES

1. J. F. Owen, P. W. Barber, B. J. Messinger, and R. K. Chang, Determination of opticalfiber diameter from resonances in the elastic scattering spectrum, *Opt. Lett.* 6, 272 (1981).
2. J. F. Owen, R. K. Chang, and P. W. Barber, Internal electric field distributions of a dielectric cylinder at resonance wavelengths, *Opt. Lett.* 6, 540 (1981).
3. B. Schlicht, K. F. Wall, R. K. Chang, and P. W. Barber, Light scattering by two parallel glass fibers, *J. Opt. Soc. Am.* A 4, 800 (1987).
4. H. A. Youssif and S. Khohler, Scattering by two penetrable cylinders at oblique incidence, *J. Opt. Soc. Am.* A 5, 1085 (1988).
5. D. Marcuse, Investigation of coupling between a fiber and an infinite slab, *J. Lightwave Tech.* 7, 122 (1990).
6. M. A. Taubenblatt, Light scattering from cylindrical structures on surfaces, *Opt. Lett.* 15, 255 (1990).
7. K. G. Ayappa, H. T. Davis, E. A. Davis, and J. Gordon, Two-dimensional finite elements analysis of microwave heating, *J. Chem. Eng.* 38, 1577 (1992).
8. P. J. Valle, F. Moreno, J. M. Saiz, and F. Gonzàlez, Near-field scattering from subwavelength metallic protuberances on conducting flat substrates, *Phys. Rev.* B 51, 13681 (1995).
9. A. Madrazo and M. Nieto-Vesperinas, Detection of subwavelength Goos-Hanchen shifts from near-field intensities: a numerical simulation, *Opt. Lett.* 20, 2445 (1995).
10. P. J. Valle, F. González, and F. Moreno, Electromagnetic wave scattering from conductinc cylindrical structures on flat substrates: study by means of the extinction theorem, *Appl. Opt.* 33, 512 (1994).
11. A. Madrazo and M. Nieto-Vesperinas, Scattering of electromagnetic waves from a cylinder in front of a conducting plane, *J. Opt. Soc. Am.* A 12, 1298 (1995).
12. R. Borghi, F. Frezza, F. Gori, M. Santarsiero, and G. Schettini, Plane-wave scattering by a perfectly conducting circular cylinder near a plane surface: cylindrical-wave approach, *J. Opt. Soc. Am.* A 13, 483 (1996).

13. G. Cincotti, F. Gori, M. Santarsiero, F. Frezza, F. Furnò, and G. Schettini, Plane wave expansion of cylindrical functions, *Opt. Commun.* 95, 192 (1993).
14. C. A. Balanis, *Advanced Engineering Electromagnetics,* (Wiley, New York, 1989).
15. H. A. Ragheb and M. Hamid, Scattering by N parallel conducting circular cylinders, *Int. J. Electron.* 59, 407 (1985).
16. A. Z. Elsherbeni and A. A. Kishk, Modeling of cylindrical objects by circular dielectric and conducting cylinders, *IEEE Trans. Antennas Propag.* 40, 96 (1992).
17. A. Z. Elsherbeni, A comparative study of two-dimensional multiple scattering techniques, *Radio Science* 29, 1023 (1994).
18. F. Frezza, G. Gerosa, F. Gori, M. Santarsiero, F. Santini, G. Schettini, and M. Sgroi, Gaussian beam diffraction by a quasi-optical grating for coupling lower-hybrid plasma waves, *Int. J. Inf. Mill. Waves* 16, 1009 (1995).
19. R. Borghi, F. Frezza, F. Gori, M. Santarsiero, and G. Schettini, Plane-wave scattering by a set of perfectly conducting circular cylinders in the presence of a plane surface, *J. Opt. Soc. Am.* A 13, 2441 (1996).

# STEADY STATES IN SPHERICAL MICROPARTICLES AND THEIR INSTABILITIES

L. A. Kotomtseva and G. P. Lednyeva

B.I.Stepanov Institute of Physics of the Academy of Sciences
68 F. Skaryna prosp., Minsk 220072, Belarus

## 1. INTRODUCTION

For various physical and engineering applications, for example in ultrafast optical data processing systems, many types of spatial lasing configurations have been recently proposed.

The advantages of the spherical microresonator have been pointed out [1-5] because of high quality factor of morphology-dependent resonances (MDRs) in the dielectric microsphere.

Among the advantage of lasing microspheres are a low threshold for lasing and optical bistability due to extremely high quality of spherical microcavity, easy matching of a small sphere to an optical fiber or other element in an optical system, and the opportunity to obtain radiation on various space modes simultaneously.[6-7] Lasing microparticles in some respects may be preferable to other small light sources. The conventional semiclassical theory of stimulated processes has been modified to accommodate a description of lasing of spherical particles.[7-10, 18-21]

On the basis of a proposed theoretical model for consideration of the time dependent stimulated emission of radiation of spherical microparticles, we describe the steady states of this system and the possibilities of their instabilities for some typical cases. Examples of the distribution of modes in a spherical microcavity are given below.

## 2. STEADY STATES OF THE BASIC SYSTEM

The system of semiclassical equations for the interaction of radiation with a spherical microparticle is:

$$\frac{\partial q_\ell}{\partial t} + 2\pi\sigma q_\ell + 2\pi\omega w_\ell \sin(\alpha - \beta) = 0,$$

$$\frac{\partial w_\ell}{\partial t} + \gamma w_\ell + \frac{N|d|^2}{3h} q_\ell y \sin(\alpha-\beta) = 0$$

$$\frac{\partial y}{\partial t} = D(y_0 - y) - \frac{1}{Nh} q_\ell w_\ell (S_\ell^2 + Q_\ell^2) \sin(\alpha-\beta), \qquad (1)$$

$$\frac{2\omega}{v^2}\frac{\partial \alpha}{\partial t} q_\ell + \frac{4\pi\omega^2}{v^2} w_\ell \cos(\alpha-\beta) + \frac{\omega^2}{v^2} q_\ell + \nabla^2 q_\ell + \ell(\ell+1) q_\ell = 0,$$

$$\frac{\partial \beta}{\partial t} + \omega - \omega_a = \frac{N|d|^2 q_\ell}{3hw_\ell} y \cos(\alpha-\beta).$$

We consider steady states, corresponding to zero values of time derivatives for amplitudes

$$\frac{dw_\ell}{dt} = 0, \qquad \frac{dq_\ell}{dt} = 0, \qquad \frac{dy}{dt} = 0$$

and definite frequency of radiation of the steady state $\omega_s = \omega + d\alpha/dt$ with

$$d\alpha/dt = \delta_1, \qquad d\beta/dt = \delta_2, \qquad \alpha - \beta = \delta_c.$$

The following relationships apply in this case:

$$q_\ell = -2\pi\omega w_\ell \sin\delta_c,$$

$$y = y_0 \Big/ [1 + \frac{|d|^2 q_\ell^2 (S_\ell^2 + Q_\ell^2)}{3Dh^2(\delta_2 + \omega - \omega_a)} \sin\delta_c \cos\delta_c],$$

$$\text{tg }\delta_c = \gamma \Big/ (\omega_a - \omega - \delta_2),\qquad (2)$$

$$w_\ell = -\frac{N|d|^2 q_\ell y}{3h} \sin\delta_c.$$

The equation

$$\nabla^2 q_\ell + \frac{\omega^2}{v^2} q_\ell + \ell(\ell+1) q_\ell = -\frac{2\omega}{v^2}\delta_1 q_\ell - \frac{4\pi\omega^2 N y q_\ell}{3hv^2} \cos\delta_c \sin\delta_c \qquad (3)$$

with value y from (2) can be solved with the help of the Green function.

For any parameters, the steady state with zero intensity and inversion of populations $y_{tr} = y_0$ exists and it is known as the trivial state.

For the four level model of the material, which is appropriate both for glass and dye-doped spheres, we can describe the case of weak intensity, when saturation does not occur, and the strong saturation case due to high intensity of radiation. In the first case, for resonant interaction $\omega = \omega_a$ and polarisation following the field modes, morphology dependent resonances can be considered as the modes of the laser. The distribution of radiation in the equatorial plane of the sphere for the magnetic wave modes is given in

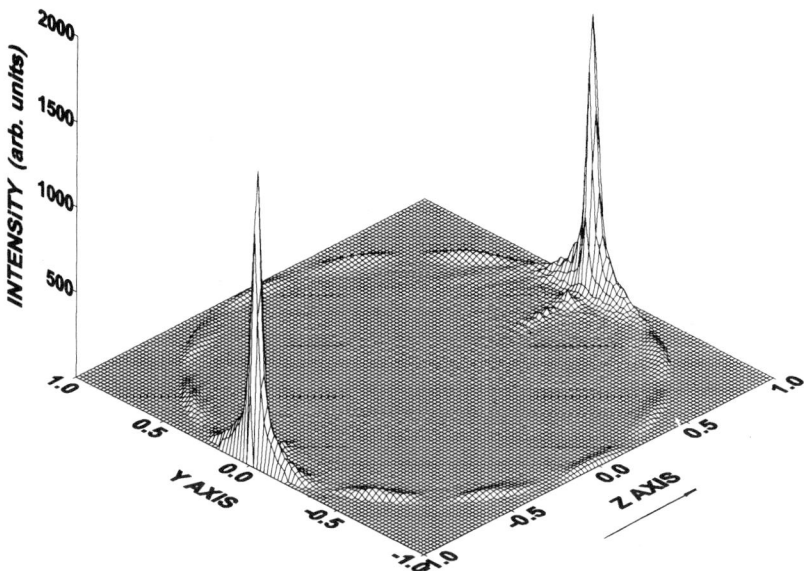

Fig. 1 The lasing magnetic wave intensity distribution in a spherical microlaser for $l = 96$, $n = 1$; the real part of the refraction index is 1.33 and its imaginary part is $3 \cdot 10^{-8}$; the particle radius is 7.56 μm.

Fig.s 1, 2 and 3 in arbitrary units. The distribution for the electric wave is in Figs. 4 and 5 in arbitrary units. From Figs. 1, 2 and 3 we can see for modes with number $l = 96$ that the highest intensity has a magnetic wave with $n=2$ in Fig. 2, which is two orders higher than the wave with $n = 1$ (Fig. 1) and $n = 3$ (Fig. 3). The maximum intensity for the magnetic wave is on the principal diameter, and the order $n$ determines the extreme intensity along the radius. For the electric wave, maximum intensity is observed in two areas slightly shifted from the

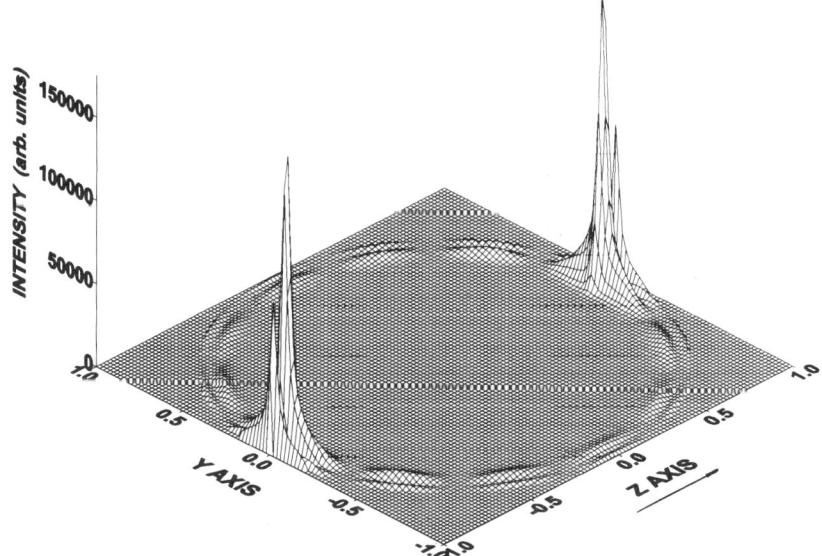

Fig. 2 The lasing magnetic wave intensity distribution in a spherical microlaser for $l = 96$, $n = 2$; the real part of the refraction index is 1.33 and its imaginary part is $3 \cdot 10^{-8}$; the particle radius is 7.56 μm.

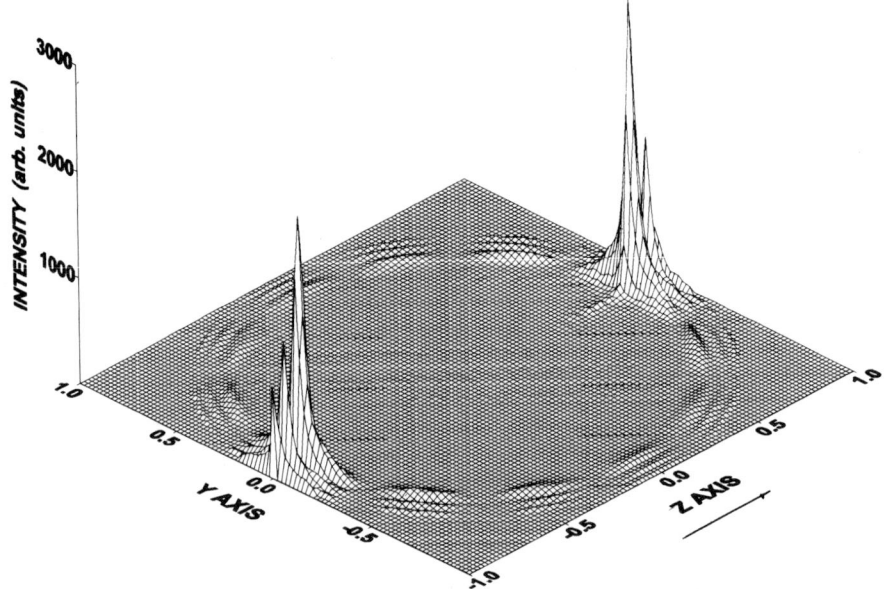

Fig. 3 The lasing magnetic wave intensity distribution in a spherical microlaser for $\ell = 96$, $n = 3$; the real part of the refraction index is 1.33 and its imaginary part is $3 \cdot 10^{-8}$, the particle radius is 7.56 µm.

principal diameter, as is seen in Figs. 4 and 5. For this example, the intensity of the mode with higher number and order in Fig. 5 is higher than for the mode with lower number and order as in Fig. 4. For $n = 2$ we see two clearly expressed maxima in distribution of intensity along the radius.

For the second case of high intensity, with highly saturated radiation exiting the sphere, we have

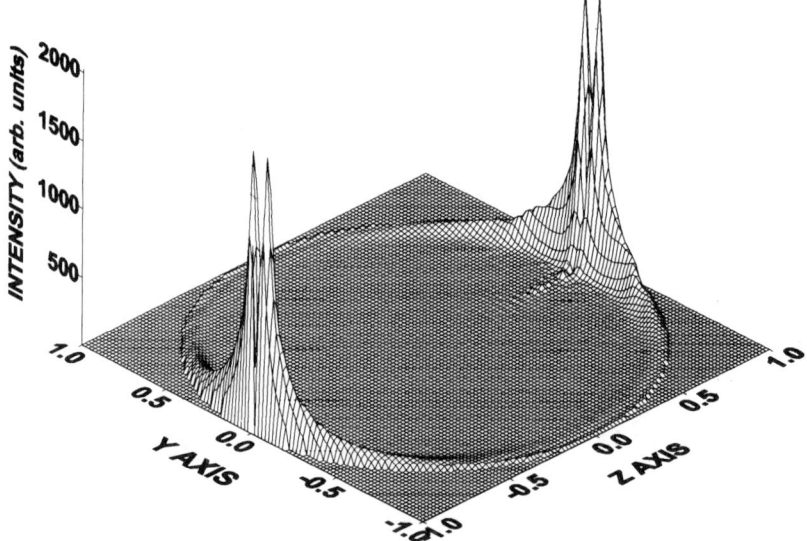

Fig. 4 The lasing electric wave intensity distribution within the equatorial plane of the microparticle for $\ell = 90$, $n = 1$; the real part of the refraction index is 1.33 and its imaginary part is $3 \cdot 10^{-8}$; the particle radius is 7.56 µm.

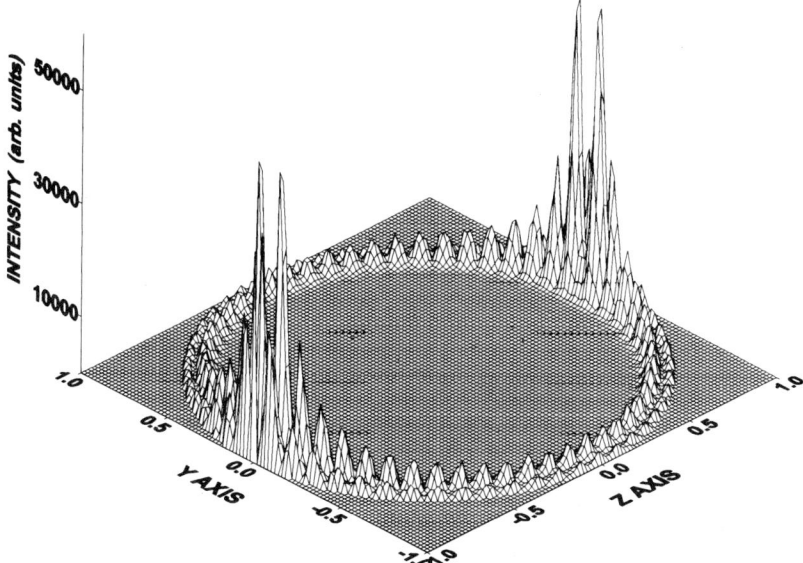

Fig. 5 The lasing electric wave intensity distribution within the equatorial plane of the microparticle for $l = 119$, $n = 2$; the real part of the refraction index is 1.33 and its imaginary part is $3 \cdot 10^{-8}$; the particle radius is 7.56 μm.

$$q_\ell = b_\ell \xi_\ell(kr) \{1 - \frac{2\omega}{v^2}\delta_1 F_\ell^2 \int_0^a \psi_\ell^2(mkr)dr - \frac{4\pi\omega^2 Nh(\delta_2 + \omega - \omega_a)}{v^2 |d|^2 (S_\ell^2 + Q_\ell^2)} \times \quad (4)$$

$$\times \sum_{j=1}^\infty \int_0^a \{c_j^2[-\frac{j(j+1)}{r^2}\psi_j^2(mk_p r)Q_j^2 + (\psi_j')^2(Q_j^2 + S_j^2)] + F_j^2\psi_j^2(Q_j^2 + S_j^2)\}dr\}.$$

Here $k_p = 2\pi v_p$, $v_p$ is the pumping frequency, $\Psi_{\ell,j}$ are the Ricatti-Bessel functions, $\xi_\ell$ are the Ricatti-Hankel functions, $F_\ell$ is the MDR coefficient for magnetic wave for field inside the particle, $c_j$ is the MDR coefficient for the electric wave for the field inside the particle, $k = 2\pi/\lambda$, $a$ is the radius of the particle, $b_\ell$ is the coefficient for solution of the linear Mie problem for magnetic wave outside the particle, index $\ell$ is the number of the lasing mode under consideration, and the summation on $j$ gives us pumping mode numbers with the frequency of pumping.

It is known that experimental lasing spectra exhibit intensity peaks associated with morphology-dependent resonance wavelengths.[6,9-12] For a given spherical microparticle MDR's occur at specific values of $\rho_{n,\ell}$, where $\rho = 2\pi a / \lambda$ is the size parameter for the particle of radius $a$ and the free space wavelength of light $\lambda$, and $n$ and $\ell$ are integers, the number and the order of the mode respectively.

It is known,[22] that spacing between modes for a series with the mode order $n$ with sequentially increasing mode number $l$ in the approximation of large argument for the Bessel functions can be written:

$$\Delta\rho = \{\arctan[(m^2-1)^{1/2}]\} / [(m^2-1)^{1/2}]. \quad (5)$$

For a microparticle with radius $a=7.56$ μm, index of refraction $m = 1.33$; and $\lambda$ in the interval from 570 to 610 nm, the value $\Delta\rho \approx 0.7369$. This is in good agreement with the results of our simulations, which give $\Delta\rho \approx 0.78$, and in value corresponds to experimentally observed frequencies of modes in dye droplets.

## 3. INSTABILITIES IN SPHERICAL MICROPARTICLES

Linear analysis for small deviations from the steady state for the definite mode (2), with different values of the resonant frequency of a group of particles and of the frequency of light for our system of Eq.(1) can give us a characteristic quasipolynomial. For the resonant case when the frequency of light coincides with the frequency of transition in an active medium and for the adiabatic elimination of polarisation from equations as in glass particles, the quadratic characteristic equation for a definite mode has been considered in Refs. 18- 21. It appears that the steady state of Eq.(2) is stable for any number of mode $\ell$. . Roots of these equations are

$$\lambda_{1,2} = -M/2 \pm \sqrt{M^2 - \nu\kappa y_s(M-D)} \qquad (6)$$

where

$$M = D + B\psi_\ell^2 f_\ell^2 [Q_\ell^2(\vartheta) + S_\ell^2(\vartheta)]/2\pi \ .$$

Here $\kappa$ is extremal theoretically achievable gain for all particles in the upper state, B is Einstein coefficient for stimulated emission, $y_s$ is the steady state value of Eq.(2) of inversion of population, $f_\ell$ is the coefficient for the expansion of the laser field inside the particle [21], and $l$ is the number of the mode under consideration. For the magnetic wave with $\ell=50$, $\rho = 37.451$, $\kappa = 5$ cm$^{-1}$, the roots (6) are complex conjugate with real part $\alpha = -0.01$ s$^{-1}$ and imaginary part $\beta = 2.2 \cdot 10^6$ s$^{-1}$. This means that under steady state pumping the microparticle can radiate transient microsecond pulsations with a rate of decay of about milliseconds. From the analytical expressions that are given in Ref. 21, the shift of the lasing frequency from the morphology dependent resonance frequency for this mode can be determined. It appears that in our approximation of noninteracting modes, any mode over the first threshold of generation is stable and multistability can take place.[23]

## 4. CONCLUSION

A theoretical model for consideration of the dynamics of interaction of light with spherical microparticles for arbitrary material parameters is proposed. Using systems of equations presented in this chapter, we have considered steady states for different modes and their instabilities for various parameters of the microparticle.

For glass and dye-doped microspheres a simpler system of equations has been proposed by one of the authors. On this basis, the authors have tried to understand the role of the interaction between modes and to consider multistable and degenerative states. Various kinds of simplifications can be introduced for the specifice type of sphere under consideration. The model described here is supposed to give an opportunity to obtain new interesting results, and to understand the mode structure and potential opportunities of lasing in spherical microparticle.

# REFERENCES

1. H.M.Tzeng, K.F.Wall, M.B.Long, and R.K.Chang, Laser emission from individual droplets at wavelengths corresponding to morphology-dependent resonances, *Opt. Lett.* 9:499 (1984).
2. S.C.Hill, and R.E.Benner, Morphology-dependent resonances associated with stimulated processes in microspheres, *J. Opt. Soc. Am.* B 3:1509 (1986).
3. H.-B.Lin, A.L.Huston, B.L.Justus, and A.J.Campillo, Some characteristics of a droplet whispering gallery mode laser, *Opt. Lett.* 11:614 (1986).
4. S.X.Qian, J.B.Snow, H.M.Tzeng, and R.K.Chang, Lasing droplets: highlighting the liquid air interface by laser emission, *Science* 231:486 (1986).
5. V.B.Braginsky, M.L.Gorodetsky, and V.S.Ilchenko, Quality-factor and nonlinear properties of optical whispering-gallery modes, *Phys. Lett.* A 137:393 (1989).
6. H.-B.Lin, A.L.Huston, J.D.Eversole, and A.J.Campillo, Double resonance stimulated Raman scattering in micrometer-sized droplets, *J. Opt. Soc. Am.* B7:2079 (1990).
7. G.Kurizki, and A.Nitzan, Theory of stimulated emission processes in spherical microparticles, *Phys. Rev.* A 38:267 (1988).
8. V.V.Dacjuk, I.A.Izmajlov, and V.A.Kochelap. Whispering modes gallery light lasing in droplet, *Kvantovaja Elektronika*, Kiev. No 38:56 (1990).
9. H.-B.Lin, J.D.Eversole, and A.J.Campillo, Spectral properties of lasing microdroplets, *J. Opt. Soc. Am.* B9:43 (1992).
10. G.Chen, D.Q.Chowdhury, R.K.Chang, and W.-F.Hsieh, Laser-induced radiation leakage from microdroplets, *J. Opt. Soc. Am.* B 10:620 (1993).
11. H.Taniguchi, Time difference in stimulated emission from fluorescent dye-doped small-size liquid droplets, *Japan. J. Appl. Phys.* 32:L1615 (1993).
12. H.Taniguchi, and S.Tanosaki, Three-color whispering-gallery-mode dye lasers using dye-doped liquid spheres, *Japan. J. Appl. Phys.* 32:L1421 (1993).
13. H.Misava, R.Fujisawa, K.Sasaki, N.Kitamura, H.Masuhara, Simultaneous manipulation and lasing of a polymer microparticle using a CW 1064 nm laser beam, *Japan J. Appl. Phys.* 32:L788 (1993).
14. A.S.Kwok, and R.K.Chang, Suppression of lasing by stimulated Raman scattering in microdroplets, *Opt. Lett.* 18:1597 (1993).
15. G.Chen, R.K.Chang, S.C.Hill, Wavelength variation of laser emission along the entire rim of slightly deformed microdroplets, *Opt. Lett.* 18:1993 (1993).
16. A.N.Oraevsky, and D.K.Bandy, Ball bistability, *Quant. Electr.* 22:211 (1995).
17. D.Q.Chowdhury, S.C.Hill, Md.M.Mazumder, Quality factor and effective-average modal gain of loss in inhomogeneous spherical resonators: application to two-photon absorption, *IEEE J.Quant.Electr.* 29:2553 (1993).
18. G.P.Lednyeva, and L.G.Astafieva, Time-dependent stimulated emission of glass microparticle, *J. Aerosol Sci.* 26(Suppl.1):S255 (1995).
19. G.P.Lednyeva, Nonstationary oscillations in spherical microparticle, *Optika i Spektroskopija.* 76:506 (1994).(in Russian)
20. G.P.Lednyeva, Regime of oscillations of microlaser, *Vesti Akademii Nauk Belarusi, ser. fiz.-mat. n.* No 1:91 (1995) (in Russian).
21. G.P.Lednyeva, and L.G.Astafieva, To the theory of a neodimium glass laser with spherical microcavity, *Optika i Spektroskopija* 80:858 (1996).
22. P.Chylek, Partial-wave resonances and the ripple structure in the Mie normalized extinction cross section, *J. Opt. Soc. Am.* 66:285 (1976).
23. A.Mekis, J.U.Nockel, G.Chen, A.D.Stone, R.K.Chang, Ray chaos and Q-spoiling in lasing droplets, *Phys.Rev.Lett.* 75:2682 (1995).

# MODELLING OF A MICROLASER BASED ON SPHERICAL MICROPARTICLE

L. A. Kotomtseva and G. P. Lednyeva

B.I.Stepanov Institute of Physics of the Academy of Sciences
68 F. Skaryna prosp., Minsk 220072, Belarus

## 1. INTRODUCTION

Optical microsystems as compact sources of coherent light are very promising for applications in many areas of science and engineering. For this reason, great attention is being paid to their potential opportunities and to understand their specific properties that arise due to geometry, the properties of materials, and the role of various physical linear and nonlinear phenomena. Systems with spherical symmetry with sizes leading to diffraction effects when illuminated belong to the same interesting class of objects. We know that in homogeneous transparent or weakly absorbing spherical microparticles, morphology dependent resonances appear under illumination by light. Such resonances have been intensively studied since the 1970s by Chang et al. and other scientists.[1-3] During last ten years experiments with microparticles have shown that for *whispering gallery modes* a very low threshold exists for the appearance of intense radiation, and lasing in glass or dye-doped spheres with a radius of several micrometers has been obtained.[4-10] Both theoretical considerations[11-15] and experiments[4-10] show many interesting characteristics for such objects.

The theoretical problem of the diffraction of light on a sphere was first solved by Mie in 1908 on the basis of the Maxwell equations for diffraction of a plane monochromatic electromagnetic wave on a homogeneous sphere of arbitrary radius.[16] A little later Debye published his equivalent solution for this problem.[17] It worth noting that the Mie solution can be used for a number of identical spheres, randomly distributed in space, when the distance between them is much greater than the wavelength of light, through summation of the contributions of separate spheres.

In the Institute of Physics of the Academy of Sciences of Belarus, optical and laser physics is the main subject of investigations. One of the scientific groups, where Lednyeva works, considers theoretical aspects of electromagnetic radiation interaction with aerosols (substances in a dispersed state). They have developed a theoretical model of nonstationary heating of dye-doped polystyrene and neodymium glass microspheres and microlasers, taking into account the strength of scaling effects. The problem of heating of spherical particles requires the solution of the two-dimensional thermal diffusion equation in spherical co-ordinates with appropriate initial and boundary conditions. Nonuniform heat release inside the particle and temperature dependencies of thermophysical properties of the material of particle are taken into account.[18-23] Kotomtseva considers nonlinear dynamics in laser and optical systems.[24-27] The results of efforts to formulate a theoretical model of microlasers based on spherical particles are summarized in this Chapter.

The phenomenon of diffraction inside a sphere can be used for pumping of this sphere to create population inversion in areas of concentration of the light energy. Mie scattering theory permits us to calculate the distribution of light inside the sphere and determine the pumping. For a particle substance with appropriate disposition of energy levels this intense light leads to transition of molecules from the ground state to an excited level. The microparticle description of the excitation of glass doped with neodyuium ions and of light emission in dye molecules must take into account not only the action of pumping, but also the nonlinear interaction of molecules of the droplet substance with spontaneous emission and stimulated emission leading to saturation of gain. This problem demands modification of the semiclassical system of equations for the spherical particle. Below we show examples of the solution of this problem. In the second part the results of traditional Mie scattering theory are adapted to determine the pumping distribution inside a spherical particle. In the third part the basic system of equations for the interaction of light with the substance of a spherical particle is presented.

## 2. MIE THEORY OF PUMPING OF A SPHERICAL PARTICLE

To describe the field distribution inside a spherical microparticle when a plane electromagnetic wave is incident on the surface of the sphere, a curvelinear, namely spherical, system of coordinates is introduced. A detailed consideration of this problem is given in Ref. 28. Below we give brief a introduction to this problem. The complete field inside and out side of the sphere is considered as sum of two subfields. The electrical vector of one field contain no radial component; it is a transverse electric field and is called the magnetic wave. For the second field, the magnetic vector has no radial component; it is a transverse magnetic field and is called the electric wave. Maxwell's equations for a monochromatic wave with time dependence of field exp (- i$\omega$t) in a medium with dielectric permeability $\varepsilon$ and conductivity $\sigma$ may be written

$$\nabla \times \vec{H} = -\frac{i\omega}{c}(\varepsilon + i\frac{4\pi\sigma}{\omega})\vec{E}, \nabla \times \vec{E} = \frac{i\omega}{c}\vec{H} . \tag{1}$$

With boundary conditions on the surface of a spherical microparticle in a spherical system of coordinates, the system of ordinary differential equations can be solved. We the remember that the total electric field outside the sphere consists of the sum of the incident field $E^{(i)}$ and of the scattered or diffracted field $E^{(s)}$. The field inside the sphere is $E^{(w)}$. Similar symbols can be introduced for the magnetic field. Solutions of Maxwell's equations for either type of wave (electrical or magnetic) can be expressed with help of scalar Debye potentials $^e\Pi$ ( $^m\Pi$ ), which are solutions of the wave equations in spherical coordinates. The usual representation of the solution in the form

$$\Pi(r,\vartheta,\varphi) = R(r)\Theta(\vartheta)\Phi(\varphi) \tag{2}$$

leads to a system of ordinary differential equations for the new functions. For single valued solutions some additional conditions must be fulfilled. In the case of $\Phi$ we obtain:

$$\Phi = a_m \cos(m\varphi) + b_m \sin(m\varphi). \qquad (3)$$

where m is integer number. The differential equation for $\Theta$ is the well known euation for spherical harmonics, whose solutions are associated Legendre polynomials $P_\ell^{(m)}(\xi)$ (spherical harmonics of the first kind), namely

$$\Theta = P_\ell^{(m)}(\xi) = P_\ell^{(m)}(\cos\vartheta). \qquad (4)$$

These functions are equal to zero for $|m| > \ell$, and $2\ell+1$ such functions exist for each value $\ell$, for $m = -\ell, -\ell+1, ..., \ell-1, \ell$.

Solutions of the equation for R(r) may be written using cylindrical functions of the general type as linear combinations of two cylindrical functions of ordinary type, for example the Bessel function $J_{\ell+1/2}(kr)$ and the Neumann function $N_{\ell+1/2}(kr)$:

$$rR = c_\ell \psi_\ell(kr) + d_\ell \chi_\ell(kr), \qquad (5)$$

where

$$\psi_\ell(\rho) = \sqrt{\frac{\pi\rho}{2}} J_{\ell+1/2}(\rho), \chi_\ell(\rho) = -\sqrt{\frac{\pi\rho}{2}} N_{\ell+1/2}(\rho).$$

Accordingly to Eq.(2) the partial solution $\Pi_\ell^{(m)}$ is the product of Eqs. (3), (4) and (5). The general solution of the wave equation is equal to

$$r\Pi = r\sum_{\ell=0}^{\infty}\sum_{m=-\ell}^{\ell}\Pi_1^{(m)} = \sum_{l=0}^{\infty}\sum_{m=-\ell}^{\ell}\{[c_\ell\psi_\ell(kr) + d_\ell\chi_\ell(kr)] \times$$

$$\times [P_\ell^{(m)}(\cos\vartheta)][a_m\cos(m\varphi) + b_m\sin(m\varphi)]\}. \qquad (6)$$

Here $a_m, b_m, c_\ell$ and $d_\ell$ are constant, which must be chosen to fulfill the boundary conditions.

The general solution can be written for the electric field inside the spherical particle, $E^{(w)}$, and the field outside it as sum of the incident field $E^{(i)}$ and of the scattered or diffracted field $E^{(s)}$. For example, we exhibit below the components of the electric field of the scattered wave $E^{(s)}$:

$$E_r^{(s)} = \frac{1}{(k^{(1)})^2}\frac{\cos\varphi}{r^2}\sum_{\ell=1}^{\infty}\ell(\ell+1)^e B_\ell \zeta_\ell^{(1)}(k^{(1)}r) P_\ell^{(1)}(\cos\vartheta),$$

$$E_\vartheta^{(s)} = -\frac{1}{k^{(1)}}\frac{\cos\varphi}{r}\sum_{\ell=1}^{\infty}\{^e B_\ell \zeta_\ell^{(1)'}(k^{(1)}r) P_\ell^{(1)'}(\cos\vartheta)\sin\vartheta -$$

$$-i^m B_\ell \zeta_\ell^{(1)}(k^{(1)}r) P_\ell^{(1)}(\cos\vartheta)\frac{1}{\sin\vartheta}\}, \qquad (7)$$

$$E_\varphi^{(s)} = -\frac{1}{k^{(I)}} \frac{\sin\varphi}{r} \sum_{\ell=1}^{\infty} \{^e B_\ell \zeta_\ell^{(1)'}(k^{(I)}r) P_\ell^{(1)}(\cos\vartheta) \frac{1}{\sin\vartheta} -$$

$$- i^m B_1 \zeta_1^{(1)}(k^{(I)}r) P_1^{(1)'}(\cos\vartheta)\sin\vartheta\}, \qquad \zeta_\ell^{(1)} = \psi_\ell - i\chi_\ell.$$

Here

$$^e B_\ell = i^{\ell+1} \frac{2\ell+1}{\ell(\ell+1)} \times \tag{8}$$

$$\times \frac{k_2^{(I)} k^{(II)} \psi_\ell'(k^{(I)}a) \psi_\ell(k^{(III)}a) - k_2^{(II)} k^{(I)} \psi_\ell'(k^{(III)}a) \psi_\ell(k^{(I)}a)}{k_2^{(I)} k^{(II)} \zeta_\ell^{(1)'}(k^{(I)}a) \psi_\ell(k^{(III)}a) - k_2^{(II)} k^{(I)} \psi_\ell'(k^{(III)}a) \zeta_\ell^{(1)}(k^{(I)}a)}$$

$$^m B_\ell = i^{\ell+1} \frac{2\ell+1}{\ell(\ell+1)} \times \tag{9}$$

$$\times \frac{k_2^{(I)} k^{(II)} \psi_\ell(k^{(I)}a) \psi_\ell'(k^{(III)}a) - k_2^{(II)} k^{(I)} \psi_\ell'(k^{(I)}a) \psi_\ell(k^{(III)}a)}{k_2^{(I)} k^{(II)} \zeta_\ell^{(1)}(k^{(I)}a) \psi_\ell'(k^{(III)}a) - k_2^{(II)} k^{(I)} \zeta_\ell^{(1)'}(k^{(I)}a) \psi_\ell(k^{(III)}a)}$$

This solution describes a sphere with radius a and conductivity σ surrounded by a nonconducting medium; then

$$k_2^{(I)} = \frac{i\omega}{c} = i\frac{2\pi}{\lambda_0}, \qquad k^{(I)} = \frac{2\pi}{\lambda_0}\sqrt{\varepsilon^{(I)}} = \frac{2\pi}{\lambda^{(I)}}, \tag{10}$$

$$k_2^{(II)} = \frac{i\omega}{c} = i\frac{2\pi}{\lambda_0}, \qquad k^{(II)} = \frac{2\pi}{\lambda_0}\sqrt{\varepsilon^{(II)} + i\frac{4\pi\sigma}{\omega}}. \tag{11}$$

Here $\lambda_0$ is the wave length in vacuum, $\lambda^{(I)}$ = wave length in the medium surrounding the sphere, $\varepsilon^{(I)}$ is the dielectric permeability of the surrounding medium and $\varepsilon^{(II)}$ is the dielectric permeability of the substance of the sphere.

Detailed consideration of the solutions for the radiation inside a spherical microparticle, for example the electric field $E^{(w)}$, gives for size parameter $\rho = 2\pi a/\lambda > 1$ and $\ell$ much greater than unity, very intense radiation in two areas of the diametral zone of the sphere. Such resonances have been intensively studied and are known as morphology dependent resonances (MDR).

## 3. DISTRIBUTION OF LIGHT IN A SPHERICAL MICROPARTICLE

Numerical simulations for the parameters corresponding to a dye-doped polystyrene sphere with radius $a = 7.56$ μm with complex refraction index m - iκ = $1.33 - i \cdot 3 \cdot 10^{-8}$ give the inhomogeneous distribution of light in the equatorial plane of the sphere seen in Figs. 1-3. The relative intensity of radiation under irradiation at $\lambda = 0.532$ μm in the direction of the z axis in Fig. 1 is $E^{(w)}$, expressed as a sum of spherical harmonics. It is seen that the region with the highest intensity is near a principal diameter in the shadow area close to the exit of radiation from the sphere; a second peak, less in intensity and square area, is concentrated

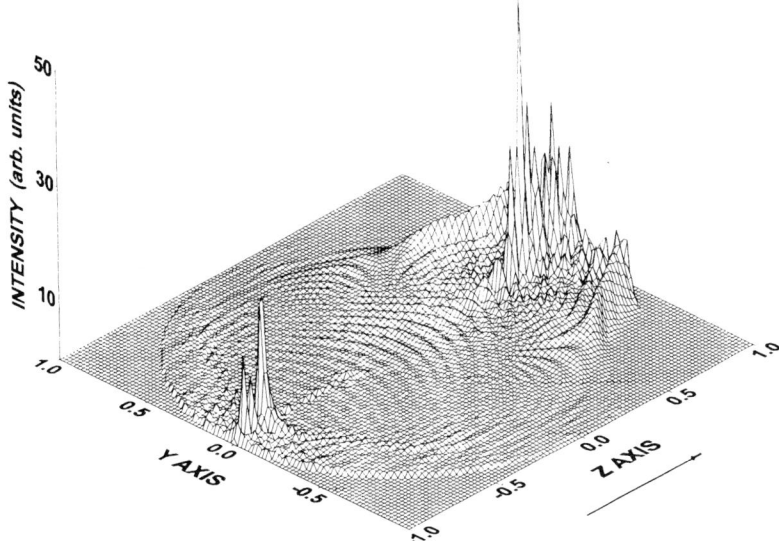

Fig. 1 The pumping intensity distribution within the equatorial plane of the sphere. The beam direction along the z axis is indicated by the arrow, and λ is 0.532 μm.

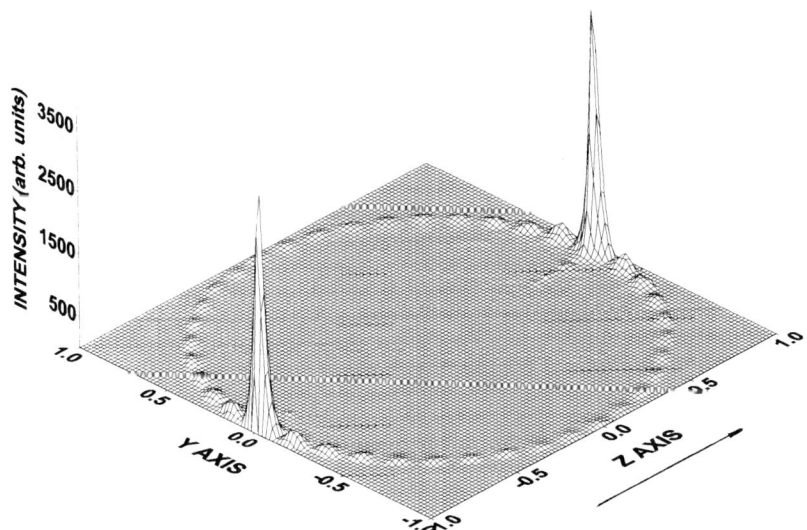

Fig. 2 The pumping intensity distribution within the equatorial plane of a particle at the morphology dependent resonance for the magnetic wave. The particle radius is 7.56 μm, $\ell = 110$, $n = 1$ and the beam direction is along the z axis.

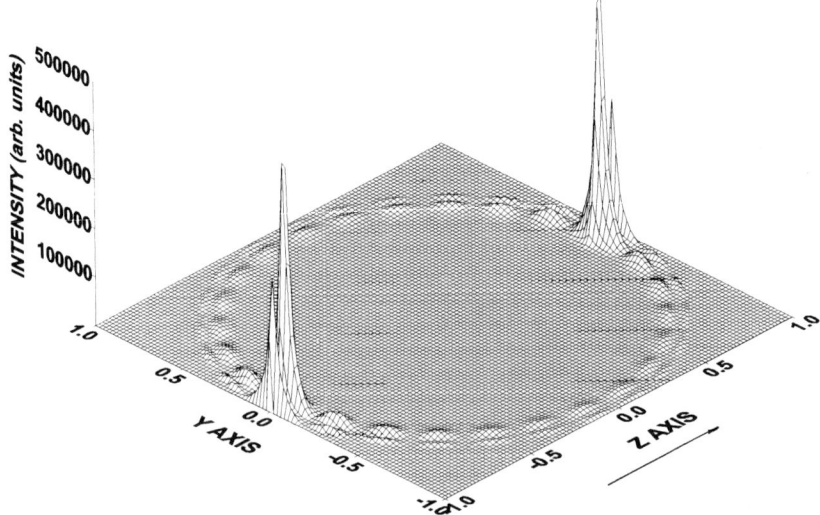

Fig. 3 The pumping intensity distribution within the equatorial plane of particle at the morphology dependent resonance for the magnetic wave. The particle radius is 7.56 μm, $\ell = 103$, $n = 2$.

near the entrance of light into the sphere. Detailed tuning of the size parameter $\rho = 2\pi a/\lambda$ with a change of this value in the fourth or the fifth significant digit permits us to get the morphology dependent resonances given in Figs. and 3 for resonance number $\ell$ ( = 110 in Fig. 2 and = 103 in Fig. 3) and for solution order $n = 1$ (Fig. 2) and $n = 2$ (Fig. 3) for the magnetic wave. The relative intensity of the resonances in Fig.3 is two orders higher than in Fig. 2 and much higher than in Fig. 1. Regions of such high resonances are very small and their selectivity in space is very high. In order for pumping light to produce laser action in such a microparticle, the case of Fig. 1 may be the most appropriate due to the relatively high intensity over a wide area.

## 4. BASIC OF EQUATIONS FOR A SPHERICAL MICROCAVITY LASER

Theoretical consideration of laser action in a spherical microparticle on the basis of semiclassical equations can be carried out with the help of the system of coupled equations for the interaction of light with an active medium. The Maxwell equations must be modified to take account of the active medium. Equations for the medium interacting with light must describe relaxation and spectral properties of the substance for transition from the upper excited energy level to the lower level at the lasing frequency. The pumping of the active medium that has been discussed in the previous section leads to excitation of an upper laser level due to the presence of additional higher energy levels, as is usually encountered in traditional lasers using a three or four energy level scheme.[24] For dipole interaction of the substance with light we use a semiclassical system of equations for the electric field E, polarisation P and inversion of population $y$ in the active medium:

$$\nabla \times \nabla \times \vec{E} = \frac{4\pi\sigma}{v^2}\frac{\partial \vec{E}}{\partial t} + \frac{1}{v^2}\frac{\partial^2}{\partial t^2}(\vec{E} + 4\pi\vec{P}) , \qquad (12)$$

$$\frac{\partial^2 \vec{P}}{\partial t^2} + 2\gamma \frac{\partial \vec{P}}{\partial t} + (\gamma^2 + \omega_a^2)\vec{P} = -\frac{2N\omega_a |d|^2}{3\hbar} \vec{E} y ,\qquad(13)$$

$$\frac{\partial y}{\partial t} = \frac{2\vec{E}}{\hbar \omega_a N} \left(\frac{\partial \vec{P}}{\partial t} + \gamma \vec{P}\right) - D(y - y_0) .\qquad(14)$$

Here $\nabla$ is the vector differential operator, $v$ is the phase velocity of light in the medium of the particle, $D$ and $\gamma$ are the rates of decay of populations and of polarisation of an active medium, $\omega_a$ is the transition frequency in an active medium, $N$ is the concentration of active particles, $y_0$ is the nonsaturated inversion of population achievable under pumping action, and $|d|$ is the modulus of the matrix element of the transition dipole moment. With the help of the angular momentum operator $L = (1/i)\,\vec{r} \times \nabla$ we introduce new variables

$$\eta = L \bullet \nabla \times \vec{E} , \qquad\qquad \Phi = L \bullet \nabla \times \vec{P} \qquad(15)$$

and we can write instead of Eqs. (12) and (13)

$$\nabla^2 \eta = \frac{4\pi}{v^2}\sigma \frac{\partial \eta}{\partial t} + \frac{1}{v^2}\frac{\partial^2}{\partial t^2}(\eta + 4\pi\Phi) ,\qquad(16)$$

$$\frac{\partial^2 \Phi}{\partial t^2} + 2\gamma \frac{\partial \Phi}{\partial t} + (\gamma^2 + \omega_a^2)\Phi = -\frac{2N|d|^2 \omega_a}{3\hbar} \eta y ,\qquad(17)$$

For electrical field and polarisation that are slowly varying in time (but not in space)

$$\vec{E} = \vec{e}\, e^{i\omega t} + \vec{e}^*\, e^{-i\omega t}, \vec{P} = \vec{p}\, e^{i\omega t} + \vec{p}^*\, e^{-i\omega t} \qquad(18)$$

Eq. (14), describing the inversion of populations, may be written

$$\frac{\partial y}{\partial t} = D(y_0 - y) + \frac{i}{2\hbar N}(\vec{e}\,\vec{p}^* - \vec{p}\,\vec{e}^*).\qquad(19)$$

For

$$\eta = \Pi(\vec{r},t)\, e^{i\omega t} + \text{c.c.} , \qquad\qquad \Phi = \Gamma(\vec{r},t)\, e^{i\omega t} + \text{c.c.} ,\qquad(20)$$

we obtain instead of Eqs. (16) and (17):

$$\nabla^2 \Pi = i\frac{4\pi\omega\sigma}{v^2}\Pi + \frac{2i\omega}{v^2}\frac{\partial \Pi}{\partial t} - \frac{\omega^2}{v^2}\Pi - \frac{4\pi\omega^2}{v^2}\Gamma ,\qquad(21)$$

$$\frac{\partial \Gamma}{\partial t} + \gamma\Gamma + i(\omega - \omega_a)\Gamma = i\frac{N|d|^2}{3\hbar}\Pi y .\qquad(22)$$

We assume that the angular distribution of modes in the nonlinear system remains the same as in the previous linear problem for the sphere. Then we can write

$$\Pi = \sum_{\ell} \sum_{m=-\ell}^{\ell} q_\ell(r,t) Y_{\ell m}(\vartheta,\varphi) , \qquad (23)$$

$$\Gamma = \sum_{\ell} \sum_{m=-\ell}^{\ell} w_\ell(r,t) Y_{\ell m}(\vartheta,\varphi) . \qquad (24)$$

Macroscopic polarisation fields and population inversion are created by all excited modes. To be able to understand the solution for a single mode, we assume that different angular modes are separated in space and don't interact one with an other. Then we can consider one angular mode with definite number $\ell$ and definite order $n$. For example, let the magnetic wave or the transversal electric wave $E_r = 0$. For this mode we can write

$$e_\vartheta = iq_\ell(r,t) S_\ell(\vartheta) e^{i\alpha(t)}, \qquad e_\varphi = -iq_\ell(r,t) Q_\ell(\vartheta) e^{i\alpha(t)}, \qquad (25)$$

$$p_\vartheta = iw_\ell(r,t) S_\ell(\vartheta) e^{i\beta(t)}, \qquad p_\varphi = -iw_\ell(r,t) Q_\ell(\vartheta) e^{i\beta(t)}. \qquad (26)$$

Here

$$Q_\ell(\vartheta) = P_\ell^{(1)}(\cos\vartheta)/\sin\vartheta, \; S_\ell(\vartheta) = -P_\ell^{(1)'}(\cos\vartheta)\sin\vartheta.$$

Then instead of Eqs. (19), (21), (22) we obtain:

$$\frac{\partial q_\ell}{\partial t} + 2\pi\sigma q_\ell + 2\pi\omega w_\ell \sin(\alpha-\beta) = 0 , \qquad (27)$$

$$\frac{\partial w_\ell}{\partial t} + \gamma w_\ell + \frac{N|d|^2}{3h} q_\ell y \sin(\alpha-\beta) = 0 , \qquad (28)$$

$$\frac{\partial y}{\partial t} = D(y_0 - y) - \frac{1}{Nh} q_\ell w_\ell (S_\ell^2 + Q_\ell^2)\sin(\alpha-\beta) , \qquad (29)$$

$$\frac{2\omega}{v^2}\frac{\partial \alpha}{\partial t} q_\ell + \frac{4\pi\omega^2}{v^2} w_\ell \cos(\alpha-\beta) + \frac{\omega^2}{v^2} q_\ell + \nabla^2 q_\ell + \ell(\ell+1) q_\ell = 0 , \qquad (30)$$

$$\frac{\partial \beta}{\partial t} + \omega - \omega_a = \frac{N|d|^2 q_\ell}{3h w_\ell} y \cos(\alpha-\beta) . \qquad (31)$$

This is the basic system of equations for a single mode laser in a spherical microparticle. The equations include the pumping radiation and the population relaxation rate of the active medium with help of corresponding correlations.[24]

## REFERENCES

1. P.Chylek, J.T.Kiehl, and M.K.W.Ko, Narrow resonance structure in the Mie scattering characteristics, *Appl. Opt.* 17:3019 (1978).
2. R.E.Benner, P.W.Barber, J.F.Owen, and R.K.Chang, Observation of structure resonances in the fluorescence spectra from microspheres, *Phys. Rev. Lett.* 44:475 (1980).

3. A.P.Prishivalko, L.G.Astafieva, G.P. Lednyeva, and M.S.Veremchuk, Internal and near surface optical fields on resonances of microparticles, *J. Appl. Spectrosc.* 41:1184 (1984).
4. S.X.Qian, J.B.Snow, H.M.Tzeng, and R.K.Chang, Lasing droplets: highlighting the liquid air interface by laser emission, *Science* 231:486 (1986).
5. H.-B.Lin, A.L.Huston, B.L.Justus, and A.J.Campillo, Some characteristics of a droplet whispering gallery mode laser, *Opt. Lett.* 11:614 (1986).
6. H.-B.Lin, J.D.Eversole, and A.J.Campillo, Spectral properties of lasing microdroplets, *J. Opt. Soc. Am.* B9:43 (1992).
7. M.Kuwata-Gonokami, K.Takeda, H.Yasuda, and K.Ema, Laser emission from dye-doped polystyrene microsphere, *Japan. J. Appl. Phys.* 31:L99 (1992).
8. A.S.Kwok, and R.K.Chang, Suppression of lasing by stimulated Raman scattering in microdroplets, *Opt. Lett.* 18:1597 (1993).
9. G.Chen, M.M.Mazumder, V.R.Chemla, A.Serpenguzel, R.K.Chang, and S.C.Hill, Wavelength variation of laser emission along the entire rim of slightly deformed microdroplets, *Opt. Lett.* 18:1993 (1993).
10. H.Taniguchi, and S.Tanosaki, Three-color whispering-gallery-mode dye lasers using dye-doped liquid spheres, *Japan. J. Appl. Phys.* 32:L1421 (1993).
11. G.Kurizki, and A.Nitzan, Theory of stimulated emission processes in spherical microparticles, *Phys. Rev.* A 38:267 (1988).
12. V.V.Dacjuk, I.A.Izmajlov, and V.A.Kochelap. Whispering modes gallery light lasing in droplet, *Kvantovaja Elektronika,* Kiev. No 38:56 (1990).
13. G.P.Lednyeva, Nonstationary oscillations in spherical microparticle, *Optika i Spektroskopija.* 76:506 (1994).(in Russian)
14. G.P.Lednyeva, Regime of oscillations of microlaser, *Vesti Akademii Nauk Belarusi, ser. fiz.-mat. n.* No 1:91 (1995) (in Russian).
15. G.P.Lednyeva, and L.G.Astafieva, To the theory of a neodimium glass laser with spherical microcavity, *Optika i Spektroskopija* 80:858 (1996).
16. G.Mie, Beiträge zur optik trüber medien, speziell kolloider metallösungen, *Ann. Physik* 25:377 (1908).
17. P.Debye, Der Lichtdruck auf kugeln von beliebigem material, *Ann. Physik* 30:57 (1909).
18. L.G.Astafieva, and G.P. Lednyeva, Influence of pumping intensity inhomogeneity on active medium of neodimium-glass microlaser, *Pure and Applied Optics* (at press).
19. L.G.Astafieva, and G.P.Lednyeva, Heating of a dye-doped polystyrene microlaser by a pumping radiation, in:*Tech. Digest. Int. Quant. Electr. Conf.* Minsk, p.58 (1996).
20. G.P.Lednyeva, and .L.G.Astafieva, Neodymium-glass microlaser, *Techn. Digest. Int Conf. Tunable Solid State Lasers*, Minsk p.26 (1994).
21. A.P.Prishivalko, L.G.Astafieva, and S.T.Leiko, Heating and destruction of metallic particles exposed to intense laser radiation, *Appl. Opt.* 35:965 (1996).
22. L.G.Astafieva, and G.P.Lednyeva, Heating of dye-doped polystyrene microspheres by a UV $N_2$-laser, *Teplofizika Vysokikh Temperatur* 34:No 6 (1996).
23. L.G.Astafieva, and A.P.Prishivalko, Heating of homogeneous and hollow particles of aluminum oxide by intensive laser radiation, *Teplofizika Vysokikh Temperatur* 32:230 (1994).
24. A.M.Samson, L.A.Kotomtseva, and N.A.Loiko. *Selfpulsings in Lasers.* Nauka i Tekhnika, Minsk (1990).
25. L..A.Kotomtseva, A.M.Samson, and S.I.Turovets, Dynamics of lasers with passive and active modulation of losses, in: *High Power Lasers. Science and Engineering.* Eds. R.Kossowsky, M.Jelinek, R.F.Walter. Kluwer Acad. Publisher, Dordrecht, (1996).
26. L..A.Kotomtseva, Longitudinal modes and dynamics of a laser with a saturable absorber, *Quantum Electronics* (at press).
27. L..A.Kotomtseva, and A.M.Samson, Steady states and their stability in a laser with a saturable absorber, *J. Appl. Spectrosc.* 62:117 (1995).
28. M.Born, and E.Wolf. *Principles of Optics*, Pergamon Press, Oxford, London (1964).

**MATERIALS AND PROCESSES**

# MICROFABRICATION TECHNOLOGIES FOR INTEGRATED OPTICAL DEVICES

G.C. Righini and M.A. Forastiere

Optoelectronics and Photonics Department (IROE CNR)
"Nello Carrara" Electromagnetic Waves Research Institute
Via Panciatichi 64, 50127 Florence, Italy

## 1. INTRODUCTION

Microelectronics and Integrated Optics (IO) share most of the microfabrication technologies, in particular micropatterning. Fabrication processes of IO devices, however, are subject to specific requirements, imposed by the nature of light confinement, which are often tighter than the ones usually adopted in microelectronics.

As a matter of fact, optical waveguiding structures are usually a few centimeters long, but at the same time their transverse dimensions must be of the order of the guided-light wavelength (i.e. ~ 1 µm); this generates a number of problems, mainly connected with the necessity of keeping propagation losses as low as possible. As an example, low propagation loss and highly tolerant fabrication processes are generally not easily compatible with high-$\Delta n$ structures (i.e. structures where the refractive index difference between the guiding layer and the surrounding media is larger than 0.1), because this implies a further reduction in waveguide dimensions (e.g. down to sub-micrometer size). While this may be an advantage from the point of view of compatibility with microelectronics IC technology (especially in view of integrated optoelectronics silicon wafers), it is clear that high-$\Delta n$ guides suffer very high coupling loss (even larger than 3 dB/facet) with input/output fibers. On the other hand, in low-$\Delta n$ structures (where the index change may vary between 0.001 and 0.01) the bending loss can be very high unless the curvature radius of the guide is kept above a certain value, typically of the order of centimeters: this puts therefore a strong limit to the integration capability.

Much work is therefore currently devoted to the development of fabrication processes which could at the same time guarantee high compatibility with electronic LSI and yield integrated optical devices with low insertion loss; low cost and high flexibility are other two key factors for a process to be adopted in industrial production of IO components and devices.

The choice of the material structure and of the fabrication process of course depend also on the functional requirements of the device to be realized; since planar structures are

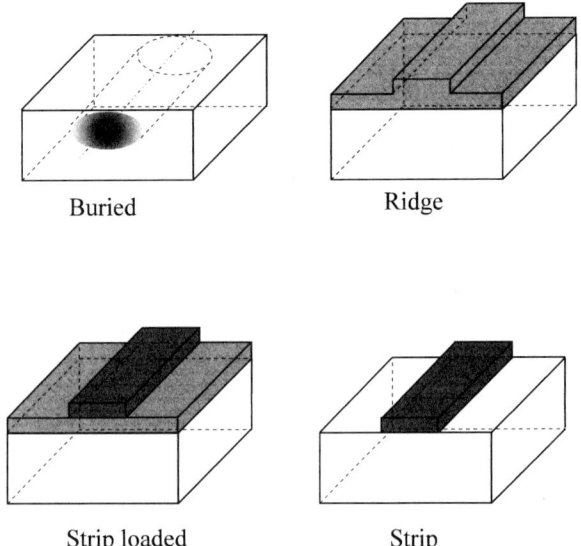

Fig. 1 Different types of channel waveguides. Light is confined transverse in the shaded regions.

being used only in very few cases, however, a common need is that of patterning the guiding layer and/or the cladding layer in order to create 2D guides. Various index or topographic configurations can be used, such as the ones sketched in Fig. 1, where darker shading indicates higher refractive index. Buried waveguides are generally obtained by diffusion processes (such as ion-exchange in glass, Ti-diffusion or proton exchange in lithium niobate, ion implantation or electron bombardment), while the other structures (ridge, strip, or strip-loaded waveguides) are produced in deposited layers. Accordingly, transverse light confinement is achieved by increase of the effective refractive-index of a limited dielectric region with respect to the surrounding regions: in most cases this is achieved by coating the substrate (or the guiding layer) with a mask material, which is then patterned lithographically with waveguide features. The following step consists of diffusion through the mask openings or of material removal, by wet or physical etching, respectively. In some cases, direct fabrication of 2D guides may be obtained, e.g. by laser or electron-beam writing.

In this Chapter a brief overview of the most important fabrication processes is presented, with particular attention devoted to glass material systems [1,2] and to silica-on-silicon integrated optics.[3,4]

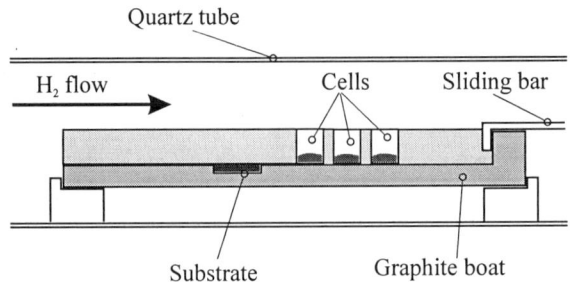

Fig. 2 Fabrication of thin films by Ar-plasma sputtering: behaviour of refractive index as a function of pressure and RF power (inset).

## 2. THIN FILM DEPOSITION TECHNIQUES

Most guided-wave optical devices are based on dielectric multilayer structures, obtained by successive deposition of thin films having tightly controlled thickness and refractive index. At this time, several deposition techniques are available. Some of them (such as e-gun evaporation, radio-frequency and magnetron sputtering) are definitely mature, having been borrowed from the classical area of optical coatings. Others were originally developed for semiconductors or optical fibers, and include for example Chemical Vapor Deposition (CVD), Flame Hydrolysis Deposition (FHD), and epitaxial growth techniques like Liquid Phase Epitaxy (LPE), Metal-Organic Chemical Vapor Deposition (MOCVD), and Molecular Beam Epitaxy (MBE). Recently, production processes have been implemented which exploit the power and selectivity of laser sources: as an example, Pulsed Laser Deposition (PLD) is becoming a widespread technique.[5] Finally, a deposition process which appeared only recently in IO is the so-called "sol-gel" technique:[6] its potential has yet to be fully exploited, but the prospects are quite appealing.[7]

### 2.1. Radio-frequency (RF) sputtering

Among the various vacuum thin-film deposition techniques, RF-sputtering is probably the most popular method for producing low-loss glassy planar waveguides. Sputtering is based on the use of positive highly accelerated ions to eject particles - usually atomic clusters or neutral atoms - from a target made of the material to be deposited. The ions are generated by injecting an electron current into an inert gas plasma (usually argon) at a pressure of $10^{-3} \div 10^{-2}$ Torr. RF-sputtering was first used in IO to deposit films of Corning 7059 glass (bulk refractive index about 1.53) onto ordinary soda-lime slides.[8] Several authors have since investigated the effect of various operating parameters on the deposition rate and composition of deposited films.[9-12]

A drawback of RF-sputtering is its slow deposition rate: as an example, in a diode sputtering system with a single 6"-diameter target the average deposition rate of Corning 7059 glass is about 1 nm/hour per Watt of RF electric power applied to the electrodes.

The quality of films is generally quite good, with propagation losses repeatedly lower than 0.8 dB/cm, for applied power less than 400 W. At higher sputtering powers, however, waveguide quality worsens and some yellow-brown discoloration of the films may occur.

The composition and optical characteristics of sputtered films vary as a function of the power level and gas pressure at which the deposition is made. This is likely due to the loss of oxygen caused by dissociation of the oxides of glass during the impact of ions onto the target. Introduction of a percentage of oxygen in the plasma (reactive sputtering) and application of a little potential to the anode (bias sputtering) can help reduce these effects. On the other hand, the possibility of varying the refractive index according to the sputtering parameters is also interesting because it can be used to increase flexibility in the design of the deposited films. As an example, a variation in the index of a 7059 glass film from 1.53 to 1.585, at 0.633 µm wavelength, was measured [9] for a corresponding change of the RF sputtering power density from 0.5 W/cm² to 4.0 W/cm². For the same type of glass, a series of measurements were carried out on the refractive index as a function of pressure in a pure argon plasma at 300 W: the results are reported in Fig. 2, where a significant decrease of the index with increasing pressure can be observed. In the same figure, the inset shows how the refractive index increases with increasing RF power, becoming much larger than the index of the bulk material itself. In the same working conditions, the addition of oxygen produces a further decrease of the refractive index, of the order of $-2 \times 10^{-3}$ for a 60% Ar - 40% $O_2$ mixture. One can easily conclude that the RF-sputtering process permits us to tailor quite accurately the optical properties of the deposited film.

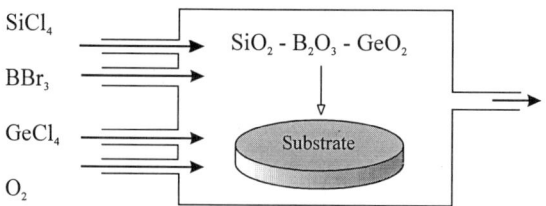

Fig. 3 Schematic diagram of apparatus for chemical vapour deposition (CVD).

RF-sputtering has also been applied for the deposition of active and/or nonlinear guiding layers: for instance, rare-earth-doped [13, 14] and semiconductor-doped [15] glass waveguides have been produced using a co-sputtering technique.

## 2.2. Chemical Vapor Deposition (CVD)

The CVD method was originally developed for the production of preforms of optical fibers, but it can be advantageously applied to planar waveguides as well. The classical CVD plant is sketched in Fig. 3: the silica (or silicon) substrate is placed in a reaction chamber, where accurately controlled flows of oxygen (carrier gas), Si Cl$_4$, BBr$_3$ and Ge Cl$_4$ gases are introduced. The reaction with oxygen at a temperature higher than 1200 °C causes reagent oxidation and the deposition of a thin oxide layer (with SiO$_2$-B$_2$O$_3$-GeO$_2$ composition) onto the substrate. The waveguide is then obtained by vitrification of the soot at a temperature around 1700 °C. A lower-index cladding layer can be finally deposited on the waveguide by reducing the flowing rate of Ge Cl$_4$.

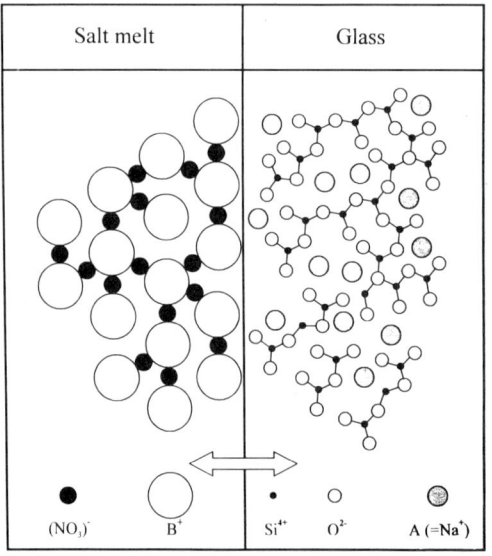

$$A_{glass} + B_{source} \leftrightarrow A_{source} + B_{glass}$$

Fig. 4 Diagram of a setup for liquid phase epitaxy (LPE).

The resulting CVD-deposited glass films usually show absorption loss lower than 0.1 dB/cm at 0.633 μm wavelength. Low-loss channel waveguides can also be formed by this technique, using suitable grooves previously formed by any reactive etching procedure.

The CVD process lends itself to many modifications, which have been developed in different laboratories with the aim of optimizing both the quality and the yield of the process. Among the most common types of CVD-derived fabrication processes used to make IO devices we can cite: flame hydrolysis deposition (FHD), [16-19] low-pressure (LPCVD), [20] plasma enhanced (PECVD), [4,21] and atmospheric pressure (APCVD) [22] processes.

## 2.3. Liquid Phase Epitaxy (LPE)

This technique is the simplest one among the epitaxial growth processes, and consists of the deposition of a solid phase from an over-saturated solution of the material to be grown. In the case of III-V semiconductors, the solvent is the same metal which is present in the compound (for example, Ga in GaAlAs). A sketch of the equipment for epitaxial growth is shown in Fig. 4. Schematically, it consists of a graphite boat placed in a quartz tube where $H_2$ flows. Several cells are present in the boat itself, each of them containing a different solution. The bottom of each cell is a slide containing a slot for the substrate, which therefore can be put in contact successively with the various solutions. Temperature is around 750 to 900 °C and must be tightly controlled (within ±0.01 °C) in order to assure the required uniformity. Mechanical tolerances are obviously very strict, but the equipment is altogether very simple.

The LPE technique has mostly been used to produce waveguiding structures for laser diodes operating at wavelengths in the range from 0.7 μm to 1.5 μm, such as $Ga_{1-x}Al_xAs$ layers grown on GaAs,[23] but LPE has also been used to grow waveguides in ferromagnetic materials, e.g. $Y_3Sc_{0.7}Fe_{4.3}O_{12}$ grown on $Gd_3Ga_5O_{12}$,[24] and in magneto-optical materials.

Channel waveguides, multilayer structures and various passive integrated optical components can be obtained by using the LPE technique combined with photolithographic and reactive etching procedures.

## 2.4. Sol-gel

The sol-gel process is a glass deposition technique based on the hydrolysis and polycondensation of metal alkoxides.[6,25] The procedure undergoes three steps: 1) hydrolysis of the different metal alkoxides in a mutual solvent (formation of the *sol*); 2) polycondensation and subsequent polymerization to form a highly porous solid network (the *gel*); and 3) drying of the gel and final heat-treatment to produce a stable and compact glass.

A number of features make the sol-gel method inherently attractive. First, this technique is based on liquid solutions, and therefore allows easy fabrication of fibers and waveguiding films. Moreover, it is a low-temperature process, so that special multicomponent compositions can be obtained without risk of phase-separation and crystallization; thus, many dopants can be introduced in the matrix, allowing for production of non-linear or active optical materials. Finally, the optical characteristics of the final glass can be varied by simply changing the initial sol composition.

Fig. 5 schematically illustrates how bulk, fiber and thin film materials can be produced by the sol-gel technique, through the careful control of process parameters (viscosity, drying and heating conditions).

Channel waveguides can be realized, as in most film deposition techniques, by the use of photolithography and a suitable etching procedure. However, since the sidewalls may remain quite rough, thus heavily contributing to scattering loss, a reflow treatment may be necessary.[26]

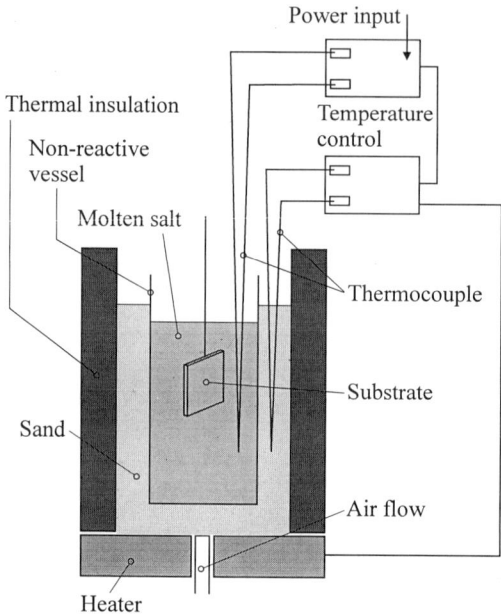

Fig. 5 Different material configurations obtainable by the sol-gel technique.

Recently, a great deal of work has been reported concerning sol-gel preparation of semiconductor-doped [27,28] and rare-earth-doped waveguides.[29-31]

## 3. ION DIFFUSION AND ION IMPLANTATION TECHNIQUES

Although thin-film deposition techniques are widely practiced, and some of them are also easy to implement, nevertheless it is sometimes more convenient to exploit the properties of a bulk material and create a guiding structure in it by locally modifying its refractive index. In this case one mainly uses diffusion techniques: a very common and effective one for glass and some crystals (such as lithium niobate, KTP and a few others) is ion-exchange, while a more general approach is that of inducing the thermal diffusion of a dopant in the bulk material. Another possibility is that of using high energy ion beams to implant a large variety of ions into a material.

Here, we will principally discuss the ion-exchange technique in glass; only a short description of the ion implantation approach will be given.

### 3.1. Ion implantation

The apparatus for ion implantation consists of a generator of ions, an accelerator, an ion separator and a deflector. Ions having energy ranging from 20 to 300 KeV can be implanted in various substrates, causing lattice modifications and the formation of impurities. These in turn produce an increment in the refractive index within a thin layer close to the surface, and therefore the creation of an optical waveguide.

As an example, the first experiments with this kind of technology were carried out on fused-silica substrates, bombarded by $H^+$ ions at 1.5 MeV [32] and by $Li^+$ ions at energies between 32 and 200 KeV.[33] The most reproducible results were obtained by using $Li^+$, for

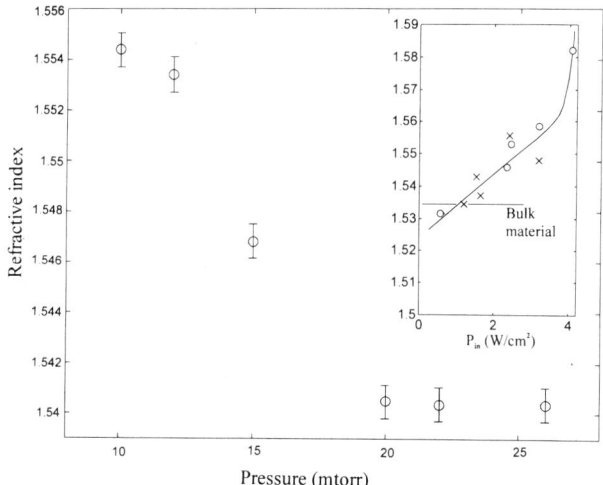

Fig. 6 Diagram of the ion-exchange process in a molten salt. The ionic species A is Na⁺ in soda-lime glasses.

which a proportionality between refractive index increase and dose of implanted ions was verified. Propagation loss in ion-implanted waveguides can be as low as 0.2 dB/cm at 0.633 μm wavelength.

Unfortunately, ion implantation requires very large and expensive equipment, although it has the advantage of allowing very precise control of fabrication parameters. The interested reader is referred to a recent review of current developments for waveguide formation as used in electro-optics, lasers or nonlinear optics applications; economic factors related to implantation have also been analyzed there.[34]

## 3.2. Ion exchange in glass

The suitability of ion-exchange technology to produce integrated optical components in glass has been recognized since the early 1970's, when a few pioneering studies demonstrated how to take advantage of the refractive index increase produced in the glass by replacing the sodium ions with other ions having higher electronic polarizability (such as silver or potassium).[35,36] One of the greatest advantages of the ion-exchange technique is the extreme simplicity and low cost of the process, which can also be easily transferred to batch production.[1,37]

In order to understand the principle of the method, one needs first of all to remember that the so-called "network modifier" ions, i.e. alkali ions such as Na⁺ which may occur in the silicate network of a glass, have a temperature-dependent mobility following an exponentially decreasing behavior with inverse temperature. At sufficiently high temperatures (typically a few hundred degrees Celsius), and in the presence of a melt solution containing other ions with similar chemical properties, the alkali ions of the glass can leave the silicate lattice, being readily substituted by similar ions from the melt solution. A sketch of the thermal ion-exchange process is shown in Fig. 6.

In a qualitative way, the exchange proceeds as follows. The glass substrate, containing for example Na⁺ ions, is immersed in a molten solution of $MeNO_3$ in $NaNO_3$, where $Me$ can be Ag, K or any other metal chemically similar to Na. Since a gradient exists in both ion concentrations, a diffusion process will take place, being essentially driven by thermal agitation. Random collisions will therefore cause replacement of Na⁺ ions by the $Me^+$ ones in the glass matrix, which gradually forms a thin layer close to the surface. The process terminates when the substrate is allowed to cool down to room temperature.

The exchange temperature is usually slightly higher than the melting point of the salt (ranging from about 200 °C to 550 °C). Excessive heating may in fact cause damage of the surface due to nitrate decomposition and thermal relaxation of glass.

The index change resulting from ion exchange can be easily determined from the fact that the ions taking part in the exchange have different electronic polarizabilities and that they occupy a different volume in the glass lattice.[38] Quantitatively, the refractive index variation $\Delta n$ can be expressed as

$$\Delta n \approx \frac{\chi}{V_0} \Delta R - \frac{R_0}{V_0} \Delta V \tag{1}$$

where $\chi$ is the concentration of the new ions in glass, $V_0$ is the volume of glass per mole of oxygen atoms, $R_0$ the refraction per mole of oxygen atoms and $\Delta V$, $\Delta R$ are the changes of these quantities due to the ion exchange.

The above model, though a very simple one, holds very well for bulk changes of composition. For surface ion-exchange, as is the case with IO, one should also take into account the effects of stress, due to the substrate resisting the localized volume change. As a consequence, the estimation of $\Delta V$ may not be accurate. Nevertheless, the model provides very useful information, giving at least the correct order of magnitude of the index change.

As an example, the predicted maximum index change in $Ag^+$-$Na^+$ exchange is 0.09, in good agreement with experimental results. In some cases, such as that of $K^+$-$Na^+$ ion-exchange, a birefringence effect also arises, due to anisotropic stress. Therefore, the index change for TM polarized modes is different from that of TE polarized modes. By taking into account volume variations induced by stress one can calculate a correction to Eq. (1) and find, for the $K^+$-$Na^+$ exchange, $\Delta n_{TE} = 0.0089$, $\Delta n_{TM} = 0.011$. The experimental results agree fairly well with the theory, being approximatly $\Delta n_{TE} \approx 0.008$ to $0.009$ and $\Delta n_{TM} \approx 0.0095$ to $0.011$.[39] The refractive index profile resulting from the ion-exchange method depends on the particular ions involved. For example, the $Ag^+$-$Na^+$ index profile for planar waveguides is well approximated by a complementary error function:

$$n(x) = n_{sub} + \Delta n \; \text{erfc} \frac{x}{\sqrt{D_e t}} \tag{2}$$

where $n_{sub}$ is the substrate index, $\Delta n$ the maximum index change, $D_e$ an "effective diffusion coefficient" and t the exchange time. Since the process is diffusion driven, the $D_e$ coefficient has the following temperature dependence:

$$D_e = C_1 \exp\left(-\frac{C_2}{T}\right) \tag{3}$$

where $C_2$ is proportional to the activation energy of the process. For the $Ag^+$-$Na^+$ ion-exchange process the salt used as the source of ions, $AgNO_3$, is usually diluted in $NaNO_3$, with concentrations varying from 5% to 20%; the temperature is kept around 300 °C (the melting point of $AgNO_3$ is 212 °C), while exchange times are of the order of a few minutes to obtain mono-modal waveguides. Dilution is advisable in the case of $Ag^+$-$Na^+$ ion-exchange because it both increases the reproducibility of the process and relaxes the tight constraints necessary for undiluted melts.

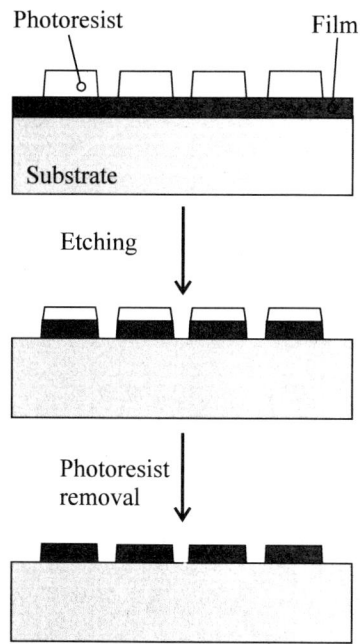

Fig. 7 Schematic setup for the production of glass waveguides by the ion-exchange technique.

A typical, economical laboratory setup for the production of waveguides by the ion-exchange technique is shown in Fig. 7 ; it consist mainly of an oven (a fluidised-bath furnace, in the case depicted in the figure, where the air flow through the sand gives rise to very good heat convection and therefore produces homogeneous heating of the reactor vessel) with accurate temperature controls in order to guarantee reproducibility of the process.

## 4. MICROSTRUCTURE FABRICATION

Fabrication of a guided-wave device, as already mentioned, requires not only having the proper refractive-index distribution but also in most cases creating a topographic structure. This means that some material has to be deposited or removed through a suitable mask. Etching, either chemical or physical, is the best process to produce these microstructures. The minimum size of the features which can be produced depends mainly on the resolution of the photoresist, which is the basic material in almost any lithographic process: currently, the minimum line width with negative resists (i.e. resists where exposure to light causes polymerization of the material, making the exposed regions insoluble in the developer) is larger than 1 µm, and a bit smaller (around 0.5 µm ) in positive resists (where exposed areas become soluble and therefore removable in the proper developer).[40]

While microlithography is still the most widespread and effective tool to produce nanostructures,[41] in less demanding applications locally-selective processing can be done by direct laser-writing: the applicability of the method obviously depends also on the choice of the material system. As an example, 2D waveguides may be produced in sol-gel films by direct laser densification, or other structures may be obtained by removing material through laser ablation.

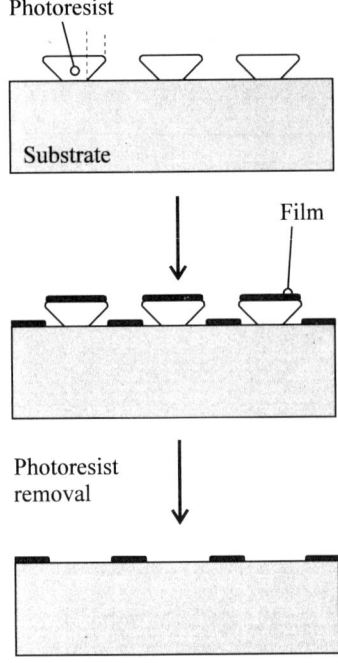

Fig. 8 Patterning of thin films by the lift-off technique (schematic).

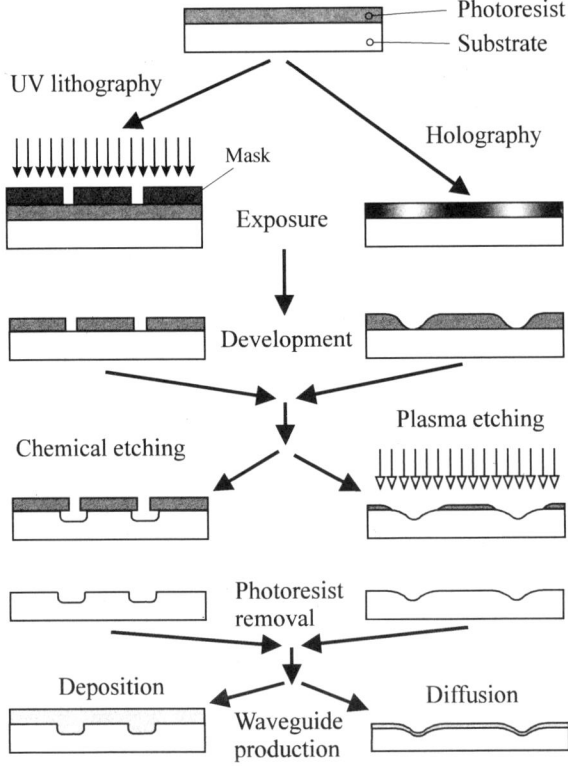

Fig. 9 Patterning of thin films by the etching technique (schematic).

## 4.1. Lift-off technique

The principle of operation of the lift-off technique, which is an alternative to the conventional etching process, is indicated in Fig. 8. A pattern is formed in the photoresist layer, previously deposited onto the substrate by spin-coating, by means of the photolithographic technique. As illustrated in the figure, an overhang is formed in the resist during development, if suitable thickness, exposure and development conditions are provided. Moreover, it is necessary that the resist thickness be larger than the thickness of the film to be deposited; the masking film is in most cases a metal film, which is deposited by means of vacuum evaporation or sputtering. Finally, the sample is immersed in a suitable solvent which dissolves the resist and thus causes the removal of the film directly deposited on it. In this way, the desired pattern is replicated in the masking film.

A few requirements have to be met in order to obtain good results by the lift-off technique: 1) an appropriate overhang must exist; 2) the resist layer must be completely removed from the patterning windows before film deposition; and, 3) good film adhesion must be guaranteed.

In order to enhance film adhesion, one can slightly ash the resist with oxygen plasma after development; another possibility is to post-bake the substrate at a temperature which does not cause resist deformation (100 to 150 °C). When using Al films for the mask, their adhesion to the substrate can be improved by previously depositing an underlayer of Cr or Ti on the substrate.

## 4.2. Etching techniques

Fig. 9 shows a schematic representation of a conventional etching procedure. The resist may operate as a masking material itself (in such a case the film indicated in the figure is the guiding layer), or instead it may be used to transfer the features from the master mask to the masking film for the waveguide fabrication process. If the resist cannot withstand the etching process, a masking underlayer must first be deposited on the substrate.

Etching processes can be divided into two classes: *chemical* (or *wet*) *etching*, which makes use of liquid chemicals, and *dry etching*, which uses gaseous etchants.

Wet etching can be achieved by a variety of substances, such as aqueous solutions of acids (HF, $H_3PO_4$, $HNO_3$), alkalis (NaOH), and salts (NaCN, $FeCl_3$). The sample is immersed in the appropriate etchant after development, kept at a certain temperature and stirred in order to remove air bubbles. The process can be monitored by observation through a microscope. Although wet etching requires extremely simple equipment and is thus very attractive, good results are rarely obtained, primarily because of the difficulty in achieving acceptable reproducibility of etch rate.

The term dry etching includes in fact a number of quite different techniques. Many of them are commercially available, being extensively used in the semiconductor industry. As an example, the basic *plasma etching* consists of an apparatus which produces plasma by RF discharge in a reactive gas (e.g. $CF_4$) at a 0.1 to 10 Torr pressure; chemical reactions with the resultant neutral radicals produce the etching of both silicon and silicon compounds.[42] *Sputter etching* is based on an ordinary sputtering apparatus, the only difference being that the samples are placed at the target position and undergo physical bombardment by $Ar^+$ ions, which cause the etching of the substrates. Variations or combinations of these two basic processes, based on physical-chemical and purely physical actions, respectively, include dry-etching configurations known as *reactive ion etching (RIE)*, *ion beam etching (IBE)*, and *reactive ion beam etching (RIBE)*.[43]

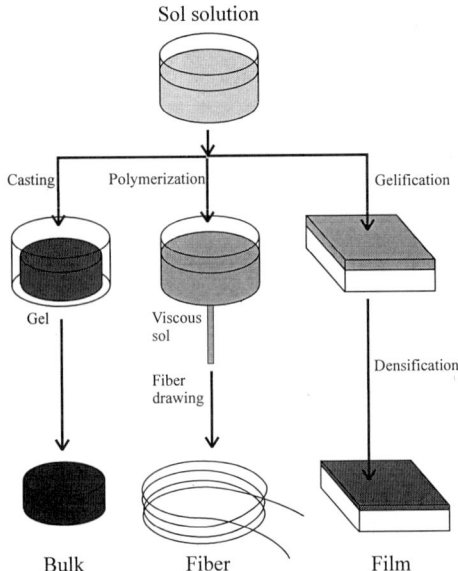

Fig. 10 Illustration of the procedure for realisation of waveguide patterns (such as gratings).

Fig. 10 summarizes some of the possible procedures to be followed for the fabrication of topographic waveguides and IO devices. When the aim is to fabricate a periodic structure, such as a grating coupler or filter, instead of using the conventional photolithographic process one can also exploit holographic techniques to produce the desired modulated pattern in the photoresist; using this approach it is easier to produce sinusoidal gratings.

### 4.3. 2D-waveguide fabrication by direct laser writing

Writing of integrated circuit masks is often done by electron-beam exposure of suitable resists (usually, PMMA), because this method offers very high resolution and dimensional precision. Laser writing, however, may also be used as an effective tool for mask production in all the cases where geometrical tolerance constraints are relaxed.[44] Both methods can also be used for direct writing of the waveguides through an induced change of refractive index or of material thickness.

Early demonstrations of direct laser writing included fabrication of guides in Kodak KPR photoresist films,[45] in sensitizer-doped polymers by the so-called *photolocking* mechanism,[46] and in $SiO_2$-$TaO_2$ films by $CO_2$-laser irradiation.[47] More recently, laser writing has been widely used in conjunction with the sol-gel deposition technique, either through local densification of the still-porous film or through UV-irradiation of a sensitizer introduced in the starting solution.

The formation of topographic structures in sol-gel films is possible by laser-induced selective densification.[48-50] Strip waveguides with propagation loss below 1 dB/cm have been produced in silica-titania sol-gel films by exploiting the localized heating of sol-gel material due to the absorption of laser light by the film itself and by the substrate.[50] A schematic representation of the selective densification process induced by a focused $CO_2$ laser is given in Fig. 11: the sol-gel film is spin-deposited and baked at a temperature below 200 °C, in order to leave it quite porous. The strip of the material which is irradiated by the

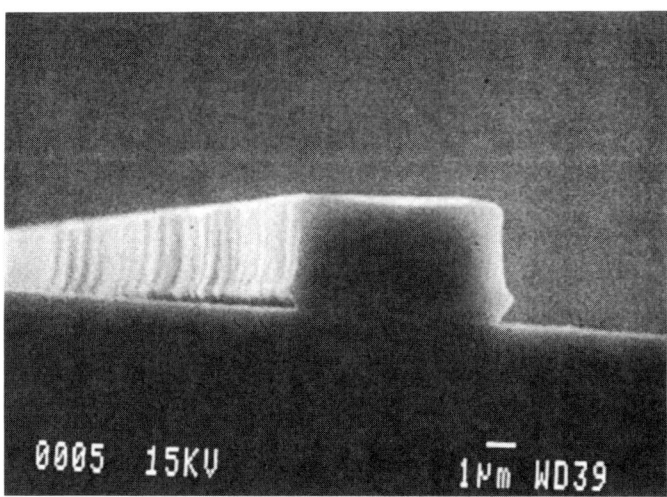

Fig. 11 Schematic of the laser densification method. A: deposition of the sol-gel layer; B: laser densification; C: densified strip; D: removal of the undensified layer by chemical etching.

focused scanning beam from a cw $CO_2$ laser (the power actually delivered onto the strip is in the range 50 to 60 mW) heats up and shrinks; the unexposed, and therefore undensified areas of the film are removed by immersing the sample in a buffered and diluted HF solution, which almost has no effect on the densified strip, due to the much slower etching rate of the dense material.

Another recently developed procedure uses UV light imprinting in a photosensitive hybrid organo-silicate film produced by a dip-coating sol-gel process. The good optical quality of a ridge optical waveguide fabricated by this procedure can be seen in Fig. 12, which shows a scanning electron microscope photograph of a 5-μm wide and 4-μm high ridge waveguide deposited onto a silicon wafer.[51] Propagation losses as low as 0.1 dB/cm at $\lambda = 1.55$ μm have been measured. A set of passive integrated optical devices, including directional couplers,[52] power splitters and wavelength-division-multiplexers, have been fabricated by this process.

## 5. CONCLUSIONS

A quick overview of the most common fabrication processes in integrated optics has been given, with particular attention devoted to practical problems in the production of passive components in glass materials and in silica-on-silicon structures.

The basic requirements for any process are the capability for very accurate control of the refractive index and thickness of the guiding layer, freedom from scattering defects and smoothness of surfaces. Moreover, in order to facilitate the commercialization of fabricated components and devices, the process should not be highly expensive nor very critical, and should be suitable for batch production. A significant role in the optimization of the processes and in the development of IO devices is obviously played on the one hand by modeling and design techniques, and on the other hand by the characterization and quality-assessment techniques, some of which have been developed specifically for use in optical waveguides.[53]

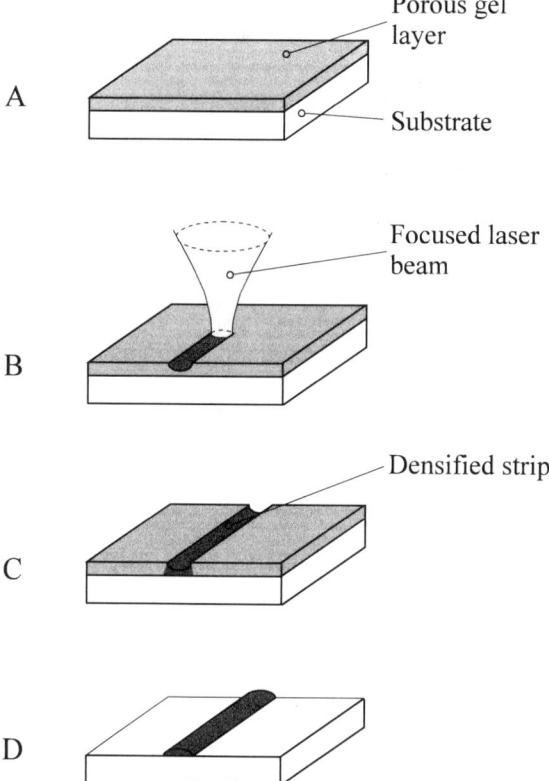

Fig. 12 SEM photograph of a strip waveguide in sol-gel glass.

ACKNOWLEDGMENTS. The authors wish to thank Dr. S. Pelli and Dr. A. Verciani for useful discussions. Financial support by the CNR Strategic Project on *Materials and Devices for Optoelectronics* and by the CNR Special Project on *Glass Waveguides for Non-Linear Optics* is also acknowledged.

## REFERENCES

1. S.I. Najafi, Ed., *Introduction to Integrated Optics* (Artech House, Boston, 1992).
2. P. Mazzoldi and G.C. Righini, Glasses for Optoelectronic Devices, in *Insulating Materials for Optoelectronics*, F. Agulló-López, Ed. (World Scientific, Singapore, 1995) 367-392.
3. R.R. Syms, Silica on Silicon Integrated Optics, in *Advances in Integrated Optics*, S. Martellucci et al. Eds. (Plenum Press, New York, 1994) 121-150.
4. S. Valette, Integrated Optics on Silicon: IOS Technologies, in *Advances in Integrated Optics*, S. Martellucci et al. Eds. (Plenum Press, New York, 1994) 151-164.
5. C.N. Afonso, Pulsed Laser Deposition of Fibers for Optical Applications, in *Insulating Materials for Optoelectronics*, F. Agulló-López, Ed. (World Scientific, Singapore, 1995) 1-28.
6. C.J. Brinker and G.W. Scherer, *Sol-Gel Science* (Academic Press, San Diego, 1990).
7. M.P. Andrews, An Overview of Sol Gel Guest-Host Materials Chemistry for Optical Devices, *Proc. SPIE* vol. 2997, 48 (1997).
8. J.E. Goell, and R. D. Standley, Sputtered glass waveguide for integrated optical circuits, *Bell System Techn. J.* 48, 3445 (1969).
9. C. W. Pitt, F.R. Gfeller, and J.R. Stevens, RF-sputtered thin films for integrated optical components, Thin Solid Films 26, 25 (1971).

10. W. M. Paulson, F.S. Hickernell and R.L. Davis, Effects of deposition parameters on optical loss for RF-sputtered $Ta_2O_5$ and $Si_3N_4$ waveguides, *J. Vac. Sci. Technol.* 16, 307 (1979).
11. H. Terui, M. Kobayashi, Refractive-index-adjustable $SiO_2$-$Ta_2O_5$ films for integrated optical circuits, *Appl. Phys. Lett.* 32, 666 (1978)
12. G. C. Righini, G. Margheri and L. Mancinelli Degli Esposti, Glass waveguides for integrated optical circuits, *Rivista Staz. Sper. Vetro* no. 6, 109 (1986).
13. B. Chen and C.L. Tang, Nd-glass thin-film waveguide: An active medium for Nd thin-film laser, *Appl. Phys. Lett.* 28, 435 (1976).
14. J. Shmulovich, A. Wong, Y.H. Wong, P.C. Becker, A.J. Bruce, and R. Adar, $Er^{3+}$ glass waveguide amplifier at 1.5 μm on silicon, *Electron. Lett.* 28, 1181 (1992).
15. H. Nasu, K. Tsunetomo, Y. Tokumitsu, and Y. Osaka, Semiconducting CdTe Microcrystalline-Doped $SiO_2$ Glass Thin Films Prepared by Rf-Sputtering, *Jap. J. Appl. Phys.* 28, L862 (1989).
16. H. Yanagawa, S. Nakamura, I. Ohyama, and K. Ueki, Broad-Band High-Silica Optical Waveguide Star Coupler with Asymmetric Directional Couplers, *IEEE J. Lightwave Technol.* 8, 1292 (1990).
17. G. Barbarossa and P.J.R. Laybourn, High-silica low-loss three waveguide couplers on Si, *Proc. SPIE* vol. 1513, 37 (1991).
18. C.J. Sun, W.M. Myers, K.M. Schmidt, S. Sumida, and K.P. Jackson, High Silica Waveguides on Alumina Substrates for Hybrid Optoelectronic Integration, *IEEE Photon. Technol. Lett.* 4, 630 (1992).
19. S. Kobayashi, Recent Development on Silica Waveguide Technology for Integrated Optics, *Proc. SPIE* vol. 2997, 264 (1997).
20. C.H. Henry, G.E. Blonder, and R.F. Kazarinov, Glass Waveguides on Silicon for Hybrid Optical Packaging, *IEEE J. Lightwave Technol.* 7, 1530 (1989).
21. A.M. Fiorello, E. Giannetta, M. Valentino, A. Vannucci, and M. Varasi, Co-doped silica-on-silicon waveguides fabricated by PECVD technique, *Proc. SPIE* vol. 2954, 124 (1996).
22. T. Hanada, M. Kitamura, and S. Nakamura, *Proc. OECC'96*, 18C2 (July 1996).
23. E. Garmire, Optical waveguides in single layers of $Ga_{1-x}Al_xAs$ grown on GaAs substrates, *Appl. Phys. Lett.*, 23, 403 (1973).
24. P. K. Tien, R. J. Martin, S. L. Blank, S. H. Wemple, and L. J. Varnerin, Optical waveguides of single-crystal garnet films, *Appl. Phys. Lett.* 21, 207 (1972).
25. B. Zelinski, C. Brinker, D. Clark, and D. Ulrich, Better Ceramics Through Chemistry IV, *Mat. Res. Soc. Symp. Proc.*, vol. 180 (1990).
26. A.S. Holmes and R.R.A. Syms, Fabrication of low-loss channel waveguides in sol-gel glass on silicon substrates, in *Advanced Materials in Optics, Electro-optics and Communication Technologies*, P. Vincenzini and G.C. Righini Eds. (Techna, Faenza, 1995) 73.
27. E.J.C. Dawnay, J. Fick, M. Green, M. Guglielmi, A. Martucci, S. Pelli, G.C. Righini, G. Vitrant, and E.M. Yeatman, Nonlinear properties of semiconductor-doped sol-gel thin films for photonic application, in *Advanced Materials in Optics, Electro-optics and Communication Technologies*, P. Vincenzini and G.C. Righini Eds. (Techna, Faenza, 1995) 15.
28. T.A. King, D. West, D.L. Williams, C. Moussu, and M. Bradford, Nonlinear optics in thin films and waveguide sol-gel composites, in *Advanced Materials in Optics, Electro-optics and Communication Technologies*, P. Vincenzini and G.C. Righini Eds. (Techna, Faenza, 1995) 21.
29. D. Barbier, X. Orignac, X.M. Du, and R.M. Almeida, Spectroscopic properties of Neodymium doped sol-gel planar waveguides, in *Advanced Materials in Optics, Electro-optics and Communication Technologies*, P. Vincenzini and G.C. Righini Eds. (Techna, Faenza, 1995) 33.
30. G. Milova, S.I. Najafi, A. Skirtach, D.J. Simkin, and M.P. Andrews, Erbium in photosensitive hybrid organoaluminosilicate sol-gel glasses, *Proc. SPIE* vol. 2997, 90 (1997).
31. X. Orignac and D. Barbier, Potential for fabrication of sol-gel-derived integrated optical amplifiers, *Proc. SPIE* vol. 2997, 271 (1997).
32. E.R. Schineller, R.P. Flam, and D.W. Wilmot, Optical waveguides formed by proton irradiation of fused silica, *J. Opt. Soc. Am.* 58, 1171 (1968).
33. R.D. Standley, W.M. Gibson, and J.W. Rodgers, Properties of ion bombarded fused quartz for integrated optics, *Appl. Opt.* 11, 1313 (1972).
34. P.D. Townsend, Application of Ion Implantation for Optoelectronics and Photonics, in *Insulating Materials for Optoelectronics*, F. Agulló-López, Ed. (World Scientific, Singapore, 1995) 393-420.
35. T. Izawa and H. Nakagome, Optical waveguide formed by electrically induced migration of ions in glass plates, *Appl. Phys. Lett.* 21, 584 (1972).
36. T.G. Giallorenzi, E.J. West, R. Kirk, R. Ginther, and R.A. Andrews, Optical waveguides formed by thermal migration of ions in glass, *Appl. Opt.* 12, 1240 (1973).
37. G.C. Righini, Ion-exchange process for glass waveguide fabrication, in *Glass Integrated Optics and Optical Fiber Devices*, S.I. Najafi Ed., vol. CR53 (SPIE, Bellingham, 1994) 24.

38. S.D. Fantone, Refractive index and spectral models for gradient-index materials, *Appl. Opt.* 22, 432 (1983).
39. J. Albert and G.L. Yip, Stress-induced index change for $K^+$- $Na^+$ ion exchange in glass, *Electron. Lett.* 23, 737 (1987).
40. R.A. Bartolini, Photoresists, in *Holographic Recording Materials*, H.M. Smith, Ed. (Springer-Verlag, Berlin, 1977) 209-228.
41. S.P. Beaumont, Today's microlithography, in *From Galileo's occhialino to optoelectronics*, P. Mazzoldi Ed. (World Scientific, Singapore, 1993) 419.
42. D.M. Manos and D.L. Flamm, *Plasma Etching* (Academic Press, New York, 1989).
43. see for instance H. Nishihara, M. Haruna and T. Suhara, *Optical Integrated Circuits* (McGraw-Hill, New York, 1987), pp. 172-184.
44. G.M. Lad, G.M. Naik, and A. Selvarajan, Laser patterning system for integrated optics and storage applications, *Opt. Eng.* 32, 725 (1993).
45. H.P. Weber, R. Ulrich, E.A. Chandross, and W.J. Tomlinson, Light-Guiding Structures of Photoresist Films, *Appl. Phys. Lett.* 20, 143 (1972).
46. E.A. Chandross, C.A. Pryde, W.J. Tomlinson, and H.P. Weber, Photolocking - A new technique for fabricating optical waveguide circuits, *Appl. Phys. Lett.* 24, 72 (1974).
47. H. Terui and M. Kobayashi, Fabrication of channel optical waveguide using $CO_2$ laser, *Electron. Lett.* 15, 79 (1979).
48. B.D. Fabes, Laser processing of sol-gel coatings, in *Sol-Gel Optics: Processing and Applications*, L.C. Klein, Ed. (Kluwer Academic Publishers, USA, 1994) 483.
49. M. Guglielmi, P. Colombo, L. Mancinelli degli Esposti, G.C. Righini, S. Pelli, and V. Rigato, Characterization of laser densified sol-gel films for the fabrication of planar and strip optical waveguides, *J. Noncryst. Solids* 147&148, 645 (1992).
50. S. Pelli, G.C. Righini, A. Scaglione, M. Guglielmi, and A. Martucci, Direct laser writing of ridge optical waveguides in silica-titania glass sol-gel films, *Opt. Materials* 5, 119 (1996).
51. S.I. Najafi, M.P. Andrews, M.A. Fardad, G. Milova, T. Tahar, and P. Coudray, UV-light imprinted surface, ridge and buried sol-gel glass waveguides and devices on silicon, Proc. SPIE vol. 2954, 100 (1996).
52. C.Y. Li, J. Chisham, M.P. Andrews, S.I. Najafi, J.D. Mackenzie, and N. Peyghambarian, Sol-gel integrated optical couplers by ultraviolet light imprinting, *Electron. Lett.* 31, 271 (1995).
53. S. Pelli and G.C. Righini, Introduction to Integrated Optics: Characterization and Modeling of Optical Waveguides, in *Advances in Integrated Optics*, S. Martellucci et al. Eds. (Plenum Press, New York, 1994) 1-20.

# FABRICATION OF DIFFRACTIVE OPTICS: SURFACE RELIEFS AND ARTIFICIAL DIELECTRICS

C. Arnone, C. Giaconia, and G. Lullo

Electrical Engineering Department, University of Palermo
Viale delle Scienze, 90128 Palermo, Italy

## 1. INTRODUCTION

After a period of fundamental theoretical research, the field of diffractive optics is now reaching a time of assessment, when we can define the useful limits of each design procedure. Selecting a suitable procedure derives from a compromise between several factors, including design time, desired optical efficiency, and precision. A major consideration in finding such a compromise has been, and still is today, the availability of technologies which can realize the design by generating the diffracting structure with assigned tolerances.

Here, most of the technologies involved in the fabrication of phase-only diffractive optical elements (DOEs) are discussed and evaluated. No attempt is made to present and compare the multitude of design techniques, since this lies beyond the purpose of this work. Moreover, each fabrication technique is evaluated only with respect to its fidelity to the geometric design, whether or not this is an optimal design.

At present, two approaches can be need to obtain the desired DOE: the well developed *surface relief* approach and the novel *artificial dielectric* approach. Both techniques benefit from the available microelectronic fabrication technologies. However, while the former uses currently available and often low cost processes, the latter requires the most advanced nanotechniques employed and is still dependent on their further improvement. In the following sections the fabrication of *classical* surface reliefs is described. In the final section the technological requirements for the new approach are discussed.

## 2. FABRICATION OF SURFACE RELIEFS

Making diffractive elements by creating a surface relief pattern is presently a widespread practice. If moderate diffraction efficiencies are acceptable, low cost bi-level structures can be used. Otherwise, when efficiencies above 90% must be attained,

Table I. Diagram of materials, patterning techniques and fabrication tools (the applicable optical wavelength range is indicated in parentheses).

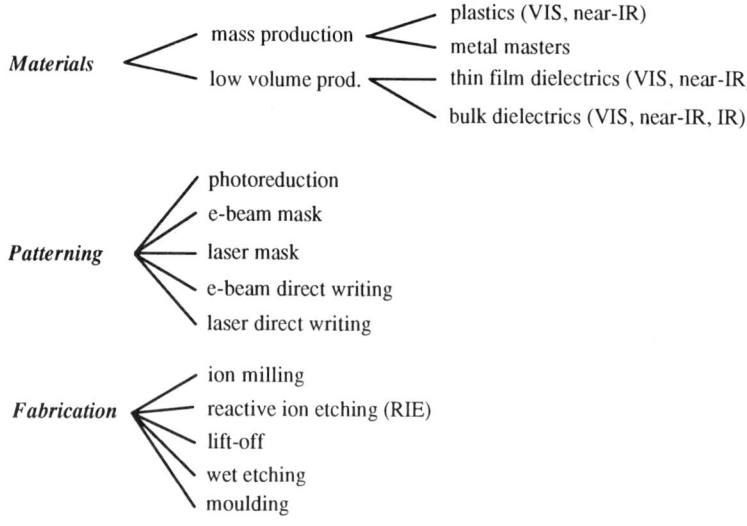

multi-level or continuous relief patterns are needed, with increased technological difficulties and fabrication costs.

Because of the multitude of applications for diffractive reliefs, a range of materials, patterning techniques and actual relief fabrication tools are available, aa shown in Table I.

In this Chapter attention is focused on fabrication technologies involving the use of planar micropatterning tools as typically encountered in the microelectronics field. For this reason, other techniques such as holographic recording, direct diamond turning and patterning of self developing photopolymers are not treated here.

## 3. MATERIALS

The selection of the most suitable material for each specific application depends on the reguired final cost of each diffractive device. If such cost must be limited to a few cents per device, as is the case for high volume production, only plastic materials - polymers - can be selected, thus setting an upper limit to optical precision and stability. If minimum cost is not a major issue or low volume productions is involved, a broad range of materials is available, whether used as thin films on a transparent substrate or as a bulk device. As a consequence, virtually no limitation occurs in the usable wavelength range, as long as a transparent material is available. If this is not the case, metal coated reflective diffracting devices can be also fabricated.

So far, diffractive devices made on polymeric substrates is expected to gain a leading role in the field, with applications ranging from consumer products to telecommunication systems. Polymethylmethacrylate (PMMA), allyl-diglycol-carbonate (CR-39), polycarbonate and polystyrene are popular choices among the wide range of optical plastics. In order to take advantage of the unique characteristics of such materials, precision embossing or injection moulding techniques have been developed, as detailed below, and the selection of the best suited material reguires also the fabrication of an accurate and durable mould.

Nickel is a common material for such application. The mould is obtained by electroforming[1]

Table II. Plastic and silica glass for DOEs fabrication.

|  | PLASTIC | SILICA GLASS |
|---|---|---|
| * cost | minimum | high |
| * impact resistance | high | low |
| * possibility of alignment features | yes | no |
| * weight | low | medium |
| index sensitivity to temperature | high | low |
| scratch resistance (no coating) | poor | high |
| thermal mechanical stability | low | high |

* Plastic exhibits a clear advantage on glass.

from a diffractive structure fabricated on resist, as detailed in the section on fabrication.

In Table II a comparison is made between plastic and silica glass for DOE fabrication. The asterisks indicate characteristics for which plastic exhibits a clear advantage as compared with glass.

For applications requiring high optical precision, IR transparency, tolerance to temperatures higher than 100°C or resistance to a harsh environment, the use of plastics must be ruled out in favour of traditional thin film or bulk inorganic dielectrics. The choice between thin film and bulk structure depends on the following factors:
- the thin film can be fabricated with tight thickness tolerances, allowing for a precise optical thickness to be specified and obtained;
- a broad range of materials is available for different refractive indexes, resistance to radiation damage, and resistance to environmental conditions;
- full dielectric reflecting structures can be fabricated, with negligible absorption loss;
- for applications where thermal or mechanical stress is induced on the device, film detachment or damage can occur, and the use of the same material for substrate and diffracting layer is preferred;
- a simpler and less expensive fabrication process is involved if the diffractive structure is made by engraving the substrate itself.

In fact, the first point is the most crucial. As discussed in the section on fabrication, the problem of obtaining minimum tolerances for the relief thickness is still incompletely solved, and this has a direct consequence on the achievement of optical efficiencies close to the theoretical design. By thin film technology, thickness errors around 1% can be routinely achieved and this is consistent with the tolerances required by a design with high diffraction efficiency.

Several examples of relief fabrication on deposited thin films are reported in the literature.[2-5] Typical film materials of industrial interest are SiOx (whose refractive index depends on the exact film stoichiometry), $SiO_2$ (n = 1.46), $MgF_2$ (n = 1.139), ZnS (n = 2.2), $TiO_2$ (n = 2.2), a-Si:N:H (amorphous silicon nitride), $Al_2O_3$ (n = 1.75), as well as specialised low-scattering or hard coatings.

A relevant type of relief fabrication on a thin film is related to the use of a resist film, intended either for electron beam (e-beam) or optical exposure, as the diffracting material. If an optical resist is used, the useful operating wavelength range is limited to the red to near infrared spectrum, because the refractive index of the resist would change with time if wavelengths are used to which the resist is sensitive. Moreover, since it is a polymer, its softness can make it unsuitable for many applications regniring durability. However, the direct use of optical resist is of special interest when novel diffracting designs are to be rapidly evaluated, involving bi-level or multilevel structures as well as continuous reliefs (kinoforms). For such applications, standard microlithographic photoresists are commonly used, with exposure wavelengths ranging from deep UV, for exposure through a high resolution mask, to blue-green, for direct patterning by laser writing.

If a resist for electron-beam exposure is used, usually polymethylmethacrylate (PMMA), the full visible to near-infrared spectrum is available for the diffracting structure, although the same limitations apply as regards softness and durability, as in the case of optical resists.

When bulk materials are used, amorphous silicon dioxide (silica) is practically the only choice for applications in the visible range. For fabricating near-IR devices - a very relevant market, including fiberoptic communications, VLSI or board optical interconnects, and optical LANs - other typically microelectronic materials join silica, such as silicon, gallium arsenide and silicon nitride. This means that well established technologies are available for their processing.

The case is different for infrared bulk materials, mostly dedicated to operations around a wavelength of 10.6 µm, for $CO_2$ laser applications. Here Ge or ZnSe are the materials of choice, but they are not typical microelectronic materials, so specific fabrication techniques have been developed for making diffractive structures on them.

## 4. PATTERNING

Whatever the complexity and accuracy of the design, we need to implement the desired features on a substrate, with minimum tolerances with respect to the design. Most past research in the field of diffractive optics has been aimed at both finding diffractive solutions that could be implemented with available technologies and improving them, in order to reduce the technological constraints that limited designers. Presently, this is again the situation when dealing with structures based on artificial dielectrics, while for *classical* surface reliefs a fair balance between technological capabilities and design needs has been reached.

The usual procedure for defining the shape, position and size of each diffracting element of a complex device relies on the fabrication of a microlithographic mask or on direct patterning of the final substrate. Generally, in both cases an e-beam pattern generator is used. However, for economical reasons or to achieve fast turnaround, laser writing systems can be convenient. This is the case, for instance, of large area devices (hundreds of square centimetres). The following diagram summarises the present patterning scenario. The practically achievable spatial accuracy is indicated in parentheses.

Either for bi-level or multilevel DOEs, patterning through a mask or a mask set follows standard procedures borrowed from microelectronics. According to published results, up to 32 levels can be fabricated, corresponding to 5 masks ($2^5 = 32$). Each mask (chromium on glass or pure silica) is produced in an electron-beam or optical pattern generator and exposure is made by UV light through the mask into a resist film deposited on the surface to be patterned. Contact exposure or photoreduction through a wafer stepper is a standard procedure.[4, 6-10]

When an e-beam machine is not available, a modern version of the old fashioned photoreduction technique can be used for patterning bi-level devices.[11,12] First, a photographic master is produced using modern desk-top or typographic publishing tools, reaching a minimum feature size in the 15 to 50 µm range. Then the actual mask is made through a photoreduction camera. Due to the optical aberration of the reducing lens, a resolution not better than 5 µm usually results (of course a much higher resolution can be reached if the lens of a stepper for microelectronics is used and the pattern size is limited to 1 or 2 cm$^2$, but this is certainly beyond the aim of using the inexpensive photoreduction technique).

Table III. Diagram of the present patterning scenario.

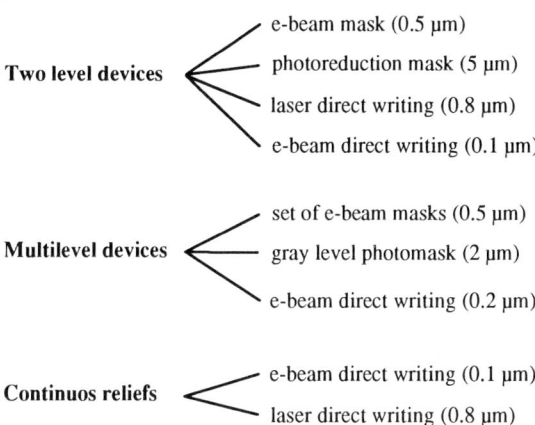

While the use of a chromium mask is a standard and simple practice for bi-level devices, it becomes inherently expensive and time consuming when more masks are used, for multilevel DOEs. Moreover, the unavoidable mask alignment errors (around 0.1 µm per mask, and these may accumulate with multiple masks) and exposure errors due to diffraction of UV light in the resist layer [6] can sensibly decrease the expected diffraction efficiency. For the above reasons other techniques have been considered too.

An interesting alternative to the multimask procedure for the fabrication of multilevel DOEs is the use of a single grey level mask.[13, 14] The relief is created on the resist layer by grey-level illumination and then it can be used directly or transferred to the substrate by RIE (reactive ion etch) or ion milling. Since a continuous-tone photomask must be used instead of a standard high contrast chrome mask, a microlithographic resolution of the process around to 1 µm can be expected, but this is sufficient for many applications. However a specific patterning system must be used and an accurate calibration is needed for defining the exact exposure energy for each level.

A better established solution to multilevel fabrication is given by direct laser patterning. Although practically limited to 1 µm accuracy, laser writing finds an interesting role when the mask-based industry standard approach is difficult this occurs mainly when continuous relief DOEs are made and when large area devices have to be fabricated in short time. Different laser system architectures have been proposed. Usually the beam from a He-Cd laser ($\lambda$ = 442 nm) is focused on the photoresist coated substrate, in a raster-scan mode, while the substrate is translated under computer control.[15-17] For patterning large area devices and optimising the drawing time, the substrate is placed on a fast rotating platform and processed in a manner similar to CD-ROM mastering.[18] In both cases, the laser beam is modulated synchronously with the scanning movement. After resist development and baking the pattern can be used directly or transferred into the substrate.

Indeed, the ultimate tool for bi-level or multilevel DOE patterning is certainly an e-beam direct writing system. By this system 0.2 µm or better resolution and accuracy in the substrate plane can be routinely achieved. The maximum practically achievable resolution is limited only by the proximity effect, due to scattering of electrons in the resist and backscattered electrons from the substrate during the exposure. Such effects can be compensated, but not completely eliminated, as they are inherent in the use of electrons for exposing the resist.[6, 19] The cost of direct patterning is higher than the conventional mask based process, due to the use of a dedicated and specialised e-beam machine and the long

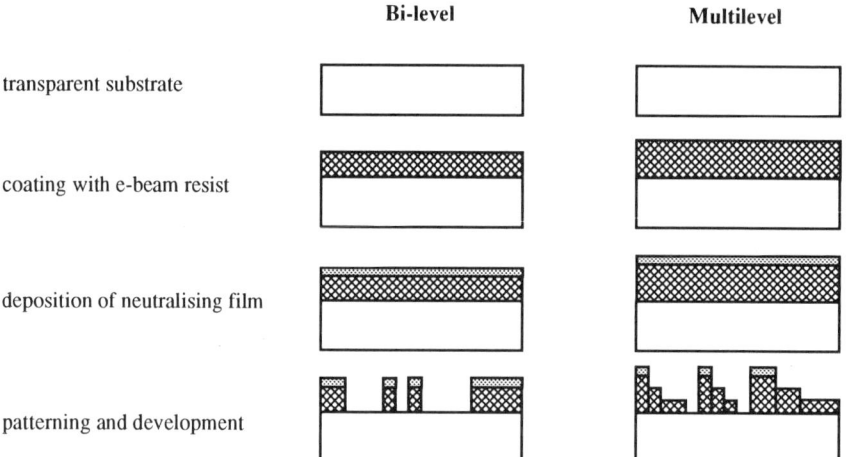

Fig. 1 Electron beam patterning process. Note that a thin metal film (usually 10-50 nm of chromium or gold) is deposited on the resist before the exposure. This avoids charge build-up due to electron bombardment, with consequent patterning distortion, while still allowing electrons to reach the resist film. For transmission DOEs, the metal layer is eventually removed. Alternatively, a transparent ITO (indium-tin-oxide) can be used between resist and substrate. For reflective type DOEs a chromium mask can be used as a substrate, and no further conductive coating is needed. If the diffracting structure has to be transferred from the resist film to the bulk substrate, a RIE or ion milling final step follows.

exposure time. Since no mask is involved in direct patterning, each device must be processed separately, and each run can require several hours or days, depending on pattern complexity Fig. 1 shows the direct patterning sequence.

Besides the need of developing specific algorithms for driving the e-beam machine in order to pattern DOEs instead of microelectronic geometries,[20,21] the bi-level structures are totally compatible with standard microelectronic practice. Conversely, multilevel patterns require a variable and yet accurate electron dosing for each step in the relief (usually in a thickness range of 100 to 500 nm). Variable dosing is obtained by locally changing the dwell time of the beam or by scanning the same area more times. Accuracy depends on a previous calibration, keeping into account electron beam intensity, resist type and further processing procedures.[22-25] Due to such difficulties and to the necessity of processing one device at a time, the achievement of the best possible patterning precision is often sacrificed in favour of the multi-mask approach, unless a completely dedicated microlithographic line (e-beam machine and related processing systems) and technical staff is available.

## 5. FABRICATION

Once the diffractive pattern has been defined on PMMA or photoresist (deposited on the final substrate or on one or more masks), the most difficult task must be performed: pattern transfer to the actual diffracting surface. The choice between the available tools is primarily driven by the precision required in the pattern transfer. Moulding, wet etching, lift-off, reactive ion etching and ion milling are techniques with respectively increasing precision as well as increasing complexity and cost. Obviously, if the patterned resist layer itself is the diffracting structure no further step is needed.

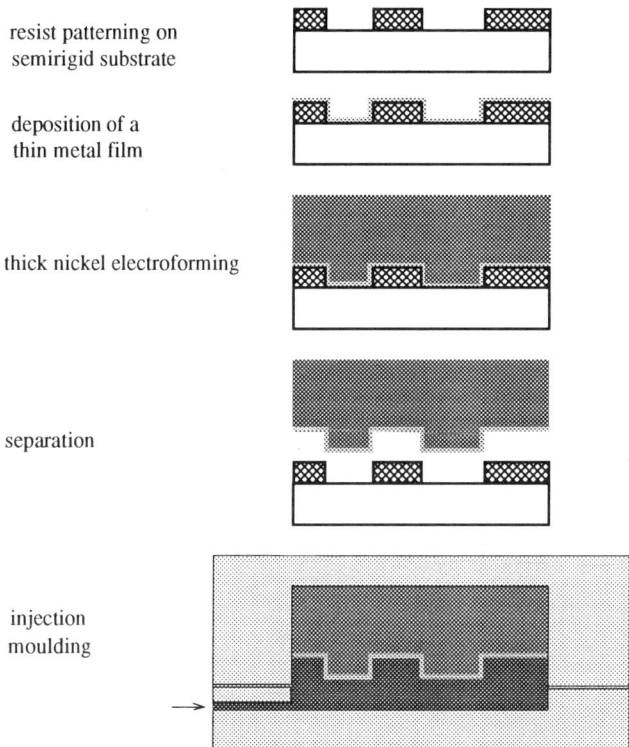

Fig. 2 Mould fabrication and DOE replication in a plastic substrate. First, the patterned resist film is coated with a metal film, in order to obtain a conductive surface. Next, a thick nickel mould is electroformed and separated from the original DOE pattern. The mould is eventually used for making replicas on the final polymeric material in an injection moulding apparatus.

## 5.1. Moulding

If DOE fabrication technologies are evaluated by the number of produced devices, moulding of plastic substrates is by far the dominant one. Modern precision injection moulding systems allow the faithful reproduction of minute details in the mould, with very high production volumes and minimum cost. The cost of moulding is improved upon only by roll-to-roll hot embossing on thermoplastic film (the same technique used for making holograms on credit cards and consumer goods). However, for quality (better than 1 µm detail reproduction capability) and time stability, injection moulding is the only choice, thus allowing plastic DOEs to be reliably incorporated into optical equipment. Fig. 2 shows a typical fabrication sequence for the mould, followed by DOE fabrication on the injected polymer[26].

Note that the moulding process is practically insensitive to the type of DOE that is being replicated. Once a precision mould is obtained, the cost of the replicas is the same for bi-level, multilevel or continuous relief structures.

For low volume or experimental applications, injection moulding is sometimes replaced by the slow photopolymerization moulding process. The UV-curing resin is placed between the nickel mould and a transparent substrate through which UV light is sent.[27]

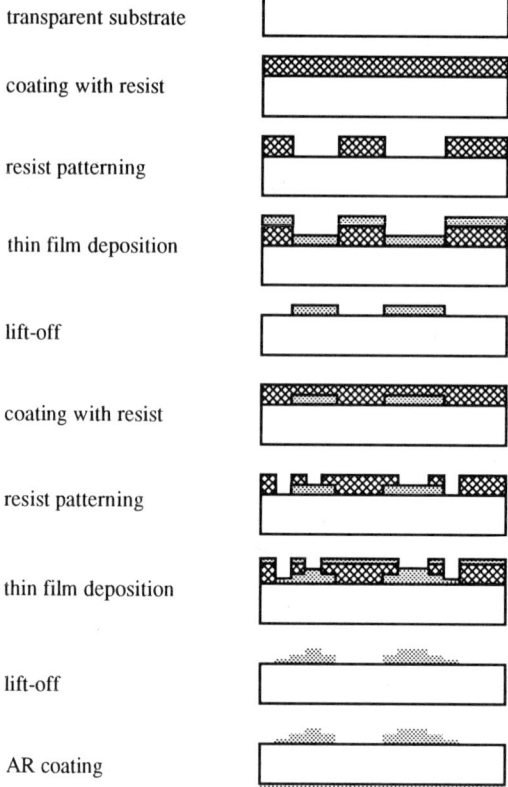

Fig. 3 Sketch of a 4-level lift-off process. For DOEs with $2^n$ levels each microlithographic cycle is repeated n times, each time with the appropriate mask and a film thickness one-half that of the previous angle. An optional antireflection coating (AR) is usually deposited on the flat side of the substrate, made of $\lambda/4$ thick index matching material.

### 5.2. Wet etching

In the field of DOE technology, wet etching is almost exclusively used for the anisotropic etching of silicon for the fabrication of V-shaped or pyramidal features for infrared applications.[28-30] Silicon wafers with (100) or (110) orientation can be easily etched, generating respectively V-shaped or U-shaped surface reliefs.[31] Then, the patterned wafer can be used directly as an infrared DOE or used as a mould for making replicas in polymeric material, for visible or near-IR applications.

### 5.3. Lift-off

When an additive fabrication process is used, that is when the active layer of the DOE is made of a thin film deposited on a transparent substrate, the lift-off technique can be applied for bi-level and multilevel structures.[3] Moreover, lift-off can be used for patterning etch masks for bi-level DOEs (see next Section). Fig. 3 shows the fabrication steps of a 4-level lift-off process. For DOEs with $2^n$ levels each microlithographic cycle is repeated n times, each time with the appropriate mask and a film thickness one-half that of the previous angle. An optional antireflection coating (AR) is usually deposited on the flat side of the substrate, made of $\lambda/4$ thick index matching material.

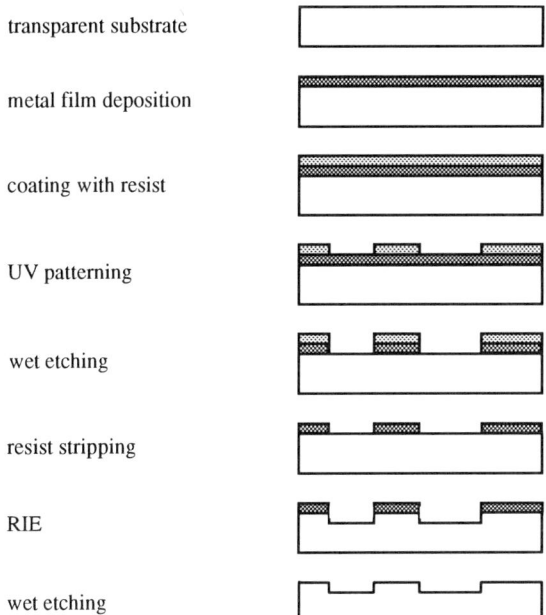

Fig. 4 Fabrication process for bi-level DOEs patterned on photoresist. The RIE metal mask is patterned first, through an e-beam generated mask and by UV exposure on photoresist, followed by wet etching (plasma etching or lift-off could be used as well).

## 5.4. Reactive ion etching (RIE)

RIE is a standard technique for accurate patterning of geometric details with submicron size. By using a chemically reactive plasma, fast and directional etching is achieved for a wide range of materials.[32-35] Pattern transfer on $SiO_2$ is usually performed by reactive ion etching in $CHF_3$ plasma, while $Cl_2$ is often used with GaAs or ZnSe substrates.

A typical RIE fabrication process is the final step of the patterning procedure described in Fig. 4. If patterning is made through a mask in UV resist, a metal layer must be used as the actual etch mask, as photoresist durability and stability under RIE processing is usually poor. The etch mask, made of chrome, nichrome or aluminium, protects the areas which are not to be etched. For chromium on silica, a 30:1 etch ratio between film and substrate is reported in the literature [36]. No metal mask is usually needed if e-beam resist is used, since this resist is very resistant to ion bombardment.

Whatever RIE process is used, accurate control of the etch depth is a major point. An error of just a few percent of the total DOE height can significantly affect the optical efficiency or generate unwanted diffracted patterns.[37, 38] For this reason, pre-calibration techniques are usually inadequate and *in situ* etch monitoring is required [39]. Another relevant issue, which however is not addressed here, is the risk of errors intrinsic in the RIE process, leading to small modifications in the profile of the etched features.[40]

## 5.5. Ion milling

Pattern transfer to the substrate by ion bombardment through a pre-patterned layer is an effective alternative to the RIE process, particularly when a strongly directional etching process is needed. In a conventional milling process, an inert beam of gas atoms (usually argon) is accelerated by an ion gun toward the target substrate, where mechanical erosion

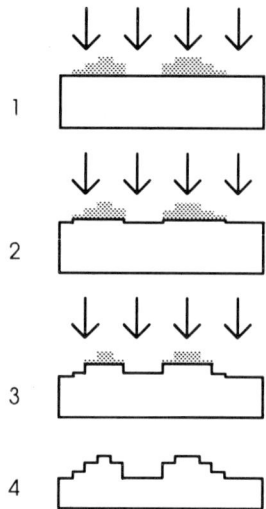

Fig. 5 Pattern transfer sequence from resist to substrate, by ion milling.

occurs. This process can be used both for simple bi-level fabrication [41] and, more interestingly, for multilevel or continuous reliefs, as shown in Fig. 5.

In some cases the erosion process can be speeded up by mixing argon with a reactive gas. Examples of this technique are given in the literature,[42, 43] where an argon/chlorine mixture is used for fast etching of GaAs, or Ar/$CHF_3$ for silica etching.

An important issue for this advanced technique concerns compensation for the uneven etch rate of the substrate. Due to the dual nature of the process - directional milling and isotropic etching - resist and substrate may exhibit different etching behaviours. In silica substrates, for instance, directional etching prevails, while isotropic etching is predominant in e-beam resist. Since directional etching locally depends on the surface slope (the higher the slope the lower the etching rate), geometrical distortions can occur.[44] However, this can be estimated in advance and compensated through a correction in the local resist thickness and slope.[43]

## 6. FABRICATION OF ARTIFICIAL DIELECTRICS

The typical artificial dielectric approach to diffractive optics fabrication is schematically shown in Fig. 6. While a surface relief is shaped in such a way as to locally affect the phase of the incoming wavefront, by varying the optical path length, the artificial dielectric controls the phase by locally varying the effective refractive index. This is obtained by means of a structure made of alternating high and low index materials (the low index material is usually air, while the high index comes directly from the substrate or from a deposited thin film). Provided that the operating wavelength is larger than the width of the alternating index features, an average index is seen. Hence, by properly specifying the material repeat distance, the desired effective index can be locally obtained.

Besides index averaging, subwavelength structures (also named zero order gratings, as no diffraction is generated) can be used for the fabrication of antireflective or artificially birefringent layers.[45 - 47] In this Chapter we are concerned only with diffractive optics fabrication. Therefore the following technological considerations are oriented to such devices only.

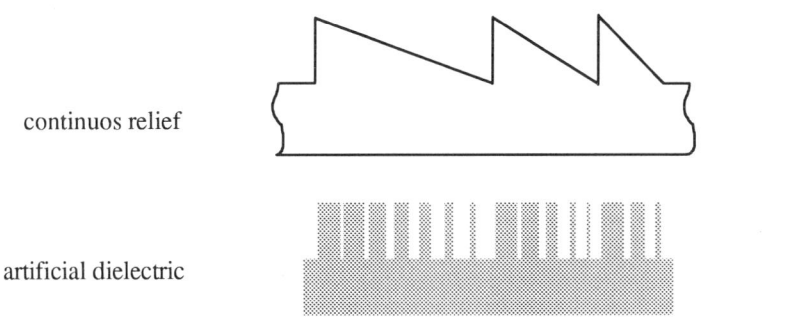

Fig. 6 Correspondence between a continuous relief DOE and the equivalent artificial dielectric structure.

In DOE fabrication using artificial dielectrics, a single lithographic step replaces the multimask fabrication procedure or the variable exposure technique, allowing for simpler and more precise patterning. However, due to the requirement that the width of each feature is smaller than the wavelength, carrying out such single lithographic step is certainly one of the most challenging tasks in the whole field of nanotechnology.[48, 49] The challenge is not only related to the width of each *pillar* in the artificial structure, but also to its height. Since the final goal is the fabrication of diffractive optics, the height of the structure is determined by the wavelength. As a result, each feature must usually exhibit aspect ratios of 5:1 or higher, well above those needed in microelectronic fabrication.

The consequence of the above considerations is that specific technological approaches are being developed, as they can be no longer borrowed from standard microelectronic technology. Certainly, electron-beam direct patterning and reactive ion etching are the basic tools for the fabrication of artificial dielectrics, for pattern resolutions well below the operating wavelength. Although the direct use of an e-beam is time consuming, as discussed in the section on patterning, it is presently the only practically available technique, since x-ray or synchrotron-radiation lithography is still not readily available to most research laboratories or manufacturing companies.

Referring to the patterning and fabrication processes, previously discussed for traditional surface reliefs, the most precise application of each technique is required for artificial dielectrics. Hence, in e-beam direct patterning the proximity effect must be always kept into account and minimised. Well calibrated electron dosing must also be assured, in order to reach 100 nm scale resolution. Moreover, *in situ* etch depth monitoring during the RIE process is strictly necessary. If thin film structures are etched, monitoring can be done by mass spectrometric analysis of the etching atmosphere. The process is stopped when the film material is no longer detected. Obviously, if the diffracting structure is etched in the substrate, mass spectrometry cannot be used and optical interferometry must be adopted.[50]

The main concern in RIE processing of high aspect ratio structures is the occurrence of the ion etch lag.[51] Due to this effect, the etch rate decreases while the trench aspect ratio increases. This is not a relevant issue if uniform gratings are being made,[46] however it leads to uneven final depth in a DOE structure, and consequently to the occurrence of inexact wavefront processing. It should be noted that the problem is not related to the DOE operation wavelength, as it is the aspect ratio and not the feature width that limits the etch rate. For instance, infrared devices for 10 µm wavelength, made with a few micron wide and several micron deep trenches, suffer similar technological difficulties.

The problem of ion etch lag is currently being faced both from the technological side and the design side. Improved RIE, ion milling and reactive ion milling are expected to contribute to a solution, together with low aspect ratio designs.[52]

# REFERENCES

The list of references is divided into two parts. The first, from Ref. 1 to 44, is mainly related to the fabrication of *surface reliefs*, and the second, from Ref. 45 to 52, contains papers on advanced submicron patterning for optical *artificial dielectrics*. Most references cover the 1990-1996 period, as preference has been given to papers which describe with current technology. However, several classical papers written in previous years are indicated in many of the references reported here.

1. R.W. Schneek, Process factors for electroforming video discs, *Plating and Surf. Finishing* 71:38 (1984).
2. E. Pawlowski, B. Kuhlow, Antireflection-coated diffractive optical elements fabricated by thin film deposition, *Opt. Eng.* 33:3537 (1994).
3. J. Jahns and S.J. Walker, Two-dimensional array of diffractive microlenses fabricated by thin film deposition, *Appl. Opt.* 29:931 (1990).
4. M.R. Taghizadeh, J.I.B. Wilson, J. Turunen, A. Vasara, and J. Westerholm, Optimization and fabrication of grating beamsplitters in silicon nitride, *Appl. Phys. Lett.* 54:1492 (1989).
5. M. E. Motamedi, B. Anderson, R. De La Rosa, W.J. Gunning, R. Hall, and M. Khoshnevisan, Binary optics thin film microlens array, *Proc. SPIE* 1544:22 (1992).
6. W. M. Moreau, "Semiconductor Lithography", Plenum Press, New York (1988).
7. M.P. Dames, D.J. Dowling, P. McKee, and D. Wood, Efficient optical elements to generate weighted spot arrays: design and fabrication, *Appl. Opt.* 30:2685 (1991).
8. W. Goltsos, M. Holtz, Agile beam steering using binary optics microlens array, *Opt. Eng.* 29:1392 (1990).
9. S. Walker and J. Jahns, Array generation with multilevel phase gratings, *J. Opt. Soc. Am. A* 7:1509 (1990).
10. G. J. Swanson, W. B. Veldkamp, Diffractive optical elements for use in infrared systems, *Opt. Eng.* 28:605 (1989).
11. D.C. O'Shea, J.W. Beletic, M. Poutous, Binary-mask generation for diffractive optical elements using microcomputers, *Appl. Opt.* 32:2566 (1993).
12. I. Moreno, C. Gorecki, J. Campos, M.J. Yzuel, Comparison of computer-generated holograms produced by laser printers and lithography: application to pattern recognition, *Opt. Eng.* 34:3520 (1995).
13. H. Andersson, M. Ekberg, S. Hård, S. Jacobsson, M. Larsson, and T. Nilsson, Single Photomask, multilevel kinoforms in quartz and photoresist: manufacture and evaluation, *Appl. Opt.* 29:4259 (1990).
14. Y. Opplinger, P. Sixt, J.M. Stauffer, J.M. Mayor, P. Regnault, and G. Voirin, One-step 3-D shaping using a gray-tone mask for optical and microelectronic applications, *Microelectron. Eng.* 23:449 (1990).
15. M.T. Gale, M. Rossi, J. Pedersen, H. Schütz, Fabrication of continuous-relief micro-optical elements by direct laser writing in photoresists, *Opt. Eng.* 33:3556 (1994).
16. C. Arnone, The laser-plotter: a versatile lithographic tool for integrated optics and microelectronics, *Microelectron. Eng.* 17:483 (1992).
17. U. Krackhardt, J. Schwider, M. Schrader, N. Streibl, Synthetic holograms written by a laser pattern generator, *Opt. Eng.* 32:781 (1993).
18. P. Perlo, private communication, Centro Ricerche FIAT, Turin (1996).
19. M. Ekberg, F. Nikolajeff, M. Larsson, and S. Hård, Proximity-compensated blazed transmission grating manufacture with direct-writing, electron-beam lithography, *Appl. Opt.* 33:103 (1994).
20. J. Fan, D. Zaleta, K.S. Urquhart, and S.H. Lee, Efficient encoding algorithms for computer-aided design of diffractive optical elements by the use of electron-beam fabrication, *Appl. Opt.* 34:2522 (1995).
21. T. Shiono, K. Setsune. O. Yamazaki, and K. Wasa, Computer-controlled electron-beam writing system for thin film micro-optics, *J. Vac. Sci. Technol. B* 5:33 (1987).
22. M. Ekberg, M. Larsson, and S. Hård, Multilevel phase holograms manufactured by electron-beam lithography, *Opt. Lett.* 15:568 (1990).
23. K.S. Urquhart, R. Stein, and Sing H. Lee, Computer-generated holograms fabricated by direct write of positive electron-beam resist, *Opt. Lett.* 18:308 (1993).
24. T. Shiono and K. Setsune, Blazed reflection micro-Fresnel lenses fabricated by electron-beam writing and dry development, *Opt. Lett.* 15:84 (1990).
25. H. Zarschizky, A. Stemmer, F. Mayerhofer, G. Lefranc, W. Gramann, Binary and multilevel diffractive lenses with submicrometer feature sizes, *Opt. Eng.* 33:3527 (1994).
26. M.T. Gale, M. Rossi, H. Schütz, P. Ehbets, H.P. Herzig, and D. Prongué, Continuous-relief diffractive optical elements for two-dimensional array generation, *Appl. Opt.* 32:2526 (1993).
27. M. Tanigami, S. Ogata, S. Aoyama, T. Yamashita, and K. Imanaka, Low wavefront aberration and high temperature stability molded micro Fresnel lens, *IEEE Photon. Technol. Lett.* 1:384 (1089).
28. M. Collischon, H. Haidner, P. Kipfer, J.T. Sheridan, J. Schwider, Diffractive optical elements for high power $CO_2$ laser applications, *IEE Conference Publication No.* 379:243 (1993).

29. N. Rajkumar and J.N. McMullin, V-groove gratings on silicon for infrared beam splitting, *Appl. Opt.* 34:2556 (1995).
30. J. Turunen, E. Noponen, V-groove gratings on silicon for infrared splitting: comment, *Appl. Opt.* 35:807 (1996).
31. K.E. Bean, Anisotropic etching of silicon, *IEEE Trans. Electron Devices* ED-25:1185 (1978).
32. J.A. Bondur, Dry process technology, *J. Vac. Sci. Technol.* 13:1023 (1976).
33. J.W. Coburn, "Plasma Etching and Reactive Ion Etching", American Vacuum Society, New York (1982).
34. H.F. Winters, J.W. Coburn, and T.J. Chuang, Surface processes in plasma-assisted etching environment, *J. Vas. Sci. Technol.* B 1:469 (1983).
35. J.M. Finlan, K.M. Flood, R. J. Bojko, Efficient $f/1$ binary-optics microlenses in fused silica designed using vector diffraction theory, *Opt. Eng.* 34:3560 (1995).
36. B. Kuhlow, E. Pawlowski and M. Ferstl, Two-dimensional arrays of diffractive microlenses for optical interconnects, *IEE Conference Publication No. 379*:41 (1993).
37. V.V. Wong and G.J. Swanson, Design and fabrication of a gaussion fan-out optical interconnect, *Appl. Opt.* 32:2502 (1993).
38. J. M. Miller, M.R. Taghizadeh, J. Turunen, and N. Ross, Multilevel-grating array generators: fabrication error analysis and experiments, *Appl. Opt.* 32:2519 (1993).
39. S.E Hicks, W. Parkes, J.A.H. Wilkinson, and C.D.W. Wilkinson, Reflectance modeling for *in situ* dry etch monitoring of bulk $SiO_2$ and III-V multilayer structures, *J. Vac. Sci. Technol.* B 12:3306 (1994).
40. D.A. Pommet, E.B. Grann, and M.G. Moharam, Effects of process errors on the diffraction characteristics of binary dielectric gratings, *Appl. Opt.* 34:2431 (1995).
41. K.Rastani, A. Marrakchi, S.F. Habiby, W.M. Hubbard, H. Gilchrist, and R.E. Nahory, Binary phase Fresnel lenses for generation of two-dimensional beam arrays, *Appl. Opt.* 30:1347 (1991).
42. J. Bengtsson, N. Eriksson, and A. Larsson, Small-feature-size fan-out kinoform etched in GaAs, *Appl. Opt.* 35:801 (1996).
43. W. Däschner, M. Larsson, and S.H. Lee, Fabrication of monolithic diffractive optical elements by the use of e-beam direct write on an analog resist and a single chemically assisted ion-beam-etching step, *Appl. Opt.* 34:2534 (1995).
44. G.M. Gallatin and C.B. Zarowin, Unified approach to the temporal evolution of surface profiles in solid etch and deposition processes, *J. Appl. Phys.* 65:5078 (1989).
45. Y. Ono, Y. Kimura, Y. Ohta, and N. Nishida, Antireflection effect in ultrahigh spatial-frequency holographic relief gratings, *Appl. Opt.* 26:1142 (1987).
46. D.C. Flanders, Submicrometer periodicity gratings as artificial anisotropic dielectrics, *Appl. Phys. Lett.* 42:492 (1983).
47. F. Xu, R. Tyan, P. Sun, and Y. Fainman, Fabrication, modeling, and characterization of form-birefringent nanostructures, *Opt. Lett.* 20:2457 (1995).
48. F.T. Chen, H. G. Craighead, Diffractive phase elements based on two-dimensional artificial dielectrics, *Opt. Lett.* 20:121 (1995).
49. J.R. Wendt, G.A. Vawter, R.E. Smith, and M.E. Warren, Nanofabrication of subwavelength, binary, high efficiency diffractive optical elements in GaAs, *J. Vac. Sci. Technol.* B 13:2705 (1995).
50. W. Parkes, S. Thoms and C.D.W. Wilkinson, Application of electron beam lithography to pattern sub-micron features in a beam forming grating, *Microelec. Eng.* 23:465 (1994).
51. O. Joubert, G.S. Oehrlein, Y. Zhang, Fluorocarbon high density plasma. V. Influence of aspect ratio on the etch rate of silicon dioxide in an electron cyclotron resonance plasma, *J. Vac. Sci. Technol.* A 12:658 (1994).
52. J.M. Miller, N. de Beaucoudrey, and P. Chavel, Synthesis of a subwavelength-pulse-width spatially modulated array illuminator for 0.633 µm, *Opt. Lett.* 21:1399 (1996).

# LOW COST HIGH QUALITY FABRICATION METHODS AND CAD FOR DIFFRACTIVE OPTICS AND COMPUTER HOLOGRAMS COMPATIBLE WITH MICRO-ELECTRONICS AND MICRO-MECHANICS FABRICATION

S. H. Lee and W. Däschner

Electrical and Computer Engineering Department
University of California, San Diego
9500 Gilman Drive, La Jolla, CA. 92093-0407 U.S.A.

## 1. INTRODUCTION

State-of-the-art diffractive optical elements (DOEs) and computer generated holograms (CGHs) are fabricated using e-beam lithography and dry etching techniques to achieve multi-level phase elements with very high diffraction efficiencies. Electron beam lithography allows DOEs/CGHs with smaller feature sizes (e.g. for off-axis diffractive lenses with small f-number) to be fabricated. Dry etching techniques, especially chemically assisted reactive ion beam, allows DOEs/CGHs be fabricated in a variety of materials. Multi-level phase DOEs/CGHs, especially those of sixteen or more phase levels, mean their surface profiles are fabricated with high accuracy / quality. It is not only the e-beam system which dictates the feature size limitations, but also the alignment systems (mask aligner) and the materials (e-beam or photo resists). In order to allow DOEs/CGHs to be used in new optoelectronic systems, it is necessary also to fabricate the elements in different materials economically. Since the cost of a multi-level phase DOE/CGH is determined by the e-beam writing time, the number of etching steps for fabricating the master, and the technology for replication, we need to decrease the writing time and etching steps for the master and to develop a low cost replication technology for mass production without affecting the quality of the DOEs/CGHs. In this Chapter, we will review three DOE/CGH fabrication techniques and CAD development required, considering the overall design and fabrication time for multi-level phase DOEs.

## 2. CONVENTIONAL FABRICATION TECHNIQUE

Most multi-level phase surface relief structures are fabricated by utilizing n binary amplitude masks in n standard photolithography processes to achieve 2n relief levels. A

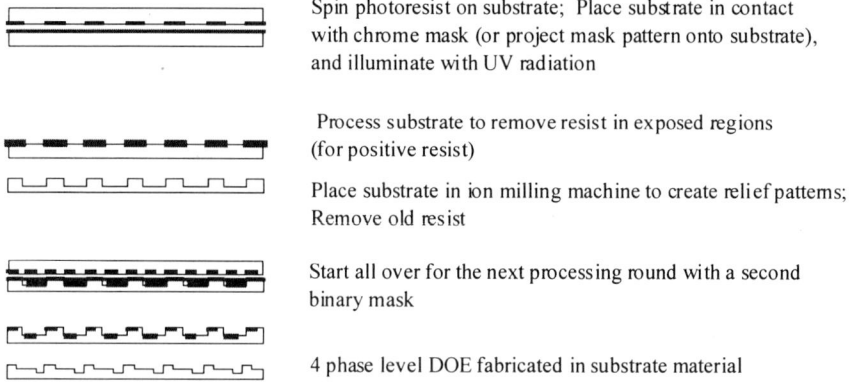

Fig. 1 Conventional fabrication technique using multiple binary masks and etching steps.

standard photolithography process cycle for a single mask consists of the following steps: 1) Spin photoresist on substrate; 2) Place substrate in contact with chrome mask (or project mask pattern onto substrate), and illuminate with UV radiation; 3) Process substrate to remove resist in exposed regions (for positive resist); 4) Place substrate in ion milling machine to create relief patterns; and, 5) Remove old resist. Thus for a 16 phase level DOE/CGH, 20 (=4x5) processing steps are required to complete the fabrication (Fig. 1). Typically, this method allows fabrication of structures down to 1 μm (mainly limited by mask aligner type and the number of re-alignments required between masks). It is easy to see that this fabrication technique is very laborious and time consuming, and subject to many processing errors, which directly increase the cost.

## 3. DIRECT WRITE ON E-BEAM SENSITIVE ANALOG PHOTORESIST

In order to eliminate the mis registrations between mask alignments and to cut down the number of processing steps, we use e-beam direct write on an e-beam sensitive analog photoresist, where the e-beam exposes the desired pattern.[1,2] In standard e-beam lithography, an e-beam exposure contains enough electron dosage in the exposed regions of the resist to fully clear the resist during the development process (for a positive e-beam resist). If the electron dosage and/or the development process is reduced, it is possible for the resist to not be fully developed. This is possible because the solubility of the resist in the developer varies with the electron dose, i.e. the higher the dose, the faster the resist dissolves in the developer. This allows different thickness of resist to be produced simply by varying the electron dose. The registration errors are decreased to the e-beam mis-registration in field addressing, which is typically of 0.025 μm (Fig. 2).

Two types of e-beam sensitive resists have been investigated: EBR-9 (Toray) and PMMA (KTI 959K 9%). Since EBR-9 could not be coated thicker than approximately 500 nm, its use was limited to reflection mode diffractive optics. PMMA was used for thicker layers (up to 2 μm thick), and can be applied to transmissive mode diffractive optics; however, high dose can cause surface defects. Both EBR-9 and PMMA have fair to poor etch resistivities. Two photoresists have also been investigated: SAL-110 and Hoechst 5200 series. They both have good to excellent etch resistances; however, they have fair to poor electron beam sensitivities. Novolac based resist OeBr-514 (Olin Ciba Geigy) was the photoresist recently found to be both e-beam sensitive and to have good etch resistance.

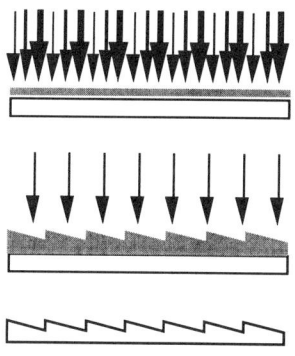

Exposure of analog photoresist with electron beam of varying dosages spatially

Transfer of structure into susbstrate using CAIBE

Resulting DOE fabricated in substrate material

Fig. 2 Direct write on e-beam sensitive analog photoresist.

After OeBr-514 is written by e-beam and developed, its surface microstructures can be transferred to its substrate by chemically assisted ion beam etching (CAIBE). CAIBE combines the mechanical etching process of ion milling and the chemical etching process of reactive ion etch (RIE); it can be applied to transfer shallow features in the resist to accommodate different optical wavelengths and substrate materials. Fig. 3 shows a DOE fabricated by e-beam direct write on analog resist.

Although the combined direct write on analog resist and CAIBE process has the advantages of reducing the DOE/CGH fabrication procedures on different substrate materials to single writing and single etching steps, it experiences difficulties in e-beam proximity effect and long exposure / writing time. Proximity effect is caused by electron

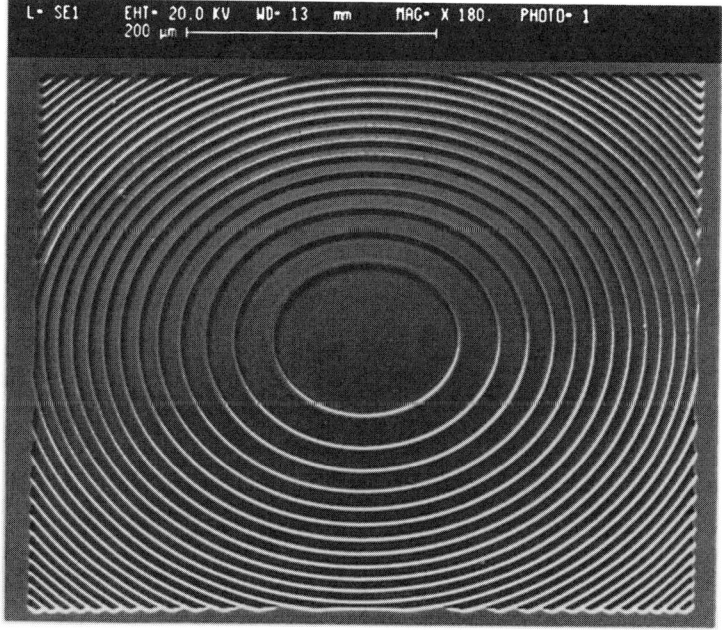

Fig. 3 DOE fabricated by direct write on analog resist.

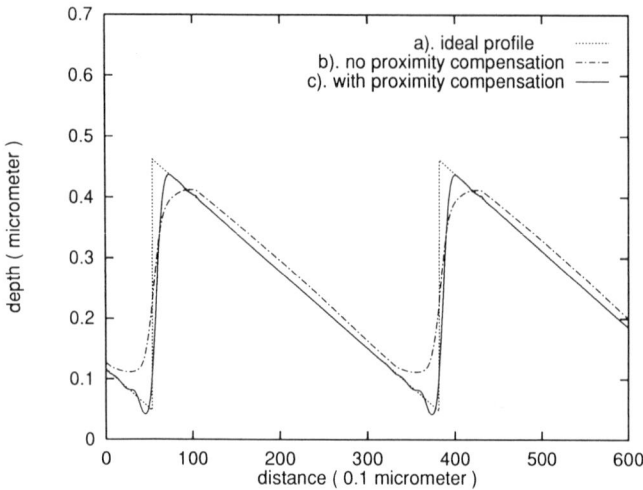

Fig. 4 Proximity compensation.

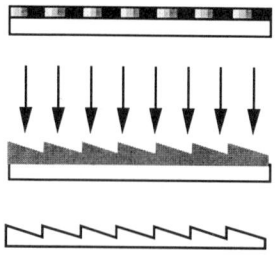

Exposure of resist using gray scale mask and contact aligner or projection stepper system

Transfer of structure into susbstrate using CAIBE

Resulting DOE fabricated in substrate material

Fig. 5 DOE fabrication using gray scale mask fabrication.

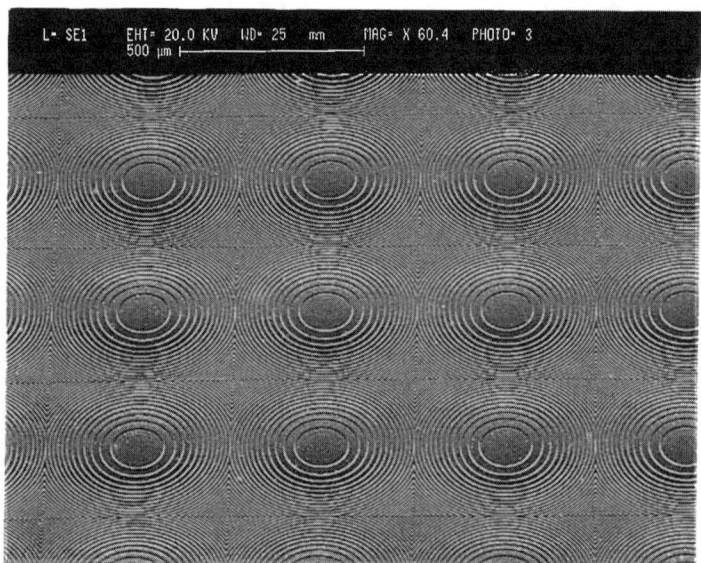

Fig.6 Doe fabricated by gray scale mask.

scattering in the resist and reduces the resolution performance of e-beam direct write. To compensate for the proximity effect the e-beam exposure data files can be adjusted (see Fig. 4) according to two alternative algorithms: iterative algorithm and de-convolution. Application of the deconvolution algorithm requires the point spread function of the proximity effect be determined by measurements. Measurements show that the proximity effect is a function of resist thickness and e-beam current. Thicker resist and higher beam current cause more electrons to scatter. When the proximity effect is properly compensated, the combined ebeam direct write on analog resist and CAIBE process is a very good method for producing high quality DOE/CGH masters on a variety of substrates (for replication). To solve the problem of long and expensive e-beam writing time, we now discuss our research into gray scale masks for DOE/CGH replication.

## 4. GRAY SCALE MASKS FOR DOE/CGH REPLICATION

Gray scale (transmission) masks, when they are generated as described below, can be used to produce many copies of multi-level or analog resist profiles in an optical stepper system by performing step-and-repeat optical exposure, single exposure for each and every copy.[3] Hence, the high cost of generating the gray scale mask (because of long e-beam exposure time) is spread over a large number of copies (Fig. 5).

Gray scale masks can be generated in high energy beam sensitive (HEBS-) glass, which consists of allcali in a low expansion zinc-boro-silicate glass. When the glass is melted, drawn, ground and polished, the base glass plates are ion-exchanged in an acidic aqueous solution containing soluble ionic silver. The ion exchange process is carried out at a temperature in excess of 320°C for a duration sufficient to cause silver ions to diffuse into the glass plates 3 µm in the thickness dimension. As a result silver ions are present in the form of silver-alkali halide $(AgX)_m(MX)_n$ complex crystals. Doping of the glass with photo inhibitors causes an increased energy band gap of the silver crystals, making the ion exchanged glasses inert to W radiation. The reduction of silver ions in the silver halide can be accomplished by exposing the HEBS material to high energy beams (e.g. e-beams >10 kV): higher electron density and higher e-beam accelerating voltages yield higher transmission density. The electron dosage needed for writing the gray scale mask vs. the resist thickness after development will be presented, along with a microlens array fabricated by HEBS-glass mask (see Fig. 6). When replica on photoresist need to be transferred onto the substrates, the CAIBE process can again be applied. It can be applied to a number of them simultaneously in batch mode to reduce production costs. The number of replica that can be etched in each batch depends on the relative size of the CAIBE system and the DOE replica.

Since this replication method employs lithographic steppers and CAIBE systems along with gray scale masks, it should be compatible with production of microelectronics and micro-mechanics, and holds the promise of hybrid / heterogeneous integration of micro-optics with micro-mechanics and microelectronics in a compact miniaturized system. Furthermore, this replication method allows DOEs/CGHs be replicated in different substrate materials for different wavelength or environmental operations.

## 5. CAD FOR DESIGN AUTOMATION

To design DOEs/CGHs one can start with some popular optical design program such as Code V for conventional designs or write one's own design programs for less conventional designs. In any case, one needs to obtain the optical path difference functions or phase profiles (in terms of aspheric or Zernike polynomials or otherwise). Then, by subtracting multiples of 2p from these phase profiles, the phase profiles for the DOEs/CGHs are obtained. For the phase profiles to be fabricated by electron beam writer and CAIBE processes, the phase profiles will have to be converted into one of the common e-beam data formats (GDS2, EBMF or MEBES) that controls the exposure function of an e-beam writer, after proximity compensation has been applied. Since the computations for design and proximity compensation are usually very laborious and the amounts of e-beam data are very large, the development of CAD is essential.[4,5]

A good CAD tool should: (a) have a user friendly interface, e.g. x-windows; (b) perform the computations of design and proximity compensation automatically, given the higher level design parameters (e.g., diffraction efficiency, operational wavelength, geometry of optical setup, and the point spread function of the e-beam writing process) by the user; (c) generate the e-beam data files efficiently; and, (d) check design requirements against limits of the fabrication technology, i.e. design rules check. In this presentation, the adaptive algorithm for efficient e-beam data generation and the CAD tool developed at UCSD will be briefly described, since efficient data generation affects e-beam writing time / cost and a good CAD tool helps to reduce the design-fabrication cycle time.

## 6. SUMMARY CONCLUSION

The foundation has been built for rapid design, high quality fabrication of DOE/CGH masters, and mass production by replication. Successful applications of the technology will most likely be found in areas where unique capabilities of DOEs/CGHs are utilized, e.g. creating wave fronts that can be defined mathematically without the physical objects, making use of the compensatory dispersion characteristics of DOEs to that of refractive optics, being able to combine the functions of several refractive optics into the same DOE/CGH (e.g. beam splitting combined with focusing), ease to produce multi-faceted DOEs/CGHs relative to multi-faceted refractive optics, etc. The future of DOE/CGH technology now rests on our creativity in finding the right applications.

## REFERENCES

1. K.S. Urquhart, R. Stein, and S.H. Lee, Computer-generated holograms fabricated by direct write of positvie electron-beam resist, *Opt. Lett.* 18, 308-310, 1993.
2. W. Daschner, M. Larsson, and S.H. Lee, Fabrication of monolithic diffractive optical elements by the use of e-beam direct write on an analog resist and a single chemically assisted ion-beam-etching step, *Appl. Opt.* 34(14): 2534-2538, May 10, 1995.
3. W. Daschner, P. Long, R. Stein, C. Wu, and S.H. Lee, One-step lithography for mass production of multilevel diffractive optical elements using High Energy Beam Sensitive (HEBS) gray-level mask, *Proceedings of the SPIE - The International Society for Optical Engineering*, vol.2689; 1996
4. K. Urquhart, S.H. Lee, C.C. Guest, M. R. Feldman, and H. Farhoosh, Computer aided design of computer generated holograms for electron beam fabrication, *Appl. Opt.* 28(15): 3387-3396 (August 15, 1989).
5. J. Fan, D. Zaleta, K.S. Urquhart, and S.H. Lee, Efficient encoding algorithms for computeraided design of diffractive optical elements by the use of electron beam fabrication, *Appl. Opt.* 34(14):2522-2529, May 10, 1995.

# DESIGN AND FABRICATION ASPECTS OF CONTINUOUS-RELIEF DIFFRACTIVE OPTICAL ELEMENTS

T. Hessler and M. Rossi

Paul Scherrer Institute
Badenerstrasse 569, 8048 Zurich, Switzerland

## 1. INTRODUCTION

Diffractive optical elements (DOEs) offer many very interesting design approaches that can became realisable due to continuous improvements in various fabrication technologies. Computer aided design and modern fabrication methods give access to optical functions which are often not realisable by one single conventional bulk optical element. Their planarity allows low cost replication in plastic material using techniques such as hot embossing or injection moulding. This offers the possibility of their integration in opto-mechanical devices to reduce assembly costs. DOEs show a chromatic and thermal behaviour which can be used for chromatic or thermal aberration correction in hybrid elements. For broad band or white light applications, their strong chromatic dispersion is usually not wanted, but in many applications, using monochromatic laser light, it is of no importance.

Typical fabrication tolerances of DOEs mainly reduce the diffraction efficiency and lead to stray light. However, the wavefront quality in the image plane can be guaranteed with high accuracy for many applications. Nevertheless, the influence of stray light and efficiency reduction needs to be investigated carefully for each specific case.

In the following, we discuss some of the principal limitations and tolerances occurring in the process of the design and fabrication of planar optical elements and their influence on optical performance, especially on the diffraction efficiency. We concentrate on continuous relief structures, fabricated by direct laser writing into photoresist and subsequent replication in plastic material. Many of the following considerations can easily be transferred to other fabrication techniques.

## 2. DESIGN OF PLANAR OPTICAL ELEMENTS

A surface relief is computed which transforms an incoming wave into a desired transmitted or reflected output wave[1]. This is done by calculating the phase function of the

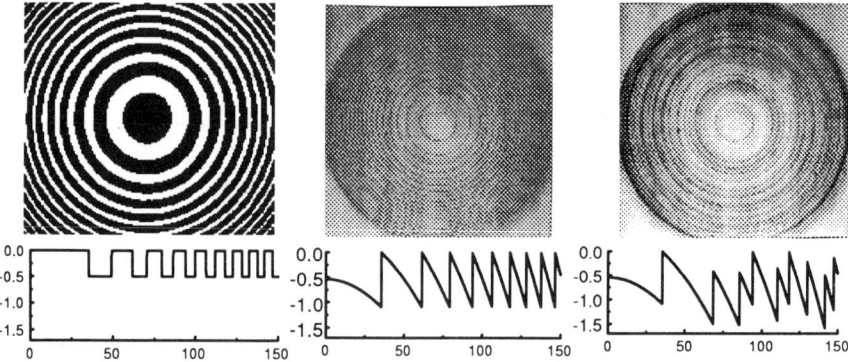

Fig. 1 Various ways of coding a lens phase function into a surface relief: a binary phase zone plate (a) and a continuous relief element (b, c). In (c) an additional design freedom was used as described in the text.

DOE by analytical, ray tracing or iterative methods and subsequent conversion of that phase function into a surface profile.

If the phase function has an amplitude of many multiples of 2 π, the phase function can be wrapped into intervals between 0 and M*2 π, where the integer number M is the so called phase-matching number.[2] This procedure maintains the planarity of the element and leads to a discontinuous surface relief, consisting of individual segments. The choice of the phase-matching number is a degree of freedom which can be used to optimise the performance of the DOE under non-ideal, real conditions. In order to fulfil the phase-matching condition, only the height of the phase steps has to be an integer multiple of 2π; the position of the phase steps can be chosen arbitrarily, as is indicated in Fig. 1c. This is an additional degree of freedom in wrapping the phase function.

DOE performance is based on diffraction at the grating given by the position of the phase transitions. This lateral pattern can be fabricated by today's microfabrication methods with high accuracy. The continuous relief of the microstructure (blaze) distributes the light among the various diffraction orders. Usually it is the aim to concentrate all the light in one single order, the phase-matching order.

By using the phase-matching order M in the coding process as a design freedom, the number of segments of a DOE can be decreased by increasing the phase-matching order. For micro-optical elements, the phase variation over the aperture $a$ is often an the order of a few multiples of 2π, so the Fresnel number ($N = a^2/(\lambda \cdot f)$) is therefore low. In this case the number of segments might be decreased to one by increasing M, leading to a lenslet without phase transitions, as shown in Fig. 2. The optical performance under the exact design conditions is nearly identical for the different surface profiles shown in Fig. 2. However real lenses are always operated at conditions which do not exactly match the design conditions. A lens with phase transitions ('diffractive lens') and a lens without phase transitions ('refractive lens') will show differences in their optical performance. In the case of development errors the refractive lens will shift its focal position, whereas the focal length of the diffractive lens will stay stable, but additional, unwanted foci appear. Often, in the first case, it is the wavefront quality in the far field which is affected by fabrication tolerances whereas in the case of diffractive lenses, the straylight increases. The use of direct laser writing into photoresist now allows the realisation all the cases of surface reliefs shown in Fig. 2 . Depending on the application's demand, one or the other design will be chosen. It is interesting to note, that even in the case of two illuminated zones, a stabilisation of the focal length is seen in the presence of depth errors. [1, 3]

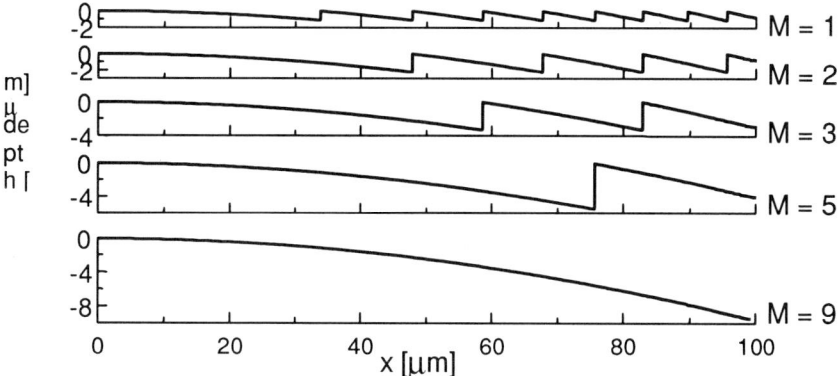

Fig. 2 By increasing the phase matching order M, the number of phase transitions is reduced.

For elements exhibiting only low deflection angles (< 5), the coding of the wrapped phase function into a continuous surface profile is straightforward. The depth d is given according to the Sweatt model[4] by $d = M\lambda/(n-1)$ and will lead to theoretical diffraction efficiencies close to 100%. The basic assumption of the underlying scalar theory is that the local grating period $\Lambda$ (the distance between two phase transitions) is much greater than the illumination wavelength $\lambda$. This is no longer fulfilled for elements employing higher deflection angles, resulting in high numerical apertures. A phase function design and coding algorithms based on rigorous diffraction theory would be needed. This is very computationally intensive and still not possible for large two dimensional structures. As an approximation, the surface relief can be calculated by using a locally blazed grating model. The optimum blaze depth is calculated by phase sensitive ray-tracing methods.[5] The depth is then given by the phase-matching condition as.[6]

$$d(\Lambda) = \frac{M\lambda}{n - \sqrt{1 - \left(\frac{M\lambda}{\Lambda}\right)^2}} \qquad (1)$$

The profile depth is now dependent on the local grating period $\Lambda$. Therefore, the segment depth of a DOE with a focusing function will decrease for increasing radii. Eq. 1 predicts the optimum depth for maximum diffraction efficiency very well down to very small grating periods such as $\Lambda/\lambda = 2$.

Furthermore, the ray-tracing approach, combined with scalar diffraction theory, can be used to give an intuitive picture for the principale limitation on maximum achievable diffraction efficiency for small grating periods. This can be explained by the shadow effect. The propagation of the rays within the surface relief results in a deflected wavefront where the rays are compressed in certain regions and where no rays are present in-between (see Fig. 3a). In the near field, the deflected wavefront contains the same amount of energy as the incident wave, but intensity modulated. This intensity modulation of the deflected wave leads to diffraction during propagation with a resulting diffraction efficiency in the far field given by.[5]

$$\eta_{Shadow} = 1 - \frac{\delta'}{\Lambda'} \qquad (2)$$

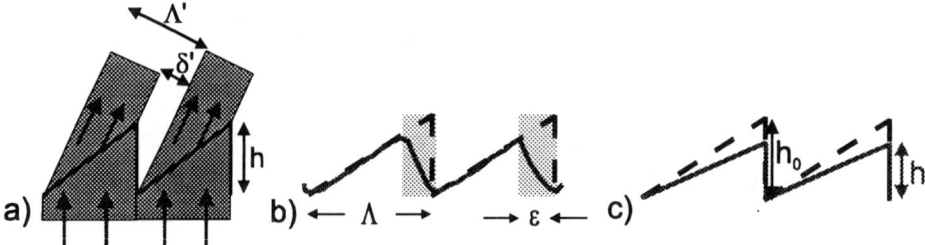

Fig. 3 The main sources for efficiency reduction. Shadow effect (a), spot size convolution (b) and depth errors (c).

where δ' is the width of the dark region.

Comparison with rigorous diffraction theory calculations showed good agreement with the simple and fast approach based on ray-tracing.[5,7]

## 3. MASTER FABRICATION

Diffractive optical elements are fabricated either by direct write methods[8] or by photolithographic multi-exposure techniques. Direct write methods include laser[9] or e-beam writing[10] and single-point diamond turning.[11] We concentrate on laser writing, but many of the conclusions also to the other fabrication techniques.

The laser writing system built up at PSI (see Fig. 4) operates as follows: An intensity-modulated focused laser beam (wavelength $\lambda$ = 442 nm) is scanned over a photoresist (Shipley AZ 1828) coated substrate (thickness 3 - 10 µm). The scan could either be a raster or a polar scan, both having their advantages depending on the element symmetries. We use a x-y roller bearing translation stage for the raster scan.

For the scan, the depth values of the desired surface relief have to be converted into a two-dimensional bitmap of exposure intensities, taking into account the sensitivity of the photoresist and the raster scan geometry. The intensity data are calculated in advance or in real-time for large elements to reduce the amount of data handled. They are loaded line by line into a buffer and are clocked out by the translation stage pulses during each line scan. The surface relief is obtained after development of the exposed resist using Shipley's Microposit 303 Developer. For a raster scan based system, the writing time for large area DOEs such as lens arrays can be many hours, therefore a stable environment is an important requirement for the writing system. Typical writing parameters are an interline spacing of 400 nm to 2 µm, a focused spot size of 1.2 µm to 6 µm and a writing speed of 50 mm/s. The number of intensity pixels is currently limited to 64 k along the writing direction and in principle unlimited in the other direction.

The fabricated profile will deviate from the designed profile. The reasons for this can be separated into limitations and tolerances of the fabrication process. We will discuss the individual contributions and show how degrees of freedom in the coding process can be used to maximise the diffraction efficiency.

### 3.1. Fabrication limitations

**3.1.1. Spot size convolution.** The fabricated profile is a result of the convolution of the intensity data with the intensity profile of the writing beam. The desired profile usually has

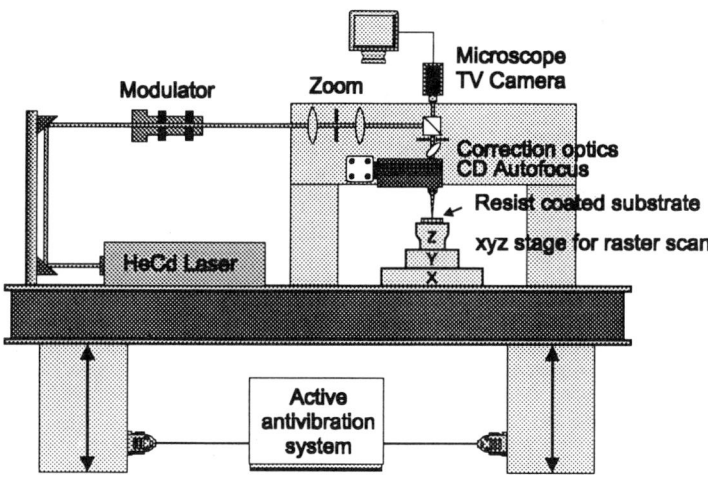

Fig. 4 Schematic of the PSI laser writing system.[12]

vertical steps at the segment boundaries. These transitions will be rounded, therefore part of the structure will be lost for the required optical function, see Fig. 3 b (dead blaze). It can be partly corrected for by deconvolution of the desired profile with the writing spot. The fraction of the dead blaze $\varepsilon$ increases strongly when the segment size decreases below a few times (5-10) the writing spot diameter. Any deviation from the ideal blaze profile leads to additional unwanted diffraction orders and a decrease of efficiency in the desired order. The resulting diffraction efficiency decreases with the square of the of the active blaze fraction.[13]

$$\eta_{Conv.} = \left(1 - \frac{\varepsilon}{\Lambda}\right)^2 \qquad (3)$$

An additional way to reduce the dead blaze fraction $\varepsilon$ is to reduce the number of segment boundaries. The use of higher phase-matching orders leads to wider and deeper segments. For a writing spot size well above the writing wavelength of 442 nm, the focus depth is a few micrometers. This allows the exposure of resist layers with a thickness of several microns and therefore the use of higher phase-matching orders. Fig. 5 a shows the

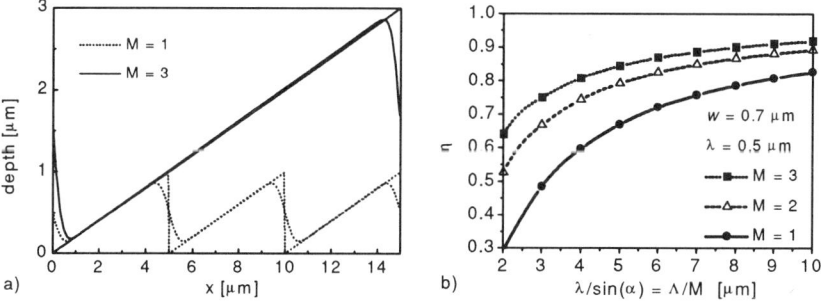

Fig. 5 The effect of spot size convolution can be decreased by using higher phase-matching orders. Desired profile and the convoluted profile for M = 1 and M = 3 (a). Expected diffraction efficiency for the convoluted profiles in dependence of the inverse deflection angle (b).

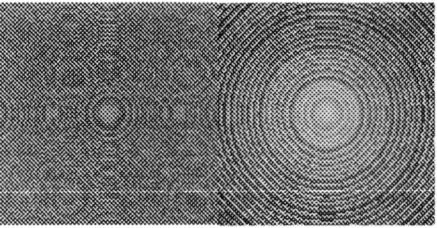 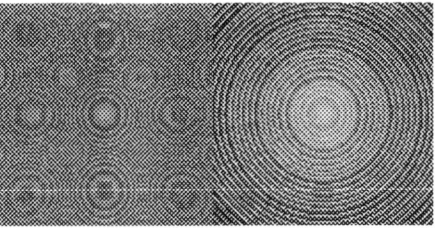

Fig. 6 Spurious lenslets may appear when a circular structure is sampled in a cartesian co-ordinate system. The exposure data bitmap for two f/1 lenses (M = 1) and (M = 1,2,3,4) for pixel sizes of 1.2 µm (left) and 400 nm respectively (right) is shown.

desired and the convoluted (writing beam waist $w = 0.7$ µm) surface profiles of a 5 µm blazed grating for two different phase-matching orders. The efficiency can be increased from 67 % to 84 % by using a phase-matching order of 3 instead of 1. The resulting efficiency gain is most pronounced for small grating periods Λ (Fig. 5 b). A drawback of using higher phase-matching orders is an increased sensitivity to depth errors[14] and the necessity to use non-standard thicker resist films.

**3.1.2. Pixel effects.** The exposure data needs to be sampled in the co-ordinate system of the scan. If the element does not have the same symmetry as the scan process, sampling errors may occur (Fig. 6). For high numerical aperture (NA) circular lenslets and an x-y raster scan, spurious lenslets will appear, eventually reducing the diffraction efficiency.[15] For the parameters used in the PSI system (pixel size of the exposure data bitmap is 400 nm), this effect is negligible for most cases. It can be further reduced by decreasing the exposure pixel size, which however leads to long writing times.

## 3.2. Fabrication tolerances

**3.2.1. Raster scan.** The dynamic position accuracy is crucial for the optical performance of the fabricated elements. Since the exposure is generated by overlapping Gaussian beams, a modulation in the profile with a grating period of the interscan distance is created.[16] The amplitude is small if the overlap of the Gaussian beams is optimal. However, if the interscan distance varies, this modulation can become significant and the resulting grating structure leads to stray light and a reduction in the optical efficiency of the element. The writing spot size for a specific element is therefore chosen as large as possible. For the current system, the dynamic positioning accuracy is approximately ± 30 nm. The fraction of the scattered light $\eta_{Scatter}$ is on the order of a few percent.

**3.2.2. Development errors.** Even if the development of the exposed resist is controlled carefully, remaining tolerances in the process give deviations from the ideal profile depth on the order of a few percent. These depth errors influence the light distribution among the possible diffraction orders. The resulting efficiency is given from Fourier analysis by [14]:

$$\eta_{Depth} = \operatorname{sin c}^2(\alpha \mu M - N) \tag{4}$$

The depth is usually controlled within a fraction of its depth and not to an absolute value. Development errors therefore can be modelled as leading to a depth $d = \mu \cdot d_0$ (Fig.

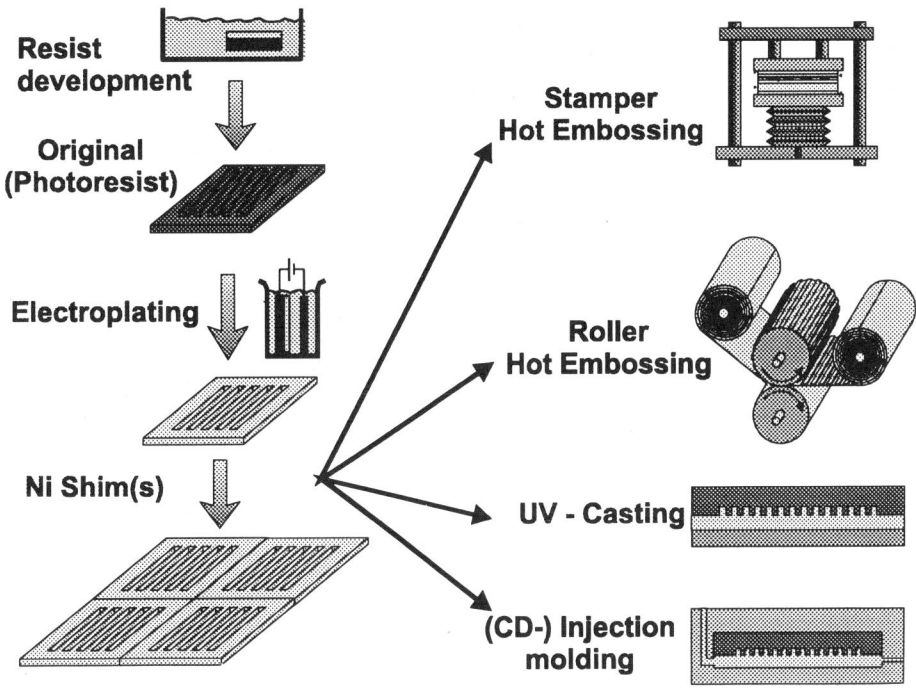

Fig. 7 Replication process: A negative copy is formed from the original structure in an electroplating process. This Ni shim can be used for various replication methods, such as hot embossing, casting or injection moulding.[17]

3 c). As can be seen from Eq. 4, the efficiency reduction due to a depth error is most pronounced with increasing depth and therefore increasing phase-matching order. For elements working in first order, a depth error of 10 % will lead to an efficiency of 98.4 % whereas for a third order element, the same depth error will reduce the efficiency to 85.8 %.

## 4. REPLICATION

The planar structure of Fresnel lens arrays is suitable for low-cost mass production by replication techniques.[17] Fig. 7 shows an overview of the major replication technologies. From the original resist structure, a Ni shim is formed by electroplating the gold-coated resist relief. From this first Ni shim, second and third generation shims may be formed. They can also be combined for hot embossing using large area, industrial roller machines, a well established technology for the replication of surface relief elements with depths up to ~1 µm. Deeper structures may also be replicated by uv-casting or injection moulding, which has demonstrated good results for elements with a minimum segment size of 5 µm and a relief depth of 5 µm.

The replication process can introduce additional surface roughness on the order of a few nanometers, which is still small compared to the roughness of the original resist profile originating from inaccuracies of the laser writing system. The replication fidelity is usually high; the main problem in the replication process is the low surface planarity of the Ni shim, which is transferred into the replicated element.

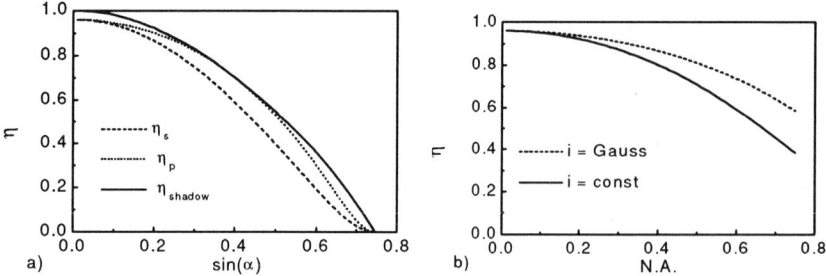

Fig. 8 Principal limitation on efficiency of an uncoated lens without considering fabrication tolerances and limitations: Local efficiencies for the two polarisations and the contribution due to the shadow effect as a function of the local deflection angle α (a). Integrated efficiency for a lens as function of its numerical aperture (NA). The resulting value depends on the illumination conditions (b).

## 5. MODELLING

For the design of an optical microsystem, it is important to get an estimation of the diffraction efficiency of an individual element before it is fabricated and tested. The expected diffraction efficiency of a DOE can be calculated by locally decomposing the DOE into blazed gratings and integrating the local diffraction efficiencies over the entire element. For a radial symmetric lens with aperture radius $a$, the overall efficiency is then given by:

$$\eta = \frac{\pi}{I_0} \int_0^a (\eta_s(r) + \eta_p(r)) \cdot i(r) \cdot r \, dr \tag{5}$$

with the integrated incident intensity

$$I_0 = \int_0^a i(r) \cdot 2\pi r \, dr \tag{6}$$

and $i(r)$ the radial intensity distribution of the incident wave. The local diffraction efficiencies $\eta_s$ and $\eta_p$ for s and p polarisation respectively, can be described by a function $f$ composed of the individual contributions:

$$\eta_{s/p}(r) = f(\eta_{Shadow}, \eta_{Fresnel}, \eta_{Conv.}, \eta_{Depth}, (1-\eta_{Scatter}), \eta_{Pixel}) \tag{7}$$

as they have been described above. Apart from the convolution and shadowing term, the individual efficiency terms will enter multiplicatively in $f$. It is important to notice that most of these contributions are strongly dependent on the local grating period which changes with the radius $r$. For a given fabrication method with fixed parameters, there is a principal limitation of achievable efficiency which is in practice lowered by the tolerances in the fabrication process. This is especially critical for elements with a high numerical aperture, employing small grating periods. The contribution due to Fresnel losses at the interfaces could in principle be accounted for by applying an antireflection coating, if the application requirements allow this additional fabrication step.

Care has to be taken in defining efficiency values for lenses, as the resulting numbers are strongly dependent on the illumination conditions. The variation of the local diffraction efficiencies with the local grating period $\Lambda$ ($\Lambda = \lambda/\sin(\alpha)$) is shown in Fig. 8a. Fig. 8b shows the expected, integrated efficiency for lenses as a function of the lens NA. Gauss illumination in Fig. 8b is defined as an Gaussian intensity distribution with 99% aperture transmission.

## 6. CONCLUSIONS

Although the use of diffractive optical elements can offer significant advantages for certain optical systems, up to now only a very few commercially available products have made use of them. Examples include aberration corrected hybrid lenses,[18] employing rather large grating periods, and beam shaping optics in laser machining systems and laser pointers. One of the main limitations in using DOEs in products is the difficulty in obtaining high efficiency and low stray light performance.

We have discussed and summarised the sources of DOE efficiency reduction from fundamental, theoretical considerations and as they are imposed by the limitations and tolerances for a given fabrication technology. The performance of the elements can be optimised by the use of certain design freedoms such as the phase-matching order. A simple and fast efficiency estimate can be made based on a ray tracing approach and scalar diffraction theory if the fabrication parameters and tolerances are known. Recent published results[5] describe elements fabricated by diamond turning that show measured efficiencies close to the theoretical maximum for an element having zone sizes as small as 13 µm working in $5^{th}$ order. The achievement of high efficiencies in many kinds of elements using different fabrication technologies remains a key requirement for the successful application of DOEs.

The authors gratefully acknowledge M.T. Gale, R.E. Kunz, H. Schütz, J. Pedersen, R. Stutz and H. Teichmann of PSI Zurich for their support in the laser beam writing, Ni Shim generation and helpful discussions.

## REFERENCES

1. H. P. Herzig, Design of refractive and diffractive micro-optics, in: *Micro-Optics - Elements, Systems, and Applications*, H. P. Herzig, ed., Taylor & Francis, London, (1997)
2. R.E. Kunz and M. Rossi, Phase-matched Fresnel elements, *Opt. Comm.* 97:6 (1993)
3. T. Hessler and R.E. Kunz, Relaxed fabrication tolerances for low Fresnel number lenses *J. Opt. Soc. Am. A*, 14, 1599 – 1606 (1997)
4. W. C. Sweatt, Describing holographic optical elements as lenses, *J. Opt. Soc. Am.* 67:803 (1977)
5. M. Rossi, C.G. Blough, D.H. Raguin, E. K. Popov and D. Maystre, Diffraction efficiency of high-NA continuous-relief diffractive lenses, in *Diffractive Optics and Micro-Optics*, Vol. 5, OSA Technical Digest Series, 233 (1996)
6. G.J. Swanson, Binary optics technology: Theoretical limitations on the diffraction efficiency of multilevel diffractive optical elements, *MIT Technical Report* 914 (1991)
7. E. Noponen, J. Turunen and A. Vasara, Electromagnetic theory and design of diffractive lens arrays, *J. Opt. Soc. Am. A* 10:434 (1993)
8. M.T. Gale, Direct writing of continuous-relief elements, in: *Micro-Optics Elements, Systems, and Applications*, H. P. Herzig, ed., Taylor & Francis, London, (1997)
9. M.T. Gale, M. Rossi, J. Pedersen and H. Schütz, Fabrication of continuous-relief micro-optical elements by direct laser writing in photoresist, *Opt. Eng.* 33:3556 (1994)
10. See the Chapter of S. H. Lee or E. Di fabrizio in these proceedings
11. C. G. Blough, S. K. Mack, R. L. Michaels, and M. Rossi, Diamond turning and replication of high-efficiency diffractive optical elements, in *Diffractive Optics and Micro-Optics*, Vol. 5, OSA Technical Digest Series, 342 (1996)
12. M. T. Gale, T. Hessler, R.E. Kunz, and H. Teichmann, Fabrication of continuous-relief micro-optics: Progress in laser writing and replication technology in *Diffractive Optics and Micro-Optics*, Vol. 5, OSA Technical Digest Series, 335 (1996)

13. T. Fujita, H. Nisihara, and J. Koyama, Blazed gratings and Fresnel lenses fabricated by electron-beam lithography, *Opt. Lett.* 7:578 (1982)
14. M. Rossi, R.E. Kunz, and H. P. Herzig, Refractive and diffractive properties of planar micro-optical elements, *Appl. Opt.* 34:5996 (1995)
15. E. Carcolé, J. Campos, I. Juvells, and S. Bosch, Diffraction efficiency of low-resolution Fresnel encoded lenses, *Appl. Opt.* 33:6741 (1994)
16. M.T. Gale and K. Knop, The fabrication of fine lens arrays by laser beam writing, *Proc. SPIE* 398:347 (1983)
17. M.T. Gale, Replication, in: *Micro-Optics - Elements, Systems, and Applications*, H. P. Herzig, ed., Taylor & Francis, London, (1997)
18. Melles Griot Dapromat™

# FABRICATION OF DIFFRACTIVE OPTICAL ELEMENTS BY ELECTRON BEAM LITHOGRAPHY

E. Di Fabrizio, L. Grella, M. Baciocchi, and M. Gentili

C.N.R. Solid State Electronics Research Institute
Via Cineto Romano 42, 00156 Rome, Italy

## 1. INTRODUCTION

Advanced optical functions, such as optical interconnects, beam splitters, fan-out elements, read-out optics, device displays, miniature arrays of microlenses and holograms can be realised by means of Diffractive Optical Elements (DOE). Many DOEs exploit micrometer and sub-micrometer surface relief patterns to modulate the phase of the incoming radiation and to perform the redistribution of the electromagnetic field arising from the DOE itself.[1] Computer generated phase profiles are directly transferred onto the optical active media by means of standard microfabrication technologies borrowed from microelectronics. Many of the surface details of a DOE are comparable in size with those of current microelectronics integrated circuits (ICs), and several technologies used in IC production can be adapted for DOE fabrication. Lithography is the first and fundamental step of any microfabrication process; by means of lithography the fine and often complex features composing a DOE are delineated. Electron Beam Lithography (EBL) is ideally suited as a pattern generation technique in the fabrication of DOE. Its intrinsic flexibility, along with its exceptional resolution, makes EBL the natural choice for the pattern generation of a wide range of DOE. In this paper, EBL is applied to the fabrication of DOE for a wavelength range spanning from the infra-red to x-rays. EBL is used both for binary shaping, and for more complex element generation. The process cal have developed allows the fabrication of nanometer size diffractive elements and, in combination with a custom electron scattering algorithm and proximity effect correction package, permits the shaping of a suitable resist in just one shot of exposure (continuous profiling). The description of the algorithm will be given, as well as examples of some of the fabricated DOEs, which include x-ray Zone Plates, beam transformers and beam splitters for infra-red and visible applications.

This Chapter discusses the application of Electron Beam Lithography to the fabrication of Difractive Optical Elements for a wide range of wavelengths. The e-beam process we have developed is used for binary patterning as well as for continuous profiling of resist

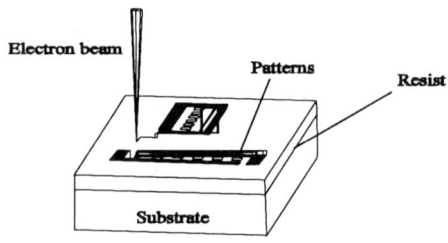

Fig.1 Principle of Electron Beam Lithography.

features for phase element definition. Three-dimensional shaping of resist features with 200 nm resolution has been achieved. The extensive use of Monte Carlo electron scattering simulations, as well as of a newly developed proximity effect correction algorithm, allows better understanding of the complex scattering processes occurring in the resist and in the substrate. A description of such effects in practical diffractive optical element patterning will be given. Examples of fabricated micro-optical elements include: Fresnel Zone Plates for x-ray focusing, blazed profile components, and beam shapers for lasers.

## 2. E-BEAM LITHOGRAPHY PROCESS SET UP

The principle of EBL is schematically described in Fig.1. A finely focused e-beam is directed onto the substrate, which has been previously coated with an e-beam sensitive film. The film is usually deposited by spinning a thin film from a liquid solution which contains the e-beam sensitive compound, the so called «resist».

Complex e-beam scattering effects, occurring in the resist itself and within the subtrate, both affect the deposition of the energy.[2] In a simplified model, the energy is transferred by scattering, from the electrons to the resist molecular bonds, which are broken (positive resist) or created (negative resist); as a consequence, the polymer mean-molecular-weight is locally changed, so that the exposed regions become more (positive-tone resist) or less (negative tone resist) soluble to the developer. It is now possible to locally remove the desired part of the resist.

The created pattern can be subsequently used to etch the optically active material beveath, or to deposit a suitable film on it (additive process). This is described schematically in Fig. 2. The function of the resist is therefore to record the information of the layout and to make it available for the subsequent etching or deposition process steps. Two pattern transfer processes have been optimised for DOE fabrication, namely metal plating and fused silica etching. Metal plating is used primarily for x-ray absorber media formation, whereas silica etching is used for UV and visible DOE phase element fabrication.

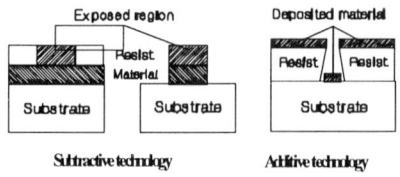

Fig.2 Technologies for pattern replica on substrate.

Table I. EBMF specifications

| Max beam deflection speed | 10 MHz |
|---|---|
| Max dimension of sample | 4 inches |
| Resolution | 35 nm |
| Max accelerating voltage | 50 kV |

All the exposures for this study were performed using a commercial e-beam lithography system, the Leica Lithography Systems EBMF 10.4 cs/120 installed at the CNR-IESS Institute of Rome. E-beam exposure was usually carried out at 50 kV with a typical beam probe currents of 500 pA; small values of beam currents are required since a small spot size is also needed to match some of the finest relief details presents on our DOEs; a typical beam-spot-size for a 50 kV exposure is about 35 nm.

After exposure, samples were developed at 20 °C in a mixture of one part of MIBK (Methyl-Iso-Butyl-Ketone) and three parts of IPA (Iso-Propyl-Alcohol). The EBMF performance characteristics are reported in Table I. Microstructure metrology was carried out by means of a calibrated SEM employing $LaB_6$ cathode.

## 3. FABRICATION OF X-RAY ZONE PLATES FOR MICROFOCUSING APPLICATIONS

It is well known that x-ray focusing is a formidable challenge, due to the fact that the refractive index of all materials is close to unity for the wavelengths involved. Refractive optics can be attempted, but difficulties in making ultra-precise and ultra-smooth multilevel x-ray mirrors make this choice particularly challenging.[3] A different way to focus x-rays is by means of Zone Plates (ZP) [4]. A ZP is essentially a circular diffraction grating made of alternating concentric rings of absorbing and transmitting materials, where the spatial modulation of the single ring follows a precise law:

$$r^2_n = n f \lambda + (n \lambda)^2/4 \qquad (1)$$

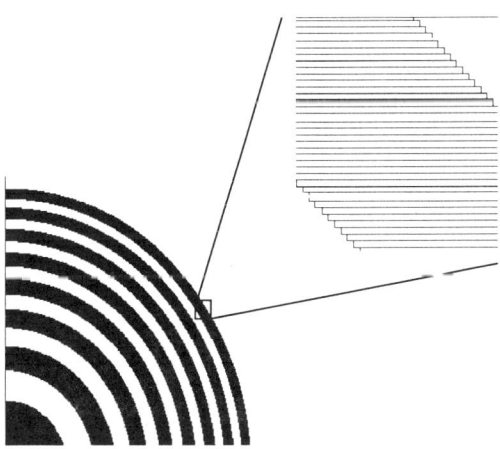

Fig.3 Fragmenting strategy for a ZP pattern.

where $r^2_n$ is the radius of the n-th zone, f is the focal length and $\lambda$ is the wavelength by which the ZP will be illuminated. Fig. 3 shows an example of part of a layout for a simple ZP; black patterns represent the absorbing media.

ZP are not ideal optics, however: they suffer from chromatic aberration and, since they are essentially diffraction gratings, also produce several diffractive focusing orders, which cause many coaxial foci to form. In addition, ZPs working in amplitude mode produce a focusing efficiency no higher than 10%. However, the highest x-ray focusing resolution obtained so far with any x-ray optics has been achieved with ZPs.

The fabrication process of ZP patterned by EBL and featuring gold absorbers for soft x-ray focusing (wavelength around 40 Å) will be discussed.

## 3.1. Fabrication of soft x-ray ZPs

**3.1.1. Software.** To prepare files for ZP patterning, one may use a CAD (Computer-Aided Design) program or custom, vertical software. The former is more flexible; however, the latter is often much faster, allows precise fine-tuning of the pattern for exposure, and also outputs files already in EBL machine binary format, thus eliminating the need for conversion. We follow both strategies, since a custom program, suitable for any 80386+ IBM-compatible PC running MS-DOS, has been developed.

Once we have defined the optical properties this ZP should have (i.e. focal length,

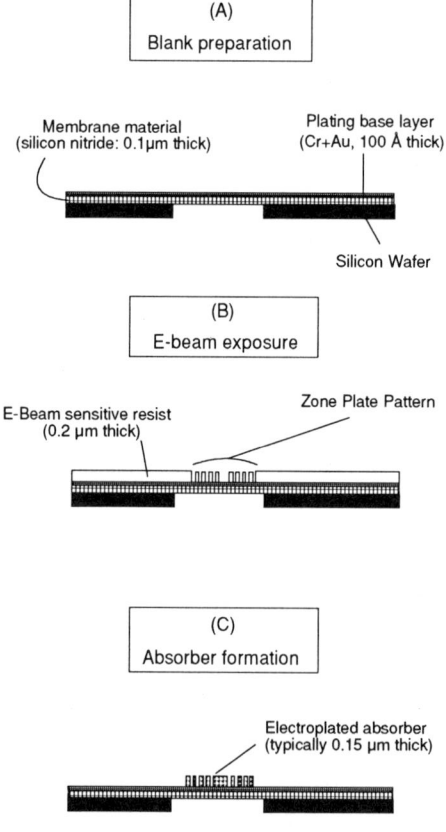

Fig.4 ZP fabrication steps.

operating wave length, zone number, etc.), the main geometrical characteristics are uniquely defined, from Eq. 1. Nevertheless, several issues arise when this curved pattern is mapped onto the orthogonal Cartesian co-ordinate space used by the E-beam deflection subsystem, and they must be taken into account.[5] The pattern must be approximated with the square «brush» defined by the EBL addressable grid cell in the same way that a computer draws a curve by using its pixels (Fig. 3). The result, if precision of the exposed pattern is an issue as expected, will be a huge file, especially in ZPs having a very high diameter/resolution ratio (i.e. a very high zone number n in Eq. 1). For a ZP having 200 µm diameter and 100 nm resolution, the typical exposure file occupies about 23 MBytes.

Fig. 4 schematically shows the procedure used for ZP fabrication.

**3.1.2. Substrate preparation.** The standard substrates used are silicon nitride (SN) membranes whose thickness is about 100 nm; these are fabricated by means of standard wet etching techniques on silicon wafers with chemically vapour deposited layers of SN. A SN membrane carrier of 100 nm thickness guarantees enough x-ray transmission at the wavelength used. The standard dimension of membrane windows depends on the desired ZP diameter, and in our case ranged from 100 to 300 µm.

Substrates are first covered with a double layer of chromium (5 nm) and gold (5 nm), so that the surface is electrically conductive as required by the subsequent electroplating process to form the absorbing layer.

**3.1.3. Resist spinning.** The resist used was Poly-Methyl-MethAcrylate (PMMA) of molecular weight of 950 K, spun on the substrate to a thickness of 200 nm. This resist thickness is adequate for the gold image transfer process, which for a good contrast amplitude ZP demands an absorber thickness of 120-150 nm.

**3.1.4. E-beam exposure.** EBL is carried out at 50 kV with beam probe currents of less than 500 pA; small values of beam currents are required since a small spot size is needed. In our experiment the writing dose was individually adjusted case by case by means of modelling and experiments and ranged 500-1200 µC/cm$^2$.

**3.1.5. Resist development and metallization.** The electrical contact necessary for electroplating metallization was defined by selectively removing the PMMA resist in O$_2$ plasma; a commercial gold-cyanide electroplating apparatus [6] was used to transfer the resist images into ZP gold rings. Current density was kept at 2.5 mA/cm$^2$ and typical growth time was one minute for 150 nm thick gold.

Due to the very thin substrate (100 nm silicon nitride), most of the incoming electrons pass through the membrane and therefore a weak effect of electron backscattering from the substrate is expected. Monte Carlo simulations [7] were carried out to evaluate this amount: only 2 % of electrons are reflected back and therefore this contribution is negligible in terms of proximity effects (PE). PE can be a limiting factor in EBL; electrons from the substrate and the resists are de-localised as a consequence of electron scattering events occurring when electrons pass into the substrate. Such a de-localisation results in a loss of image contrast since electrons deposit energy at undesired locations.[8] More important is the spreading of the electron beam passing through the resist. For the forward scattering effect (FS), a Monte Carlo calculation indicates that FS (σ value of the first gaussian term used to approximate the energy point-spread function [9]) is about 60 nm.

**3.1.6. Considerations and results.** It can be concluded that the attainable ultimate resolution in this experimental system is related to the beam spot size, the FS effect, and the accuracy of approximation of the circular features with rectangular elements. Development

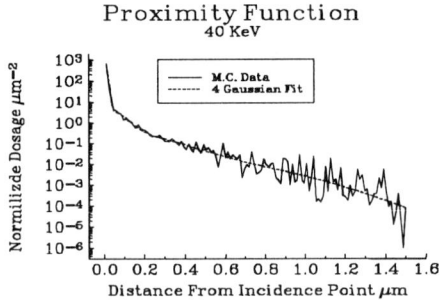

Fig. 5 MonteCarlo computed proximity function.

time, exposure dose, or very aggressive developers can significantly alter the targeted line-to-space ratio, and therefore the diffraction ZP efficiencies. Nominal line-width values must normally be smaller than the final pattern, because FS and spot size are finite; this procedure, which consists in altering (reducing) the dimension of the patterns to be written, is called «biasing». The amount of bias to be introduced for a given pattern and developer must be determined by modelling and/or experiments. If a very accurate line-to-space ratio has to be maintained along the whole diameter of the ZP, a «differential» bias and/or exposure dose must be used. This is essentially due to the «proximity effect» introduced by the overlapping of forward scattering distributions of adjacent features.

Fig. 5 shows Monte Carlo computed proximity function [10] for a dot exposure in case of ZP substrate and resist. The very «noisy» character of the large angle scattering distribution is evident: this is caused by the relatively low quantity of backscattered electrons from the substrate. This distribution extends for only 1.3-1.4 µm and is many orders of magnitude less intense than the collimated part making up the forward scattering; four gaussians have been used to model this distribution.

Fig. 6 shows the Monte Carlo computed absorbed dose for a 150 µm diameter ZP with sub-100 nm resolution and for two e-beam spot sizes: B.D. = 20 nm and 50 nm. Fig. 7 shows the computed dose to be assigned to the ZP of figure 6 after proximity effect compensation (self consistent exposure).

Deviation from 1:1 line-to-space-ratios can be exploited for ZPs working in higher diffraction orders.[11] In this case the biasing and/or proximity correction must account for a final line-width exceeding the space-width, and this ratio should be maintained along the whole ZP diameter. Fig. 8 shows a finished ZP (after gold plating and stripping) optimised for $1^{st}$ order focusing: the line-to-space ratio is 1:1 with a resolution of about 100 nm; measured efficiency for such a ZP was about 7% .[12]

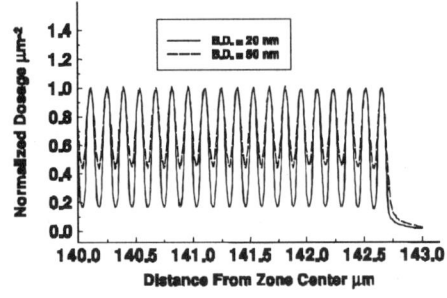

Fig. 6 Computed dose for a 150 µm diameter ZP at 40 KeV exposurew energy.

155

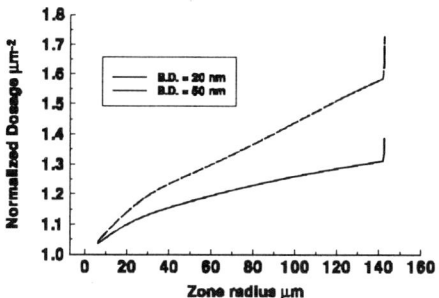

Fig. 7 Self consistent dose assignment for two different e-beam diameters at 40 KeV exposurew energy.

Resist profile is generally influenced by scattering effects and the developer; in our experimental system the extent of backscattering is negligible, and since FS is comparatively smaller than the spot size used, a substantially vertical profile should be expected. In fact, in our experiment nearly vertical profiles were always observed.

## 4. EXAMPLE OF CONTINUOUS RESIST PROFILES MADE BY A SINGLE E-BEAM EXPOSURE.

The requirement of step-like approximation of continuous surface profiles for high efficiency (up to 95 % for an eight-level approximation) DOEs, introduces multiple lithographic and etching (deposition) process steps that have to be aligned to each other at a small fraction of the minimum feature size. This multi-printing approximation technique has been previously described and has proven its feasibility for a wide class of DOEs.[13] However, the cyclic printing process requiring a number of masking levels is N (where $2^N$ is the number of steps approximating the ideal lens profile), often introduces additional defects and errors which can limit the final performance of DOEs. One alternative approach to the step approximation is the technique which shapes a suitable resist in just a single shot of

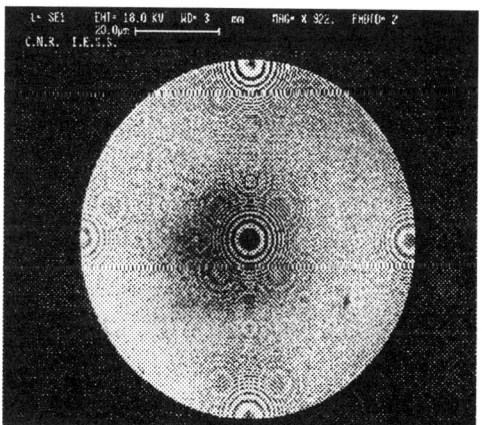

Fig. 8 Finished ZP with 100 nm resolution.

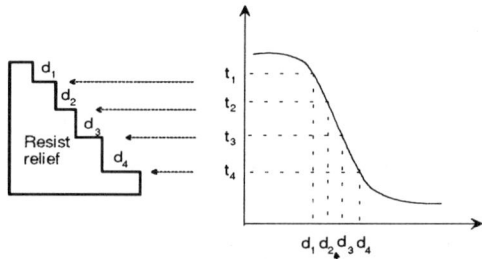

Fig. 9 Dose behaviour with contrast

exposure, using the resulting image to fabricate a durable metal copy of it for mass replication.[13] Other notable reports on resist profiling include laser techniques [14] and grey-tone optical lithography.[15] We have focused on e-beam lithography as a flexible means to achieve a high resolution 3D resist profile suitable for DOE fabrication.

The first step of the e-beam 3D profiling process is the measurement of the normalised resist thickness (NRT) vs. absorbed dosage curve. This NRT curve can be easily obtained by exposing, at increasing dosages, areas whose lateral size is larger than the electron backscattering range. Since for large areas the exposure dosage is equal to the absorbed dosage, this curve shows the experimental relationship between remaining thickness of resist (i.e. vertical value) and locally absorbed dosage. With the aid of a specially developed algorithm it is possible to control locally the resist height by reguiring any exposed area to absorb the dosage d(t) corresponding to the required height t through the NRT curve.

Fig. 9 shows that the d(t) latitude becomes wider as contrast decreases. In this condition, a low contrast developer would be an appropriate choice, but the high background etch rate may lead to vertical accuracy loss for the unexposed zones . A better trade off is to use a high contrast developer on a low contrast latent image that can be obtained by using an exposure energy greater than 40 KeV.

The second step is to self consistently expose a resist area (i.e. shape of DOE layout) with a dosage corresponding to the planned vertical value. This last constraint means the solving of the following n-dimensional linear system:

$$\mathbf{d} = \mathbf{B} * \mathbf{D} \qquad (2)$$

where n is the number of shapes contained in the electron range, **d** is the n-dimensional adsorbed dosage vector $d(t_t)$ corresponding to the planned vertical values through the NRT curve, and **D** is the n-dimensional vector corresponding to the delivered dosages during the exposure time. The matrix **B** is made up of the elements $B_{ij}$

$$\int\int_{A_j A_i} f_p(\mathbf{r}_{ij}) dA_j dA_i = B_{ij}, \qquad (3)$$

where $A_i$ is the area of the i-th elementary structure interacting with the j-th structure and $f_p$ is the proximity function; a typical mathematical expression of $f_p$ is a linear combination of Gaussian functions:

$$f_p(r) = \sum_{k=1}^{k=k_{max}} b_k \exp\left(\frac{r^2}{\sigma_k^2}\right), \qquad (4)$$

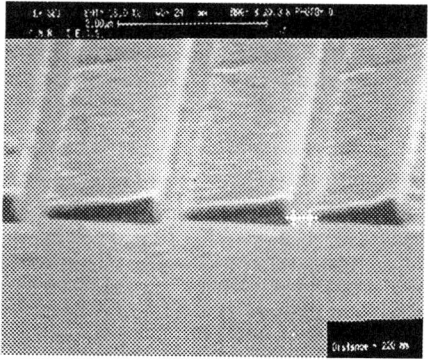

Fig. 10 Detail of eight-levels PMMA structure on $Si_3N_4$ substrate.

where the parameters $b_k$ and $\sigma_k$ are related to the scattering process as explained elsewhere.[9]

We point out that unlike conventional proximity effect corrections, not all $d(t_t)$ terms are equal to the same threshold dosage.

### 4.1. Semi-continuous zone PMMA profiles with 200 nm resolution

Continuous profiling by accurate compensation of e-beam scattering by dosage modulation can generate high-resolution blazed profiles in PMMA.

In order to demonstrate this challenging concept, a 2 μm thick silicon nitride membrane was coated with 0.5 μm of PMMA and exposed at 50 kV. In Fig. 9 we described schematically how the dose is assigned to each zone.

Fig.s 10 and 11 show a detail of an eight-level PMMA structure achieved on $Si_3N_4$ substrate, where the gap between two adjacent zones is controlled to better than 200 nm.

The lithographic resolution of the resist relief is related to the number of discretization levels for which the DOE is designed. In the present case, 200 nm resolution is determined by the eight-level structure and the short wavelength of the radiation the DOE is designed for.

### 4.2. Beam shaper for applications at λ = 0.633 μm

Generally, optical devices known as *beam shaper* realise the optical function that transforms a beam with a particular intensity into another with a different energy distribution over a different area.

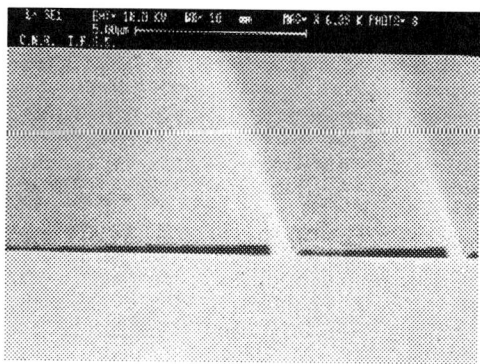

Fig. 11 Detail of eight-levels PMMA structure on $Si_3N_4$ substrate.

In this Chapter we focus attention on the fabrication of a beam shaper that transforms *a gaussian beam into a rectangular one*. The beam shaper can be seen as a hologram that causes a phase change of the incoming beam, in a such a way that the outgoing beam has, in a desired focal plane, a uniform energy distribution.

The hologram phase function is obtained by ray tracing calculations.[16] In this case there is an analytical expression for the phase function in the plane of the beam shaper:

$$\phi(u,v) = \varphi_x(u) + \varphi_y(v) \tag{5}$$

where:

$$\varphi_x(u) = -\frac{k \cdot u^2}{2 \cdot f} - \frac{k \cdot a}{f \cdot E_a} \left\{ u \cdot \text{erf}\left(\frac{u \cdot \sqrt{2}}{\sigma}\right) - \frac{\sigma}{\sqrt{2\pi}} \left[ 1 - \exp\left(\frac{2 \cdot u^2}{\sigma^2}\right) \right] \right\} \tag{6}$$

and a similar expression holds for $\varphi_y(v)$. In these formulas the (u,v) co-ordinates are referred to the beam shaper plane: $k = 2\pi/\lambda$ is the wave vector of the laser beam radiation; $\lambda$ is the wavelength of the laser; f is the focal length of the beam shaper; a is the aperture of the beam shaper in the x-direction; and, $\sigma$ is the laser beam diameter. Once the function $\phi(u,v)$ is known, the next step is to reduce it to the interval $(0,2\pi)$. This is the general procedure for approximating the phase function with a multilevel structure.

In our case we have approximated the phase function with an eight-level staircase whose theoretical efficiency is about 95%.

In order to demonstrate the flexibility of our process, an eight-level DOE for laser beam shaping at $\lambda = 0.633$ µm wavelength was fabricated. A standard silicon nitride membrane was employed as a substrate; resist thickness was 0.5 µm. Fig. 12 shows an optical micrograph of the fabricated device; minimum feature size is 2 µm and entire size is 3.2 x 3.2 mm².

## 5. DOEs ETCHED INTO FUSED SILICA LAYERS

Fused silica is widely used as an active optical layer for grating structures, Computer Generated Holograms (CGH), etc. Many of these applications demand phase encoding which is normally realised by fine structuring the surface of the silica plate.

This process involves patterning of an etch-resistant material followed by its precise transfer into the bulk silica by Reactive Ion Etching (RIE). EBL can be used to pattern the DOE but has a substantial drawback: a charge is built up when the e-beam is addressed to an

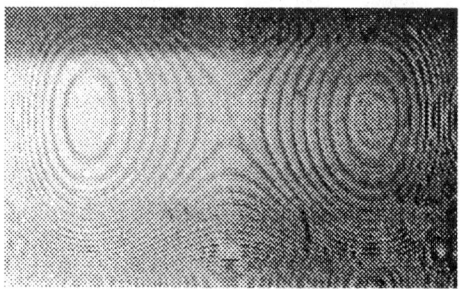

Fig.12 Optical micrograph of the eight-level beam shaper device.

Table II. Etching characteristics for fused silica layers.

| $SF_6$ (sccm) | $CF_4$ (sccm) | $O_2$ (sccm) | Power (W) | Etch rate (nm/min) | Pressure ($10^{-3}$ Torr) |
|---|---|---|---|---|---|
| 1.69 | 8.7 | 2 | 275 | 90 | 25 |

insulating substrate. The charge locally distorts the pattern and also alters its placement. In order to solve this problem we have developed a technique which uses a sacrificial Cr layer. The conductive Cr layer inhibits the process of charge build-up during writing.

Cr is first deposited to a thickness of about 500 Å on the surface of the fused silica plate by thermal evaporation. Resist is then applied and patterned by EBL. After development the chromium layer is selectively removed in the resist open region by a wet etching process based on a Cerium-Ammonium-Nitrate solution. After having carried out the chromium etching, the resulting pattern is used as a RIE durable etch stop mask, to transfer into the silica layer by a process. The RIE optimised process is based on a mixture of $SF_6$, $CF_4$ and $O_2$; main etching characteristics are summarised in Table II.

Fig. 13 shows a detail of a submicrometer grating etched into fused silica; the roughness visible in the cross section is due to the gold layer sputter coated to prevent charging during SEM inspection.

## 6. CONCLUSIONS

A wide range of diffractive optical elements have been fabricated by means of electron beam lithography. A custom e-beam scattering modelling algorithm, vertical software suited for ZP generation, and custom 3-D proximity effect correction software were successfully developed and used for patterning of practical DOEs, including *nanometer resolution Fresnel Zone Plates for x-ray focusing* and *continuous resist profiling*. The suitability of electron beam lithography in DOE fabrication is confirmed by the variety of fabricated optical devices ranging from the infra-red to the x-ray regions.

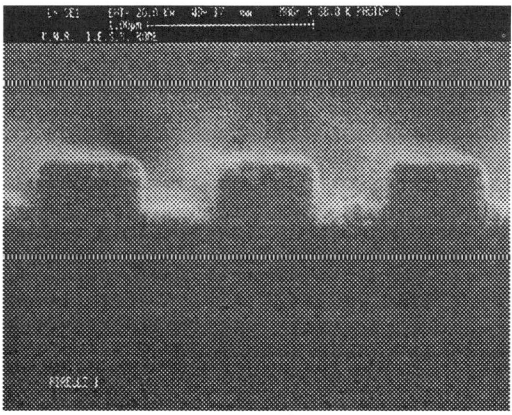

Fig. 13 Detail of a submicrometer grating etched into fused silica.

# REFERENCES

1. G.J. Swanson, Binary Optics Technology: The theory and design of multi-level diffractive optical elements, *MIT Lincoln Laboratory Technical Report* # 854 (1989).
2. M. Parikh and D. F. Kayser, Energy deposition function in electron resist films on substrates, J. App. Phys. (2), 1979, 1104
3. A.E. Rosenbluth and P.Lee, Bragg Condition in Absorbing x-ray Multilayers *App. Phys. Lett.*, 40, pp. 466-468, 1982.
4. A.G. Michette, Optical Systems for Soft X-rays, Plenum, New York, 1986.
5. D. Progue, H. Rothuizen, F. Vasey and P. Vettiger, Enhanced e-beam System for the Fabrication of Optical Elements, *Microelectronic Engineering* 27 (1995) 163.
6. M. Gentili, A. Lucchesini, P. Lugli, G. Messina, A. Paoletti, S. Santangelo, A. Tucciarone and G. Petrocco, Electron Scattering Effects in master mask fabrication by single layer process for submicron XRL, *J. Vac. Scie. Technol.* B **7** (1989) 1586.
7. M. Gentili, A. Lucchesini, L. Grella, L. Luciani, L. Mastrogiacomo and P. Musumeci, Energy Density Function determination in a very High Electron Beam Lithography, *J. Vac. Scie. Technol.* B 8 (1990), 1867.
8. N. Aizaki, Proximity effect dependence on substrate material, *J. Vac. Scie. Technol.*, 16 (6), Nov/Dec 1979.
9. S. J. Wind, M. G. Rosenfield, G. Pepper, W.W. Molzen and P. D. Gerber, Proximity correction for electron beam lithography using a three gaussian model of the electron energy distribution, *J. Vac. Scie. Technol.* B **7** (6), Nov/Dec 1989.
10. L. Grella, E. Di Fabrizio, M. Gentili, M. Baciocchi and R. Maggiora, Proximity correction for e-beam patterned sub-500nm diffractive optical elements, *Proceedings of the Micro and Nano Circuit Engineering Conference*, 22-25 September 1996, Glasgow U.K.
11. D. Morris, *Ph.D. Thesis*, University of London, June 1991, p. 43.
12. M. Kinskinova, Sincrotrone Trieste, Private communication.
13. M. T. Gale, Replication, in *Micro-Optics*, H.P. Herzig Editor, Taylor and Francis, Chapter 6, 1997.
14. M. T. Gale, Direct Writing of continuous-relief micro optics in *Micro-Optics*, H.P. Herzig Editor, Taylor and Francis, Chapter 6, 1997.
15. K. Reimer, W. Henke, H. J. Quenzer, W. Pilz and B. Wagner, One level grey tone design, *Microcircuit Engineering* 30, 569, 1996.
16. M.R. Duparre, B. Ludge, R.M. Kowarschik, M.A. Golub, E.B. Kley, and H.J. Fuchs, Investigation of computer-generated diffractive beam shapers for the task of Gauss-to-ring laser beam transformation, *SPIE* Vol. 2689, pp. 112-119, *1996*.

# PULSED LASER DEPOSITION: PERSPECTIVES AS A MICRO-OPTICS FABRICATION TECHNIQUE

S. Martellucci, M. Richetta, A. Tebano, and A. Spena

The University of Rome "Tor Vergata"
Via di Tor Vergata snc, 00133 Roma, Italy

## 1. INTRODUCTION

The PLD technique possesses many unique properties that have led, up to now, to fabrication of excellent quality films of a variety of materials. In these cases the great advantage of the method is strongly connected to its stoichiometric target reproduction. Recently PLD has attracted much effort in the deposition of Diamond Like Carbon (DLC) films. The growing interest in DLC is essentially due to the peculiar properties of this material which make it a very attractive choice in the microtechnology field.

In this Chapter we show that thin films of Diamond Like Carbon can be grown by means of the Pulsed Laser Deposition (PLD) technique, using an XeCl excimer laser. Their characterisation can be carried out by SEM, Raman Spectroscopy and Microindenter. As these materials could be applied in microsystems technology, we started a study on the etching mechanism of DLC films. In the following we will present first a brief overview on what the PLD technique means and on the main characteristics of DLC films (Sections 2 and 3). Afterwards, in Section 4, will be presented the experimental apparatus employed to grow hydrogenated and hydrogen-free carbon films and the technique applied to characterise them. In Section 5 we will analyse the effect of excimer laser irradiation on films, to study conditions for device patterning. Finally, Section 6 will be devoted to concluding remarks.

## 2. PULSED LASER DEPOSITION TECHNIQUE

Both conceptually and experimentally PLD is extremely simple, probably one of the simplest among all thin film growth techniques.[1] It is based on the interaction between a laser beam, with wavelength ranging from 193 nm to 1064 nm, and the target material. In particular, in the case of thin film growth, the useful wavelength interval lies between 200 and 400 nm. Over this range many materials used in deposition exhibit strong absorption,

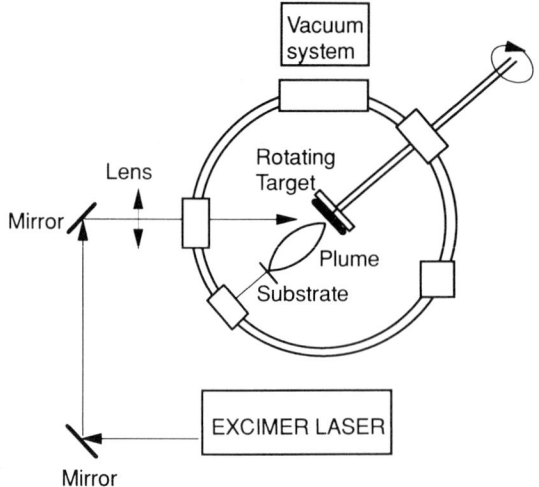

Fig. 1 Experimental set-up of the PLD apparatus for film deposition.

since their absorption coefficients tend to increase as one moves towards the shortest wavelength. This fact causes a reduction of the penetration depth into the target material and the possibility to more easily ablate thinner layers of surface. The stronger absorption at the shortest wavelength also results in a decrease of the ablation fluence threshold ($\lambda = 200$ nm is the lower limit imposed by absorption band of molecular oxygen).

From the reasons just mentioned, it has turned out that excimer lasers represent a favourite choice for PLD work. They emit radiation directly in the UV range and deliver high outputs, typically 200 mJ/pulse. They can also achieve high repetition rate (up to several hundred Hertz).

In Fig. 1 is shown a schematic diagram of a typical experimental set-up. It consists of: a target holder, which can rotate about its main axis during the irradiation, and a substrate holder, both housed in a vacuum chamber; an external laser used as power source to vaporise material and deposit thin films; and finally, a set of optical components to focus the laser beam over the target surface.

The decoupling of the vacuum hardware from the evaporation power source makes this technique so flexible that it is easily adaptable to different operational modes, without the constraints imposed by the use of internally powered evaporation sources.

Film growth can be carried out in an environment containing any kind of gas, with or without plasma excitation. It can also be conducted in conjunction with other types of evaporation sources in a hybrid approach. In contrast to the hardware simplicity, the laser target interaction is quite a complex physical phenomenon. Theoretical descriptions are multidisciplinary and combine equilibrium and non-equilibrium processes.

The mechanism that leads to material ablation depends on laser characteristics (wavelength, fluence, spot size, etc.) as well as on optical, topological and thermodynamical properties of the target. In the ablation process one can distinguish three main steps:

a) interaction of the laser beam with the target material, which causes "evaporation" of the surface layers;

b) interaction of the emitted material with incident radiation and subsequent formation of a plasma; and,

c) anisotropic plasma expansion followed by deposition of ablated materials.

As is shown by the work that has been reported, the PLD method offers many advantages, but also disadvantages. Among the former we can count simplicity and flexibility, due essentially to the fact that laser beams are easier to transport and manipulate, and allow decoupling of the vacuum chamber from the power source. PLD also offers the opportunity to grow multilayer and artificial structures by positioning different targets under the laser beam. This can be accomplished by mounting a multitarget rotating holder inside the chamber. Compared with other sputtering processes, it is rather clean, thanks to the confinement of the laser-target interaction region. All these properties also give PLD a good degree of reproducibility. In any case, its key feature is the possibility to reproduce the stoichiometric structure of the target material, under appropriate conditions.

Despite these advantages, it is essential to remind ourselves that a small deposition area and the presence of droplets are considered the two major drawbacks of PLD. A narrow angular distribution of the plume causes lack of uniformity over large areas. However, one can bypass this problem, or at least reduce its effects, by rastering the substrate, via rotation or translation. It could also be possible to raster the laser beam, but in this way off-axis evaporants could strike the substrate. On the other hand, splashing is an intrinsic problem, therefore more difficult to solve. The origin of particulates is still a subject of investigation. Possible causes of droplets include target inclusion, energy density fluctuations and target texturing. Two main approaches are applied to reduce the effects of splashing, working in different directions. The first approach tends to minimise the production of particulates, while the second tends to stop droplets from reaching the substrate.

## 3. DIAMOND LIKE CARBON FILMS

The term DLC, or amorphous Carbon (a-C), was introduced to emphasise the possibility of obtaining hard, transparent, amorphous carbon films with properties close to diamond.[2] DLC is not a well defined material because its properties can change over a wide range depending on the $sp^3$ bonds content. DLC is an amorphous aggregation state of carbon with very interesting properties varying between those of graphite and those of diamond. DLC films therefore strongly depend on their structure and their properties are essentially governed by the ratio $sp^3$ over $sp^2$ bonds, which in turn is controlled by the deposition process. The tribological, electrical and optical properties of amorphous carbon diamond-like thin films, such as chemical resistance, thermal conductivity, mechanical hardness, electrical insulation, optical transparency, low friction coefficient, good adherence to the substrate, and low roughness, make this material very unusual and attractive for technological applications. The use of diamond-like carbon thin films as a mechanical protective coating material is well known.[3,4] The hard mechanical properties arise due to the $sp^3$ hybridised tetrahedral bonds between the C-C atoms.

Unlike other Group IV amorphous materials, which only form other tetrahedrally bonded structures (a-Si, a-Ge), a-C can form tetrahedral $sp^3$, trigonal $sp^2$ and linear sp hybridised bonds. DLC has very wide electromagnetic radiation transparency, ranging from IR to part of the visible region, which can be very useful for optical systems. The thermal conductivity, several times higher than copper, combined with its electrical insulating property makes DLC an ideal heat sink for high power semiconductors. Although diamond films have properties sometimes more desirable than DLC films their compatibility with other solid state devices or materials is often limited as a result of the high temperature processing involved.

Among the several techniques actually available for DLC thin film deposition (Ion Beam Deposition, Sputtering, Chemical Vapour Deposition, etc.), PLD emerges as one of the more promising. The PLD technique offers the significant advantage of very low substrate temperature deposition: typical values are around room temperature. A strong dependence of film properties on instantaneous laser power density (fluence) and wavelength is typically reported in a number of works [5-9]. A comparison of different DLC studies shows that there is a wavelength/fluence region for PLD parameters, where the formation of DLC occurs. For the deposition of DLC it is important to achieve a certain energy level, several 100 eV, of ablated carbon species. This level can be achieved with lower fluences, using a shorter wavelength laser, due to the reciprocal dependence of energy absorbence on laser wavelength. These findings are in good agreement with energy driving models of DLC formation.[10] These models predict deposition of DLC films with high percentage of $sp^3$ bonds at kinetic energies of carbon atoms higher than 100 eV. Several works evaluate the introduction of hydrogen into the chamber to deposit hydrogenated a-C films at pressures of around 1 Torr and substrate temperature below 100 °C.[11]

## 4. EXPERIMENTAL SET-UP AND CHARACTERISATION OF FILMS

The radiation produced by an XeCl excimer laser (Lambda Physik Compex 102) with wavelength of 309 nm was focused by a 40 mm focal length lens on a slowly rotating target. The pulse energy was 170 mJ, with 10 ns pulse duration and repetition rate of 10 Hz. The laser power density on the target was around $5 \times 10^8$ W/cm$^2$. The target was a high purity pyrolitic graphite pellet (99.99% purity). The films were deposited onto Si (100) boron doped substrates (1 cm$^2$ surface area). All of the substrates were ultrasonically cleaned in acetone, air dried at room temperature (R.T.), glued by silver paint onto a resistively heated holder and then placed into the stainless steel vacuum chamber at 5 cm distance from the rotating target. The chamber is evacuated to a base pressure of $10^{-5}$ Torr prior to the deposition, and to de-gas the substrate the temperature is set for 20 min at 500 °C. The substrate temperature during deposition was kept constant between R.T. and 700 °C. For the a-C:H film deposition a 2 Torr hydrogen pressure was used.

The morphology of the films was studied by Scanning Electron Microscopy (Leica Cambridge 260). The grown DLC films were very homogeneous. The surface for a-C and a-C:H films was very smooth. The surface roughness was estimated to be less than 0.1 µm. The presence of some particles was due to the laser ablation process and is typical of materials deposition using PLD. This phenomenon is more evident for materials deposited using longer laser-wavelengths, because melted particles are ejected from the target directly to the substrate. In our case the presence of these particles does not strongly affect the film morphology. By the analysis of the film cross sections it was possible to measure quite accurately a thickness between 100 nm and 1 µm, corresponding to a growth rate of 0.8 Å/shot. The growth rate is directly dependent on the laser beam focusing.

The films as deposited were highly resistant to scratching attempts. The mechanical hardness for a-C sample was determined by Vickers method utilising a Microindenter. We analysed several samples grown in vacuum and hydrogen atmosphere. The hydrogen-free films showed a much higher hardness than the hydrogenated ones for all the deposition temperatures. The substrate temperature was an important parameter in determining the hardness of the hydrogen-free films. The film hardness starting from a value of 1300 HV (kg$_f$/mm$^2$) for deposition temperature of 25 °C increased to 3200 HV (kg$_f$/mm$^2$) for the films deposited at 60 °C and then decreased to 1400 HV (kg$_f$/mm$^2$) for the films deposited at 300 °C.

By Raman spectroscopy it was possible to study the structure of the DLC films. This is one of the most common techniques to evaluate DLC film quality, by studying the variations in the different types of carbon bond by analysis of the Raman spectra. The Raman spectra show a strong dependence on the $sp^2$ component and can be used to characterise the

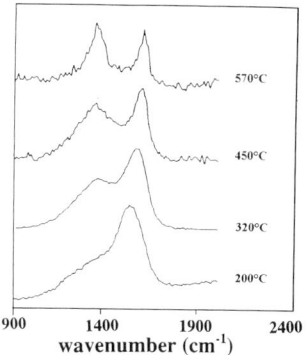

Fig. 2 Raman spectra of the film grow in $H_2$ ambient at different deposition temperature.

varying "diamond-like" quality of the films. Thus Raman spectroscopy is a powerful technique for studying amorphous-diamond phases together with the formation of other phases in a carbon film [3, 12, 13]. The Raman spectrum of a large single crystal of graphite consists of a sharp peak (G peak) located at about 1580 cm$^{-1}$. The Raman spectrum of DLC films consists of a broad peak centred at about 1500 - 1550 cm$^{-1}$. An asymmetric broadening of the Raman peak and the appearance of a shoulder in the 1350 cm$^{-1}$ region (D peak) is due to an increase of the sp$^2$ bonds. The intensity ratio $I_D / I_G$ is correlated with the degree of graphitisation of the sample. The Raman study showed that the deposition temperature strongly influences the film quality, especially for the hydrogenated films. The Raman spectra (Fig. 2) show the growth of the D peak and the narrowing of D and G lines as the deposition temperature increases

The Raman spectra were fitted by a Lorenzian (D peak), a Breit Wigner Fano (G peak) and a linear background. The ratio $I_D / I_G$ obtained from the fits shows a trend with increasing deposition temperature that is related to an increase of the graphitic component in the film. In Fig. 3 we report the spectrum of the hydrogen-free sample deposited at 60 °C before and after laser irradiation.

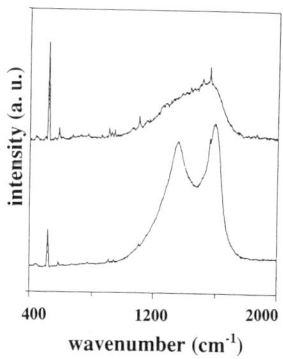

Fig. 3 Raman spectra of unhydrogenated film before and after laser irradiation.

## 5. PATTERNING OF DLC FILMS

It is possible to imagine broad applications of DLC for micromechanical devices such as accelerometers and pressure sensor structures, as have been already realised with polycrystalline diamond. DLC is one of the most interesting and promising materials for microsystem applications.

Several technologies are now available for patterning and machining of DLC. Conventional patterning is based on resist selected irradiation and etching processes, such as wet-chemical, reactive-ion plasma-assisted methods. One of the simplest and more effective, methods is laser ablation. Excimer lasers are used. For the excimer lasers, each UV laser photon possesses sufficient energy to break the bonds responsible for holding together the material. In comparison, the longer wavelength infrared Nd:YAG and $CO_2$ laser photon require the absorption of many photons before molecular dissociation can occur as the result of thermal processes. The absorption of the infrared photons excites the materials to successively higher vibrational levels which then causes melting and vaporisation. This produces the so called "heat-affected zone" in the surrounding area. On the other hand, the higher energy UV photons directly break the material bonds; this process of material removal is the photoablation process. Because this bond breaking and material ejection takes place so rapidly, there is little heat-transfer to the rest of the material and no heat-affected zone outside of the ablated area is observed. Due to this mechanism excimer lasers can be used to very precisely pattern the material. However, UV excimer lasers offer an alternative route by providing a resist-free, simple patterning process, that can be utilised either in direct-write mode or by projecting the laser light via a lens or a mask, in order to improve throughput performance. In particular the different varieties of amorphous diamond can be usually obtained with band gap energies corresponding to more than 500 nm wavelength: this is longer than the emission wavelength of excimer lasers, commonly operating in the UV region between 193 and 350 nm. Thus enhanced ablation can be obtained, via photodissociation of the carbon bond network.

In order to implement this technique under optimised conditions, we studied the effect of laser irradiation amorphous diamond. The excimer laser exposure experiments were carried out, in air, with the same XeCl laser used for the film deposition. The apparatus used for DLC film patterning consist of an attenuator, a mirror, a focusing lens and an in-plane movable sample holder. The attenuator was necessary to control the fluence that must be much lower than that usually used for DLC film deposition. The fluence was between 0.5 and 13 $J/cm^2$.

In Fig. 3 are shown the spectra of a hydrogen-free sample (a) before and (b) after irradiation performed at a fluence of 6.5 $J/cm^2$ by one laser pulse. In the Raman spectrum of the virgin material it is possible to distinguish only the broad asymmetric band typical of a-C films. The Raman spectrum of the irradiated film shows: a) two distinct peaks at 1335 $cm^{-1}$ and 1580 $cm^{-1}$; and, b) a strong decrease of the sharp Si – line at 520 cm-1. This last feature is evidently due to an increase in the absorption coefficient of the irradiated film. With this fluence it was possible to measure an etch rate of about 400 nm/pulse.

Thus Raman spectra confirm that the laser etching of DLC is a two-step process. In the first step ablation creates a graphite layer on top of the film. After this highly absorbing layer has been formed the etching may be characterised by the dependence of ablation rate on the laser fluence. We observe a strong dependence of the etch-rate on the fluence: the etch rate increases more than linearly with laser fluence. For hydrogen-free films we found at 0.5 $J/cm^2$ a fluence threshold for the etching mechanism. To compare the etch-rate of hydrogenated films we irradiated at 0.5 $J/cm^2$ a film grown in hydrogen; in this case we found an etch-rate of about 30 nm/pulse.

## 6. CONCLUSIONS

We studied the deposition by PLD of hydrogenated and unhydrogenated amorphous carbon films. Film quality, investigated by Raman and hardness measurements, is better for films grown in vacuum at deposition temperatures around 60 °C. We preliminary investigations of excimer laser patterning in air. The modification of the Raman spectra after irradiation is explained in terms of a double step process: a graphitisation followed by removal of the modified layer. Material removal was strongly dependent on laser fluence and on film quality. We observe the existence of a threshold fluence for etching mechanism which depends on film quality.

## REFERENCES

1. *Pulsed Laser Deposition of Thin Films*, Eds.: D.B. Chrisey and G.K. Hubler, John Wiley & sons, Inc., New York 1994
2. S. Aisenberg and R. Chabot, J. Ion-Beam Deposition of Thin Films of Diamondlike Carbon, *J. Appl. Phys.*, 42 (1971) 2953
3. H. Tsai and D. B. Bogy, Characterization of Diamond like Carbon Films and their application as overcoats on thin film media for magnetic recording, *J. Vac. Sci. Technol.* A, 5 (1987) 3278
4. J. Robertson, Amorphous Carbon, *Adv. Phys.*, 35 (1986) 317
5. T. Sato, S. Furuno, S. Iguchi and M. Hanabusa, Deposition of Diamond-like Carbon Films by Pulsed-Laser Evaporation, *Jpn. J. Appl. Phys.* 26, L1487 (1987)
6. T. Sato, S. Furuno, S. Iguchi and M. Hanabusa, Deposition of Diamond-like Carbon Films by Pulsed-Laser Evaporation, *Appl. Phys.* A 45, 355 (1988)
7. F. Davanloo, E. M. Juengermann, D. R. Jander, T. J. Lee and C. B. Collins, Amorphic diamond films produced by a laser plasma source, *J. Appl. Phys.* 66, 2081 (1990)
8. J. J. Cuomo, D. L. Pappas, J. Bruley, J. P. Doyle and K. L. Saenger, Vapor deposition processes for amorphous carbon films with $sp^3$ fractions approaching diamond, *J. Appl. Phys.* 70, 1706 (1991)
9. D. L. Pappas K. L. Saenger, J. Bruley, W. Krakow, J. J. Cuomo, T. Gu and R. W. Collins, Pulsed laser deposition of diamond-like carbon films, *J. Appl. Phys.* 71, 5675 (1992)
10. J. Robertson, The deposition mechanism of diamond-like a-C and a-C:H, *Diamond Relat. Mater.* 3, 361 (1994)
11. A. P. Malshe, S. M. Kanetar, S. B. Ogale and S. T. Kshirsagar, Pulsed laser deposition of diamondlike hydrogenated amorphous carbon films, *J.Appl.Phys.* 68, 5648 (1990)
12. Y. Lifshitz, G. D. Lempert, S. Rotter, I. Avigal, C. Uzan-Saguy and R. Kalish, The influence of substrate temperature during ion beam deposition on the diamond-like or graphic nature of carbon films, *Diamond Relat. Mater.*, 2 (1993) 285
14. S. Prawer, K. W. Nugent ,Y. Lifshitz, G. D. Lempert, E. Grossman, R. Kalish and Y. Avigal, Systematic variation of the Raman spectra of DCL films as a function of $sp^2$:$sp^3$ composition, *Diamond Relat. Mater.*, 3 - 5 (1996) 433.

# FABRICATION OF THIN-FILM MICROLENS ARRAYS BY MASK-SHADED VACUUM DEPOSITION

R. Grunwald, S. Woggon, and R. Ehlert

Gesellschaft zur Förderung angewandter Optik, Optoelektronik
Quantenelektronik und Spektroskopie e.V. (GOS)
Rudower Chaussee 5, 12489 Berlin, Germany

## 1. INTRODUCTION

Refractive microoptical arrays are of interest for many applications in photonics, laser beam shaping and measuring techniques. In comparison to diffractive optical elements, the minimum structure sizes are much larger and broadband transmission can be realized. The use of thin-film microstructures opens new prospects because of their specific properties. Compact and flexible systems can be designed (e.g. microlenses on thin flexible substrates or fiber optics). A wide range of focal lengths and depths of foci can be reached. Adhesive layers with high thermal and chemical stability and laser damage resistance can be fabricated. Additional spectral and spatial selectivity can be added by using multilayers (partial reflectance, apodization, dichroic properties, envelope phase and amplitude functions). When the optical efficiency has to be maximized (image sensing, optical pre-processing, homogenizers), high geometrical fill factors and low aberrations are necessary. For the collimation of anisotropic sources such as laser diode arrays and for light concentration on rectangular sensor pixels, anamorphic microlenses are needed. In all cases, the main problem of design and fabrication of thin-film microlenses is the simultaneous optimization of all critical parameters (focal length, fill factor, lens diameter, profile function, number of elements, array area).

Microoptical structures in optical materials can be formed by subtractive, modifying or additive processes (photolithography,[1] holography,[2-3] vapor deposition[4-7]) which deliver deepened or elevated reliefs or a periodically changed refractive index. Different types of processes can also be used stepwise (thin-film deposition followed by etching[8-9]). *High geometrical fill factors* can be obtained with *crossed linear interaction zones*. For this purpose, sequential line scans of direct writing laser spots or simultaneously written arrays of linear laser distributions (formed with grey-tone masks) are applied in two or more angular steps (excimer laser ablation of polymers[10]). Most of the subtractive processes exhibit nonlinear characteristics. The deposition of debris must be avoided and long-term beam quality must be controlled with high accuracy. Crossed exposure with linear interference patterns has been used to write hexagonal structures with small pitches (down to the range of a few wavelengths) into a recording medium [2-3]. In this case, shape function and statistics are strongly influenced by the beam homogeneity, process nonlinearities and misalignments. A direct additive process can be realized by shading the vapor beam with periodic diaphragms. Thin-film waveguide Luneburg lenses have been sputtered with extended sources and 3D shading masks [4-6]. Graded reflectance dielectric mirrors [12,13], micro-mirror arrays [14,15] as well as microlens arrays have been fabricated by the authors with mask-shaded vacuum deposition using a rotating system and a point-like vapor source. Two different variants of this technique (hole mask shading [16-18]; crossed deposition of cylindrical lenses with slit mask shading [19-21]) have been investigated. Recent results will be presented and both methods will be compared here.

## 2. FABRICATION METHODS

### 2.1. Mask shading in rotating systems

In a vacuum deposition system with planetary rotation of the substrate, 3D-diaphragms fixed directly on or in proximity to the substrate can be used to form thin layers of transparent materials with lenslike properties by a defined shading of the vapor beam. The geometrical conditions (substrate movement, position and angular characteristics of the beam vaporizer, distance, depth profile, tilt and orientation of the masks) have been simulated with a computer program. The trajectories of the surface points of the rotating and evolving substrate can be described by hypotrochoidal curves. The angular distribution of the vapour beam has been measured to be a $\cos^k$-distribution with k = 0.95. The calculated total durations of deposition and shading for each point of the substrate deliver the theoretical thickness distributions and have been used to optimize the depth profiles of the masks. Two different fabrication methods have been tested experimentally: (a) 1-step-deposition through *hole array masks*, (b) multi-step-deposition through *slit masks*.

### 2.2. Shading with hole array masks (one-step-method)

In the first case (Fig. 1), arrays of conical holes (preferentially etched holes in thin metal plates and foils) have been used for shading.[14-18] Large areas, high numbers of elements and small periods can be obtained by this method. Anamorphic microlenses can be formed by elliptical holes. The theoretical limits for the geometrical fill factors of hexagonal and orthogonal arrangements of circular lenses are 0.91 and 0.79, respectively. As numerical simulations show, parabolic profiles can only be expected for the central regions.

### 2.3. Shading with slit array masks and crossed deposition (multi-step-method)

As an alternative approach, a *crossed deposition* of cylindrical microlenses with shading slit mask arrays (as depicted for one direction in Fig. 2) has been tested.[19-21] Theoretically, the overlapping profiles allow one to realize geometrical fill factors up to 1 and nearly parabolic shapes. Otherwise, this method is more sensitive against mask errors because of the adding up of asymmetries. Compared with the hole mask method, fewer elements and only larger pitches can be produced simultaneously because of mechanical limitations (except when stabilizing parts without slits are contained). By varying the layer thickness and/or the slit widths, *anamorphic* microlenses can be designed with this method too. By using multilayers, spectrally as well as spatially selective properties can be obtained.

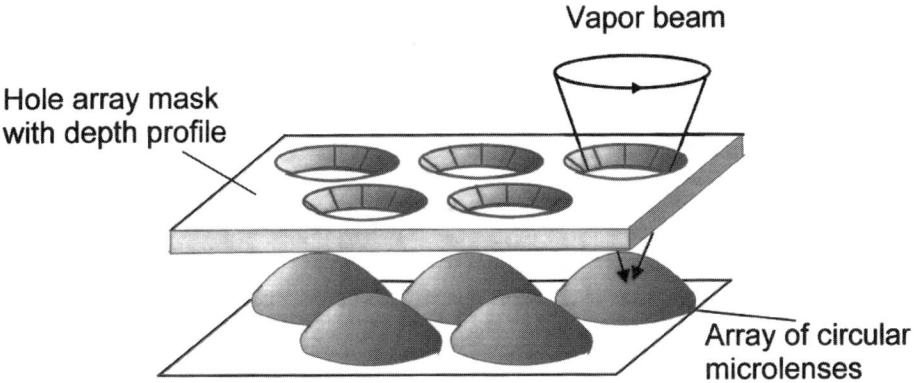

Fig. 1 Deposition of thin-film microlens arrays with shading hole array masks in a rotating system (schematically, for circular elements).

Applications of partially reflective, apodized micro-mirror arrays as outcoupling mirrors in self-imaging (Talbot) laser resonators have been published elsewhere.[14-18, 22-24]

### 2.4. Parameter stability

The vertex focal length $f$ of a spherical microlens of a diameter $d$, a refractive index $n$ and a sag h (vertex directed to the image plane, $h < d/2$) is given by

$$f = (h + d^2/4h) / [2(n-1)] \tag{1}$$

For very small $h$, the derivative $df/dh$ is high and the process is unstable. For large $h$, $df/dh$ converges to a constant value. Therefore, the best operating range may be found for moderate $h$. Furthermore, it can be concluded that a shortening of the focal lengths by enhancing the thickness h is impossible for large $h$ and can only be obtained by smaller $d$.

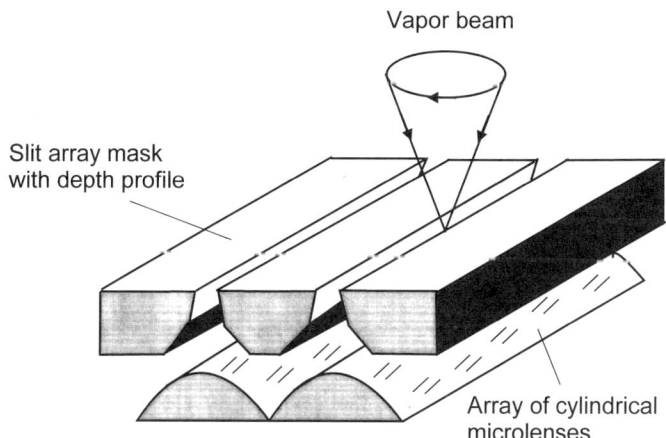

Fig. 2 Deposition of thin-film microlens arrays with shading slit array masks in a rotating system (schematically, for cylindrical elements).

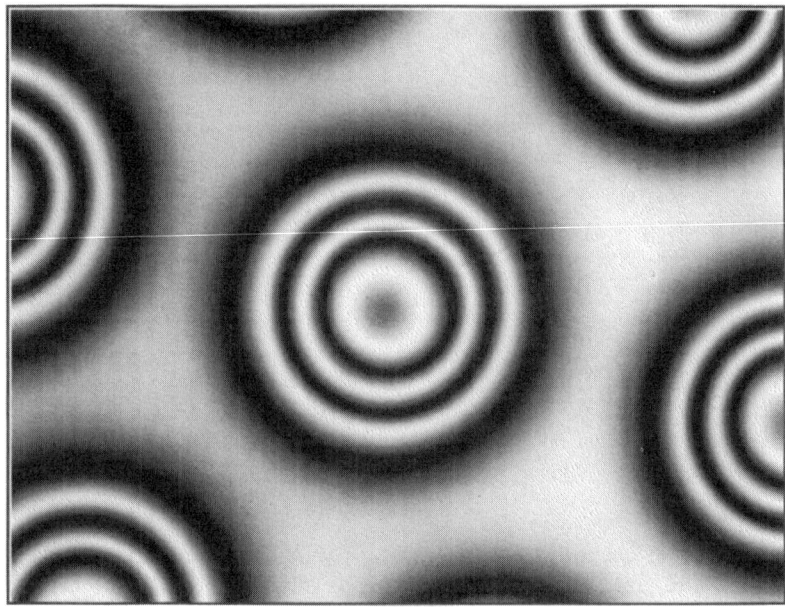

Fig. 3 Interference fringe pattern for a part of a microlens array deposited with a shading hole array mask (400 µm pitch, $SiO_2$ on quartz).

## 3. RESULTS

Different types of laser-cut and etched, monolithic and composed, hole and slit masks of conical profiles have been used as shading elements in a conventional vacuum deposition apparatus with e-beam vaporizer and a planetary rotation of the substrate mount. Thin-film microoptics have been deposited on rigid (quartz and glass) as well as flexible substrates (polymer foils). The maximum layer thickness without remarkable structural defects was about 15 µm ($SiO_2$ and $SiO_2$:$HfO_2$ on polycarbonate). Therefore, Fresnel numbers < 50 can be obtained. Thickness profiles have been measured with a commercial white-light Mirau interferometer (Zygo). Fig. 3 and Fig. 4 show typical interference fringe patterns for microlens arrays deposited with both methods (Fig. 3: shading hole array mask, 400 µm pitch, $SiO_2$ on quartz; Fig. 4: crossed deposition with slit masks, 200 µm pitch, $SiO_2$ on quartz).

Areas of > 50 $cm^2$ and numbers of elements > 80,000 have been obtained with the *hole mask method*. Hexagonal and orthogonal arrays with lens diameters of 25 to 600 µm, periods of 40 to 750 µm and focal distances of < 1 to 60 mm have been fabricated. Besides circular holes, other geometries have been tested. Fig. 5 shows a part of an orthogonal array of > $10^4$ $SiO_2$ microlenses of 40 µm pitch on quartz which has been deposited through square-shaped holes. The radii of curvature in the central part of the lenses are in the range of 50 µm. As demonstrated in Fig. 6, the structure has been transferred into a quartz substrate by plasma etching with reactive ions (1 µm step, not fully resolved by the interferometer). Thickness deviations by etching have been determined to be < 5%.

By additional masks and a variation of the mask distance from the substrate, it was possible to realize envelope phase functions. Fig. 7 shows a part of an array with the same pitch of 40 µm, deposited on a flexible polycarbonate substrate, having a linear envelope function for the maximum layer thickness (position dependent focal lengths). Super-Gaussian envelope functions have been realized by additional macroscopic masks.[24]

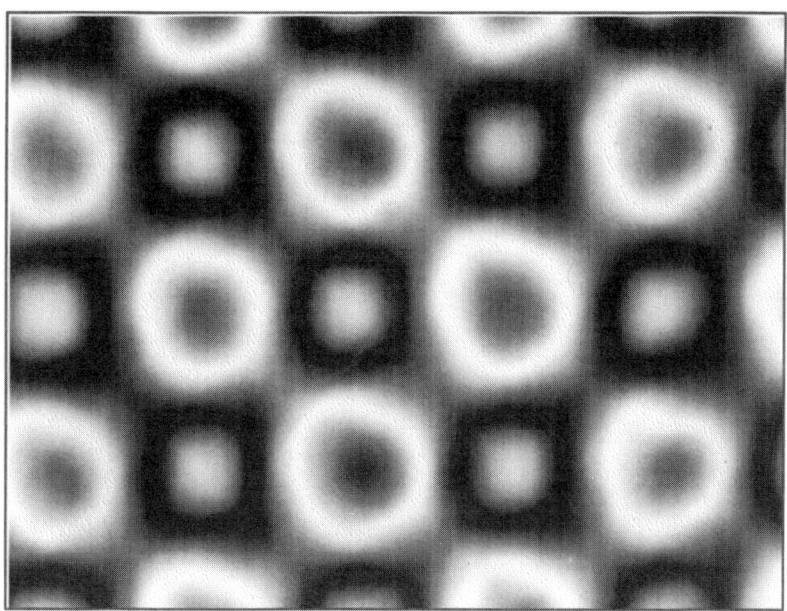

Fig. 4 Interference fringe pattern for a part of a microlens array produced with a shading slit array mask (crossed cylindrical microlenses, 200 µm pitch, SiO$_2$ on quartz).

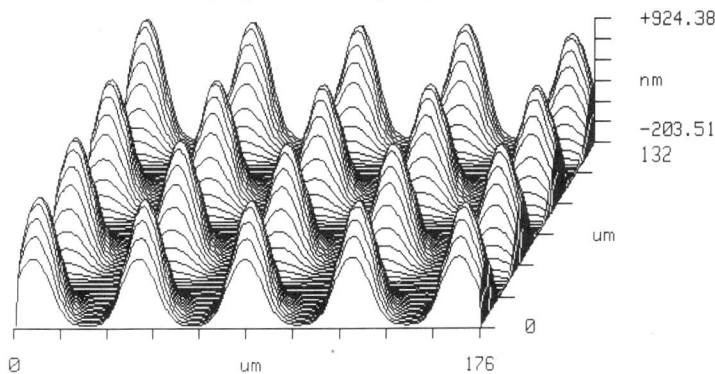

Fig. 5 Measured thickness distribution of a microlens array deposited with shading hole mask (40 µm pitch, SiO$_2$ on quartz).

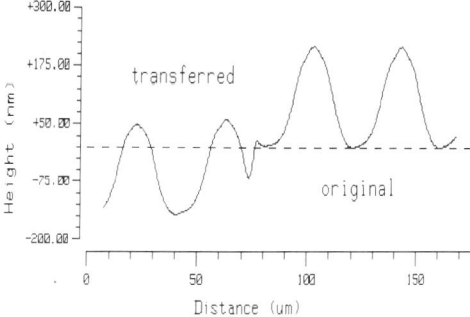

Fig. 6 Transfer of microlens structures from a SiO$_2$ layer into a quartz substrate by reactive ion etching.

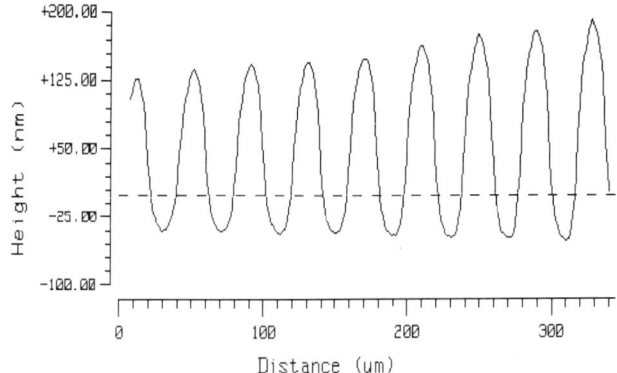

Fig. 7 Cut through a microlens array with envelope phase function (hole mask method, 40 µm pitch, SiO$_2$ on polycarbonate).

Closely packed hexagonal microlens arrays of 600 µm pitch have been fabricated with a net of hexagonal holes (Fig. 8).

With the method of crossed deposition of cylindrical microlenses (*slit masks*), arrays with up to 2,500 elements, areas of up to 6.25 cm$^2$, diameters from 40 up to 300 µm, pitches from 100 to 500 µm and focal lengths of typically some mm have been produced. Closely-packed linear arrays of cylindrical microlenses and orthogonal arrays of nearly spherically shaped microlenses could be formed (200 µm pitch). With elliptical holes, arrays of partially reflecting anamorphic microlenses have been deposited (Fig. 9, layer thickness 0.4 µm). The typical errors of the interferometry of multilayers (parasitic oscillations, depth errors) could be avoided by using thin auxiliary ligh reflectance layers. The profile of a part of an anamorphic microlens array with high fill factor (near 1) fabricated with crossed deposition of cylindrical microlenses is depicted in Fig. 10 (2 steps, angle of 90°).

Different radii of curvature have been generated by changing the deposition time for one angular step in this case. Other experiments have been carried out with different slit diameters for different angular steps. Hexagonal arrays could also be fabricated by the

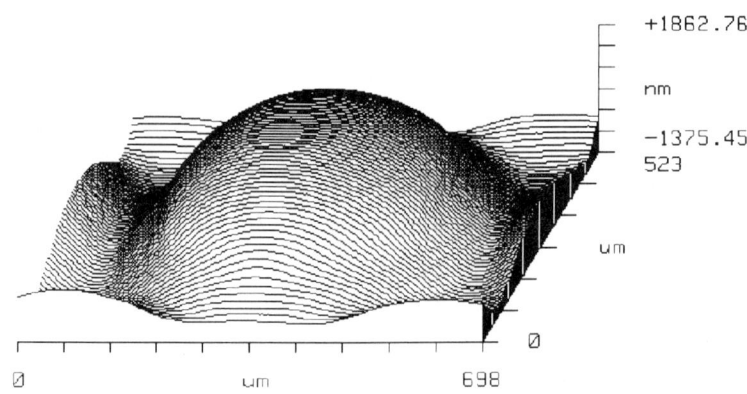

Fig. 8 Closely packed hexagonal array deposited through a hexagonal net mask (600 µm pitch, SiO$_2$ on quartz).

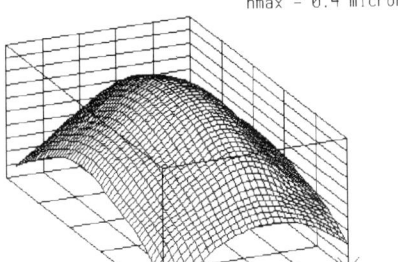

Fig. 9 Thickness profile of a single element of an array of partially reflecting anamorphic microlenses (SiO$_2$:HfO$_2$ on quartz) deposited through elliptical holes.

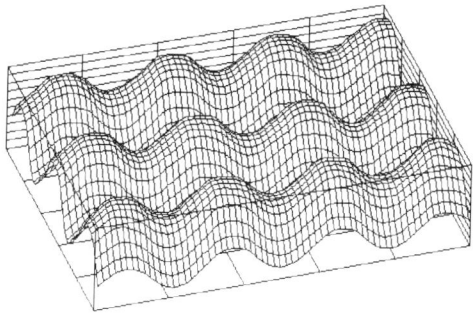

Fig. 10 Anamorphic microlenses (SiO$_2$ on quartz), produced by crossed deposition of cylindrical microlenses with different deposition times (two steps, h$_{max}$ 0.4 μm, pitch 200 μm).

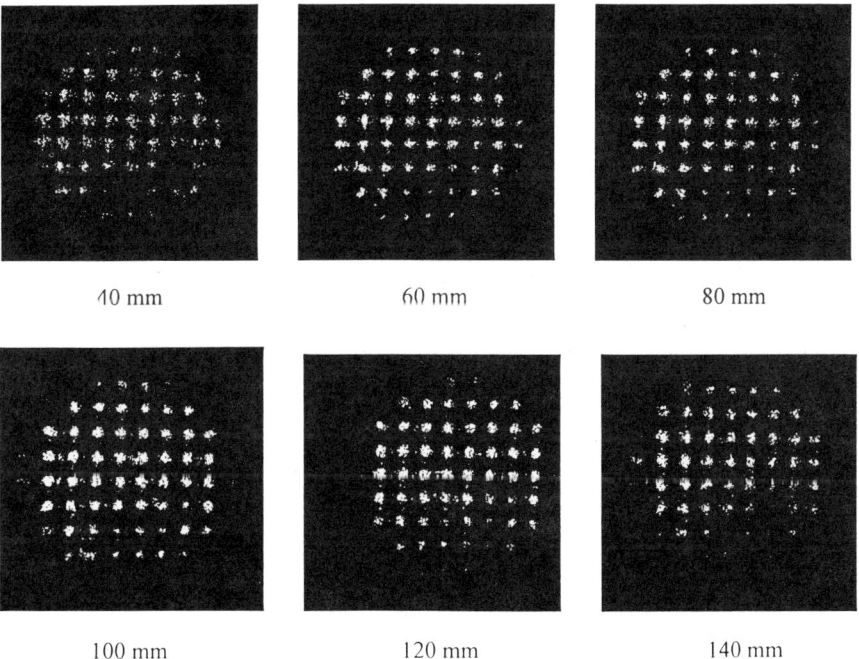

40 mm     60 mm     80 mm

100 mm     120 mm     140 mm

Fig. 11 Microlens array with 100 mm focal length: Intensity distribution at different distances from the vertices (crossed deposition of cylindrical microlenses, SiO$_2$ on quartz, lens profile slightly asymmetrical).

method of crossed deposition (3 steps, angle of 60°); however, the adjustment is critical, highly symmetrical slit masks are necessary and the maximum layer thickness per step is reduced. First experimental results with a rotatable mask holder were not satisfactory enough and will be improved in the future.

With very small layer thicknesses, extremely long focal lengths could be realized. Figure 11 shows the intensity distribution at different distances for an array with 100 mm focal length. For flexible microlens arrays, the adhesion of the layer is of crucial interest. The adhesion of $SiO_2$ on polycarbonate substrates has been measured to be very strong (> 24 N/cm). Furthermore, it should be mentioned that one of the most interesting features of the fabricated thin-film components are their broadband transmission (also in the UV region) and their ability to withstand laser radiation. The stability has been improved by the transfer of the structures into a substrate, which can be also a first step to further replication procedures.

## 4. CONCLUSIONS

To conclude, mask-shaded vacuum deposition of thin optical films has been shown to be capable of the fabrication of micro-optical components within a wide range of parameters. Both methods presented here show specific advantages and disadvantages. Small diameters (25 µm) as well as high numbers of elements (> 80.000) can be produced using *hole mask arrays* (1-step-method) whereas high fill factors can be realized by *slit mask arrays* (crossed deposition of linear arrays of cylindrical microlenses, ≥ 2 steps). The advantages of thin-film microoptics are the simplicity of the technology and the possibility of designing compact and flexible components (microlenses on thin polymer substrates [21]). Furthermore, spectral properties can be integrated by using multilayers. Both methods allow the realization of anamorphic properties. Focal distances from < 1 up to > 100 mm can be reached. Theoretical calculations indicate that aspheric microlenses which can be produced by the crossed deposition of cylindrical microlenses are stable against tilting and can be used for the circularization of line-shaped sources.[20] New types of masks (wire grids, compound masks, etched monolithic masks) are under investigation.

ACKNOWLEDGMENTS. The Authors thank Dr. W. Reinecke (Ingenieurbüro für Optikentwicklung, Berlin) for calculating optical functions, H.-H. Witzmann (GFaI, Berlin) for laser cutting of masks, Dr. D. Schäfer (BIFO, Berlin) for technical support, H. Mischke from Adolf-Slaby-Institute (ASI, Berlin) for plasma etching experiments, Prof. H. Schönnagel and U. Griebner from Max-Born-Institute for Nonlinear Optics and Short-Pulse Spectroscopy (MBI, Berlin) for many years of cooperation and Prof. E. Lüder (INS, University Stuttgart) for initializing the image sensing applications. The investigations have been supported by the Federal Ministry for Research and Technology (BMBF) under grant number 01M2035C.

## REFERENCES

1. L. D'Auria, J. P. Huignard, A. M. Roy, and E. Spitz, Photolithographic fabrication of thin film lenses, *Opt. Commun.* 5:232-235 (1972).
2. J. J. Cowan; The holographic honeycomb microlens, *Proceedings of SPIE* 523:251-259 (1985).
3. N. J. Phillips and Ch. A. Barnett, Micro-optic studies using photopolymers, *Proceedings of SPIE* 1544:10-21 (1991).
4. S. K. Yao and D. B. Anderson, Shadow sputtered diffraction-limited waveguide Luneburg lenses, *Appl. Phys. Lett.* 33:307-309 (1978).
5. S. K. Yao, D. B. Anderson, R. R. August, B. R. Youmans and C. M. Oania, Guided-wave optical thin-film Luneburg lenses: fabrication technique and properties, *Appl. Opt.* 18:4067-4079 (1979).

6. S. K. Yao, Theoretical model of thin film deposition profile with shadow effect, *J. Appl. Phys.* 50:3390-3395 (1979).
7. M. Kubo and M. Hanabusa, Fabrication of microlenses by laser chemical vapor deposition, *Appl. Opt.* 29:2755-2759 (1990).
8. J. Jahns and S. J. Walker, Two-dimensional array of diffractive microlenses fabricated by thin film deposition, *Appl. Opt.* 29: 931-936 (1990).
9. E. Pawlowski, H. Engel, M. Ferstl, W. Fürst and B. Kuhlow, Diffractive microlenses with antireflection coatings fabricated by thin film deposition, *Opt. Eng.* 33:647-652 (1994).
10. K. Zimmer and F. Bigl, 3D-Strukturierung von Polymeren durch Excimerlaserablation, *1. Int. Mittweidaer Fachtagung Qualitäts- und Informationsmanagement*, Mittweida, 1994.
11. G. Przyrembel, Continous-relief microoptical elements fabricated by laser-beam writing, H. Reichl, A. Heuberger, eds., *Micro System Technologies '94*, Berlin (1994), 219-228.
12. R. Grunwald, G. Szczepanski, I. Pinz and D. Schäfer, Variable-reflectivity IR-laser outcoupling mirrors, *2nd European Conference on Quantum Electronics*, Europhysics Conference Abstracts, Dresden (1989), Vol.13D, II, P 2.31.
13. Ch. Budzinski, R.Grunwald, I. Pinz, D. Schäfer and H. Schönnagel, Apodized outcouplers for unstable resonators, *Proceedings of SPIE* 1500:264-274 (1991).
14. R. Grunwald, U. Griebner, and D. Schäfer, Graded reflectivity micro-mirror arrays, *Proceedings of SPIE* 1983:49-50 (1993).
15. R. Grunwald and U. Griebner, Segmented solid-state laser resonators with graded reflectance micro-mirror arrays, *Pure Appl. Opt.* 3:435-440 (1994).
16. R. Grunwald, U. Griebner, and R. Ehlert, Microlens arrays for segmented laser architectures, *Proceedings of SPIE 2383*:324-333 (1995).
17. U. Griebner, H. Schönnagel and R. Grunwald, Diode-pumping of fiber array lasers via microlens arrays, in: *Advanced Solid-State Lasers*, B. H. T. Chai, S. A. Payne, eds., OSA Proceedings 24:253-256 (1995).
18. R. Grunwald, U. Griebner, R. Ehlert and S. Woggon, Micro-optical array components for novel-type lasers and artificial compound eyes, *National Workshop on Microlens Arrays*, National Physical Laboratory, Teddington, UK (1995), 85-88.
19. R. Grunwald, R. Ehlert, S. Woggon, H.-J. Pätzold and H.-H. Witzmann, Microlens arrays formed by crossed thin film deposition of cylindrical microlenses, *Diffractive Optics and Microoptics Topical Meeting, OSA Technical Digest Series* 5:27-30 (1996).
20. R. Grunwald, S. Woggon, R. Ehlert, and W. Reinecke, Thin-film microlens arrays with high fill factors produced by crossed deposition of cylindrical microlenses, *Appl. Opt.*, submitted.
21. R. Grunwald, R. Ehlert, S. Woggon, and H.-H. Witzmann, Thin-film microlens arrays on flexible polymer substrates. - Micro System Technologies, Potsdam, September 17-19, 1996, in: *Micro System Technologies 96*, VDE-Verlag GmbH, Berlin (1996), 793-795.
22. R. Grunwald, and U. Griebner, Optically segmented unstable resonators with graded reflectance micro-mirror arrays. - *Workshop Laser resonators with graded reflectance mirrors*, Technical Abstracts Florence (1993).
23. R. Grunwald, U. Griebner, and R. Koch, Phase-coupled multiple-beam solid-state laser with Talbot-resonator. - *CLEO '94*, Technical Digest, Anaheim (1994), 410.
24. R. Grunwald, and U. Griebner, Passively Q-switched high-power Nd:glass laser with different types of graded reflectance outcoupling mirrors. - *CLEO '95*, Technical Digest, Baltimore (1995), 38-39.

# NEW PHOTOLUMINESCENT MATERIALS BASED ON LiF:NaF MICROSTRUCTURES

G. Baldacchini[*], E. De Nicola[*], R. M. Montereali[*], M. Cremona[§],
M. Passacantando[†], and F. Somma[‡]

[*] Division of Applied Physics INN Department, C.R. Frascati ENEA
  Via E. Fermi 45, 00044 Frascati (RM), Italy
[§] Department of Material Science, Pontifical Catholic University
  C.P. 38071, 22452-970, Rio de Janeiro, Brazil
[†] Department of Physics, University of L'Aquila
  67010 Coppito (AQ), Italy
[‡] "E. Amaldi" Physics Department
  The Third University of Rome
  Via della Vasca Navale 84, 00146 Rome, Italy

## 1. INTRODUCTION

Alkali halides are ionic crystals with face centered cubic structure, optically transparent from the near UV to the IR. Point defects, known as color centers, can be produced by ionizing radiation (X-rays, γ rays, electron beams, etc.) or by chemical treatments, and they colour the crystals, because of broad absorption bands in the visible range. These defects have a characteristic four-level optical system which is generally suitable for the realization of optically pumped tunable solid state lasers with high spectral purity ($\Delta\nu/\nu < 10^{-9}$), which are used for high resolution spectroscopy and optical fiber signal propagation. Color center lasers usually operate at liquid nitrogen temperature.[1]

Among the alkali halides, LiF occupies a special place because of its peculiar properties. First of all the cation-anion distance is the shortest (0.2013 nm) and the $Li^+$ and $F^-$ ions have the smallest radius (0.06 and 0.136 nm, respectively) of all the alkali and halide ones. Moreover it is of particular interest for its hardness and because it is not hygroscopic, allowing easy manipulation in open air.

LiF and NaF crystals can be colored only by ionizing radiations and they can host point defects which are laser active in the visible and near infrared at RT.[2] In particular $F_2$ and $F_3^+$ centers (two electrons bound to two or three neighbouring anionic vacancies, respectively, the last one being a charged defect) in LiF have overlapping absorption bands peaked at ~ 450 nm (M band) and emit in the spectral range between 510-570 nm ($F_3^+$ center) and 650-740 nm ($F_2$ center). In NaF they absorb at ~ 500 nm with emission bands peaked at 580 nm ($F_2$) and 655 nm ($F_3^+$)

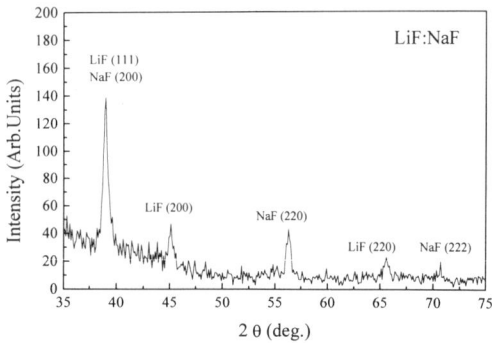

Fig. 1 X-ray Bragg-Brentano diffraction spectrum of a multilayer LiF:NaF film grown at $T_s = 230 \,°C$.

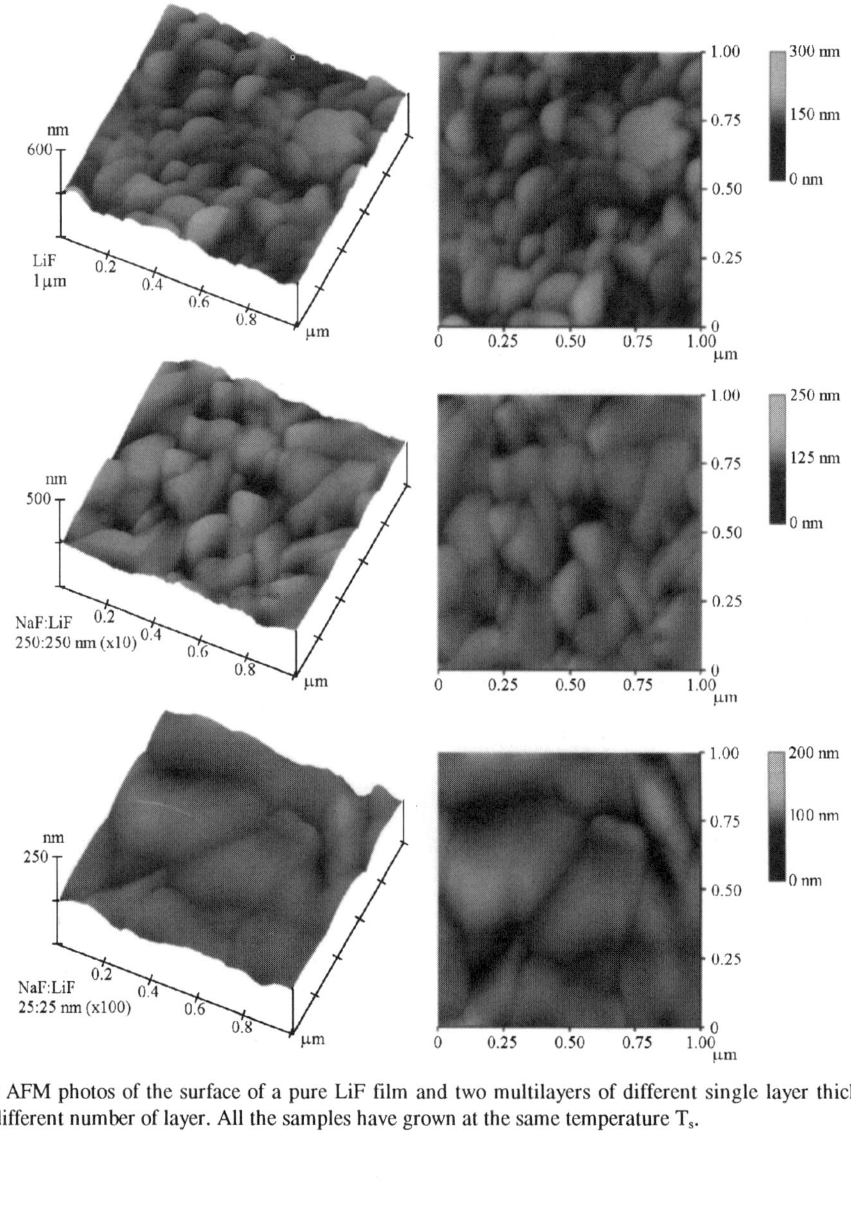

Fig.2 AFM photos of the surface of a pure LiF film and two multilayers of different single layer thickness and different number of layer. All the samples have grown at the same temperature $T_s$.

The growing interest in component miniaturization and the search for new materials and technologies in integrated optoelectronics[3] prompted us to investigate the possibility to realize a new photoluminescent device based on a thin layer of active defects. Therefore we have grown single[4] and multilayer[5] polycrystalline films, where high concentrations of color centers have been created by low energy electron beams. The opportunity to create point defects in thin films by low penetrating radiation has been already investigated for applications in the area of frequency domain optical storage using the inhomogeneous broadening of zero phonon lines,[6] and in the field of optical waveguides by varying the refractive index of the materials with the presence of defects.[7]

The limited penetration of this kind of radiation gives the advantage of controlling the coloration depth by selecting the electron energy and the possibility to focus it using electron beam lithography techniques, which also allow the production of predesigned patterns.

The aim of this Chapter is to present the results of a structural and spectroscopic investigation on polycrystalline multilayer LiF:NaF films with different layer thicknesses, grown by physical evaporation on amorphous substrates, and to compare them with those obtained on pure LiF and NaF thin films. These materials were chosen because the defects are efficiently created by low energy electron beams and have a room temperature stable visible photoemission, which has been already utilized in bulk crystals for realizing solid state lasers. Since it is impossible to produce mixed LiF:NaF crystals with a controlled proportion of the two components,[8] the multilayered films represent one efficient way to obtain this new material.

## 2. EXPERIMENTAL DETAILS

LiF, NaF and multilayer LiF:NaF films were grown by electron beam assisted physical evaporation on silica substrates by heating superpure powders. During the growth the substrate temperature $T_s$, was kept at a fixed value in the range between 30 and 350 °C. The pressure in the evaporation chamber was less than $10^{-4}$ Pa and the growing rate was around 1 to 2 nm/s. The total film thickness, measured by a stylus profilometer was ~ 2.5 µm. It was possible to control the single layer thickness of the multilayer samples using a quartz oscillator. The upper layer of the LiF:NaF samples is always LiF because it is less hygroscopic therefore it protects the multilayer structure.

Several complementary techniques have been used for the structural and morphological characterization of the film: XRD (X-Ray Diffraction), SEM (Scanning Electron Microscopy), AFM (Atomic Force Microscopy) and XPS (X-Ray Photoelectron Spectroscopy) depth profile technique.[3,9]

Color centers were produced by irradiating the films with low energy electron beams (3 to 12 keV) at RT. Since at these energies the electrons have a limited penetration depth (d = 0.15 to 1.6 µm),[6] the substrate is not colored. At the energy of 3 keV a continuous electron gun with current of 100 µA and spot area of 1 cm$^2$ has been used. At electron energies greater than 3 keV the irradiation has been performed in a scanning electron microscope equipped with an electron beam lithography system (I = 2 nA, spot area = 3x3 mm$^2$).

The spectroscopic properties of the colored films have been investigated at room temperature (RT). Absorption measurements have revealed the presence of several kinds of color centers. Emission spectra have been taken in a collinear geometry of pumping source and detector, exciting the samples with the 458 and 496 nm Ar$^+$ laser lines. These lines were chosen because they match the absorption band of colored pure LiF ($F_2$ at 448 nm, $F_3^+$ at 458 nm)[4] and NaF ($F_2$ at 498 nm, $F_3^+$ at 520 nm).[10] The photoluminescence was detected by a monochromator, 22 cm focal length, and a photomultiplier, S20 response, using a lock-in technique.

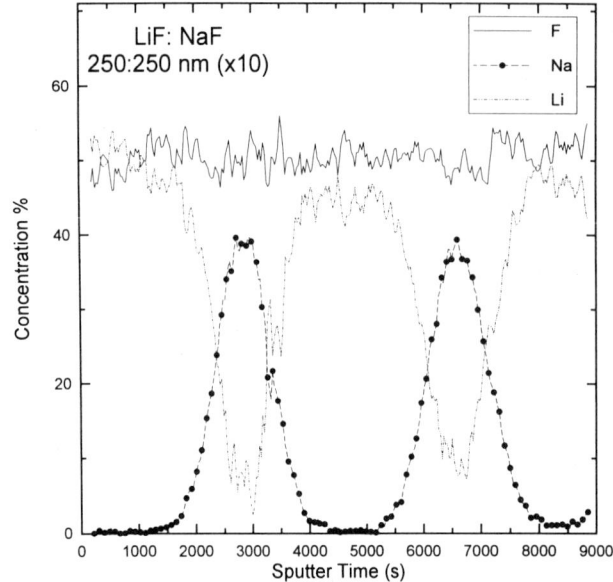

Fig. 3 XPS depht profile of a LiF:NaF multilayer consisting of 10 layer with single layer thickness of 250 nm.

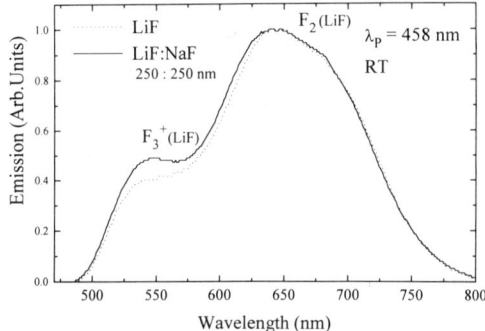

Fig. 4 RT emission spectra of a pure LiF film and a multilayer LiF:NaF made of 10 layer with single layer thickness of 250 nm excited with the 458 Ar $^+$ laser line. The samples have been grown at the same $T_s$ and irradiated at RT by a 12 KeV electron beam with the same dose.

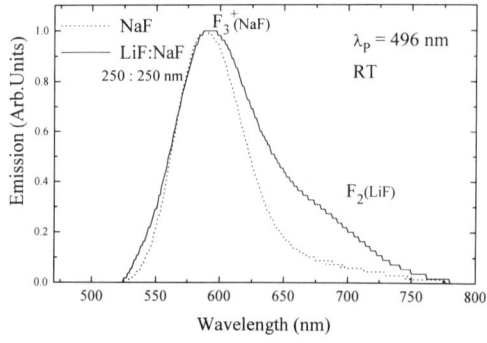

Fig. 5 RT emession spectra of a pure LiF film and multilayer LiF:NaF made of 10 layer woth single layer tickness of 250 nm excited with the 496 Ar $^+$laser line. The samples have been grown at the same $T_s$ and irradiated at RF by a 12 KeV electron beam with the same dose.

## 3. RESULTS AND DISCUSSION

### 3.1. Structural and morphological characterization

The X-ray Bragg-Brentano diffraction spectrum of a multilayer LiF:NaF film, grown at high substrate temperature, is shown in Fig. 1. Both the LiF and NaF diffraction lines are present and there is not any spurious diffraction peak, showing that the single layers of the multilayer film preserve their individual structure. In other words, the materials used do not form other phases. Therefore the film is composed of pure LiF and pure NaF layers which are polycrystalline, as shown by the presence of several diffraction peaks.

The morphological structure of the surface of the LiF:NaF multilayers has been analyzed by AFM and SEM techniques. The AFM technique, due to the possibility of observing the sample without treating the surface, is more suitable than the SEM for these insulating materials.

Fig. 2 shows the surface of a pure LiF and two LiF:NaF films grown at 230 °C and with the same total thickness. The LiF:NaF multilayers are composed of 10 and 100 layers with single layer thickness d = 250 and 25 nm, respectively. From a comparison of the AFM photos, it is possible to note that the grain size and the formation of interstitial empty spaces at the grain boundaries are strongly influenced by the single layer thickness.

The XPS depth profile technique allows us to investigate the chemical composition of the multilayers. Fig. 3 shows the concentration of F, Na and Li atoms as a function of the sputtering time, i.e. of the depth inside the film. While F is always present at 50 % inside the sample, Li and Na alternate, showing the expected sequence of LiF and NaF layers inside the sample. The relative concentrations of Li and F confirm the correct stoichiometric composition for the LiF layers, while the presence of smaller LiF grains in the interstitial empty spaces present between the larger NaF grains is probably the cause of the reduction of the sputtering time for the NaF deposit and of the fact that the concentration of Na never reaches 50 %. We have also measured the binding energies of the 1s core energy level of the three atoms of the multilayer and we have found the following values which agree with those reported in literature:[11] 1071.4 eV for Na, 684.7 eV for F and 55.6 eV for Li. We have also investigated the chemical purity of the multilayers, analyzing the possible presence of other materials inside the samples; we have found that on the surface there is a slight contamination from oxygen and carbon, while they are absent inside the films.

### 3.2. Spectroscopic properties

In order to relate the structural and morphological characteristics of the films with their spectroscopic properties and to distinguish the film growth parameters influence on the photoluminescence spectra, we have performed absorption and emission measurements on the pure LiF, the pure NaF and on the multilayer LiF:NaF films colored by low energy electron beams. These measurements have confirmed the formation of $F_2$ and $F_3^+$ centers in each of the samples. The measured emission bands coincide in position and half-width with the $F_3^+$ and $F_2$ bands of LiF and NaF, peaked respectively at 535 and 670 nm for LiF and at 580 and 655 nm for NaF. In Fig. 4 the RT normalized photoluminescence spectrum of the LiF:NaF multilayer with single layer thickness of 250 nm colored by 12 keV electrons and excited with the 458 nm $Ar^+$ laser line is compared with that of a LiF film irradiated in the same conditions and pumped with the same wavelength at the same power. Since the form of the emission is the same, it follows that the photoluminescence of NaF layers of the LiF:NaF multilayer is not excited by the 458 nm wavelength. Indeed, $F_2$ and $F_3^+$ emission signals of pure NaF film measured in the same experimental conditions are much smaller than the $F_2$ and $F_3^+$ ones measured in the LiF film.

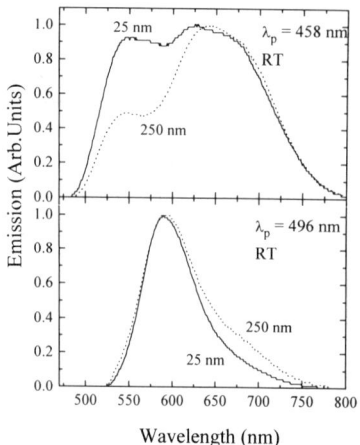

Fig. 6 RT emission spectra of two LiF:NaF multilayers made of 10 and 100 layer with single layer thickness of 250 and 25 nm, respectively. The excitation has been performed with the 458 and 496 nm Ar$^+$ laser line. The samples have been grown at the same $T_s$ and irradiated at RT by 12 keV electron beam with the same dose.

Under excitation at 496 nm, in the LiF:NaF multilayer the $F_3^+$ photoluminescence of NaF prevails, while the LiF emission is less pronounced than in the spectrum excited at 458 nm, as shown in Fig. 5, where the photoluminescence of the NaF film colored in the same conditions is also reported. Indeed, in this case, the pure LiF film emission signal is lower than the pure NaF one.

The emission spectra of the two multilayers which differ in the single layer thickness, excited at 458 and 496 nm, are compared in Fig. 6. The photoluminescence pumped at 458 nm is different in the two samples and presents a more flat profile when the single layer thickness is smaller. Also, the 496 nm pumped emission spectra depend on the morphological properties of the films. When the single layer thickness is 25 nm, only $F_3^+$ photoluminescence of NaF is present, while in the other sample there is a hint of $F_2$ luminescence of LiF. From a comparison of the emission spectra of the LiF:NaF samples, which differ in the single layer thickness and in the number of layers, with their AFM photos, the formation efficiency of the different kinds of centers in the multilayers seems to be related to the uniformity of the surface and to the packing of the microcrystals. Indeed, irradiating two multilayer LiF:NaF films having the same morphological and structural characteristics with electron beams of different energy, 3 keV and 12 keV, the formation efficiency of different kinds of centers is the same, as derived by the ratio of their emission intensities.

## 4. CONCLUSION

It has been possible to grow multilayer films of LiF and NaF. The described technique consists of successively adding layers of the two original materials. The morphology and the structural and optical properties of the resulting samples have been studied and compared with those of pure LiF and pure NaF films, and it has been shown that the formation efficiency of color centers seems to be related to the film compactness.

It has been possible to produce optically active defects in a single multilayer deposit which combines the emission properties of the defects present in each of the two alkali halides forming the multilayer. Such a combination of properties cannot be achieved by using optical materials in bulk form.

The possibility of producing color centers of the two alkali fluoride components in a single multilayer structure allows us to obtain new photoluminescent materials with predesigned photoluminescent properties. The quality of the surfaces and the stability of the color centers is of interest for potential applications, because the colored regions could become light sources and amplifiers, if appropriately pumped.

## REFERENCES

1. L.F.Mollenauer, Color center lasers, *in:* "Tunable Lasers", L.F. Mollenauer and J.C. White, Eds., Springer Verlag, New York (1987), p.225.
2. T.T. Basiev, S.B. Mirov and V.V. Osiko, Room temperature color center laser, *IEEE J. Quantum Electron.* 24: 1052 (1988).
3. G. Baldacchini, M. Cremona, RM. Montereali, E. Masetti, M. Montecchi, S. Martelli, G. C. Righini and S. Pelli, Colored layers of LiF for integrated optical devices, *in:* Advanced Materials in Optics, Electrooptics and Communication Technologies, P. Vincenzini and G. C. Righini Eds., Techna s.r.l. , Faenza (1995), p. 42S.
4. RM. Montereali, G. Baldacchini and L.C. Scavarda do Carmo, LiF films: absorption and luminescence of colour centres, *Thin Solid Films* 205: 106 (1991).
5. F. Somma, M. Cremona, R.M. Montereali, M. Passacantando, P. Picozzi, S. Santucci, Structural and optical properties of low energy electrons irradiated KCl:LiF multilayer films, *Nuclear Inst and Meth. in Phys. Res. B.* 116 : 212(1996) .
6. C. Ortiz, RM. Macfarlane, R.M. Shelby, W. Lenth and G. C. Bjorklund, Thin-film aggregate color centers as media for frequency domain optical storage, *Appl. Phys.* 25: 87 (1981).
7. P.J. Chandler, F.L. Lama, P.D. Townsend and L. Zhang, Buried double waveguide by ion implantation in quartz,*Appl Phys. Lett* 53: 89 (1988).
8. A. Smakula, N.C. Maynard and A. Repucci, Color centers in mixed crystals of alkali halides, *Phys Rev.* 130: 113 (1963).
9. F. Somma, A. Ercoli, S. Santucci, L. Lozzi, M. Passacantando, and P. Picozzi, Production and characterization of multilaya KCl:LiF thin film on glass, *Journal oJff Vacuum Science & Technology A* 13: 1013 (1995).
10. M. Cremona, A.P. Sotero, R.A. Nunes, M. H. do Pinho Mauricio, L.C. Scavarda do Carmo, R.M. Montereali, S. Martelli and F. Somma, NaF films: growth properties and electron beam induced defects, *Rad. Eff and Defects in Solids* 136: 163 (1995).
11. Handbook of X-Ray Photoelectron Spectroscopy, J. Chastain, Ed., Perkin Elmer, Eden Prairie, MN (1992).

**COMPONENTS AND MEMS**

# ADVANCES IN OPTICAL MICROSYSTEMS COMBINING MICROTECHNOLOGIES AND BATCH PROCESSING FABRICATION

S. Valette

LETI, Commissariat a l'Energie Atomique
17 rue des Martyrs 38054 Grenoble Cedex 9, France

## 1. INTRODUCTION

The use of microelectronics technologies to fabricate optical components and devices really began in the seventies with the development of semiconductor light sources and integrated optics.

Most of the thin film technologies currently used in microelectronics have been applied to achieve planar optical components in different materials including III-V semiconductors,[1-4] electro-optic materials such as lithium niobate[5] $LiNbO_3$ or tantalate $LiTaO_3$, various glasses [6], polymers [7] and of course silicon.[6, 8-13]

Although a complete parallel cannot be established between microelectronics and integrated optics, the main objective is the same : to achieve more attractive and more miniaturized devices compatible with technical specifications in order to decrease the required driving voltages and the power consumption for active components, to get novel and more reliable architectures and assembly and to take advantage of batch processing to reduce fabrication costs.

Planar optics is of course well suited to integrate microelectronics know-how in the field of optical devices because it is built on the same intellectual concept and uses very similar thin film processes. Although new technological problems must be solved to take into account the specificity of optical components and the use of new materials for multilayer structures, integrated optoelectronics can be seen as a normal development of microelectronics.

The extension of these techniques to free space optics is much less obvious.
This extension requires, because it has to deal with a 3D environment, both intellectual and technological jumps[14] and depends on a lot of technological advances in various technological fields including miniaturized light sources and detectors, novel optical passive elements and micromechanical structure which, have to be combined while satisfying the objectives of low cost and reliable packaging.

Much progress has been made in this field during the last 10 years, and various key advances have been obtained which open the way for Micro Electro-Mechanical and

Optical Systems (MEMOS).

During the last twenty years, the development of integrated optics has made the microelectronic technological approach and batch processing fabrication methods familiar to the optical community. At first, these new developments only concerned planar devices ; more recently new concepts in miniaturized light sources and different technological advances open the way for batch processing fabrication of free space optics. They allow the achievement of real optical microsystems providing both attractive performance and potential low cost.

In this Chapter we give a review of this new landscape by summarizing recent work in the field of optical and optomechanical components and devices.

## 2. NOVEL MINIATURIZED LIGHT SOURCES

Miniaturized light sources are key components in optical microsystems because they have to provide photons in a suitable form. Miniaturized light sources must provide various characteristics in terms of performance, emission geometry, reliability and cost.

They must also be fully compatible with low cost assembly of the complete microsystems in which they are used.

From this point of view Vertical Cavity Surface Emitting Lasers VCSEL, solid state microchip lasers and new light emitting diodes with resonant cavities appear as key emissive components.

### 2.1. Vertical Cavity Surface Emitting Laser (VCSEL) [15-20]

The development of semiconductor heterostructures and the demonstration of the laser effect in compound semiconductors at the beginning of the seventies have revolutionized the optical world.

Edge emitting semiconductor laser diodes offered the first efficient miniaturized laser sources able to give high electrical to optical conversion yield.

However, their configuration presents some drawbacks for the achievement of advanced microsystems because the edge emission of the light requires a mechanical dicing operation before laser operation, and prevents testing at the wafer level.

Although different approaches can be used to overcome this problem (for instance 45° reflectors), they lead to additional complex technical steps which can reduce the fabrication yields.

From this point of view, Vertical Cavity Surface Emitting Laser (VCSEL) offers very attractive features and appears more and more as one of the ideal light sources for many optical microsystems.

The VCSEL structure is a multilayer structure obtained by Molecular Beam Epitaxy (MBE) or Metal Organic Chemical Vapor Deposition and which consists of top and bottom quarter-wave mirror stacks surrounding a spacer region ; the spacer region, usually of a thickness equal to an optical wavelength within the material in order to reduce cavity losses, contains the active quantum wells at its center.

Because of their very thin active layer, VCSELs require a very high cavity Q factor. Therefore, the quarter wave stack (generally referred as a Distributed Bragg Reflector, DBR) must be optically very efficient to decrease the cavity losses, and thus the current threshold, but must also exhibit low serial electrical resistance to reduce the driving voltage.

These requirements are not easily compatible and the challenge is to find the best multilayer arrangement and the most convenient VCSEL design in order to obtain, for a given emissive area, the lowest current and voltage threshold and the best emission characteristics.

Since the first demonstrations of VCSEL action at the end of the eighties [15, 16] some impressive progress [17-20] has been made, both with regard to increasing the performance within a given material system, and to enlarging the number of semiconductor families in which VCSEL structures can be implemented.

The VCSEL effect has now been demonstrated at room temperature in continous wave operation with many different structures and quantum well materials ; the best results are obtained when GaAlAs layers can be used for the reflector because of the larger refractive index difference attainable with this material family. This is the case for VCSELs emitting around 0.85 µm [18-21] and 0.78 µm [22] (GaAlAs quantum wells), 0.98 µm [23] (GaInAs quantum wells), 0.66 - 0.68 µm [24] (GaInAlP quantum wells) and even at 1.3 µm [25] (GaInAsP quantum wells) in spite of the difference in the thermal expansion coefficient of InP ($4.6.10^{-6}$ $K^{-1}$) and GaAs ($6.3 \times 10^{-6}$ $K^{-1}$). Threshold currents and output powers are typically in the range of few mA and few mW (except at 1.3 µm where the available power remains smaller), depending of course on the VCSEL structure and on the diameter of the emission area.

The best results obtained with the more extensively investigated structures around 0.85 µm and 0.98 µm are impressive, including very low thresholds current (few 10 µA to few 100 µA) [26] and low voltage (less then 2 V and 1.5 V respectively), high output powers corresponding to differential quantum efficiency at or above 50 %, and high speed modulation above 15 GHz.[17, 21, 27]

Moreover, even better performance can be expected with additional technical improvements leading both to better current and lateral optical confinement.

Although room temperature operation has been demonstrated, results are still poor at long wavelength around 1550 nm.[28] This is due to the high recombination rate near room temperature and to the increased difficulty of achieving low optical loss Bragg reflectors at this wavelength (more than 30 pairs of λ/4 mirror layers are needed because of the small refractive index difference in the material couple InP/InGaAsP).

On the other hand, VCSEL pulsed operation at 490 nm under threshold conditions at 77 K of 3 mA and 17 V has been recently reported with ZnSe based structures (ZnSe cladding on GaAs, CdZnSe active region, $SiO_2/TiO_2$ mirror and current injection by annular ring) by Matsushita Electric.[29] More recently ultraviolet emission (363 nm) has also been observed at room temperature from a gallium nitride VCSEL structure, optically pumped and using GaAlN Bragg reflectors.[30]

Thus VCSEL technologies are progressing very quickly. Thanks to their nice optical characteristics (symmetrical emission, low beam divergence, versatility of near field pattern including single transverse and longitudinal mode emission), their very good temperature behaviour compared to edge-emitter laser diodes and of course their intrinsic availability in one or two dimensional addressable arrays, they are drastically changing the face of optical techniques in the field of communications,[21] optical storage and more generally optical microsystems.[31]

## 2.2. Microchip Lasers

Although semiconductor lasers and now VCSELs provide a wonderful miniaturized light source, they cannot provide all of the optical features required in some applications of optical microsystems.

In particular, the very short radiative lifetime (in the nanosecond range) of semiconductor materials prevents the achievement of efficient pulsed lasers using for instance Q switching.

On the other hand, high power operation can be only obtained with array configurations which lead to the loss of spatial coherence.

Microchip solid state lasers overcome these last drawbacks and appear from this point of view complementary to laser diodes.[32-35]

The concept of the laser chip is basically very simple. It uses the capability of optical techniques to realize very flat wafers of different materials with a very precise control of side-angle. After suitable coating with multi-layer dielectric films and generally standard optical mounting at the wafer level, the microchip laser is then realized by dicing the final assembly in little square pieces (generally on the order of a square millimeter).

The simplest assembly to realize an array of microchip lasers requires only a double side polished wafer of an amplifier material (for instance Neodynium doped YAG or Erbium doped glass) coated with suitable dielectric dichroic mirrors (input mirror transparent for the pump wavelength and highly reflective for the solid state laser emission ; output mirror slightly transparent for the laser emission and generally highly reflective for the pump).

These kinds of structures have been extensively investigated with different amplifier materials in continuous wave or gain-switched operation according to the pumping mode, and microchip lasers have been thus achieved at various emission wavelengths :

- around 1 µm with different Neodynium doped crystals such as YAG (Yttrium Aluminium Garnet), LMA (Lanthanum and Magnesium Aluminate), YLF (Yttrium Lanthanum Fluroride), $YVO_4$ (Yttrium Vanadate) and YSO (Yttrium Silicate) using high power gallium aluminum arsenide laser diode pumps, [32-35]

- around 1.5 µm with Erbium-Ytterbium codoped phosphate glass and using a GaInAs high power diode laser pump emitting at 0.98 µm,[34-36]

- around 2 µm with Thulium or Thulium-Holmium doped crystals such as YAG-YSO, also with a GaAlAs diode laser pump.[34,35]

In the simple flat-flat configuration, threshold pump powers typically range from a few tens of milliwatts (for instance 15 mW with an optimized Er-Yb phosphate glass microchip laser ; 35 mW for Nd/YAG...) to about 100 mW with Nd/YLF.[34,35]

Differential quantum efficiencies are in the range 10 to 60 %, and output powers in a TEMoo mode of several hundred milliwatts have been commonly observed with, for instance, standard GaAlAs or GaInAs diode laser pumps of 1 W.

However, many different schemes and improvements can be developed from the basic microchip configuration in order to obtain novel laser functions such as passive Q-switching,[34,35] frequency modulation,[37] internal or external frequency doubling,[32,34,35] or to increase the laser performance in terms of low threshold, high power stability or high frequency pulsed operation.

The first drastic improvement brought to the microchip laser was the use of a stable cavity scheme by introducing microlenses.[34,35]

This kind of improvement is compatible with batch process fabrication by adding a microlens array to the basic configuration.

It gives an intrinsically better resonator stability which is only obtained in the flat-flat scheme by favorable thermal effects.

Depending on the laser material and on the packaging, the gain in performance can be very impressive, reaching a factor 20 in the threshold current with Nd/YAG microchip laser (1,8 mW compared to about 40 mW in a comparable flat-flat scheme).[35]

However, the main advantage of solid state microchip lasers is their ability to be Q-switched because of the long lifetime of the excited state (few 100 µs with Nd but milliseconds with Erbium for instance).

The first Q-switched operation of a YAG/Nd microlaser was published by Zayhowski using a coupled cavity configuration which associated the laser material with an electro-optic modulator.[32,37]

Depending on the coupled cavity parameters, both Q-switched and wavelength modulation operation can be achieved with this scheme.

However the use of an electro-optic modulator (lithium tantalate, $LiTaO_3$) requires driving voltages in the range of 1000 Volts to sufficiently modify the cavity length and observe efficient pulsed operation or frequency shifting.[37]

Even if favourable configurations using microlenses can lead to lower driving voltages, this is a limitation of that approach.

More recently, passive Q-switched microlasers have been demonstrated with different saturable absorbers such as polymers and thick epitaxial layers of YAG/Cr$^{4+}$ obtained by liquid phase epitaxy, directly grown on the YAG/Nd amplifier wafer.[34,35]

In this last work, peak powers and pulse duration are in the range of a few kilowatts (5 to 6 typically) and 0,3-2 ns respectively, leading to pulse energies ranging between 0,5-5 µJ per pulse with a standard pumping diode around 1 Watt.

With higher pumping diode power up to 10 W, a quasi proportional increase of peak powers and of the pulse energy (which can then reach 50-100 µJ) is observed, however associated with a transverse multimode behaviour.[35]

Of course other Q-switched schemes using different methods (mechanical choppers, acousto-optic modulation...) are currently under study to achieve pulsed microlasers at wavelengths where no attractive saturable absorbers are presently available.[34]

Finally, microchip lasers, because of their excellent mode properties even at high power emission, are very convenient to observe efficient nonlinear effects in suitable external or internal cavity configurations.[32,34,35]

A very nice demonstration of this capability has been reported by LETI with the achievement of a pulsed green microlaser obtained by frequency doubling of a passive Q-switched microlaser with a KTP crystal.[34,35]

With no special optimization, peak power pulses of several hundred Watts in the TEM$_{00}$ mode were observed at 532 nm with a 300 mW pump at 808 nm.

This miniaturized green laser is now commercially available from NANOLASE company.

Much work has been recently done in these fields and additional new results are expected: for instance, using an optimized cavity scheme in order to increase the conversion effiency or to enlarge the wavelength spectra.

UV microlasers have been recently achieved by generating the fourth harmonic from a Nd/YAG microlaser with promising conversion efficiency.[32,35]

## 2.3. Novel Light Emitting Diodes

The laser effect is very attractive for obtaining high power continuous or pulsed light sources with high space and time coherence.

However, it is well known in optics that time coherence can be a drawback in many imaging systems because it can lead to parasitic interference and increased optical noise.

Moreover, different undesirable effects associated with laser diodes as mode hopping, frequency shift, etc. have to be overcome in practical microsystems.

In some cases, light emitting diodes can be an attractive alternative if they can deliver reasonable power with increased space coherence.

Resonant cavity light emitting diodes seem able to reach these requirements.[38-41]

Compared with conventional light emitting diodes, resonant cavity LED's offer higher efficiency because of the cavity effect similar to that used in VCSEL's. However, because of the smaller number of Bragg periods required (5-6) instead of the 20 or 30 typical with a VCSEL, these LEDs are easier to build. They can be very useful in the infrared range for many applications including chemical analysis and pollutant detection, remote measurement and short distance communication. Such LEDs have been demonstrated with CdHgTe alloys grown by molecular beam expitaxy on ZnCdTe substrate, and they can operate in the range 3 to 5 µm, which is especially attractive for optical spectrometry.

Devices working at 3.2 µm wavelength present a reduced electroluminescence spectrum width of 8 meV at room temperature, an improved directivity emission angle down to 55° (FWHM) and exhibit an external quantum efficiency $0.2 \times 10^{-3}$ [38-40] for a 10 mA current. Similar results have been recently obtained on devices emitting at 4.2 µm wavelength.[39,40]

Further progress in device technology and in cavity design should both improve the efficiency [41,42] (a 16 % efficiency resonant cavity LED has been obtained with a GaInAs structure emitting at 0.94 µm [42]) and of course increase the wavelength spectrum beyond 4.2 µm or in the short wavelength region, with simpler technical steps than those involved in VCSEL fabrication.

## 3. SMART IMAGE SENSORS

Detectors must exhibit high conversion efficiency, low noise, room temperature operation and must be available in complex array configurations compatible with efficient electrical signal processing for advanced compact imaging systems. Progress in smart imaging sensors and room temperature infrared detectors is addressing those requirements.

### 3.1. Active Pixel Sensors APS

In the field of detector arrays, CCD arrays continue to progress and remain an attractive choice for many imaging systems because of their sophisticated level of performance. However, smart sensors with integrated detectors and image processors on a stacked and compact assembly [43,44] appear more and more as a very strong competitor able to overcome the drawbacks associated with CCD technology,[45,46] specifically:

- The need for near-perfect charge transfer efficiency, which leads to sophisticated process steps, limits the size of the array and prevents the achievement of radiation-hard devices.
- The difficulty in integrating CCD with CMOS integrated circuits for digital timing and signal processing on the same chip without undesirable compromise (additional fabrication steps at the price of reduced yield and increased manufacturing cost) and to realize sensors in a very compact form.
- Limited readout rates (~ 50 frames/s) due to the inherent high capacitance of CCDs, which is too slow for many future applications.
- High power dissipation due to the need for large drive currents and voltages less compatible with standard CMOS technologies.
- No random access.
- Limited spectral range.

Several manufacturers are working on a different approach tuown as APS,[46] making rapid progress fowards the achievement of stacked chips which associate high sensitivity detection, high and versatile rate readout, low power consumption and more or less sophisticated signal processing including blooming and fixed pattern noise suppression, but also contour extraction, windowing, etc.

### 3.2. Room Temperature Infrared Detectors [47-49]

In the infrared range, a lot of progress has been made with HgCdTe arrays bumped to silicon readout integrated circuits in the atmospheric windows 3-5 µm and 8-12 µm for military applications. For other application windows, arrays of InSb, PtSi and extrinsic Si doped with gallium or germanium have been developed, for instance for space and astrophysics applications.

With the HgCdTe hybrid approach on Si, complex arrays up to 512 x 512 (and soon

1024 x 1024) have been demonstrated with low NETD (Noise Equivalent Temperature Difference of about 10 mK) and high reliability.

However, all these detectors suffer from the need for cryogenic operation. Although significant progress on signal processing has allowed increasing the operating temperature in such case without significant performance degradation, these devices cannot work at room temperature and are not well suited for many military or civilian applications.

The recent demonstration of efficient uncooled microbolometers [47, 48] is drastically changing the situation. In the near future, low cost, compact size infrared imaging systems working at room temperature will be available. Today NETDs around 40 mK (normalized to f/1 and a 30 Hz frame rate) have been already measured for a pixel and could be expected soon for complex focal plane arrays with further improvements in the readout electronics design, the pixel structure and sensing materials. Eventually, an NETD close to 10 mK appears reasonably attainable.

At the present stage of the technology, arrays of about 80,000 pixels with high fill factor (~ 50 %) have been demonstrated with different suspended structures supporting a thermal detecting element (microbridges) over an integrated CMOS readout circuit,[48, 49]

Although the response time of thermal effects tends to be low compared to quantum devices, it is nevertheless fast enough for operating in the range 40-70 frames per second sufficient for many applications (civilian : surveillance cameras, thermography, thermal viewer, medical imaging ... or military : helmet mounted sights, reconnaissance FLIRs, missile seeker sensors ...).[48, 49]

Moreover, the capability of microbolometers for multispectral response could be a significant advantage for large spectral range infrared analysis with or withaut the use of the optical filters described below.

## 4. LOW COST DIFFRACTIVE AND REFRACTIVE OPTICAL ELEMENTS

The achievement of optical systems requires of course the availability of suitable imaging elements such as lenses, beam splitters, polarizers, prisms or mirrors, compatible with the objective of miniaturization and low cost, and that present very high optical performance.s in terms of optical aberrations, noise level, numerical aperture, etc.

### 4.1. Diffractive Optical Elements : "DOE"

Although Diffractive Optical Elements (DOE) cannot be considered as a very novel optical element, they are beginning to exhibit high performance and to penetrate commercial markets.[50-52]

DOE have not only enabled very compact optical components at reduced cost, but also the development of industrial products that are difficult to produce with standard technologies.

The main difficulty with realizing DOEs is directly related to the wavelength the light, which requires dealing with very small lithographic patterns.

With the rapid progress of VLSI techniques and the access to submicron feature sizes (0,25 µm, now 0,18 µm and soon 0,12 µm for industrial integrated microelectronic circuits), tighter alignment tolerances and more accurate etching depth given by the sophisticated but now readily available equipment of the microelectronic industry, many DOE drawbacks will soon be overcome.[52, 53]

The newer VLSI techniques will permit DOEs with high diffraction efficiency at moderate and low f/numbers, increased spectral width, and better focusing properties even at high optical power. These significant improvements in DOE performance will be available from suppliers with full production capabilities.

At the same time, progress in DOE design and the replacement of the traditional analytic quantization method by more sophisticated encoding methods (such as the radially symmetric iterative discrete on-axis method based on a rigorous scalar diffraction model) have also greatly contributed to increase performance.

For example, f/1 collimating elements can provide very high diffraction efficiencies (greater than 90 %) over a spectral range of several tens of nanometers.[52, 53]

Low cost DOEs can be fabricated either by a photolithographic approach using a master mask, or by replication techniques such as injection molding or embossing.

Of course, if DOEs can make efficient focusing devices, they are also very attractive for fabricating high performance beam splitters or beam deflectors [53, 54] compatible with advanced optical microsystems in many fields of application, including fiber optic telecommunication, head mounted displays, optical instrumentation, and future optical interconnections between microelectronic chips.

## 4.2. Miniature Refractive Optical Components

It is also clear that progresses in DOE fabrication can be very useful for achieving low cost refractive optical elements exhibiting high performance and reproducible characteristics.

With standard lithography and processing of photoresist followed by etching or ion milling, aspheric microlenses can be easily fabricated; [55 - 57] however, focal lengths cannot be obtained over a wide range of values. These microlenses can be convenient in some applications such as stabilizing microlaser cavities [57] because only low numerical apertures are required, but they cannot cover all the needs of optical microsystems.

On the other hand, the use of anisotropic wet or dry etching techniques leads to a restricted range of geometrical shapes.

The use of one level gray tone lithography overcomes most of these limitations and opens a new age for batch process fabrication of various refractive optical elements.[58 - 60]

## 5. MICRO ELECTRO-MECHANICAL OPTICAL SYSTEMS (MEMOS)

Progress in silicon technologies, strongly pushed thirty years by microelectronics and VLSI integrated circuits, allows the achievement of new components and devices which are more and more complex and economically attractive.

This is particularly true in the field of microsensors and information storage, where new integrated read-write heads will soon replace more traditional components.

It is also the case in optical systems where many new components combining optical and mechanical elements driven by suitable electronical circuits are beginning to be commercially available or are in an advanced development stage.

These new components take advantage of various improvements in silicon based technologies.

The most relevant advances which allow the achievement of new components in the field of optics include:
- deep etching using anisotropic or isotropic processes in silicon, and the availability of high aspect ratio structures in silicon based technologies,
- the use of sacrificial layers to release mechanical structures and achieve movable elements,
- the development of new active material fabrication processes such as the "smart cut" techniques already developed to realize new SOI or SiC substrates.

Of course, all these new techniques for achieving high performance MEMOS must be

## 5.1. Micromirror Arrays For Digital Signal Processing [61, 62, 63]

One of the best example of a new opportunity is proveded by the Digital Micromirror Device (DMD) developed by Texas Instrument. The DMD light switch is a MEMS completely achieved from CMOS-like processing steps implemented over a CMOS memory chip. It is an aluminum mirror which can reflect an incident light beam in two different directions, -10° or +10°, depending on the state "0" or "1" of the underlying memory cell.

By combining the device with suitable projection optics, each mirror is able to reflect its incident light beam either into or out of the pupil of the projection lens, giving a bright or a dark spot at the imaging plane.

Gray scale is obtained by binary pulsewidth modulation of the "1" state so that incident light beams are switched into light bundles having suitable duration.

Colors can be accommodated by suitable combination of color filters, either rotating or stationary according to the complexity of the projection systems, which may use three, two, or only one DMD chips.

Each aluminum mirror of the DMD chip is rigidly attached to an underlying yoke connected by mechanically compliant torsion hinges to the substrate. Electrostatic forces developed by a suitable voltage cause rotation of the yoke and the mirror in the negative or positive direction, with limits at -10° and +10° due to the use of mechanical stops.

Many improvements have been made since the first proposed structures in order to overcome different practical problems leading to the use of six photomask levels.

Mirrors and mechanical hinges were first fabricated on separate layers in order to increase the reflecting area, decrease light diffraction, improve the contrast ratio up to more than 100:1, and improve optical efficiency.

Since then, new designs have been developed in order to maximize the electrostatic attraction area, and to use both the yoke and the mirror as active elements, leading to lower driving voltages compatible with CMOS technology.

The DMD chip is currently used in various Digital Light Processors, DLP, including projection displays and color printing. The first DLP product, designed to serve as a portable business display with VGA resolution (640 x 480), has been commercialized and various DLP systems are about to be manufactured or in an advanced demonstration stage.[61, 64, 65]

The several years effort by Texas Instruments researchers and the impressive results finally obtained demonstrate the real potential of silicon based technologies to achieve very complex, efficient and reliable optical microsystems.

## 5.2. Optical Scanners and Choppers

DMD provides concrete proof that optical microsystems can work perfectly and drastically modify an industrial sector.

The large markets opened up by Digital Mirror Devices has jutified the technical effort devoted to their fabrication.

Of course, this is not yet the case for many optomechanical components which cannot be considered as key components for a large range of consumer products but only as one piece of the optical microsystems puzzle.

Therefore, many attractive realizations presently under study in advanced laboratories can still appear very immature and far from an industrial product compared with DMD.

Nevertheless, the of work carried on this field, along with the impressive progress on

DMD described previously, shows that the landscape of optical MEMs is considerably changing.

It is obviously difficult to give a complete surveyof all published work, but different examples can be chosen to illustrate properly the state of the art and trends.

Active micromirrors and deflectors for light beam steering or scanning are one of the more useful components for various applications.
Although the DMD belongs to this family, its characteristics are of course mainly suited for image projection, and present drawbacks for beam steering of laser beams.

The first of these drawbacks is a deflection angle limited to a few degrees because of the need for small electrode gaps compatible with low driving voltages which in turn leads to a small range of motion.

Different solutions can be developed to overcome this difficulty by separating the electrode plate, on which the driving voltage is applied, from the mirror plate ; both plates are connected by a suitable torsion bar and the mirror acts as a dynamic damper. This configuration allows one to increase arbitrarily the clearance between the mirror plate and the substrate, leading to a wider amplitude angle without increasing the driving voltage when working at resonance.[66]

Various laboratories work on such approaches using different technological tricks. Another attractive approach has been proposed by H. Goto, using a silicon micromachined two dimensional scanning mirror with a piezoelectric actuator for excitation of the resonator.[67-69]

In this device, a scanning mirror supported by a tiny torsion bar is etched on a silicon substrate and excited at its resonance frequencies for bending and torsional vibration modes by a piezoresistor.

This scanner is able to work as a two dimensional optical scanner by driving the actuator at both resonance frequency modes ; large scanning angles greater than 60° for bending mode and 30° for torsional mode can be obtained with an amplitude excitation of 5 µm.[68,69]
It has been employed as a miniature 3D vision sensor for pipe inspection by a mobile robot.

However, even if all approaches concerning moving mirrors described previously were perfectly convenient for achieving various types of miniaturized scanning components suitable for low cost application, they present an inherent drawback when they are integrated in a complete optical microsystem. In all cases, the incident light is reflected backward and returns in the same half plane ; this configuration prevents the realization of a direct assembly in which the scanning element and the light source for instance could be combined at the wafer level.

To overcome this problem, a double mirror configuration realized by chemical anisotropic etching of silicon has been recently proposed.[70] In this scheme, one or both mirrors can be moved by electrostatic or piezoelectric excitation after suitable deposition and etching stages, leading to a 2D scanner where the light is well deflected forward.
This is also the case for scanners using moving lenses or microlenses arrays driven by linear electrostatic or piezoelectric actuators.[71,72] Although this approach requires short focal lengths to reach large deflection angles because of the limited displacement excursion available, it leads to simpler structures which are easy to insert in a microsystem assembly.

More generally, microfabricated actuators[73] and particularly electrostatic comb drivers can be very useful to achieve a wide variety of active micro-optical elements, taking advantage of considerable progress in the field of micro-accelerometers.

Optical choppers and micro shutters can be made easily in a similar way, and have been demonstrated by several laboratories. They can be inserted in various measurement microsystems to increase the signal to noise ratio, and can operate up to Mhz frequencies.[74,75]

## 5.3. Optical Filters

Optical filters are other basic components for many optical systems, especially in the fields of infrared spectrometry and optical fiber communication.

Several approaches have already been explored both in planar and 3D optics to develop compact, low cost and efficient filters using micromachined fixed or movable gratings,[8, 9, 76, 77] phase array integrated elements [78] and different kinds of Fabry Perot interferometers.[79-80]

For example, a tunable grating filter has been demonstrated by combining a LIGA fabricated grating and a linear magnetic drive actuator [76] in order to obtain a spring flexure system which allows expansion or contraction of the filter spacing and therefore the changes the spectral range of analysis.

However one of the most attractive approaches for realizing optical filters to uses the ability of silicon based technologies to fabricate thin and transparent moving membranes in order to form a tunable Fabry-Perot interferometer.

This approach, which uses several versions of the well-known sacrificial layer technique to release the membrane from the substrate and electrostatic actuation to move it, is derived from the fabrication of miniature and low cost capacitive pressure sensors developed automotive engine for controls. In optical filters, a $\lambda/4$ membrane is generally suspended above the substrate with a $3\lambda/4$ thick air gap (where $\lambda$ is the center wavelength of the desired operating spectral window), providing high reflection of the light. When the moving membrane is electrostatically attracted towards the substrate by a suitable voltage, the air gap is reduced to $\lambda/2$ for which low reflection conditions are obtained. High contrast ratios (greater than 24 dB) have already been measured with such devices, with low driving voltages and very reduced size allowing high frequency modulation up to a few MHz.[79, 83]

Tunable filters exhibiting a transmission spectrum linewidth in the range of 3-5 nm (full width half maximum) in the 1550 nm window and a tuning range 40-60 nm wide have been also reported with maximum driving voltages of 60 V.[80, 81]

With further improvements, these components can be incorporated in wavelength division multiplexed passive optical network systems where after suitable modulation the same light is transmitted downstream and upstream between the subscriber and the central office.[79, 83]

These components can be also very useful alone as a single element or in an array configuration, in spectral analysis microsystems associated with such broad band light sources as LEDs, infrared emissive microfilaments or directly integrated with semiconductor VCSELs [85] in order to realize a widely tunable light source (over 19 nm with tuning voltage around 14 V).

A tunable Fabry-Perot filter has also been demonstrated, using AlGaAs distributed Bragg reflectors associated with a cantilever structure, exhibiting continuous tuning over 70 nm around 940 nm, 4.9 V tuning voltage, contrast ratio of 19 dB, loss of 5.5 dB and response frequencies close to 10 Khz.[86]

## 5.4. Micro Optical Focusing and Adaptive Optics Components

Structures very similar to that of the Fabry-Perot filter can be applied to fabricate controlled flexible mirrors or lenses in order to develop stand-alone adaptative optical component,[87-90] as proposed long ago,[87] or dynamic focusing microsystems using electrostatic or piezoelectric actuators.[91, 92]

An adaptive mirror of 10,5 x 10,5 mm² aperture controlled by nine rectangular actuators has been recently demonstrated at Delft University.[88, 89] The deformable membrane is realized by chemical etching of a silicon bulk substrate, leaving a 0,5 µm thick silicon nitride layer covered by aluminium and suspended by the edges of a rectangular window.

The flexible mirror array is then mounted over the actuator structure, with an air gap ranging from 20 to 100 µm, and driven by electrostatic force with voltage on the order of 40 Volts. Of course various improvements can be implemented in similar structures leading to a great versatility in adaptative optics configurations.

In the same way, micro focusing devices have been demonstrated using silicon bulk micromachining and piezoelectric actuators involving PZT thin films. In these devices the light provided by a surface emitting laser is focused at different positions by a deformable mirror associated with a micro Fresnel lens.[92]

These devices provide the ability to change the focal position over 100 mm with very low applied voltages (≈ 5 Volts) and potentially good optical characteristics. This offers a new opportunity to develop optical choppers and modulators, various kind of displacement sensors, control of optical heads in compact disc drives, and selffocusing miniature cameras.

## 5.5. MEMOS for Optoelectronic Coupling

The development of surface micromachined mirrors using sacrificial layer and microhinge fabrication techniques, combined with integrated electrostatic comb drive actuators, can be now generalized to implement more complex out of plane optical beam steering systems.

Such devices have been recently developed in order to achieve precise alignment of optoelectronic components such as laser diodes and optical fibers.[93-97]
The structures developed at Berkeley [94,95] use three different sets of microhinges connected with two movable electrostatically driven sliders and a mirror. Linearly translating the sliders causes the mirror to foldup and rise off the substrate. Translation of the mirror without tilt is obtained when both sliders are moved together in the same direction and the mirror angle is changed if one slider is moved with respect to the other, leading, with appropriate design, to a mirror system with two degrees of freedom.

Associated with collimating lenses, the coupling module can position a laser beam or a fiber with a 0,17 µm standard deviation and present a range of motion greater than 200 µm. Coupling efficiency of 40 % has been measured, and 75 % is expected when various identified loss mechanisms are corrected with an improved silicon bench.

Although the use of such modules for accurate active coupling on chip can appear today nonsensical in terms of cost, this experiment shows clearly the potential of microtechnologies.

It opens new possibilities in the field of optical microsystems in various applications such as optical disk pick-up heads [98] or external cavity semiconductor lasers [99] for laser linewidth narrowing, frequency tuning or mode locking.[100,101]

## 5.6. MEMOS and Micromotors

The availability of micromotors can be an opportunity for optical systems because the required mechanical forces are generally lower than those necessary in other fields of application.

Although other approachesare possible, micromotors are very useful for microscanners, microspectrometers, auto focusing devices, and mechanical choppers, as described in several publications.[102-105]

Although linear micromotors have already demonstrated [106] their ability to work properly and will soon be integrated in optical systems, rotating micromotors are not yet mature for this purpose. Up to now various approaches to electrostatic or piezoelectric micromotors have been proposed.

The electrostatic approach requires high rotation speed to give acceptable torques, and this is generally incompatible with the present state of the art of micromachined devices [107]. Therefore, although very good photographs have been published in the literature showing miniature gears and different electrostatic micromotor configurations, no very attractive results in terms of both suitable mechanical forces and reliability have been yet demonstrated.

Piezoelectric and more generally friction force micromotors seem a better way to realize miniature rotating elements and to overcome problems associated with miniaturization.[108, 109]

## 5.7. Integrated Optics MEMOS

Considerable work has been carried out in integrated optics for more than twenty years on various substrates : III-V semiconductors,[1-4] LiNbO3,[5] glasses [6] or silicon, applied to optical communications,[8-13] sensors [8, 9, 11] and read-write optical or magneto-optical [110-113] heads, and very efficient components and devices have been demonstrated.

However mechanical structures are new additions, having been proposed and introduced very recently.

These mechanical structures address the need for low speed switches in optical fiber systems for network safety or network reconfiguration, but also in optical sensor systems to permit new devices or more efficient optical signal processing.

In all these cases, high frequency response is not necessary and mechanical switches, as compared with electro-optic or acousto-optic devices, offer suitable solutions in terms of polarization and wavelength insensitivity. In addition, mechanical switches can allow lower response time and drastically lower power consumption than devices using thermal effects.[114]

Silicon based integrated technologies are well suited to achieve micro-mechanical components, and the first opto mechanical structures were proposed seven years ago by S. Valette and J.S. Danel.[115]

In these devices a doped silica waveguide is buried in a movable cantilever driven by electrostatic forces. Different switching configurations can be realized using this basic idea, from a simple $1 \rightarrow 2$ switch to complex M x N matrices. Practical devices have been recently achieved and reinforce interest in the approach.[116-118]

The best results have been those published by E. Ollier [117, 118] with polarization insensitive switches in phosphorus doped silica waveguides working at 28 V, exhibiting less than 3 dB fiber to fiber loss and operating in the two wavelength windows of optical communication, 1.3 and 1.55 µm.

Although such devices are simple in principle and not very difficult to available silicon based technologies, various tricks must be implemented in order to decrease the driving voltage, for instance by using an electrostatic comb driver taking advantage of the bistability effect, to reach low insertion losses. The design is critical because of the stress gradient in the silica waveguide structure, which leads to undesirable vertical deflection when releasing the cantilever.

This effect can be greatly minimized by suitable thermal treatments and by adding compensating mechanical arms which prevent the cantilever from moving in the vertical direction without a significant effect on the in plane displacement.[117, 118]

These realizations demonstrate the potential of electromechanical structures for integrated optics devices and open the way for new components in this field.[119]

## 6. TECHNIQUES FOR MEMOS ASSEMBLIES

The last but not the least field of microsystem activities concerns MEMOS assemblies. Although all-monolithic approaches can be used and have been already demonstrated to realize advanced microsystems, the great number of different functions which will have to be implemented in future MEMS suggests that hybrid approaches will be also widely applied.

This is particularly true in optical microsystems because so many different technologies need to be employed : semiconductor technologies for optical sources, quantum detectors or high speed amplitude modulators, silicon or quartz technologies for electromechanical devices, various glasses or polymers for diffractive or refractive elements etc.

Thus, it may be impossible to achieve complex MEMOS without using a hybrid approach, which supposes the availability of efficient and low cost assembly techniques.

Assembly of several individual optical components is a well-established way to realize complex optical systems and that know-how is of course also useful in the case of microsystems. These techniques must be associated with a more general know-how coming from advanced microelectronic packaging which is well suited to build a complex miniaturized device involving various complementary functions.

The Multichip Module family offers many advantages for realizing reliable and compact assemblies able to overcome mechanical, electrical and thermal problems. These packaging techniques, combining solder bumps, wire bonding and connecting vias through metallized holes, have already demonstrated their potential with the achievement of advanced mass memory modules (64 x 16 Mbit DRAMS interconnected on 8 stacked MCMs equivalent to a 1 Gbit memory).[120]
Although some adaptation will be required to take into account the specific problems involved in opto-mechanical packaging, this technical approach will be useful in the future.

Another attractive technique for the development of optical microsystems is wafer bonding. It can be used to realize rugged mechanical assemblies at the wafer scale, as already demonstrated in the field of micromachined sensors.
In addition, these techniques are attractive for achieving new structures, as illustrated by the so called "smart-cut" process already used to build SOI structures.[121, 122]

This process, which involves slicing a thin film material created by hydrogen implantation, followed by heat-treatment and wafer bonding, offers new opportunities to realize new substrates and active thin film devices in the fields of optoelectronics, microelectronics, microsensors and micro-actuators.

## 7. CONCLUSION

New optical components fabricated using silicon technologies and batch processing are currently under study in most major research laboratories.

Progress in miniaturized light sources exhibiting various optical properties, in detectors, in high performance diffractive optical elements and in micro machined optical components and devices for free space or planar optics, will permit a variety of industrial optical microsystems.

These systems can combine attractive features, compact size and low cost to address a wide range of applications including chemical analysis by infrared spectrometry, range finders for obstacle avoidance, various displays for digital light processing and communication networks, smart imaging sensors etc.

The more advanced optical microsystems are entering the market place and already demonstrate the potential of these technical approaches.

## REFERENCES

1. D. Hofstetter, H.P. Zappe, P. Riel, J.E. Epler, and O.J. Homan, III-V based integrated optical chip for metrology : device and integration technology, *Proceedingss of the 7$^{th}$ European Conference on*

*Integrated Optics ECIO 1995*, pp 109 - 112, DELFT, (April 1995).
2. D.Trommer, Photonic integration on InP, *Proceeding of the 7$^{th}$ European Conference on Integrated Optics ECIO 1995*, pp 93 - 98, DELFT, (April 1995).
3. M.R. Amersfoort, WDM devices in InP/InGaAsP, *Proceeding of the 7$^{th}$ European Conference on Integrated Optics ECIO 1995*, pp 499 - 504, DELFT, (April 1995).
4. A. Carenco, Advances Semiconductor Integrated Optics, in *Advances in Integrated Optics*, Plenum Press (New York and London), edited by S. Martellucci, A.N. Chester and M. Bertolotti, (1994).
5. Y.M. Yurek, P.G. Suchoski, S.W. Merrit and F.J. Leomberger, Commercial $LiNbO_3$ Integrated Optics Devices, *Optics and Photonics News*, pp 26 - 30, (June 1995).
6. S. Kobayashi, F. Kiger, M. Meyers, and A. Spector, Prospects for silicon and glass based integrated optics, *Proceedingss of the 7$^{th}$ European Conference on Integrated Optics ECIO 1995*, pp 309 - 314, DELFT, (April 1995).
7. W.H.G. Horsthuis, and R. Lytel, Prospects for Integrated Optics polymer components, *Proceedings of the 7$^{th}$ European Conference on Integrated Optics ECIO 1995*, pp 67 - 71, DELFT, (1995).
8. S. Valette, Integrated photonic circuit on silicon, in *Novel Silicon Based Technologies*, R.A. LEVY editor, Kluwer Academic Publishers, pp 173 - 240, (1991).
9. S. Valette, Integrated Optics on Silicon : IOS technologies, in *Advances in Integrated Optics*, Plenum Press (New York and London), edited by S. Martellucci, A.N. Chester and M. Bertolotti, (1994).
10. C.H. Henry, Glass waveguides on silicon for hybrid optical packaging, *Journal of Lightwave Technology*, 7, pp 1530 - 1539, (1989).
11. P. Mottier, Integrated Optics at LETI, *International Journal of Optoelectronics*, Vol. 9, pp 125 - 135, (1994).
12. C. Artigue, Trends in silicon based optoelectronics hybrid integration, *Technical Digest of Integrated Photonics Research Conference*, pp 304-306, Boston, (April-May 1996).
13. R. Bozeat and A. Loni, Silicon-based waveguides offer low cost manufacturing, *Laser Focus World*, pp 97 - 102, (April 1995).
14. S. Valette, Micro-optics: a key technology in the race to microsystems, *Journal of micromechanics and micro-engineering*, Vol. 5, pp 74 - 76, (June 1995).
15. K. Iga, F. Koyama and S. Kinoshita, Surface emitting semiconductor lasers, *IEEE J. Quantum Electronics*, 24, pp 1845 - 1855, (1988).
16. J.L. Jewell, Y.H. Lee, A. Scherer, S.L. McCall, N.A. Olsson, J.P. Harbison, and L.T. Florez, Surface emitting microlasers for photonic switching and interchip connections, *Optical Engineering*, 29, pp 210 - 214, (1990).
17. R.A. Morgan, Vertical Cavity Surface Emitting Laser: The next generation, SPIE Vol. 1992, *Miniature and Micro-optics and Micromechanics*, pp 64 - 88, (1993).
18. S.W. Koch, F. Jahnke and W.W. Chow, Physics of semiconductor microcavity lasers, *Semiconductor Science Technologies*, 10, pp 739 - 751, (1995).
19. J.L. Jewell, J.P. Harbison, A. Sherer, Y.H. Lee and L.T. Florez, Vertical cavity surface emitting lasers : design, growth, fabrication, characterization, *IEEE Journal of Quantum Electronics*, 27, pp 1332 -1346, (June 1991).
20. J.L. Jewell, G.R. Olbright, R.R. Bryan and A. Scherer, Surface emitting lasers break the resistance barrier, *Photonic spectra*, pp 126 - 130 (November 1992).
21. G.R. Olbright, VCSEL's could revolutionize optical communication, *Photonic spectra*, pp 98 - 101, (February 1995).
22. H.E. Shin, Y.G. Ju, J.H. Shin, J.H. Ser, T. Kim, E.K. Lee, I. Kim and Y.H. Lee, 780 nm oxidized vertical cavity surface emitting laser with $Al_{0.11}Ga_{0.89}As$ quantum wells, *Electronics Letters*, 32, pp 1287 - 1288, (July 1996).
23. J.W. Scott, B.J. Thibealt, L.A. Coldren, Design and modeling of high speed intracavity contacted vertical cavity lasers, *Proceedingss of SPIE Vol. 2399*, pp 572 - 582, (1995).
24. M. Hagerott Crawford, R.P. Schneider Jr, K.D. Choquette, K.L. Lear, S.P. Kilcoyne and J.J. Figiel, High efficiency AlGaInP based 660 - 680 nm vertical cavity surface emitting lasers, *Electronics Letters*, 31, pp 196 - 197, (February 1995).
25. J.J. Dudley, D.I. Babic, R. Mirin, L. Yang, B.I. Miller, R.J. Ham, J. Reynolds, E.L. Hh and J.E. Bowers, Low Threshold, wafer fused long wavelength vertical cavity lasers, *Appl. Phys. Lett.*, 64, pp 1463 - 1465, (March 1994).
26. K.D. Choquette, H.Q. Hou, K.L. Lear, H.C. Chui, K.M. Geib, A.A. Mar, and B.E. Hammons, Seff pulsing oxide confined vertical cavity lasers with ultra low operating current, *Electronics Letters*, 32, pp 459 - 460, (February 1996).
27. K.L. Lear, A. Mar, K.D. Choquette, S.P. Kilcoyne, R.P. Schneider Jr and K.M. Geib, High frequency modulation of oxide confined vertical cavity surface emitting lasers, *Electronics Letters*, 32, pp 457 - 458, (February 1996).
28. T. Baba, Near room temperature continuous wave losing characteristics of GaInAsP/InP surface emitting laser, *Electronics Letters*, pp 913 - 914, Vol. 29, (1993).

29. *Compound Semiconductor*, p 36, (July/August 1995).
30. *Laser Focus World*, p 11, (August 1996).
31. Achim Strass, Surface emitting lasers cut data transfer costs, *Opto Laser Europe*, pp 23 - 27, (April 1995).
32. J.J. Zayhowski, Microchip lasers create light in small spaces, *Laser Focus World*, pp 73 - 78, (April 1996).
33. A. Mooradian, K. Wall and J. Keszen Heimer, Microchip lasers and laser arrays: technology and applications, *Optics and Photonics News*, Vol. 6, pp 16 - 19, (November 1995).
34. S. Valette, Microlasers: key components for optical microsystems, *Proceedingss of the SPIE conference*, Vol. 2783, Micro-optical Technologies for Measurement, Sensors and Microsystems, pp 16 - 21, BESANCON, (June 1996).
35. E. Molva et al, Microchip laser and micro-optics technologies, *Proceedings of CLEO Conference*, Baltimore, (May 1995).
36. Ph. Thony et al, 1.55 µm wavelength CW microchip lasers, *Proceedings of the Advanced Solid State Lasers Conference*, pp 296 - 300, San Fransisco, (January - February 1996).
37. J.J. Zayhowski and C. Dill, Coupled cavity electro-optically Q switched Nd - $YVO_4$ microchip lasers, *Optic Letters*, Vol. 20, p 716, (1995).
38. E. Hadji, J. Bleuse, N. Magnea and J.L. Pautrat, 3,2 µm infrared resonant cavity light emitting diode, *Appl. Phys. Lett.*, 67, pp 2591 - 2593, (1995).
39. R. Roux, Resonant-cavity LED's, operate at 3 to 5 µm, *Laser Focus World*, p 32, (February 1996).
40. E. Hadji, J. Bleuse, N. Magnea and J.L. Pautrat, Resonant cavity light emitting diodes for the 3-5 µm range, *Solid State Electronics*, 40, pp 473 - 476, (1996).
41. E.F. Shubert, N.E.J. Hurt, M. Micovic, R.J. Malik, D.L. Sinco, A.Y. Cho, and C.J. Zidzik, Highly efficient light emitting diode with microcavities, *Science*, Vol. 265, pp 943 - 945, (1994).
42. J. Blondelle, H. Deneve, P. Demeester, P. Yan Daele, G. Borghs and R. Baets, 16 % external quantum efficiency from planar microcavity LED's at 940 nm by precise matching of cavity wavelength, *Electronics Letters*, Vol. 31, pp 1286 - 1288, (July 1995).
43. J.C. Carson, D.R. Shostak and D.E. Ludwig, Smart sensors integrate detectors and processors, Laser Focus World, pp 113 - 115, (July 1996).
44. S.K. Medis, S.E. Kemeny, R.C. Gee, B. Pain, Q. Kim and E.R. Fossum, Progress in CMOS active pixel image sensors, *Proceedings of the SPIE Vol. 2172*, pp 19 - 29, (1994).
45. E.R. Fossum, Active pixel sensors: are CCD's dinosaurs ?, *Proceedings of the SPIE Vol. 1900*, pp 2 - 14, (1993).
46. E.R. Fossum, Assessment of image sensor technology for future NASA mission, *Proceedings of the SPIE Vol. 2172*, pp 38 - 53, (1994).
47. R.A. Wood, C.J. Han and P.W. Kruse, Integrated uncooled infrared detector imaging arrays, *IEEE Solid State Sensors and Actuator Workshop*, pp 132 - 135, Hilton Head Island, (June 1992).
48. R.A. Wood, Uncooled thermal imaging with monolithic silicon focal plane arrays, *Proceedingss of SPIE* Vol. 2020 Infrared Technology XIX, pp 329 - 336, (1993).
49. N. Butler, R. Blackwell, R. Murphy, R. Silva, C. Marshall, Low cost uncooled microbolometers imaging system for dual use, *Proceedings of SPIE* Vol. 2552, Infrared Technology XXI, San Diego, pp 583 - 591, (July 1995).
50. T.D. Milster, R.M. Trusty, M.S. Wang, F.F. Froehlich, J.K. Erwin, Micro-optic lens for data storage, *Proceedings of the SPIE* Vol. 1499, "Optical data storage", pp 286 - 292, (1991).
51. N. Streibl et al, Digital optics, *Proceedings of the IEEE*, Vol. 77, pp 1954 - 1969, (December 1989).
52. M.R. feldman, Diffractive optics move into the commercial arena, *Laser Focus World*, pp 143 - 152, (October 1994).
53. S.J. Walker et al, Design and fabrication of high efficiency beam splitters and beam deflectors for integrated planar micro-optic systems, *Applied Optics*, 32, pp 2494 - 2501, (May 1993).
54. Ph. Regnault, H. Bucsek, O. Anthamatten, G. Voirin, Ch. Zimm, and H. Gilgen, Diffractive interconnecting device for filter to chip coupling: *CSEM Scientific and Technical Report 1993*, p 38, (1993).
55. M.C. Hutley, Opticals technics for the generation of microlens arrays, *Journal of Modern Optics*, 37, pp 253 - 255, (1990).
56. M.C. Hutley, R. STEVENS and D. DALY, Microlens array, *Physic World*, pp 27 - 31, (July 1991).
57. M. Rabarot, V. Tarazona, E. Molva, C. Clement, Stabilized Mocrochip laser fabricated by micro-optical technologies, *Proceedingss of the SPIE* Vol. 2783, "Micro-optical technologies for measurements, sensors and Microsystems", pp 40 - 45, BESANCON, (June 1996).
58. Y. Oppliger, P. Sixt, J.M. Stauffer, J.M. Mayor, Ph. Regnault, and G. Voirin, Gray-tone masks for one step shaping of DOE's, *CSEM Scientific and technical report* 1993, p 37, (1993).
59 K. Reimer, H.J. Quenzer, R. Demmeler, and B. Wagner, One-level gray-tone lithography mask data preparation and pattern transfer, *Proceedingss of SPIE Vol*. 2783, Micro-optical technologies for measurements, sensors and microsystems, pp 71 - 79, BESANCON, (June 1996).

60. Y. Oppliger, P. Sixt, J.M. Stauffer, J.M. Mayer, P. Regnault and G. Voirin, One-step 3D shaping using a grey-tone mask for optical and microelectronics applications, *Microelectronics Engineering*, 23, pp 449 - 454, (1994).
61. L.J. Hornbeck, Digital light processing and MEMS: timely convergence for a bright future presented at *Micromachining and Microfabrication 95*, Plenary Session paper, Austin, (October 1995).
62. L.J. Hornbeck, Current Satus of the Digital Micromirror Device (DMD) for projection television, *Proceedingss of the International Electron Devices Conference*, pp 381 - 384, (1993).
63. L.J. Hornbeck, Deformable mirror spatial light modulators, *Proceedingss of SPIE* Vol. 1150, "Spatial Light Modulators and Application III", pp 86 - 102, (August 1989).
64. W.E. Nelson and R.L. Bhuva, Digital micromirror device imaging bar for hard copy, *Proceedingss of SPIE* Vol. 2413 Color Hardcopy and Graphic Art IV, San Jose (February 1995).
65. M.A. Mignardi, Digital micromirror array for projection TV, *Solid State Technology*, Vol. 35, p 63 - 66, (July 1994).
66. T. Usuda, H. Yamada and A. Umeda, A novel silicon torsional resonator with two degrees of freedom, *Proceedings of the $7^{th}$ International Conference on Solid State Sensors and Actuators*, Tokyo, (June 1993).
67. H. Goto, High performance microphotonic devices with micro-actuator, *Proceedings of SPIE* Vol. 1992, Miniature and Micro-optic and Micromechanics, pp 32 - 39, (1993).
68. M. Ikeda, H. Goto, M. Sakata, and S. Wakabayashi, Two dimensional silicon micromachined optical scanner integrated with photo detector, *Proceedings of SPIE* Vol. 2383, pp 118 - 124 - San Jose (February 1995).
69. H. Goto, Si micromachined 2D optical scanning mirror and its application to scanning sensor, *Digest of IEEE LEOS 96 Summer Topical Meeting*, Keystone (USA), pp 17 - 18, (August 1996).
70. S. Valette, Micro-élément de balayage pour système optique, French patent n 9508751.
71. G.F. Mc Dearmon, K.M. Flood and J.M. Finlan, Comparison of conventional and microlens-array agile beam steerers, *Proceedings of SPIE* Vol. 2383, pp 167 - 178, (1995).
72. R. Göring and S. Glöckner, Micro-optical beam deflectors and modulators - present state of development, *Proceedings of SPIE* Vol. 2783, pp 154 - 162, (1996).
73. J.J. Sniegowski and E.J. Garcia, Microfabricated actuators and their application to optics, *Proceedings of SPIE* Vol. 2383, pp 46 - 64, San Jose, (February 1995).
74. M.T. Ching, R.A. Brennen and R.M. White, Microfabricated optical choppers, *Proceedings of SPIE*, Vol. 1992, Miniature and Micro-optics and Micromechanics, pp 40 - 46, (1993).
75. H. Toshiyoshi, H. Fuhita, and T. Ueda, A self-excited chapper made by quartz micromachining and its application to an optical sensor, *Proceedings of the $7^{th}$ IEEE workshop on Micro Electro-Mechanical Systems (MEMS 94)*, pp 325 - 330, OHISO (Japan), (January 1994).
76. D. Arch, T. Ohnstein, D. Zook and H. Guckel, A MEMS based tunable infrared filter for spectroscopy, *Digest of IEEE/LEOS 96 Summer Topical Meeting*, Keystone (USA), pp 21 - 22, (August 1996).
77. P. Kpippner, J. Mohr, C. Müller, and C. Vandersel, Microspectrometer for the infrared range, *Proceedings of the SPIE* Vol. 2783, pp 277 - 282, (1996).
78. Y. Inoue, Silicon-based planar lightwave circuits for WOM systems, *Technical Digest of Integrated Photonic Research*, pp 32 - 35, BOSTON, (April-May 1996).
79. J.A. Walker, K.W. GOOSEN and S.C. ARNEY, Mechanical anti-reflexion switch (MARS) device for fiber-in-the-loop applications, *Digest of IEEE/LEOS 96 Summer Topical Meeting*, KEYSTONE (USA), pp 59 - 60, (August 1996).
80. A.T.T.D. Tran, Y.H. Lo, Z.H. Zhu, D. Horonian, E. Mozdy, Surface micromachined Fabry Perot tunable filter, *IEEE Photonic Technology Letters*, Vol. 8, pp 393 - 395, (March 1996).
81. Y.H. Lo, A.T.T.D. Tran, S.H. Shy, G.L. Christenson, Integrated micro-optical interferometer arrays, *Digest of IEEE/LEOS 96 Summer Topical Meeting*, KEYSTONE (USA), pp 25 - 26, (August 1996).
82. D.B. Fanner, R.M. Caragelo, P.J. Kung, P.G. Hanblen, and J.I. Budnick, Silicon etalon arrays for IR transformer spectroscopy, *Digest of IEEE/LEOS 96 Summer Topical Meeting*, Keystone (USA), pp 23 - 24, (August 1996).
83. O. Anthamanen, R.K. Bottig, B. Valk and P. Vogel, Packaging of a reflective optical duplexer based on silicon micromechanics, *Digest of IEEE/LEOS 96 Summer Topical Meeting*, Keystone (USA), pp 61 - 62, (August 1996).
84. N.C. Macdonald and A. Jazairy, Single crystal silicon, application to micro-opto electromechanical devices, *Proceedings of the SPIE* Vol. 2383, pp 125 - 133, (1995).
85. C.J. Chang-Hasnain, E.C. Veil and M.S. Wu, Widely tunable micro-mechanical vertical cavity lasers and detector, *Digest of IEEE/LEOS 96 Summer Topical Meeting* Keystone (USA), pp 43 - 44, (August 1996).
86. E.C. Vail, M.S. Wu, G.S. Li, L. Eng and C.J. Chang-Hasnain, GaAs micromachined widely tunable Fabry Perot filter, *Electronics Letters*, Vol. 31, pp 228 - 29, (February 1995).
87. R.P. Grosso and M. Yellin, The membrane mirror as an adaptative optical element, *J. Opt. Soc. Am.*, 67, pp 399 - 406, (1977).

88. G. Vdovin and P.M. Sarro Flexible mirror micromachined in silicon, *Applied Optics*, Vol. 34, pp 2968 - 2972, (June 1995).
89. M.J. Daneman, O. Solgaard, M.C. Tien, K.Y. Lou and R.S. Muller, Laser to fiber coupling module using a micromachined alignment mirror, *IEEE Photonics Technology Letters*, Vol. 8, pp 396 - 398, (March 1996).
90. L.M. Miller, M.L. Argonin, R.K. Bartman, W.J. Kaise, T.W. Kenny, R.L. Norton, and E.C. Vote, Fabrication and characterization of a micromachined deformable mirror for adaptative optics applications, *Proceedings of SPIE* Vol. 1945, pp 424 - 430, (1993).
91. M. Hisanaga, T. Koumura and T. Nattori, Fabrication of three dimensionally shaped Si diaphragm dynamic focusing mirror, *Proceedings of the IEEE conference on Micro Electro Mechanical Systems*, pp 30 - 35, New York, (1993).
92. H. Goto, S. Wakabayashi, M. Ikeda, M. Sakata, and K. Imanaka, Micro focusing optical device using piezoelectric thin film actuator, *Proceedings of the SPIE* Vol. 2383, pp 136 - 143, (1995).
93. M.J. Daneman, O. Solgaard, M.C. Tien, K.Y. Lau and R.S. Muller, Laser to fiber coupling module using a micromachined alignment mirror, *IEEE Photonic Technology Letters*, 8, pp 396 - 398, (March 1996).
94. N.C. Tien, O. Solgaard, M.H. Kiang, M. Daneman, K.Y. Lau, and R.S. Muller, Surface-micromachined mirrors for laser beam positioning, *Sensors and Actuators A*, 52, pp 76 - 81, (1996).
95. O. Solgaard, M. Daneman, N.C. Tien, A. Friedberger, R.S. Muller and K.Y. Lau, Optoelectronic packaging using silicon surface micromachined alignment mirrors, *IEEE Photonics Technology Letters*, 7, pp 41 - 43, (1995).
96. M.C. Wu, L.Y. Lin, S.S. Lee and K.S.J. Pister, Micromachined free-space integrated micro-optics, *Sensors and Actuators A*, 50, pp 127 - 134, (199?).
97. Y. Lin, S.S. Lee, K.S.J. Pister and M.C. Wu, Micromachined three dimensional micro-optical for free-space optical systems, *IEEE Photonic Technology Letters*, 6, pp 1445 - 144?, (1994).
98. L.Y. Lin, J.L. Shen, S.S. Lee and M.C. Wu, Realization of novel monolithic free-space optical disk pickup heads by surface micromachining, *Optics Letters*, 21, pp 155 - 157, (1996).
99. M.H. Kiang, O. Solgaard, R.S. Muller and K.Y. Lau, Silicon micromachined micromirror with integrated high precision actuators for external cavity semiconductor lasers, *IEEE Photonic Technology Letters*, 8, pp 95 - 97, (January 1996).
100. W.R. Trutna and L.F. Stokes, Continuously tuned external cavity semiconductor laser, *J. Lightwave Technology*, 11, 8, pp 1279 - 1286, (1993).
101. J. Yu, M. Schell M. Schulze and D. Bimberg, Fourier-limited, 6ps pulse with variable repetition rate from 1 to 26 GHz by passive mode-locking of a semiconductor laser in an external cavity, *IEEE Photonics Technology Letters*, 7, pp 467 - 469, (1995).
102. M. Mehregany, An overview of micromechanical systems, *Proceedingss of SPIE* Vol. 1793, Integrated optics and microstructures, pp 2 - 11, (1992).
103. J.L. Sniegowski and E.J. Garcia, Microfabricated actuators and their application to optics, *Proceedings of SPIE* Vol. 2383, pp 46 - 63, (1995).
104. H. Miyajima, K. Deng, M. Mehregany, F.L. Merat and S. Furukawa, Fabrication of polygon mirror microscanner by surface micromachining, *Proceedings of SPIE* Vol. 2291, pp 62 - 73, (1994).
105. F. Merat and M. Mehregany, Integrated micro-opto mechanical systems, *Proceedings of SPIE* Vol. 2383, pp 88 - 98, San Jose, (February 1995).
106. J. Mohr, M. Kohl and W. Menz, Micro-optical switching by electrostatic linear actuators with large displacements, *Proceedings of the 7$^{th}$ Solid State Sensors and Actuators Conference*, pp 120 - 123, Yokohama, (June 1993).
107. M. Mehregany and F. Merat, MEMS application in optical systems, *Digest of IEEE/LEOS 96 Summer Topical Meeting*, Keystone (USA), pp 75 - 76, (August 1996).
108. A.M. Flynn, L.S. Tavrow, and S.F. Bart, Piezoelectric micromotor for microrobots, *Journal of micromechanical systems*, 1, pp 44 - 51, (March 1992).
109. Ph. Robert, J.S. Danel, and P. Villard, The electrostatic ultrasonic micromotor, *Proceedings of Micromechanics Europe MME' 96*, pp 133 - 136, (October 1996).
110. T. Suhara and H. Nishihara, Integrated optic disk pick-ups devices using waveguide holographic components, *Proceedings of SPIE*, Vol. 1136, pp 92 - 99, (1989).
111. H. Nishihara, Possibilities of integrated optic in optical storage, *Proceedings of the 7$^{th}$ European Conference on Integrated Optics ECIO 1995*, pp 249 - 252, DELFT, (April 1995).
112. V. Lapras, P. Labeye and P. Gidon, Development of a reading-integrated optical circuit for a magneto-optical head, *Proceedings Optics ECIO 93*, pp 12 - 34 to 12 - 35, NEUCHATEL, (16-22 April 1993).
113. B.N. Kurdi, Integrated optics for optical data storage, *Proceedings of ECIO 1993*, pp 12 - 17 to 12 - 19, Neuchatel, (16-22 April 1993).
114. W. Riethmuller and W. Benecke, Thermally excited silicon micro-actuators, *IEEE Trans. Electron. Devices*, 35, pp 758 - 763, (1988).
115. S. Valette and J.S. Danel, Commutateur et système de commutation optique intégrée et pocédé de fabrication du commutateur, French Patent n 9003902.

116. E. Voges and M. Hoffman, Optical waveguides on silicon combined with micromechanical structures, *Proceedings of IEEE/LEOS 96 Summer Topical Meeting*, Keystone, pp 71 - 72 (August 1996).
117. E. Ollier, P. Labeye, and F. Revol, Micro-opto-mechanical switch integrated on silicon, *Electronics Letters*, Vol. 31, pp 2003 - 2005, (November 1995).
118. E. Ollier, P. Labeye, and F. Revol, A micro-opto-mechanical switch integrated on silicon for optical fiber network, *Digest of IEEE/LEOS 96 Summer Topical Meeting*, KEYSTONE (USA), Invited paper, pp 71 - 72, (August 1996).
119. E. OLLIER and P. MOTTIER, A new micro-optical vibration sensor connected to optical fiber, Proceedings of IEEE/LEOS 95 Summer Topical Meeting, KEYSTONE, pp 92 (August 1996).
120. C.G. Massit and G.C. Nicolas, High performance 3D MCM using Silicon Microtechnologies, *Proceedings of Electronic components and technologies conference*, pp 641 - 644, Las Vegas, (May 1996).
121. M. Bruel, Application of hydrogen ion beams to Silicon On Insulator material technology, *Nuclear Instrument and Method in Physics Research* B, 108, pp 313 - 319, (1996).
122. M. Bruel, Silicon on Insulator material technology, *Electronics Letters*, 31, pp 1201 - 1202, (July 1995).

# ACTUATION MECHANISMS FOR MICROMECHANICS

E. M. Yeatman

Department of Electrical and Electronic Engineering
Imperial College of Science, Technology and Medicine
London, SW7 2BT, England

## 1. INTRODUCTION

The information revolution, which is having such dramatic and continuous effects in nearly every field, is underpinned by the rapid advances that have occurred in microelectronics technology over the last decades. In particular, silicon fabrication technology offers extraordinary capabilities in terms of device scale and precision, complexity, and low cost. More recently, there is an increasing interest in extending the application of this technology, beyond the processing of information, to devices which interact with their immediate environment through sensing and actuation, and in the integration of all these functionalities in so-called "Microsystems". In this article, we will concern ourselves specifically with micromechanical actuation, reviewing the mechanisms that can be employed, their practical implementation, and specific applications in optics. This is intended primarily as a tutorial for those (particularly working in optics) who are unfamiliar with the topic, so while example references are given, a comprehensive literature review of the field is not intended.

Integrated circuit fabrication relies on the reproduction and superposition of complex patterns, over areas of 10's of $cm^2$ and with features down to 100's of nm, by the ultraviolet imaging of mask patterns onto photosensitive polymer films. These patterns are transferred to other materials by various processes of etching, thin film deposition and diffusion. Such a methodology can also be applied to the fabrication of micromechanical structures, and indeed, much of micromechanical fabrication is based on the use of standard microelectronics processes and materials. However, this change of use is not entirely straightforward, as there are key differences in the requirements for micromechanics. Most important is the need for some considerable degree of 3-dimensionality. Microelectronic devices are made up of relatively flat patterns, the layer thicknesses being typically much less than the lateral feature sizes. Mechanical structures, however, require greater layer thicknesses for strength and rigidity, and both the deposition and patterning of such multi-micron layers place new challenges on the relevant processes. Associated with the need for thicker layers is a low tolerance of in-built stress, particularly in freed structures, which is in

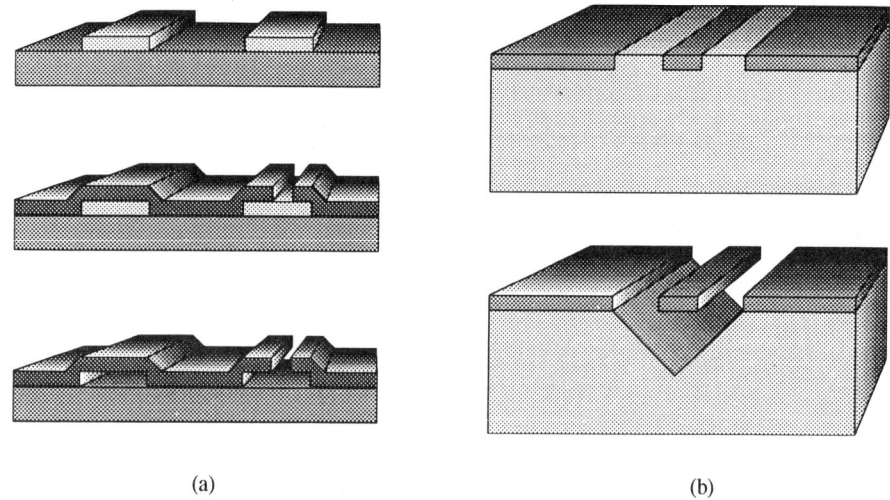

Fig. 1 (a) A simplified surface micromachining process. From top to bottom, a sacrificial layer is deposited on the substrate and patterned, a mechanical layer is then deposited and patterned, and finally the sacrificial layer is etched away. (b) A simplified bulk micromachining process. First mechanical parts are defined by a patterned diffusion, then the structure is freed by anisotropic etching.

conflict with an increased difficulty in avoiding stress in deposited layers as thickness increases. Even where these difficulties are overcome, the structures fabricated by superimposed patterned layers remain quasi-2-dimensional, which for many potential applications may be a severe limitation. However, practical structures can be made, and an increasing number of successful products are based on them.

There are two main approaches to the fabrication of micromechanical structures, and these are known as surface and bulk micromachining respectively. In surface micromachining, a series of layers are deposited and patterned onto a substrate (which is typically silicon), using two types of material; one which will form the mechanical parts, and a sacrificial material which can be etched away to free these parts (see, e.g., Mehregany et al.[1]). This process is illustrated in Fig. 1(a). Silicon nitride and polycrystalline silicon are the most commonly used mechanical materials, for which phosphosilicate glass is a suitable sacrificial material; variants of chemical vapour deposition are most commonly used for deposition. In bulk micromachining, the silicon substrate itself forms the mechanical material. A patterned diffusion is used to make the desired parts heavily p-type; this provides strong resistance to certain anisotropic etchants which can then be used to clear the parts from the substrate. These etchants stop along crystallographic planes, allowing precise definition of deep pits by mask patterns at the surface. However, this characteristic also limits the etch shapes that can be obtained. A typical process is illustrated in Fig. 1(b). The most important material for bulk micromachining is silicon; its use as a mechanical material is the subject of a classic paper by Petersen.[2]. However, quartz and gallium arsenide have also been investigated.

There are many other important techniques in micromachining used along with or in place of those above. A prominent one is LiGA, an acronym from the German words for lithography, electroplating and molding (see, e.g. Rogner et al.[3]). It uses x-ray exposure of a very thick (hundreds of microns) resist layer to obtain high quality deep projected images well beyond the range possible with conventional lithography. The developed resist is used as a master; metal structures plated around this layer are then used as molds for plastic injection molding of the final products, so that the expensive x-ray exposure step is not repeated for each replication.

The behaviour of materials and structures in the size scale of micromechanics can generally be analysed in the same way as can be done with structures of more macroscopic dimensions. The typical feature sizes of microns and greater are orders of magnitude larger than atomic dimensions so, for example, different behaviour resulting from quantum mechanical effects is not usually an issue. However, the reduced scale does have an important impact. One reason is that different effects scale at different rates, and so their relative significances alter. This has been discussed by Trimmer,[4] who compares the scaling of different forces. Gravitational and inertial forces are proportional to volume, and therefore scale as $s^3$, where $s$ indicates the linear scale variable, while electrostatic forces are shown to scale as $s^2$ or even $s^1$, as we shall see. Thus, electrostatic forces become rapidly more effective as the device size decreases. Apart from the direct effects of scaling, there are also consequences for the mechanical behaviour of the different materials used. In particular, the use of single crystal silicon results in some anisotropy in mechanical properties such as the elastic moduli, and in altered failure patterns. Silicon is a hard material which fails by fast fracture rather than by plastic yielding, and the fast fracture strength of a particular structural element depends on the largest defect size present. The very precise machining possible in silicon micromechanics results in very high strength, values of several GPa being reported.[5]

To date, many of the most successful micromechanical devices do not incorporate actuation, but are simply physical sensors. In particular, silicon diaphragms and cantilevered beams are used as pressure and acceleration sensors respectively.[6,7] However, ink-jet printer heads based on micromachined silicon have long been commercially produced, and an increasing variety of actuator structures are being investigated or developed for a wide range of applications. These applications include microfluidics, where a variety of pumps and valves are employed (the ink-jet printer is, of course, an example); fine positioning, for example in optics and in scanning probe microscopy; and even the self-testing of mechanical sensors such as those mentioned above. It seems probable that the application possibilities will continue to proliferate as the technology itself becomes more powerful, cheap and versatile.

## 2. RESISTING FORCES

Before introducing the actuation mechanisms that can be applied, let us consider the types of forces which will counteract our intended motion. One is the inertial forces of the mechanical parts themselves. However, since these scale so rapidly with size, they will tend to be relatively small in micromechanics, and will chiefly be of importance in determining the resonant behaviour of oscillating structures. Gravitational forces can generally be neglected for the same reason. In some cases, the most important resisting force will be associated with the application; for example, in liquid pumping, viscous drag may dominate. However, in many cases, the strongest forces will be those intrinsic to the mechanism. There are two main ways to allow motion of mechanical parts in a controlled fashion; they can be firmly attached to the main structure and move by flexure, or they can be unattached but confined to a fixed pattern of motion by guides and bearings. In the former case, elastic forces are obtained, and in the latter, friction.

Silicon has a stiffness similar to that of steel; while the precise value depends on crystallographic direction, 190 GPa can be taken as an average for the Young's modulus.[2] This means that to allow sufficient deflection with reasonable actuation forces, large aspect ratio structures must be employed. For an element moving in the direction orthogonal to the plane of the substrate, this is obtained by having the patterned structural dimensions (e.g. diaphragm area or cantilever length) much greater than the structure thickness. For structures moving in the in-plane direction, the challenge is greater; to have out-of-plane stiffness the mechanical parts must be made from reasonably thick layers, but this thickness limits how narrow they can be. This particular challenge of high-aspect-ratio machining is one that has attracted much attention, and much progress has been made.

Many in-plane actuators are based on a folded beam suspension stiffly joined to the substrate at both ends. For such a structure, standard mechanical analysis shows[8] that for a point load applied at the centre, the stiffness dF/dx is given by $192EI/L^3$, where L is the beam length, $E$ is Young's modulus, and I is the second moment of area. For a beam of rectangular cross section, $I = wd^3/12$, where w and d are the cross section dimensions, w being the width and d the thickness in the direction of motion. Thus we can write:

$$dF/dx = 16Ew(d/L)^3 \qquad (1)$$

For a cantilever with a point load applied at the free end, the stiffness is reduced to $3E\,I/L^3$.

The most common type of freely moving structure consists of a rotor spinning on a central fixed shaft. In conventional engineering the relative dimensions of rotor and shaft are very well defined, a lubricating fluid is introduced under controlled pressure, and the frictional forces can consequently be analysed with some precision. In micromechanics, the situation is less amenable to analysis; the bearing gap is typically much larger with respect to the shaft diameter, and usually neither bearing surface quality nor lubrication are well controlled. However, we can consider the scaling of frictional effects. The frictional force on a surface is proportional to the normal force on that surface. This force may arise from a variety of sources, for example the actuation mechanism, surface tension and Van der Waals forces, or the external load. We also have two important effects of friction to consider: one is the reduction in actuation effectiveness resulting from this countervailing force, and the other is the resultant wear. To analyse the effects of scaling on wear, let us choose as a parameter the power consumed by friction per unit area, P/A, which to a simple approximation should be proportional to the rate of surface wear, divided by the characteristic length $s$ to give the fractional shrinkage rate of the part. If $F_n$ is the normal force, v the relative velocity of the surfaces, and µ the coefficient of friction, then we have:

$$P/A\,s = \mu F_n v/As \qquad (2)$$

If we assume that v scales as $s$, then this fractional wear rate scales as $F_n/s^2$, so that if $F_n$ scales more slowly than $s^2$ the wear problem will be very serious for micromechanics. We must, however, be cautious about assuming that µ remains constant for very small contact areas, as deviations have been reported.[9] In any case, wear is found to be a problem, so that various techniques are used to reduce or eliminate the effects of friction. One is the use of elastic suspensions as we have seen; others are:

i) use of rotating rather than sliding joints
ii) use of rolling rather than sliding bearings
iii) use of non-contact bearings (e.g. magnetic, electrostatic)
iv) use of lower friction surface coating materials.

As to the restraining force of friction, one likely effect is that it will quickly dissipate the small inertial forces of micromechanical parts, which will hinder smooth motion where the driving force is not itself smooth.

## 3. ELECTROSTATIC ACTUATION

Electrostatic actuation is essentially based on the attractive force between two oppositely charged bodies. When we apply a voltage difference V between two conductors, they become oppositely charged, and consequently attracted to each other. They also act as

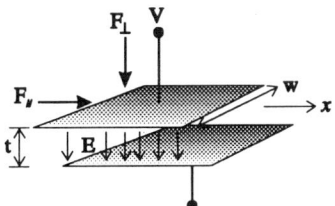

Fig 2. Schematic diagram of parallel plate capacitor.

a capacitor, the capacitance C being the ratio of the stored charge Q to the voltage, according to Q = CV. The energy stored in the resulting electric field E will then be given by U = ½ CV²; alternatively, U can be found by integrating the energy density, which is given by ½ εE², ε being the permittivity of the dielectric medium between the conductors. Let us consider the example of the parallel plate capacitor shown in Fig. 2. In this case, E = V/t, and C = εA/t, where A is the overlapping area and t is the plate separation; (in fact, this is an approximation for lateral dimensions much greater than t, such that fringing fields at the edges of the overlapping region can be neglected).

The relations described above allow the electrical behaviour of capacitors to be analysed, but they can also be used to determine the forces between the plates. If the plates are electrically isolated, then Q is constant. If the separation between the plates changes, the capacitance and voltage change in inverse proportion. Expressing U as ½ Q²/C, we obtain:

$$dU/dC = -½ Q^2/C^2 = -½ V^2 . \qquad (3)$$

The force in an arbitrary direction $s$ is given by the spatial derivative of the energy, e.g. $F_s = -dU/ds$, the sign indicating that the force tends to reduce the stored energy. Since dU/dC is negative, a decrease in U is associated with an increase in C, and thus the force will tend to decrease t and increase A, i.e. pull the plates together, as expected. Using Eq. 3 we can show that $F_s = ½ V^2 (dC/ds)$. We can now derive the normal and parallel forces, as indicated in Fig. 2. For the normal force, then, $F_\perp = ½ V^2 (dC/dt)$, and dC/dt = - C/t. In order to consider the scaling behaviour, we will write $F_\perp$ in terms of the electric field, giving:

$$F_\perp = ½ \varepsilon E^2 A . \qquad (4)$$

For a fixed field strength, the force is independent of the gap (within the limits of our approximation). For the parallel force, taking $x$ as the lateral position of the upper plate and w the plate width as indicated in Fig. 2, dC/dx = εw/t. Then:

$$F_{//} = ½ \varepsilon E^2 wt \qquad (5)$$

In both cases, the force is proportional to the area of the dielectric in the plane orthogonal to the direction of motion.

We can now consider the scaling of electrostatic forces. The maximum force for a given geometry will be obtained for the maximum field strength that can be applied without dielectric breakdown. For common dielectrics at macroscopic dimensions, the breakdown strength is a constant, and assuming this to be the case here results in a scaling rate of $s^2$ for both $F_\perp$ and $F_{//}$. However, it has been long observed[10] (although not well understood) that dielectric strength can rise substantially for gaps below 10's of µm; by 2 µm, for example, the breakdown strength of air at atmospheric pressure has risen from its 'bulk' value of 3 x 10⁶ V/m to 1.7 x 10⁸ V/m, close to the vacuum value of 3 x 10⁸ V/m. If we approximate this

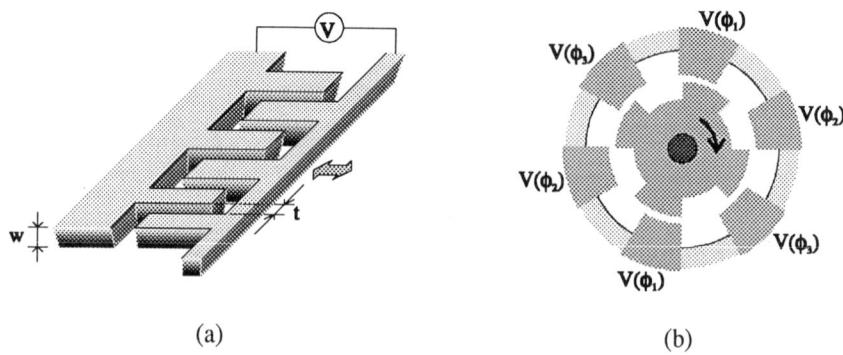

Fig. 3 Typical geometries of (a) an electrostatic comb drive, and (b) an electrostatic motor.

variation by allowing $E_{max}$ to increase as $s^{0.5}$, then the electrostatic forces scale as $s$. In both cases, but particularly the latter one, which may be more realistic, the scaling rate of electrostatic forces is slow, and therefore they may have much increased significance in the micro-world.

Scaling shows why electrostatic actuation may be more viable for micromechanics than at conventional dimensions. However, the main reason for its very wide use is more likely to be the relative ease of implementation. No special materials are required, and a wide range of structures can be used to obtain different types of motion for different application. There are two common electrostatic micro-mechanisms, corresponding to the flexing and freely moving variants we introduced earlier. The first is the inter-digitated or comb electrode structure used in conjunction with a folded suspension, as reported, for example, by Tang et al.[11] This is illustrated in Fig. 3(a). Eq. (5) for the force may be applied directly, except that the total force is multiplied by the number of gaps. To get an idea of the absolute quantities, let us take as an example an actuator having w = 10 µm, t = 5 µm, and 20 fingers on each side, giving 39 gaps. The moving side is at the centre of a suspension beam 10 µm thick (equivalent to w if the comb and suspension are a contiguous structure) and 5 µm wide, and of length 1 mm. This gives a stiffness (using Eq. 1) of 3.8 N/m. If the dielectric is air, ε = 8.85 x 10$^{-12}$ F/m, and we find that simply to get a displacement of 1 µm, we have to apply just over 100 V (if indeed the gap can withstand 20 V/µm); this despite an aspect ratio in the suspension of 200. We can see that although the force may be viable, it is still not easy to get substantial deflections. The use of smaller gaps allows large field strengths to be obtained using more practical voltages, but as we have seen this does not fundamentally increase the actuation force. Another important characteristic implied by Eq. 5 is that the force, and therefore the displacement within the linear range of the suspension, varies as the voltage squared rather than linearly with applied voltage.

One way to increase the obtained deflections is to drive the mechanism in resonance. If we can approximate the structure as a single mass m supported by a linear massless spring of stiffness k, the resonant frequency is given by $2\pi f_o = \sqrt{(k/m)}$. Silicon has a density of 2.33 g/cm$^3$, so if we take as an example a 10 x 100 x 100 µm structure, m = 2.33 x 10$^{-7}$g, and using the stiffness obtained in the previous example of 3.8 N/m we obtain a resonant frequency of 20 kHz. We can expect a large increase in the displacement at resonance; Tang et al. measured resonant quality factors of about 30 in air, but about 5 x 10$^4$ in vacuum.[10]

The second important type of electrostatic actuator is the rotating multi-phase motor, as reported, for example, by.[12] This is illustrated in Fig 3(b). This works on essentially the same principle as the comb drive, i.e. using the lateral force $F_{//}$. A multi-bladed rotor is mounted on a central axis, and is surrounded by stator elements or blades to which voltages can be applied. A voltage difference between a stator blade and a rotor blade pulls the rotor blade laterally in such a way as to increase the overlap area; when the overlap is maximum the drive voltage is transferred to the next stator blade, and so on. In the structure of Fig. 3(b), Each of three phases of drive signal is applied to a pair of opposite stator elements, such that the effective drive field rotates around the stator in the desired direction, the rotation speed being determined by the drive frequency. The reduced number of rotor blades ensures that as each pair is pulled into alignment, the next pair is in position to be attracted by the stator blades to which the drive signal is then transferred.

Just as the deflection force is modest in the comb drive, the torques obtained in this type of motor are also limited. This is partially because the cross-sectional area of the gaps between rotor and stator are small compared to the overall structure. Considerable improvement can be achieved if the rotor and stator elements are stacked vertically, such that the effective width w for Eq. 5 is the width of the blade rather than its thickness. Such structures have been demonstrated;[13] however, they are more difficult to fabricate, and as the normal force is now vertical rather than radial, there is a tendency for the rotor to be bent. This must be prevented either by making it very thick, or by having a balanced set of stators, one above and one below, cancelling out the vertical force.

Electrostatic micromotors have had much publicity, as they make fascinating technology demonstrators and are highly photogenic. They have had less success in practical application to date, however. There are several reasons, not least of which is the lack of obvious applications for the particular characteristics of this device. In addition, as mentioned above, there is a significant problem of frictional wear, leading to insufficient operating lifetimes. One approach to reducing wear, as also mentioned above, is to use a rolling rather than sliding bearing, and this is what is done in the so-called wobble motor. Here the drive is applied as in the motor described above, but in this case the rotor is not axle mounted, but is free to roll eccentrically around the inside of the larger circular cavity of the stator. Such motors have been demonstrated in 'flat' micromachining technology,[14] and are also being fabricated in full 3-dimensionality at mm size,[15] using batch processing.

Other variants of electrostatic actuation merit a mention. The techniques described above all involve pulling two conductors together by applying a potential difference across a dielectric gap. A repulsive force can also be obtained, however, if fixed buried charges are used. Such devices are known as electret actuators. Also, the field between conductors also produces a force pulling a dielectric part or fluid into the gap to replace a material of lower permittivity. This principle has been used to pump fluid through micromachined channels, and to handle biological objects such as cells, as described for instance in a review by Fujita and Gabriel.[16]

## 4. PIEZOELECTRIC ACTUATION

The piezoelectric effect is exhibited by certain crystalline materials, and consists of a coupling between the mechanical stress/strain relations and the electric field/displacement relations. This coupling is described in full tensor form by the equations:

$$S_{ij} = s_{ijkl} T_{kl} + d_{ijm} E_m , \qquad (6)$$

$$D_n = d_{nkl} T_{kl} + \varepsilon_{nm} E_m . \qquad (7)$$

Here, S and T are the strain and stress tensors respectively, D and E are the electric displacement and field, s is the mechanical compliance tensor at constant field, $\varepsilon$ is the permittivity at constant stress, and d is the piezoelectric tensor. Fortunately, symmetries reduce the number of independent elements of d to six. Let us consider a piezoelectric plate or slab of lateral dimensions L and w, the upper and lower surfaces coated with conducting

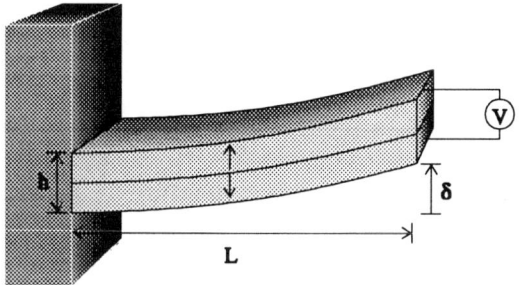

Fig. 4 A simplified piezoelectric bimorph structure.

electrodes, and thickness between the electrodes of t. If the slab is not mechanically constrained, then the change in thickness resulting from an applied voltage is simply given by $d_{11}V$, while the fractional change in lateral dimensions is given by $\Delta w/w = \Delta L/L = d_{31}V/t$, where 1 denotes the direction normal to the electrodes.

Popular piezoelectric materials are quartz and PZT; the latter is an abbreviation for lead zirconate titanate, a perovskite ceramic which in a particular stoichiometry has excellent piezoelectric properties. The material PZT5A, manufactured by Morgan Matroc Ltd., has values for $d_{33}$ and $d_{31}$ of 374 and -171 x $10^{-12}$ m/V respectively.[17] Using the equations above, a 10 μm thick slab to which 100 V is applied will contract or expand (depending on the polarisation) in thickness by about 0.04 μm, a strain of 4 x $10^{-3}$. In practice, strains greater than about $10^{-3}$ (or 0.1%) cannot be obtained in ceramic materials without damage. However, the forces available to achieve the displacement are enormous. If we consider clamped (S = 0) rather than free (T = 0) motion, then the maximum stress the plate can develop is given by $d_{33}EE$ (E being the Young's modulus, given for PZT5A as 61 GPa). Then for 100 V applied across 10 μm we obtain 220 MPa (about 2000 bar).

In order to get large displacements from these large forces without having to strain the material significantly, we need to leverage the effect in some way. One popular technique is to use a bimorph structure, where two identical piezoelectric slabs are bonded with their polarisations reversed. A field is than applied across this sandwich which, through $d_{31}$, applies opposing lateral stresses to the two slabs, and therefore causes the bimorph to bend. This is illustrated in Fig. 4. Piezoelectric bimorphs were extensively reviewed and analysed by Smits et al..[18] They find that for the structure of Fig. 4, the radius of curvature is given by $2h/(3 d_{31}E)$, while the tip displacement is given by:

$$\delta = 3 d_{31}EL/2h \qquad (8)$$

For an unconstrained bimorph cantilever as shown, with L = 100 μm and h = 10 μm, to get a tip displacement of 1 μm requires only about 4 mV to be applied. This contrasts with the much lower forces and displacements obtained in the comb drive structure discussed above.

Piezoelectric microactuators have a achieved considerable success. One of the first commercially exploited applications was the ink-jet printer head, also described by Petersen.[19] Here the ink cavity is fabricated by bulk micromachining, the nozzle by isotropic wet etching, and the cavity is closed by a glass plate. A piezoelectric crystal is then bonded to this plate; the actuation is similar to the bimorph discussed above except that one of the piezo- layers is replaced by the (passive) glass. This results in a very effective device, but there is clearly a disadvantage from a more general micromechanics point of view. Bonding a conventionally fabricated and machined crystal is an expensive process step, and in addition limits the geometries that can be achieved. It would be far more compatible with the other fabrication steps if the piezo- material could be deposited as a film, and patterned and processed like the other materials.

This is an important goal, not only for micromechanics but for other applications. PZT is also of great interest for semiconductor memory, because it also has strong ferroelectric properties and a very high relative permittivity (800 - 1000). Many methods of depositing PZT and related materials are being examined for these applications,[20] but there are many challenges, particularly in achieving the desired crystalline structure and orientation, which usually requires high temperature processing, while maintaining the integrity of the electrode layers and compatibility with the active electronics. Most applications of piezoelectric actuation in micromechanics have used the bonded crystal approach, fluidics being a particularly popular application area.[21] It should also be noted that while piezoelectric actuators can be used at high frequencies, their low frequency effectiveness is limited by dielectric leakage below about 10 Hz.

Finally, we can note that rotating micromotors can also be fabricated based on piezoelectric actuation.[22] In this case, a displacement (surface acoustic) wave is propagated along the surface of the stator (or rotor), causing cyclical motion of local surfaces; the adjacent surface is then propelled in the direction of the wave motion.

## 5. OTHER ACTUATION MECHANISMS

Thermally actuated structures have also been investigated for micromechanics. These exploit thermal expansion, so again a method of leveraging small strains is needed. One implementation is the thermal bimorph or bi-metallic strip. This behaves as the structure in Fig. 4, except that the differential stress used to cause bending comes from the two layers having different coefficients of thermal expansion, and from a temperature increase being applied, or alternatively a temperature differential being applied between two surface of the same material. Silicon has a coefficient of thermal expansion [23] of $2.6 \times 10^{-12}$/K, so that for a temperature difference of 40°C we get a strain difference of $10^{-10}$. This is several orders of magnitude less than can be achieved in the piezoelectric case, but still compares reasonably well with electrostatic mechanisms, and also has the advantage of relative ease of implementation. Fujita[24] has demonstrated thermal cantilevers based on two different polymer layers sandwiched around a thin film metal heating strip; these were arrayed to produce a ciliary motion transport mechanism. Van de Pol et al.[25] demonstrated an alternative approach in a diaphragm pump in which actuation was effected by heating the air in a sealed cavity. Thermal actuators have two related disadvantages; first, they tend to be slow, since the de-activation time is limited by the rate of heat loss to the surrounding structure; second, adding large amounts of heat can have negative effects on the surrounding materials and devices, and if the heater elements are too densely spaced or heavily used, the substrate temperature will rise, limiting cooling and thus defeating the actuation. Having a thermally well isolated actuator reduces the heat that must be introduced, but may excessively reduce the cooling rate as well.

A variation on the thermal actuator involves the use of shape memory alloy (SMA) materials.[26] These are specially treated metal alloys which 'remember' particular shape configurations, deform from them when heated, and return to them when cooled. Actuators have been fabricated using SMA springs and wires, and SMA in thin film form is being investigated.

Electromagnetic actuators form probably the most important class at larger dimensions. However, they appear to scale less well to micro-dimensions than electrostatic types, primarily because of the heat dissipation associated with magnetising currents. There are also practical difficulties of implementation; a magnetising coil is likely to be needed, which requires 3-dimensional fabrication, and a ferromagnetic material is highly desirable to guide and magnify the magnetisation. Despite the difficulties, electromagnetic microactuators have been demonstrated by a number of groups, e.g. Ang & Allen.[27] More exotic mechanisms, such as magnetostriction or the Meissner effect in superconductors, can also be exploited.

A final technique deserving of consideration is the use of surface tension. Surface tension scales as $s$, and therefore is highly effective at micro-dimensions. Indeed, the capillary forces produced by the drying of etchants around micromechanical structures tend to distort the structures, pulling them together, and may be so high as to bond parts permanently together.[28] This can be a serious problem in processing, and various methods, such as critical drying, have been exploited to avoid it. However, surface tension can also be used advantageously. In flip-chip bonding, an integrated circuit is bonded upside-down onto another, and high alignment precision is needed. This is achieved by adding patterned solder pads to each and roughly aligning the chips in placement; the solder is then melted, so that matching pads pull each other into perfect alignment by surface tension. Developed at IBM, this technique has more recently been adapted to optical assembly, e.g. by Hayashi.[29]
A similar method had also been developed for folding surface micromachined parts into 3-dimensional configurations.[29] In this case the solder is deposited as a hinge, and its final angle is determined by the position of minimum surface area, and can therefore be controlled through initial thickness. Electrically controllable surface tension actuators are also being investigated.[31]

## 6. APPLICATIONS OF MICROACTUATORS IN OPTICS

As we have seen, there are applications for microactuators in many fields, of which optics is one. In light of the optical theme of this volume as a whole, we will now review this particular application area in more detail. Optics is a particularly interesting application for microactuators for several reasons. First, a variety of structures on the micron scale are of interest in optics: waveguides, diffractive elements, micro-lenses, etc.. Also, useful action is obtained by moving the parts themselves, in the sense that the interaction with the outside world, i.e. with the light, does not usually introduce significant loading forces. Thus, applications have been investigated in optical alignment, beam scanning and modulation, adaptive optics and spectral tuning. Examples are described below.

The V-shaped grooves formed by anisotropic etching of (100) silicon can form a conveniently precise alignment feature for the interconnection of single-mode fibre, or for the connection of such fibres to integrated optical components, where sub-micron precision is desired. Kikuya et al.[32] have extended this technique by coating a fibre end with a metal film, suspending it above a widened V-groove surface, and then actively orienting it electrostatically using orthogonal electrodes on the V-groove.

Uenishi et al.[33] demonstrated the integration of opto-electronic and micromechanical functionality, by micromachining of a GaAs based structure including a cantilever beam suspended between two laser diode stripes and a photodetector. The lasers are used both to effect and to monitor the beam's oscillation. One laser is modulated at the beam's resonant frequency, and its modulated output causes local heating on one side of the beam, resulting in oscillating strain, driving the beam into vibration. This effectively varies the external cavity length of the second laser stripe, for which the cantilever acts as one of the mirrors. This causes oscillation in its output intensity as the internal and external cavities interfere, and the photodetector detects this variation. By measuring the vibration amplitude against frequency we can find the resonant frequency; this may be dependent on, for example, an adsorbed chemical species, and thus we have an integrated sensor.

Another use of an actuator for laser diode tuning was demonstrated by Uenishi et al.[34] In this case, the laser is hybrid mounted onto a silicon platform onto which a nickel comb-drive electrostatic actuator has been fabricated. This comb drive moves a cantilevered nickel mirror which acts as one mirror in an external cavity of the laser. Motion of ±4 µm was obtained using a ±20V signal, for which a total tuning range for the output wavelength of about 20 nm was obtained.

Optoelectronic tuning can also be obtained by quite a different technique, as shown by Hung et al..[35] They mounted a GaAs laser diode on a piezoelectric bimorph, and showed that the relative intensities of the longitudinal modes could be varied by straining the laser, and that the photoluminescence peak could be shifted in wavelength by several nm, indicating an shift in the energy gap.

A miniature optical bench on a chip was developed at UCLA,[36] in which a laser diode mounted on its edge on a silicon substrate is combined with planar mirrors, Fresnel lenses and gratings folded out of plane to guide the laser output. In this case the folded components were fabricated by sacrificial layer processing to incorporate staple hinges and an associated locking mechanism. Assembly is carried out simply by vibrating the elements until they lock into their out-of-plane position. They even show such an upright diffraction grating mounted on an electrostatic motor, allowing the possibility of rotational tuning. Assembly of optical components was also considered by a group at Berkeley,[37] who developed a microvibromotor for positioning such components, in order to achieve large displacements. A slider is mounted in a flanged groove, alongside which are positioned two sets of electrostatic comb drive actuators which vibrate prods that push obliquely on the slider's edge. Depending on which set of vibrators is chosen, motion in one direction or the other of the slider is achieved. This group has also demonstrated a 3-D micro-optical bench.[38]

A micromachined scanner unit was developed by OMRON Corp.,[39] in which a cantilevered silicon mirror is driven into its two vibration modes (bending and torsional) by applying the relevant frequencies to a hybrid mounted piezoelectric unit. Application in a two-dimensional bar-code reader is suggested. Deflection or distortion of arrays of mirrors also forms the basis of an important class of spatial light modulator, particularly for applications in large area video projectors.[40] The use of microactuation to distort a mirror can also be used in adaptive optics, and particularly in dynamic focusing mirrors. At Nippondenso Co. an electrostatically actuated silicon diaphragm mirror was demonstrated,[41] in which adjustment of the distortion shape was achieved by varying the diaphragm thickness radially, thus reducing spherical aberrations.

In fact, a great variety of possibilities exist for microactuation in optics, and as the two fields of micromechanics and micro-optics progress, the possibilities from combining their capabilities will continue to grow. Future developments may include a greater degree of integration of electronics with these types of structures, and a greater use of optical functions in microsystems such as those being developed for chemical analysis.

# REFERENCES

1. M. Mehregany, K.J. Gabriel and W.S. Trimmer, Integrated fabrication of polysilicon mechanisms, *IEEE Trans. Electr. Dev.* 35: 719 (1988).
2. K. Petersen, Silicon as a mechanical material, *Proc. IEEE* 70: 420 (1982).
3. A. Rogner, J. Eicher, D. Munchmeyer, R-P. Peters and J. Mohr, The LIGA technique - what are the new opportunities, *J. Micromech. Microeng.* 2: 133 (1992)
4. W.S.N. Trimmer, Microrobots and micromechanical systems, *Sensors and Actuators* 19: 267 (1989).
5. F. Ericson and J-A Schweitz, Micromechanical fracture strength of silicon, *J. Appl. Phys.* 68: 5840 (1990).
6. S.K. Clark and K.D. Wise, Pressure sensitivity in anisotropically etched thin-diaphragm pressure sensors, *IEEE Trans. Electr. Dev.* ED-26: 1887 (1979).
7. L.M. Roylance, A batch fabricated silicon accelerometer, *IEEE Trans. Electr. Dev.* ED-26: 1911 (1979).
8. J.D. Todd, "Structural Theory and Analysis," Macmillan Press, London (1974).

9. U. Beershwinger, D. Mathieson, R.L. Reuben, S.J. Yang, R.S. Dhariwal and H. Ziad, Tribological measurements for MEMS applications, *in Micro System Technologies '94*, H. Reichl and A. Heuberger, ed., Vde-Verlag gmbh, Berlin (1994).
10. B. Bollee, Electrostatic motors, *Philips Tech. Rev.* 30: 177 (1969).
11. W.C. Tang, T-C. H. Nguyen, M.W. Judy and R.T. Howe, Electrostatic-comb drive of lateral polysilicon resonators, *Sensors and Actuators* A21-A23: 328 (1990).
12. L.S. Tavrow, S.F. Bart and J.H. Lang, Operational characteristics of microfabricated electric motors, *Sensors and Actuators A* 35: 33 (1992).
13. Ziad, S. Spirkovitch, N. Milne, U. Beershwinger and S. Rigo, Pneumatic stabilisation of top drive micromotors, *in Micro System Technologies '94*, H. Reichl and A. Heuberger, ed., Vde-Verlag gmbh, Berlin (1994).
14. T. Furuhata, T. Hirano, L.H. Lane, R.E. Fontana, L.S. Fan and H. Fujita, Outer rotor surface-micromachined wobble motor, *Proc. IEEE Micro Electro Mech. Syst. 93* : 161 (1993).
15. Matsushita Res. Inst. Tokyo Inc., Nondestructive testing device, *Micromachine* No. 14: 7 (1996).
16. H. Fujita and K.J. Gabriel, New opportunities for micro actuators, *Proc. IEEE Micro Electro Mech. Syst. 91* : 14 (1991).
17. *Piezoelectric Data Book for Designers*, Morgan Matroc Ltd., Thornhill, Southampton.
18. J.G. Smits, S.I. Dalke and T.K. Cooney, The constituent equations of piezoelectric bimorphs, *Sensors and Actuators* 28: 41 (1991).
19. K. Petersen, Fabrication of an integrated, planar silicon ink-jet structure, *IEEE Trans. Electr. Dev.* ED-26: 1918 (1979).
20. A. Patel and J.S. Obhi, Ferroelectric thin films for integrated sensor and memory devices, *GEC J. Res.* 12: 141 (1995).
21. J.G. Smits, Piezoelectric pump with three valves working peristaltically, *Sensors and Actuators* A21-23: 203 (1990).
22. K.R. Udayakumar, S.F. Bart, A.M. Flynn, J. Chen, L.S. Tavrow, L.E. Cross, R.A. Brooks and D.J. Ehrlich, Ferroelectric thin film ultrasonic micromotors, *Proc. IEEE Micro Electro Mech. Syst. 91* : 109 (1991).
23. *Properties of Silicon*, London: INSPEC (1988).
24. M. Ataka, A. Omodaka, N. Takeshima and H. Fujita, Fabrication and operation of polyimide bimorph actuators for a ciliary motion system, *J. Microelectromech. Syst.* 2: 146 (1993).
25. F.C.M. Van de Pol, D.G.J. Wonnink, M. Elwenspoek and J.H. Fluitman, A thermo-pneumatic actuation principle for a microminiature pump and other micromechanical devices, *Sensors and Actuators* 17: 139 (1989).
26. M. Bergamasco, P. Dario and F. Salsedo, Shape memory microactuators, *Sensors and Actuators* A21-A23: 253 (1990).
27. C.H. Ang and M.G. Allen, A fully integrated surface micromachined magnetic microactuator with a multilevel meander magnetic core, *J. Microelectromech. Syst.* 2: 15 (1993).
28. C.H. Mastrangelo and C.H. Hsu, Mechanical stability and adhesion of microstructures under capillary forces - part II: experiments, *J. Microelectromech. Syst.* 2: 44 (1993).
29. T. Hayashi, An innovative bonding technique for optical chips using solder bumps that eliminate chip positioning adjustments, *IEEE Trans. Comp., Hybrids & Manuf. Tech.* 15: 225 (1992).
30. P.W. Green, R.R.A. Syms and E.M.Yeatman, Demonstration of three-dimensional microstructure self-assembly, *J. Microelectromech. Syst.*4: 170 (1995).
31. E.M. Yeatman, R.R.A. Syms and A.S. Holmes, Research on micromolding and microactuation using surface tension, *Micromachine* No. 16: 13 (1996).
32. Y. Kikuya, M. Hirano, K. Koyabu, and F. Ohira, Micro alignment machine for optical coupling, *Proc. IEEE Micro Electro Mech. Syst. 93*: 36 (1993).
33. Y. Uenishi, H. Tanaka and H. Ukita, AlGaAs/GaAs micromachining for monolithic integration of optical and mechanical components, *Proc. Soc. Photo-Opt. Instr. Eng.* 2291: 82 (1994).
34. Y. Uenishi, K. Honma and S. Nagaoka, Tunable laser diode using a nickel micromachined external mirror, *Electr. Lett.* 32: 1207 (1996).
35. C.Y. Hung, R. Burton, T.E. Schlesinger and M.L. Reed, Microelectromechanical tuning of electrooptic devices, *Proc. IEEE Micro Electro Mech. Syst.* 92: 154 (1992).
36. L.Y. Lin, M.C. Wu and K.S.J. Pister, Micromachined integrated optics for free-space interconnections, *Proc. IEEE Micro Electro Mech. Syst.* 95: 77 (1995).
37. M. J. Daneman, N.C. Tien, O. Solgaard, A.P. Pisano, K.Y. Lau and R.S. Muller, Linear microvibromotor for positioning optical components, *J. Microelectromech. Syst.*5: 159 (1996).

38. O. Solgaard, M. J. Daneman, N.C. Tien, A.Friedberger, K.Y. Lau and R.S. Muller, Optoelectronic packaging using silicon surface-micromachined alignment mirrors, *IEEE Photonics Tech. Lett.* 7: 41 (1995).
39. M. Ikeda, H. Goto, M. Sakata, S. Wakabayashi, K. Imanaka, M. Takeuchi and T. Yada, Two dimensional silicon micromachined optical scanner integrated with photodetector and piezoresistor, *Proc. Transducers '94* (1994).
40. J.B. Sampsell, Digital micromirror device and its application to projection displays, *J. Vac. Sci. Tech. B* 12: 3242 (1994).
41. M. Hisanaga, T. Koumura, T. Hattori, Fabrication of 3-dimensionally shaped Si diaphragm dynamic focusing mirror, *Proc. IEEE Micro Electro Mech. Syst.* 93: 30 (1993).

# DESIGN OF COMPUTER GENERATED BINARY HOLOGRAMS FOR FREE SPACE OPTICAL INTERCONNECTIONS

I. Montrosset[*], D. Cojoc[§], and F. Sartori[*]

[*] Polytechnic of Turin, Electronics Department
C.so Duca degli Abruzzi 24, 10129 Torino, Italy
[§] Precision Engineering and Optics Department
University "Politehnica" of Bucharest
Spl. Independentei 313, 77206 Bucharest, Romania

## 1. INTRODUCTION

The purpose of this Chapter is to give an introduction to the use of computer generated holograms (CGHs) as free space optical interconnection elements. We will focus mainly on the techniques for designing binary CGHs, which can be easily fabricated in photographic emulsions and can be implemented in reconfigurable devices such as binary spatial light modulators. In particular, we will briefly review the advantages offered by optical interconnections compared with electronic interconnects.[1-4]

Optical interconnections potentially offer a freedom from mutual coupling effects not afforded by conventional electronic interconnects. This potential advantage of optics becomes more and more important as the bandwidth of the desired interconnections increases, because the strength of mutual coupling associated with electrical interconnects is proportional to the frequency of the signals propagating on the interconnect lines.

A second potential advantage of optical interconnections is additional flexibility in routing. Electrical interconnect paths cannot cross and therefore must be routed over or under one another through multiple interconnect layers. Optical interconnections can indeed be routed through one another without any deleterious effect.

Same further advantages of optical interconnects are a partial freedom from certain capacitive loading effects inherent in electrical interconnect lines, and on the possibility of realizing dynamically changeable interconnect devices.[1-3]

There are in fact many optical methods that could be used as the basis for optical interconnections. One of the best known methods for realising optical interconnections is by means of optical fibres. However, optical fibres are not necessarily the ideal solution for interconnect at all levels. In particular, at the lowest levels, i.e. intrachip and very nearby chip-to chip interconnections, the problems of bending and looping fibres become severe, due to radiation losses induced by bending. In

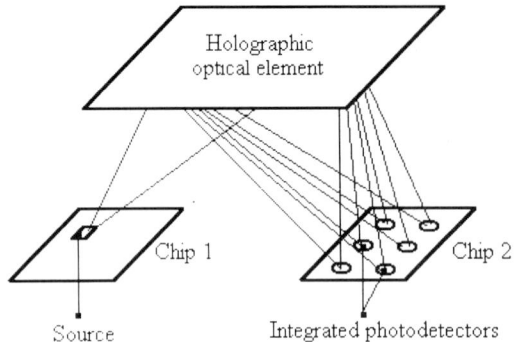

Fig. 1 Scheme of a 1 to $N$ free space imaging interconnection.

such cases it might be argued that fibres are too much like wires, requiring a material path for interconnection between every two points, and rather inflexible paths at that. An alternative approach that may be applicable to intrachip communications is the use of integrated optic technology, but an important practical problem with the integrated-optic approach is the coupling of light into and out of the waveguides.

Free space focused interconnections (imaging interconnections) differ from those discussed above in that rather general optical focusing elements are used to place nearly all the available light onto the detector sites where it is required. The focusing elements generally must be realised by means of holography and are referred to as holographic optical elements (HOEs) (Fig. 1).

Using such elements, a single source can be imaged onto one or more photodetectors with high efficiency. The practical limits to efficiency depend on the amount of fan-out required, the material from which the hologram is made, and the geometry required. With the use of dichromated gelatine as a recording material, simple reflection holograms capable of imaging one source onto one detector using visible light can be made with efficiencies in excess of 99 %. Less is known about making good holograms that work in the near infrared, where high-speed optical communication technology is extensively employed.[1] Note that the use of imaging interconnections opons up the possibility of parallelism in the interconnect network. One source can be imaged onto one photodetector, while another independent source is imaged onto another different photodetector. Each such channel can then operate independently. If holographic optical elements are to be used, it should be noted that the Bragg selectivity of a thick transmission grating is far superior to that of a thick reflection grating. Therefore, when several parallel interconnections are to be made independently, a transmission geometry is preferable to a reflection geometry.

The use of HOEs seems to be one of the most promising and vigorously researched methods for free space optical interconnection. Some advantages of HOEs, and in particular computer generated holograms, over other more conventional techniques are[1, 5-7] :large spatial bandwidth; easy and low-cost fabrication; and planar integration.

Binary CGHs represent the most extensively investigated class of CGHs. They are fundamentally attractive for the ease with which they can be fabricated by using standard VLSI lithography techniques. Morever, they can be implemented in a programmable binary spatial light modulator. Various interconnection patterns can be thus configured in real time. The main drawback of binary CGHs is the error introduced by the coding technique when computing thebinary oattern. Fortunately, this error can be substantially reduced by using various optimisation techniques.[8]

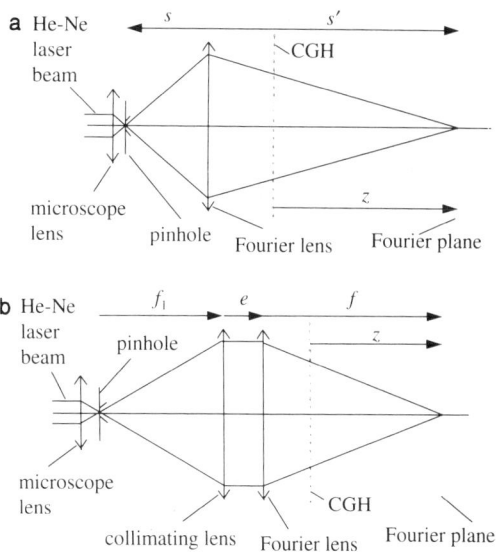

Fig. 2 Fourier transform optical setup for the 1 to $N$ interconnection. (a) Using spherical wave illumination and a Fourier hologram. (b) Using plane wave illumination and a Fourier hologram.

## 2. OPTICAL SETUP

With regard to the optical setup, there are two methods for free space optical interconnections using binary CGHs: methods that use Fraunhofer diffraction, and those that use Fresnel diffraction. Each has its advantages and disadvantages. One-to-many interconnections cause encoding problems for Fresnel zone plates,[9] which are the fundamental structure that use Fresnel diffraction. Free space optical interconnections using Fraunhofer diffraction [10, 11] are good for one-to-many interconnections but require a Fourier hologram. The length of the required optical setup is, in addition, too large for many practical applications.

Suppose that we want to connect a single point source to an array of $N$ detectors. First, we design the connection element as a Fourier hologram which is introduced in the Fourier transform optical setup depicted in Fig. 2(a). An array of $N$ spots results from the optical reconstruction of the hologram in the Fourier plane. The possibility of adjusting the scale of the resulting $N$-spot array, by shifting the hologram along with the optical axis, represents one of the advantages of this setup. The main drawback is its length. Since this setup needs a Fourier lens, the minimum length equals $4f$ ($f$ being the focal length of the Fourier lens). The length can be reduced by using plane wave illumination, as is shown in Fig. 2(b). However, the price is the limitation of the scale range. From Fig. 2(a) and 2(b), one may see that the setup could be more compact if the Fourier lens were somehow removed. This goal can be achieved replacing the Fourier lens with a Fresnel encoded lens, and superposing the Fourier CGH on it. The result is a Fresnel hologram. With a high resolution technique (e.g. laser lithography) we can obtain a Fresnel encoded lens with less than 10 mm focal length, and this leads to a considerable decrease of the setup length, required for certain applications. The main disadvantage of this setup is its low tolerance as is transverse shifts of the hologram. The reconstructed pattern is no more shift invariant when the Fourier lens and the hologram are within the same support.

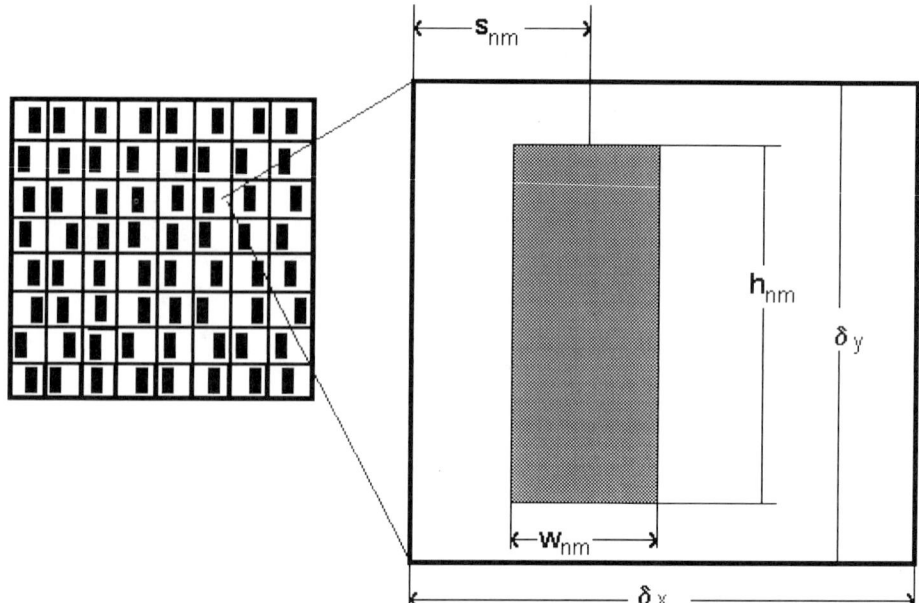

Fig. 3 Scheme of the cell-oriented coding technique.

## 3. CODING TECHNIQUES FOR FOURIER TYPE HOLOGRAMS

For the fabrication of Fourier holograms we have investigated two coding schemes: the point oriented coding (POC) and cell oriented coding (COC), which are described in this Section.

If $t(x,y)$ is the function representing the desired array of $N$ spots, $T(X,Y)$, the function describing the hologram, is given by:

$$T(X,Y) = FT^{-1}[t(x,y)] = A(X,Y) \exp[i\Phi(X,Y)] \qquad (1)$$

where $FT^{-1}[t(x,y)]$ is the inverse Fourier transform of the desired function $t(x,y)$, and $(X,Y)$ represent the spatial coordinates in the hologram plane. $T(X,Y)$ is a complex function which needs to be implemented physically. The aim is to obtain a real positive binary function, $T_B(X,Y)$ from $T(X,Y)$, that permits us to reconstruct the desired function $t(x,y)$. With POC [12] we proceed from the discrete version of $T(X,Y)$, $T_k(X,Y)$, and associate a binary value to each sample. For the case of COC,[13-16] a square cell of pixels is associated to each sample. The amplitude and phase of the complex value $T_k(X,Y)$ is coded by using an aperture inside the cell.

### 3.1. Cell Oriented Coding

In this case, a square cell is associated with each sample point of the $T_k(X,Y)$ function (Fig. 3). Each cell contains a number $n \times n$ of pixels whose transmittance can assume the value 0 or 1. The size of the aperture in each cell $h_{nm}$ is proportional to the amplitude $A_{nm}(X,Y)$ while the horizontal shift of the aperture inside the cell $s_{nm}$ is proportional to the phase $\Phi_{nm}(X,Y)$ according to the formulas:

$$\begin{cases} h_{nm} = A_{nm}\,\delta y \\ w_{nm} = w = ct \\ s_{nm} = \varphi_{nm}\,\delta x / 2\pi M \end{cases} \qquad (2)$$

In order to avoid the overlapping of an aperture in a cell with the one in the neighbouring cell, the maximum size of the aperture is limited to $n \times n/2$ pixels. This method has the advantage of providing a binary codification of the complex valued function $T_k(X,Y)$, with more than two levels of quantization for the amplitude and the phase. Its main drawback is the high number of pixels needed to describe the pattern.

### 3.2. Point Oriented Coding

One way to get a real function from $T(X,Y)$ is to consider only its real part. In this case the function $t(x,y)$ and its inverse symmetric $t(-x,-y)$ will overlap at the centre of the Fourier plane. In order to separate these two patterns, a cosine carrier signal is superimposed along with the $X$ axis, yielding:

$$T_2(X,Y) = A(X,Y)\cos[2\pi pX + \Phi(X,Y)] \qquad (3)$$

where $p$ is the carrier signal spatial frequency. Following the approach presented by VanderLugt,[17] it has been found that $p$ should fulfil the condition $p \geq L/(2\lambda z)$, where $L$ represents the extent of the pattern described by $t(x,y)$, $\lambda$ is the wavelength of the coherent illumination beam and $z$ is the distance between the CE and the Fourier plane. At the limit this condition provides a $L/2$ separation between the twin patterns $t(x,y)$ and $t(-x,-y)$. However, in this case, they are disturbed by the contribution due to the binarization. We have chosen $p = L/(\lambda z)$ in order to shift this perturbation away from the twin patterns. The function in Eq. 3 is binarized according to a threshold value $T_r$ obtaining:

$$T_B(X,Y) = \begin{cases} 1, & \text{if } \quad T_2(X,Y) \geq T_r \\ 0, & \text{if } \quad T_2(X,Y) < T_r \end{cases} \qquad (4)$$

The effect of Eq. 4 in the Fourier plane is twofold: the appearance of a central peak due to the non-zero mean value of $T_B$ and of disturbances in the reconstruction of $t(x,y)$ due to the binarization approximation.

## 4. CODING TECHNIQUES FOR FRESNEL TYPE HOLOGRAMS

By Fresnel hologram,[8, 18, 19] we mean a hologram which includes a Fresnel lens and a Fourier hologram, in the same plate. If the included Fresnel lens has a short focal length then the length of the setup can be considerably reduced.

The first coding technique presented here consists of a slight modification of the procedure to superpose the two elements of the Fresnel hologram, namely, the Fresnel encoded lens and the proper Fourier hologram. The second technique allows us to eliminate the twin image and the central dot from the reconstruction plane. To our knowledge, this represents a new approach compared with those already reported in the literature.[20, 10]

### 4.1. Fresnel Encoded Lens

It is well known that the transmittance function of a positive lens can be written as:

$$T(X,Y) = \exp\left[i\Delta\phi(X,Y)\right] = \exp\left[\frac{i\pi}{\lambda f}(X^2 + Y^2)\right] \qquad (5)$$

where $(x,y)$ are the Cartesian coordinates at the lens plane, $f$ is the focal length, and $\lambda$ is the wavelength of the light beam passing through the lens. If the discrete version of the above equation is considered, the focal length of the lens can be defined in terms of the lens diameter $D$, the number of sampling points $n \times n$, and the parameter $A$:

$$f = D^2 \frac{A}{\lambda n^2} \qquad (6)$$

The value of the parameter $A$ results from the sampling theorem ($A \geq n$). From the above equation one can see that the value of the focal length depends strongly on the resolution of the device used to fabricate the Fresnel lens. For instance, if laser lithography (2 μm resolution) is assumed, for $n = 1000$ a 2 mm diameter Fresnel encoded lens results having a focal length $f = 1.58$ mm ($\lambda = 0.63$ μm, $A = 248.85$). We have designed and realised a Fresnel lens with $n = 256$ using a laser printer (600 dpi) ($\lambda = 0.63$ μm, $A = 272.50$). In this case, we have $f = 660$ mm for a 10 mm diameter Fresnel lens. The configuration of the lens is similar to that reported by Carcole.[18] A multiple Fresnel lens can be obtained on the same support if the value of the parameter $A$ is chosen to be lower than $n$. We have designed a 7x7 multiple lens, each element having a focal length $f$ of about 100 mm, which still has good focusing properties ($n = 256$, $D = 10$ mm, $A \sim 40$, $\lambda = 0.63$ μm).

### 4.2. Fresnel Hologram of The First Type

The coding technique for a hologram of this type implies the superposition of a Fresnel lens over the Fourier hologram that was obtained using the POC method (described in Section 3). The optical reconstruction of this hologram looks similar to that obtained with a Fourier hologram, but the length of the optical setup can be easily reduced and controlled if the focal length of the Fresnel lens is adequately chosen.

### 4.3. Fresnel Hologram of The Second Type

Starting with the object function $t(x,y)$, a Fresnel transform (instead of a Fourier transform) is performed first. For objects described by a real valued function (intensity objects), this transform is performed in two steps: computation of the Fourier transform of the object function; multiplication of the FT by a quadratic phase factor, which depends on the distance between the hologram and the reconstruction plane.

Then, in order to obtain the hologram, the POC technique is used. Note that POC is the last step of the coding procedure here, while POC is an intermediate step in the coding procedure of a Fresnel hologram of the first type.

Now follows the theoretical explanation of what we have called Fresnel hologram of the second type, and how the unwanted images are removed from its reconstruction.[21-23]

If we represent the real positive object function as $t(x,y)$ (we consider only intensity objects) then the complex transmittance distribution to be realized in the hologram is given by the Fresnel transform of $t$:

$$T_{FR}(X,Y) = \text{F T}\left[t(x,y)\right]\exp[i\pi/(\lambda z)(X^2 + Y^2)] = T(X,Y)\exp[i\pi/(\lambda z)(X^2 + Y^2)] \qquad (7)$$

Now, as in the Fourier case, we are looking for a real positive function $T_H$ for making an amplitude hologram. The reconstruction is then obtained with the algorithm:

$$t_{rec}(x,y) = FT^{-1}\{T_H(X,Y)\exp[-i\pi/(\lambda z)(X^2+Y^2)]\}\exp[-i\pi/(\lambda z)(x^2+y^2)]. \quad (8)$$

As we are interested in intensity objects only, we can drop the last exponential factor which has unity modulus, obtaining:

$$t_{rec}(x,y) = FT^{-1}\{T_H(X,Y)\exp[-i\pi/(\lambda z)(X^2+Y^2)]\}. \quad (9)$$

Following the same steps as in the POC method we can begin taking the real part of $T_{FR}(X,Y)$:

$$T_R = T_{FR} + T_{FR}^* = T(X,Y)\exp[i\pi/(\lambda z)(X^2+Y^2)] + T(X,Y)^*\exp[-i\pi/(\lambda z)(X^2+Y^2)]. \quad (10)$$

The reconstruction will be:

$$t_{rec}(x,y) = FT^{-1}\{T(X,Y) + T(X,Y)^*\exp[-2i\pi/(\lambda z)(X^2+Y^2)]\}$$

$$= t(x,y) + \frac{1}{2\lambda z}\iint d\xi\, d\eta\, t(-\xi,-\eta)\exp\left\{\frac{\pi i}{2\lambda z}\left[(x-\xi)^2 + (y-\eta)^2\right]\right\}. \quad (11)$$

The original object function is reconstructed, plus a convolution of it (reversed) with a function which is the transmittance function of a positive lens with a focal length equal to $2z$. In our free space optical interconnections we are considering object functions that represent a certain number of spots. These object functions are of the type:

$$t(x,y) = \sum_{l=1}^{N}\delta(x-x_l)\delta(y-y_l) \quad (12)$$

where $(x_l, y_l)$ are the coordinates of the $N$ points that form the image pattern. Substituting Eq. (12) into Eq. (11) we obtain:

$$t_{rec1}(x,y) = \sum_{l=1}^{N}\delta(x-x_l)\delta(y-y_l) + \frac{1}{2\lambda z}\sum_{l=1}^{N}\exp\left\{\frac{\pi i}{2\lambda z}\left[(x+x_l)^2 + (y+y_l)^2\right]\right\} \quad (13)$$

As can be seen, the convolution term results in a negligible disturbance to the desired distribution $t(x,y)$. To obtain a positive $T_H$, in practice we have to add, a bias in Eq. (10). This will cause a disturbance of the form:

$$\frac{1}{\lambda z}\exp\left[\frac{\pi i}{\lambda z}(x^2+y^2)\right] \quad (14)$$

which is of the same form as the one in Eq. (13). This explains the fundamental result of obtaining only one of the twin images. Here, contrary to the Fourier case, it is not necessary to superimpose a carrier to separate the twin images because the reversed image is somehow transformed in a disturbance. In any case, the effect of a carrier would just be the shifting of the disturbance and of the image in opposite directions.

## 5. EXPERIMENTAL RESULTS

The ability of the connection element to realise the desired interconnection has been experimentally checked in the laboratory and the results are presented in what follows. The

230

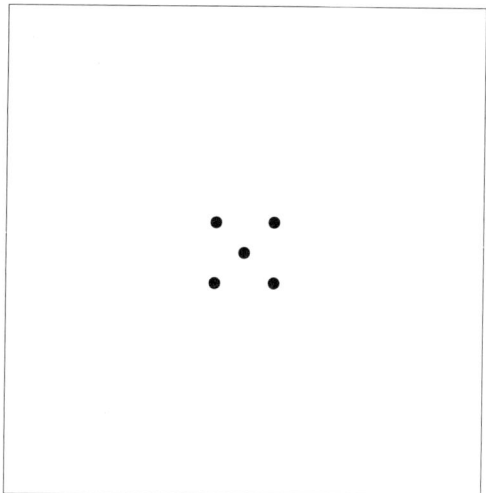

Fig. 4 Schematical representation of the 5 spots array. The actual image consists of 5 white pixels in a matrix of 256x256 black pixels.

Fig. 5 Fragment of the central region of the Fourier binary CGH obtained by the use of the COC technique.

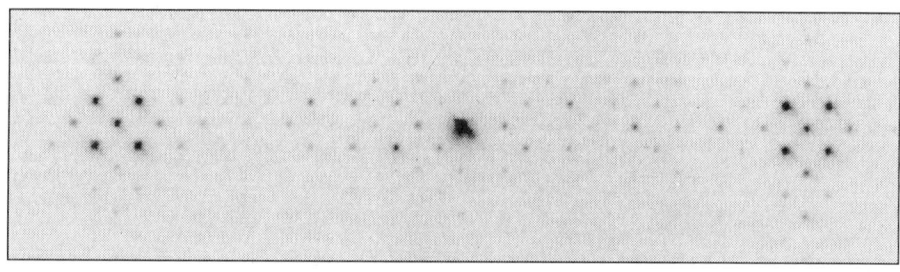

Fig. 6 Optical reconstruction in the Fourier plane (negative image) obtained with a COC Fourier hologram (Fig. 5) using the setup shown in Fig. 2(a).

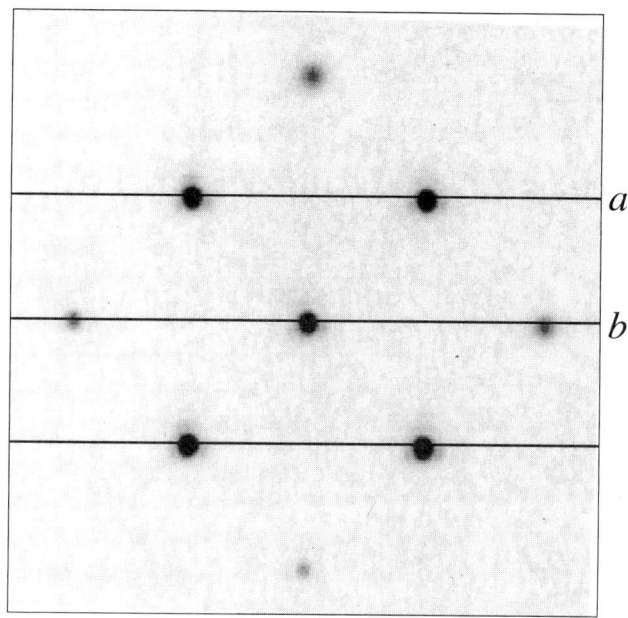

Fig. 7 Optical reconstruction in the case of COC Fourier hologram: portion of Fig. 6 representing one of the twin images.

two approaches described in Section 3 have been used to design the connection element, as a Fourier hologram. A $N$-spot array, consisting of 5 pixels inserted in a matrix of 256x256 pixels (Fig. 4), was considered for our holograms. For the case of COC, we have used a cell having 9x9 pixels to code the complex amplitude value of each sample. The amplitude values of all the cells was set to 1 (a column of 9 pixels in the cell) and only the phase of the Fourier transform was coded. In this configuration the phase was quantized with 9 levels (there were 9 positions in the cell for the 9-pixel aperture). This number of quantization levels proved to be the best compromise with respect to the errors that appear in the

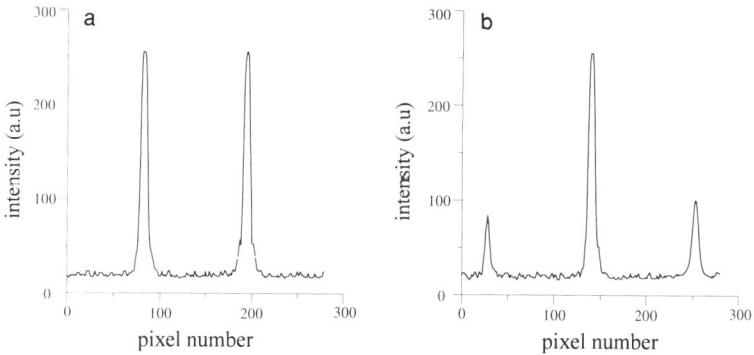

Fig. 8 Optical reconstruction in the case of COC Fourier hologram. (a) Intensity distribution along the line $a$ in Fig. 7. (b) Intensity distribution along the line $b$ in Fig. 7.

Fig. 9 The Fourier binary CGH obtained by the use of POC technique.

reconstructed image. A binary pattern of 2304x2304 pixels resulted. A fragment of the central region of the corresponding hologram is depicted in Fig. 5.

This pattern was first printed out using a laser printer with 600 dpi resolution and then photoreduced onto a lithographic film to a 1x1cm pattern. The result of the optical reconstruction in the Fourier plane of the setup shown in Fig. 2(a) is depicted in Fig. 6 (negative image).

One can see the twin arrays and the central peak that appears in the reconstruction plane. The separation between the two images depends on the frequency of the carrier superimposed on the real part of the FT, while the central peak corresponds to the constant that was added to get positive values. Because of the errors introduced by the coding technique, some undesired spots are also obtained. Nevertheless they are not very significant, because the maximum intensity of these spots is lower than half of the peak intensity of the desired spots and thus they can be easily removed. The image in Fig. 6 was captured with a CCD camera (736x560 pixels resolution). The width of the image is of 1/3". The corresponding size of the "5 spots" array is about 0.5x0.5 mm. For further considerations about the parameters that influence the optical reconstruction, we shall consider only one of the two twin images (Fig. 7).

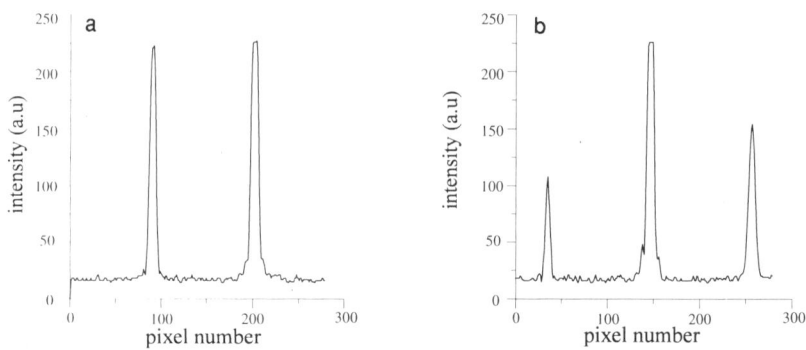

Fig. 10 Optical reconstruction in the case of POC Fourier hologram. (a) Intensity distribution along a line analogous to $a$ in Fig. 7. (b) Intensity distribution along a line analogous to $b$ in Fig. 7.

Fig. 11 Fresnel binary CGH of the first type.

The intensity distributions along the lines $a$ and $b$ in Fig. 7 are shown in Fig. 8(a) and 8(b) respectively. The width of the spot, corresponding to the half of its peak intensity, is about 0.09 mm and it is uniform for all of the five spots. This value is even a bit higher than the real one because the signal is slightly saturated. From Fig. 8(b), it turns out that the undesired spots have a maximum intensity of 100, which represents 39% of the peak intensity of the array spots. The level of noise in the reconstruction is between 20 and 30

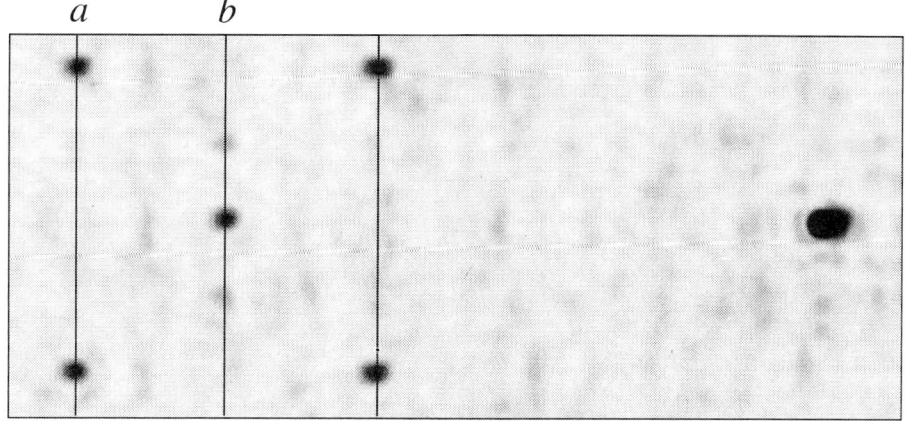

Fig. 12 Optical reconstruction in the case of Fresnel hologram of the first type (Fig. 11). Only the central peak and the left twin image are shown.

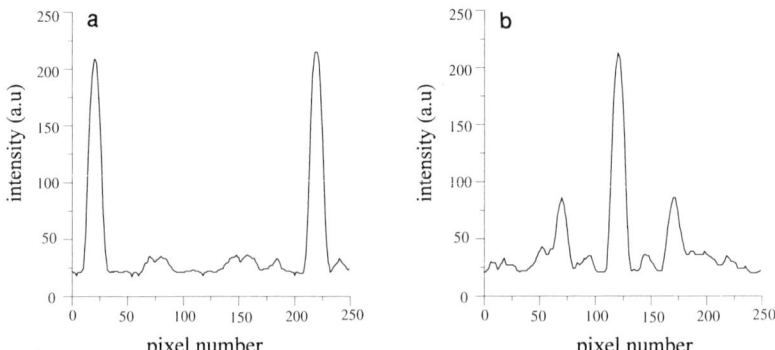

Fig. 13 Optical reconstruction in the case of Fresnel hologram of the first type. (a) Intensity distribution along the line $a$ in Fig. 12. (b) Intensity distribution along the line $b$ in Fig. 12.

(8÷12% of the maximum intensity). For the case of POC, after binarization, the 256x256 pixel pattern was printed out on a slide using a 600 dpi laser printer. A 10.8x10.8 mm size binary CGH resulted, thus avoiding the photoreduction step. This hologram is depicted in Fig. 9. The result of the optical reconstruction is similar to that in Fig. 6 and 7. The intensity distributions along lines analogous to $a$ and $b$ of Fig. 7 are shown in Fig. 10(a) and 10(b) respectively. One can see that the right side spots are larger than the left side ones. The intensity of the higher undesired spot represents 68% of that in the desired ones. The errors that are introduced with POC are greater than those with COC.

This is normally expected because the content of information about the original array is strongly reduced.Furthermore the phase variations induced by thickness nonuniformity of the slide also contributes to generate errors. Nevertheless, because of its simplicity, the technique described above can be a convenient solution for many applications. Anyway, in order to reduce the errors due to the use of the slides, the pattern can be drawn first at a larger scale and then photoreduced on a lithographic film.

In order to obtain the smallest length for the setup, we have used the Fresnel holograms described in Section 4. First, a Fresnel encoded lens was directly superposed on the Fourier binary CGH of the $N$ spots array in Fig. 4. The resulting hologram (i.e. Fresnel hologram of first type) is shown in Fig. 11. The array generated by it can be observed in Fig. 12.

The intensity distributions over the lines $a$ and $b$ of Fig. 12 are shown in Fig. 13(a) and 13(b) respectively. Comparing with the intensity distributions depicted in Fig. 10(a) and 10(b), one can see that the spots are larger now. This effect comes from the errors introduced by the substitution of the classical lens with a binary Fresnel lens. The diameter of the spots, corresponding to the half of maximum intensity in the reconstruction plane, equals 0.12 mm for the array in Fig. 13.

For the case when we want to generate only a single array, we use a Fresnel hologram of the second type (Fig. 14). The array generated in the reconstruction plane, when this hologram is introduced into the proper optical setup, is depicted in Fig. 15.

The array we have used to build up our hologram might be thought to be a special case (i.e., it is symmetric with respect to the origin).In order to eliminate this suspicion, we have also considered various other array configurations. A three spot array generated by the use of a corresponding Fresnel hologram of the second type is depicted in Fig. 16(a); a four spot array appears in Fig. 16(b) and a two spot array in Fig. 16(c).

Fig. 14 Fresnel binary CGH of the second type.

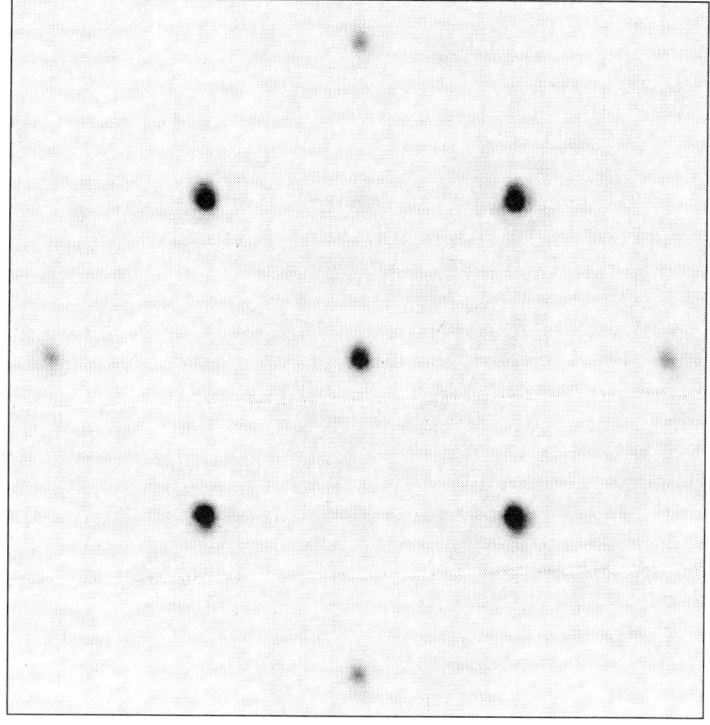

Fig. 15 Optical reconstruction in the case of Fresnel hologram of the second type.

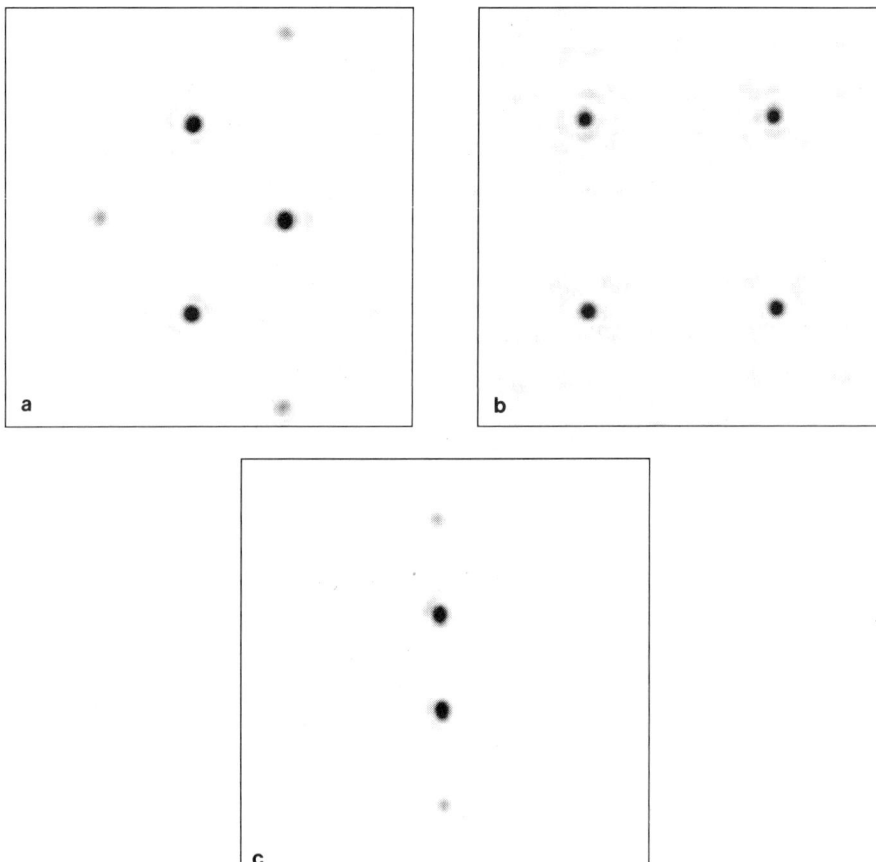

Fig. 16 Optical reconstructions in the case of Fresnel holograms of the second type for different arrays. (a) 3 spots array; (b) 4 spots array; (c) 2 spots array.

## REFERENCES

1. J.W. Goodman, Optics as an interconnect technology, in: *Optical Processing and Computing*, Academic Press. Inc., 1-32, (1989).
2. J. Wilde, R. McRuer, L. Hesselink, and J.W. Goodman, Dynamic holographic interconnections using photorefractive crystals, *Proc. SPIE*, 752, 200-208 (1987).
3. P. Chavel, and Ph. Lalanne, Parallel Algorithms for optoelectronic image processing, *Proc. EOS*, (Photonics '95, Prague, August 1995), 1-8.
4. H. Tajima, Y. Okada, and K. Tamura, A high-speed optical common bus for a multiprocessor system *Trans. Inst. electron and Commun. Eng. Jpn.* 24 (17), 1984, 850-866.
5. R. K. Kostuk, J. W. Goodman, and L. Hesselink, Design considerations for holographic optical interconnections, *Appl. Opt.* 26, 3947-3953, (1987).
6. B. Bianco, and T. Tommasi, Space-variant optical interconnection through the use of computer generated holograms, *Appl. Opt.* 34, 7573-7580, (1995).
7. W. H. Lee, Computer generated holograms: techniques and applications, in *Progress in Optics*, 1978, Vol. 16, 121-232.
8. O. Bryndgdahl, and F. Wyrowski, "Digital holography - computer generated holograms," in *Progress in Optics*, Vol. XXVIII, 28-82, (1990).
9. M. R. Feldman, C. C. Guest, Holograms for optical interconnects for VLSI circuits by electron beam lithography, *Opt. Eng.* 28, 915-921, (1989).

10. H. H. Suh, C. H. Kwak, and E. Lee, Combined binary-phase holograms for free-space optical interconnection, *Optics Letters* Vol. 20 (20), 2131-2133, (1995).
11. J. A. Davis, G. W. Bach, D. M. Cottrell, and R. A. Lilly, Suppression of selected diffraction orders with programmable masks written on spatial light modulators, *Appl. Opt.* 27, 2949-2953, (1988).
12. W. H. Lee, Binary computer generated holograms, *Appl. Opt.* 18, 3661-3669, (1979).
13. B.R. Brown, and A. W. Lohmann, Computer-generated Binary Holograms, *IBM Journal of Research and Development* Vol. 13, 160-168, (1969).
14. A. W. Lohmann, and D. P. Paris, Binary Fraunhofer Holograms, Generated by Computer, *Appl. Opt.*, Vol. 6(10), 1967, 1739-1748.
15. W. H. Lee, Sampled Fourier Transform Hologram Generated by Computer, *Appl. Opt.*, Vol. 9(3), 1970, 639-643.
16. C.B. Burckhardt, A simplification of Lee's method of generating holograms by computer, *Appl. Opt.*, 9, 1970, 1949.
17. A. VanderLugt, "Optimum sampling of Fresnel transforms," *Appl. Opt.* 29, 3352-3361, (1990).
18. E. Carcole, J. Campos, and S. Bosch, Diffraction theory of Fresnel lenses encoded in low-resolution devices, *Appl. Opt.* 33 (2), 162-174, (1994).
19. E. Carcole, J. Campos, I. Juvells, and S. Bosch, Diffraction efficiency of low-resolution Fresnel encoded lenses, *Appl. Opt.*, Vol. 33 (29), 1994, 6741-6746.
20. M. S. Kim, C.C. Guest, Block quantized binary-phase holograms for optical interconnection, *Appl. Opt.* **32**, 678-683, (1993).
21. Y. H. Wu, and P. Chavel, Cell-oriented on-axis computer-generated holograms for use in the Fresnel diffraction mode, *Appl. Opt.* 23 (2) pp. 228-238 (1984).
22. W. J. Dallas, A.W. Lohmann, Phase quantization in holograms-depth effects, *Appl. Opt.* 11 (1) pp. 192-194 (1972).
23. D. Cojoc, F. Sartori, and I. Montrosset: 1xN free space optical interconnection using computer generated binary amplitude holograms, submitted to *Appl. Opt.*, "Diffractive and Micro-Optics" feature to be published on July 1997.

# HOLOGRAPHIC DIFFRACTIVE COMPONENTS FOR BEAM COUPLING

M. Miler

Institute of Radio Engineering & Electronics Academy of Sciences
Chaberskà 57, 18251 Prague-Kobylisy, Czech Republic

## 1. INTRODUCTION

Optical devices for non-parallel transmission and processing of information have been designed as 3-dimensional structures up to now. In such arrangements optical beams as streams of photons carry information. Beams are transformed by means of classical or diffractive optical elements in 3-D space. However, such optical structures occupy a lot of volume, their discrete components must be positioned using an appropriate mechanical cage or frame, and adjustment is often very tedious and complicated. Therefore there is an evident effort to convert 3-D structures, at least in part, to 2-dimensional components where there are opportunities for technology integration. This 2-D technology better allows automated production of the components in massive quantities. 2-D or planar optics can be divided in to two branches according to the nature of light propagation inside a dielectric layer or a plate where the light field is confined, unlike propagation in 3-D free space. If the thickness of the plate is much larger than the wavelength of light, zigzag propagation takes place due to substrate modes. On the other hand, if the thickness of the layer is comparable with the light wavelength, guided waves arise and guided modes must be taken into account. In both cases, there must be a way to provide coupling between freely propagating beams in 3-D space and beams propagating in the planar component. Classical coupling elements for this purpose are prisms. However, prism couplers are discrete elements which must be fixed to the surface of 2-D optics. Their production is tedious and they are not compatible with the planar integration. On the other hand, gratings [1] offer many more advantages for coupling, because they are compatible with planar optics and can be produced large numbers.

## 2. DIFFRACTION COUPLING BETWEEN FREE-PROPAGATED BEAMS AND GUIDED WAVES

Each surface corrugation of the planar waveguide l eads to a leakage of the guided

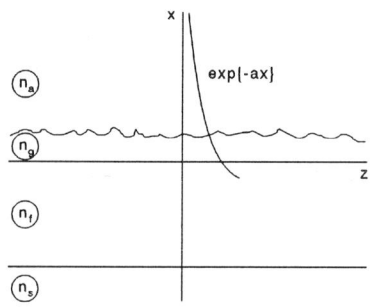

Fig. 1 The guided wave tail.

energy into the neighboring medium. This leakage is caused by scattering of the guided wave tail, located in the vicinity of the boundary between the waveguide film and neighboring medium, on the corrugation (Fig. 1). The guided wave tail decays exponentially away from the film. If the corrugation is random, the leaky waves are randomly scattered. On the other hand, if the corrugation is periodic, the leaky field concentrates in discrete directions and diffraction orders arise. This process can be also realized from opposite side of the film: if light is incident on the corrugated surface of the waveguide, it will be scattered. Some scattered rays can be identified with guided modes and propagate along the waveguide. If the corrugation is periodic, discrete directions exist from which light must impinge on the waveguide surface for a guided wave to begenerated.

## 2.1. Surface-Relief Phase Gratings

A periodic corrugation of the boundary surface can be called a grating diffraction structure having phase surface-relief modulation. If this modulation is formed holographically, i.e. by exposure of a light sensitive layer to an interference field, then, under the linear conditions of exposure and subsequent etching, a sinusoidal modulation arises (Fig. 2). The spatial period of the structure is $\Lambda$ and the amplitude is $h/2$, where $h$ is the depth of modulation. However, the conditions of etching can often be non-linear, and the profile is consist either of shallow "sea-waves" with sharp tops and round bottoms, or of deep "wells" having practically rectangular shape. The latter shape is obtained especially often in the case of ion-beam etching through a holographically prepared photoresist mask.

The theory of diffraction by surface-relief sinusoidal gratings [2] gives results in an analytic form only for shallow profiles, $h/\Lambda < 0.14$, whose period is large compared to wavelength of light, $\Lambda/\lambda > 5$. The angle of incidence must be small, so the diffraction angles are then small as well, i.e. a high number of diffraction orders exists, and the polarization of light need not be taken into account. T his type of diffraction is called Raman-Nath, and the

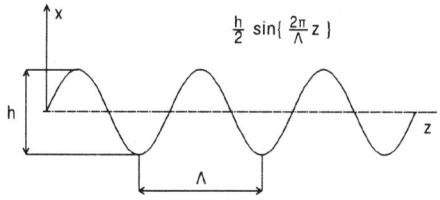

Fig. 2 The sinusoidal surface-relief modulation.

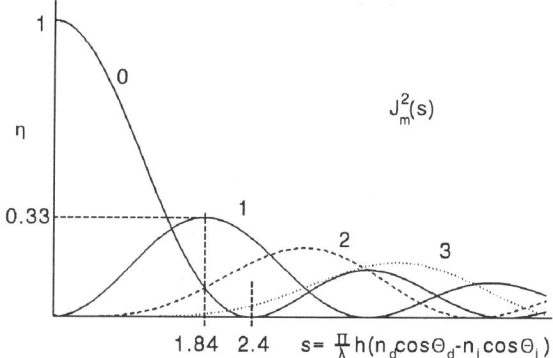

Fig. 3 Diffraction efficiency of the phase sinusoidal grating.

amplitudes of the diffracted waves are proportional to the Bessel functions of the first kind:

$$A_m = (-1)^m J_m \{\pi h(n_d \cos\theta_d - n_i \cos\theta_i)/\lambda\} \quad (1)$$

where $\theta_i$, $\theta_d$ are the angles of incidence and diffraction, respectively, and $n_i$, $n_d$ are the indices of refraction of the media in which the incident and diffracted beams propagate, respectively. The diffraction efficiency, $\eta_m$, is then proportional to the square of the $m$-th order Bessel function, where $m$ stands for the diffraction order (Fig. 3).

For the mathematical analysis of deeper profile gratings numerical methods must be used, and physical insight is problematical. Only very deep profiles [3] can be treated as the Bragg gratings where energy can be exchanged only between the zero and first orders according to the square of sine and cosine functions (Fig. 4):

$$\eta_1 = \sin^2\{\pi n_1 d/\lambda \cos\theta_d\}, \qquad \eta_0 = 1 - \eta_1, \quad (2)$$

where $n_1$ is the amplitude of modulation and $d$ is the thickness of the grating layer.

Besides the distribution among diffraction orders, one needs to of energy know the directions of propagation of the diffraction orders. These directions are determined by the so called *grating equation* which interrelates the projections of propagation vectors depending on the order of diffraction :

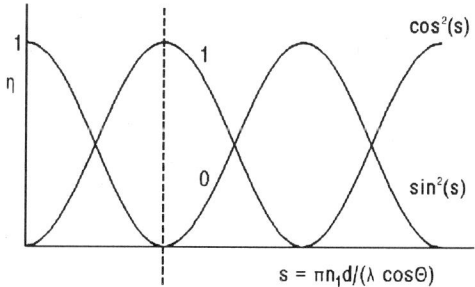

Fig. 4 Diffraction efficiency of the volume phase grating.

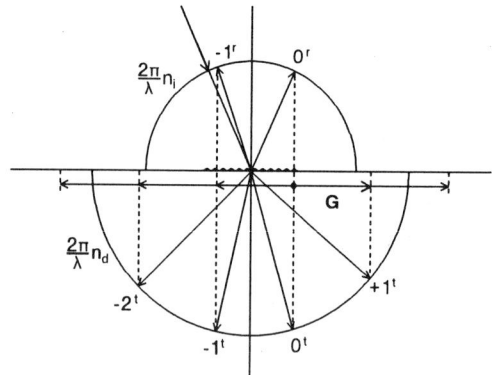

Fig. 5 Diagram of wave vectors for diffraction by a plane grating.

$$p_{m\,x} = p_{i\,x} + mG \tag{3}$$

where $\mathbf{p}_{ix}$, $\mathbf{p}_{mx}$ are the projections of the propagation vectors of the incident and $m$-th order diffracted beams, respectively, and $\mathbf{G}$ is the grating vector ($|\mathbf{G}| = 2\pi/\Lambda$).

A more illustrative form of the grating equation is shown in the diagram (Fig. 5), which gives directly the angles of diffraction for different orders in the transmission and reflection modes, and the possible number of orders. The sign convention for diffraction orders follows from Fig. 5.

For very deep (or thick or volume) gratings, the direction of propagation of the diffracted order is given by the Bragg law, which says that the propagation vectors and the grating vector must form closed triangle (Fig. 6) which is symmetrical with respect to the grating planes, i.e. the grating vector must be divided into halves by the bisector of the propagation vectors. In another words, light rays must undergo reflection on the grating planes. The second and higher diffraction orders can be obtained if a double or multiple of the grating vector is smaller than the diameter $(4\pi/\lambda)n_g$ of the circle of wave vectors.

## 2.2. Grating Couplers for Guided Waves

If the grating is placed on the surface of the waveguide film, it can serve as a coupler which makes it possible to convert a guided wave into a freely propagating beam in the

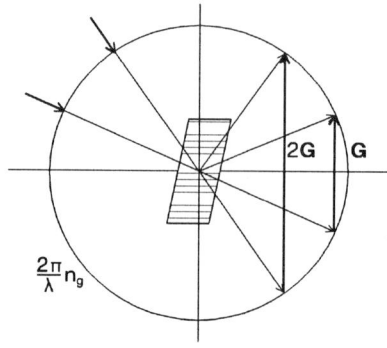

Fig. 6 Diagram of wave vectors for diffraction by a volume grating.

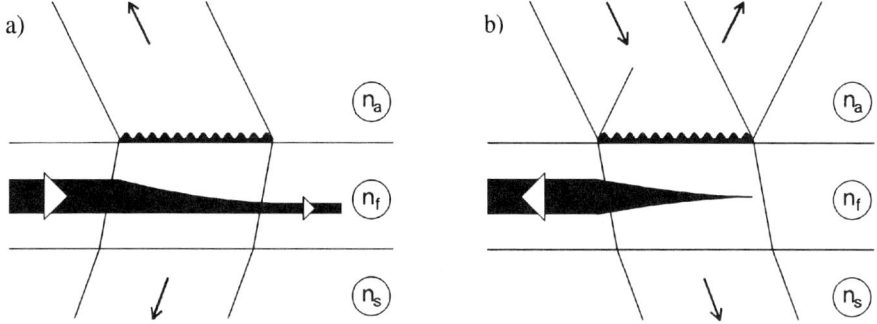

Fig. 7 Schematic sketch of: outcoupling (a); incoupling (b).

surrounding medium or vice versa. The schematic sketch of both cases is plotted in Fig. 7a,b. In the case of outcoupling the guided wave feels, by its exponential tail, the grating structure and is scattered into space-harmonic fields. The energy of the guided wave diminishes as is transformed into the energy of the freely propagating beams. The incoupling can appears much more complicated. A beam incident at a suitable angle onto the grating is coupled into the guided wave, but is partially outcoupled again while it propagates, contributing thus to the reflected or transmitted field. After attaining equilibrium the guided wave propagates in the waveguide.[4]

The diagram of wave vectors involved is shown in Fig. 8. Waveguide modes are shown as discrete points on the boundary between the substrate and superstrate (air) media at a distance from the center of the diagram greater than the radius of the circle representing the locus of the final points of all possible wave vectors in the substrate. The propagation factor $\beta$ of the guided mode, the grating vector **G** (or its multiple) and the projection of the propagation vector $\mathbf{p}_{mx}$ must satisfy the relation

$$p_{m\,x} = \beta + m \quad G. \tag{4}$$

From the diagram one can see the values of the coupling angles, the number of outcoupled orders etc. Coupling exists only for negative values of $m$. Not only can more

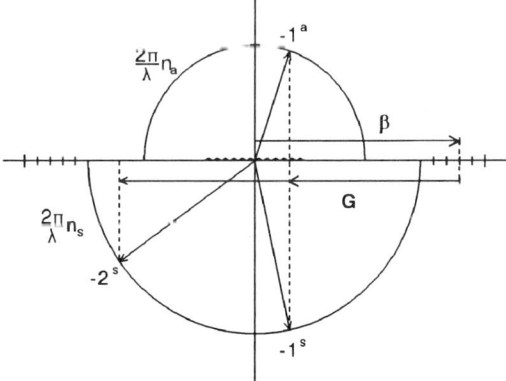

Fig. 8 Diagram of wave vectors for the grating coupler.

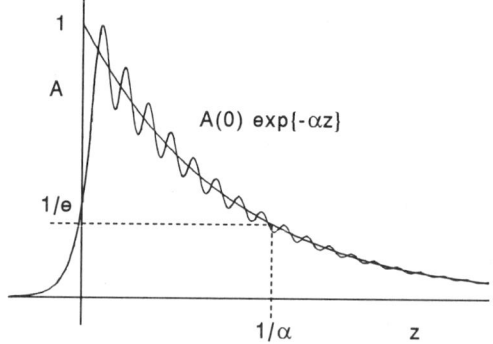

Fig. 9 The cross-sectional distribution of the amplitude in the outcoupled beam.

than one outcoupled beam appear in the case of outcoupling; also in the case of incoupling, other diffracted beams, besides the useful ones can be formed.

When the beam is being outcoupled from the waveguide, the energy in the guided wave decreases according to the familiar law

$$dA(z) = -\alpha A(z) dz, \tag{5}$$

where $A(z)$ is the amplitude of the electric field and $\alpha$ is the decay factor. The cross-sectional amplitude distribution of the out-coupled beam is exponential (Fig. 9), where the width of the beam, determined from the intensity decrease in to $1/e$ of its maximum, is approximately $1/\alpha$. This latter result actually comes from geometric-optical treatment, which does not give the detailed behavior of the curve. As follows from the wave optical analysis, there is an oscillatory ripple on the curve due to Fresnel diffraction at the edge of the grating, particularly at the beginning of the curve.[5]

If such a grating is used for incoupling, full conversion is achieved merely on condition that the impinging beam has the same intensity profile as the outgoing beam would have hads were the grating used as the outcoupler. However, incident beams have profiles which are usually quite different from the exponential form, so that there are energy losses caused by profile mismatch. For example, at most 80% of the incoming energy can be coupled into a guided wave if the incident beam profile is Gaussian.

The decay factor $\alpha$ is a complicated function of the grating parameters. It depends first of all on the profile of the grating grooves. The best kind of the grating for these purposes is the Bragg type[6] with slanted grating planes (Fig.10a). However, in most cases it is not

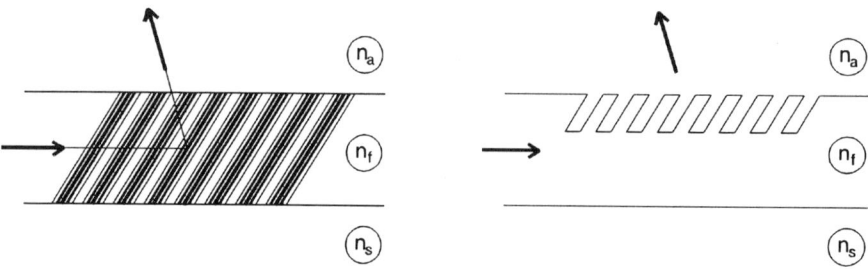

Fig. 10a The thick volume grating coupler.   Fig. 10b The volume grating coupler with deep surface modulation.

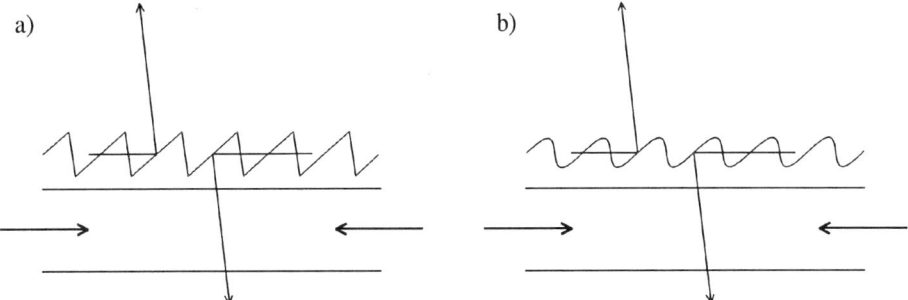

Fig. 11 Grating coupler with blazed triangular profile (a) blazed profile (b).

possible to fabricate a thick grating in the volume of the waveguide as the phase grating with the refractive index modulation. The grating is often fabricated as a deep surface corrugationby means of ion-beam etching [7] (see Fig. 10b). For mass production one must prefer gratings fabricated by embossing or related techniques. In such cases, the grating cannot have an extremely deep profile, but only a profile whose modulation depth is comparable with its period. At the same time, the largest possible part of the grating groove should be formed by an inclined working facet whose orientation makes it possible for the rays reflected or refracted on this facet to propagate at the diffraction (coupling) angle.

The most suitable form for the profile is a rectangular triangle, but by holographic techniques one can obtain a smooth profile wich is a close approximation \to an inclined sinusoidal curve (Fig. 11) in shape. In this case a part of the energy is reflected back from the perpendicular facet and can contribute to diffraction beams of higher orders if there are any. This so called blazed profile [8] increases the diffraction efficiency of the coupler and therefore also the decay factor $\alpha$. The blazing also prevents the splitting of energy into two outcoupling beams: one into the substrate and one into the air. This restriction is now given by the direction of propagation of the guided wave.

From the theoretical analysis, the approximate dependence of the decay factor $\alpha$ on the modulation depth $h$ of the grating can be derived. For small values of $h$ the factor $\alpha$ varies as $h^2$ until the depth reaches the critical value $h_c$, after which $\alpha$ increases less rapidly and tends to be limited by an asymptotic value depending on the profile shape. The variation of $\alpha$ with $h$ for a rectangular grating profile having inclined side facets can be seen in Fig. 12.

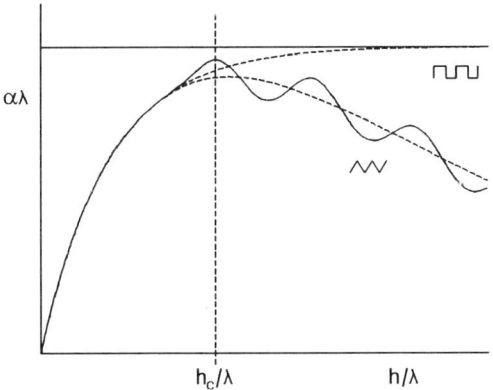

Fig. 12 Dependence of the decay factor on the modulation depth.

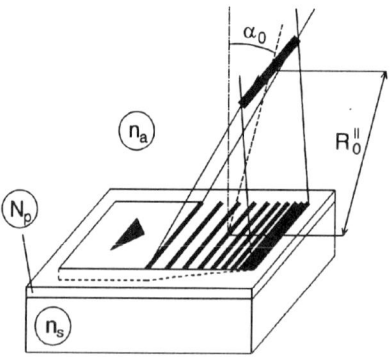

Fig. 13 The chirped grating coupler.

However, oscillations appear at large *h*, caused by interference patterns a within the thickness of the waveguide system adjoining the grating. The amplitude of these oscillations depends on how much the refractive indices of the substrate, waveguide, grating and superstrate media differ. For small differences the oscillations are rather weak (for example, the case of glass diffused waveguides) while for larger differences they are stronger (silicon-oxide-silicon). Due to the character of the $\alpha$ curve, the diffraction efficiency also exhibits oscillatory behavior. Untill now, we have assumed a constant modulation depth along the grating. If one wants to obtain a cross-sectional energy distribution which differs from exponential, the modulation depth must vary in the longitudinal direction. It can be shown that linear growth of the modulation amplitude is necessary for the cross-sectional distribution of the outgoing beam to approach a Gaussian shape. In this case, a Gaussian beam incident on a grating coupler of this type also has maximal coupling efficiency.

## 2.3. Holographic Focusing Grating Couplers

Grating couplers have not only the advantage of their compatibility with planar structures but they can also provide other optical functions such as focusing, which is the subject of this section. The common grating coupler, which couples a free by propagating plane wave into a guided wave having a planar wave-front, must have straight and equidistant grating lines; therefore, it can be formed by interference of two freely propagating plane waves. Any curvature of the grating lines and inequalities in their spacing would lead to curvature of the wavefront of the coupled waves.

First, we will analyze gratings having nonequidistant but straight grating lines.[9] If distances between neigh-boring lines regularly increase or de-crease, the grating is called a *chirped grating*. The basic chirp is linear,

$$\Lambda(z) = \Lambda_0 + \Delta\Lambda \cdot z, \tag{6}$$

where $\Lambda_0$ is the period at the beginning of the grating and $\Delta\Lambda$ is the increment of the period. The angle of the outcoupling ray will change with the local period according to the grating equation, and the rays will converge or diverge depending on the sign of the increment $\Delta\Lambda$ (see Fig. 13). The linear chirp produces a freely propagating outcoupled wave having, in the thin beam approximation, a cylindrical parabolic wavefront the focal line of which is perpendicular to the plane of outcoupling. This wave can be written in the form

$$u(z) = u_0 \exp\left\{-ik\left(\frac{z^2 \cos^2\alpha_0}{2R_0^{\|}} - z\sin\alpha_0\right)\right\}, \tag{7}$$

where $R_0^{\|}$ is the distance to the focal line from the center of the coupler and $\alpha_0$ is the angle of outcoupling of the central ray. The rate of the path change in the argument of the exponential expression of the parabolic wave $(d/dz)(z^2\cos^2\alpha_0/2R_0^{\|} - z\sin\alpha_0)$ is proportional to the grating period at the given point.

If a grating of this type is to be recorded holographically using two interfering freely propagating waves, one of the waves should approximate the guided wave and the other one a the outcoupled wave. The first wave is a plane wave impinging at a large angle of incidence on the light sensitive layer placed on the waveguide, and the other one is a cylindrical wave having its focal line perpendicular to the plane of incidence. *The holographic equation* may then be written in the form

$$n_{a,s}\frac{\cos^2\alpha_0}{R_0^{\|}} = \pm\mu n_a \frac{\cos^2\alpha_1}{R_1}, \tag{8}$$

where $n_a$, $n_s$ are indices of refraction of the superstrate and substrate respectively, $R_0^{\|}$, $R_1$, $\alpha_0$, $\alpha_1$ are distances and angles of the focal lines of cylindrical waves of the outcoupled and recording waves respectively, and $\mu = \lambda_0/\lambda_r$ is the ratio of wavelengths of the working and recording waves respectively. Terms containing plane waves, either freely propagating or guided, are not present because their $R \to \infty$. Eq. (8) must be solved simultaneously with *the grating equation*

$$n_{a,s}\sin\alpha_0 = N_p \pm \mu n_a(\sin\alpha_1 - \sin\alpha_2), \tag{9}$$

where $N_p$ is the effective refractive index of the guided wave and $\alpha_2$ is the angle of incidence of the second recording (plane) wave, which should be as large as possible. In both equations the upper sign is valid for primary imaging and the lower one for secondary imaging. This approach is valid only for the case of very thin beams. For larger beam widths aberration must be considered.

If the cylindrical recording wave has its focal line lying in the plane of incidence, the grating lines remain equidistant but they are curved.[10] In the plane of incidence (meridional plane) imaging does not take place while in the perpendicular (sagittal) plane the beam is

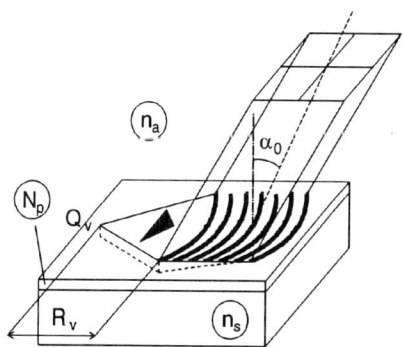

Fig. 14 The grating coupler focusing the guided wave.

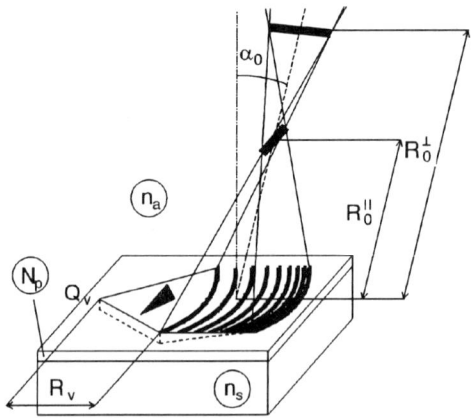

Fig. 15 The astigmatic outcoupled beam.

$$\frac{N_p}{R_v} = \pm \mu n_a \frac{1}{R_1}, \qquad N_p = n_a \sin\alpha_0 \pm \mu n_a (\sin\alpha_1 - \sin\alpha_2), \qquad (10)$$

transformed. Using this kind of grating coupler, focusing of the guided wave after incoupling can be realized (see Fig. 14). The holographic and grating equations are then
where the angle $\alpha_1$ is as large as possible.

Note that in both cases, the case of the chirped straight grating and that of the equidistant curved grating, the ratio of wavelengths $\mu$ can be compensated by the ratio of refractive indices. For fabrication of a fully (stigmatically) outcoupled beam focusing grating coupler, cylindrical recording waves cannot be generally applied.[11] By using rotationally symmetric wavefronts, astigmatic point imaging [12] is obtained (Fig. 15) because for large angles in the meridional plane a different holographic equation holds, which is different from the equation for small angles in sagittal plane. Generally, for holographic imaging in the meridional and the sagittal plane, respectively,

$$n_{a,s} \frac{\cos^2\alpha_0}{R_0^{\|}} = \pm \mu n_a \left( \frac{\cos^2\alpha_1}{R_1} - \frac{\cos^2\alpha_2}{R_2} \right), \qquad (11)$$

$$\frac{n_{a,s}}{R_0^{\perp}} = \frac{N_p}{R_v} \pm \mu n_a \left( \frac{1}{R_1} - \frac{1}{R_2} \right), \qquad (12)$$

and, for the grating equation,

$$n_{a,s} \sin\alpha_0 = N_p \pm \mu n_a (\sin\alpha_1 - \sin\alpha_2), \qquad (13)$$

where focusing takes place both for the guided and the freely propagating waves. A stigmatic outcoupled beam can be obtained if $R_0^{\|} = R_0^{\perp}$. In this case, from the last three equations it follows that

$$\left[ \frac{N_p}{n_a} \frac{1}{R_v} \pm \mu \left( \frac{1}{R_1} - \frac{1}{R_2} \right) \right] \cdot \left[ \frac{N_p}{n_a} \pm \mu (\sin\alpha_1 - \sin\alpha_2) \right]^2 +$$

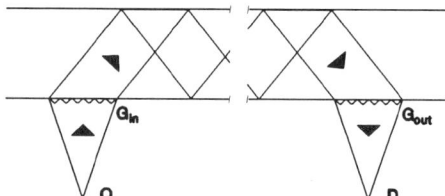

Fig. 16 Schematic sketch of the backplane inter-connection.

$$-\left[\frac{N_p}{n_a}\frac{1}{R_v} \pm \mu\left(\frac{\sin^2\alpha_1}{R_1} - \frac{\sin^2\alpha_2}{R_2}\right)\right] = 0. \tag{14}$$

For rough estimation we may use the following values: $\alpha_1 = 0°$, $R_1$, $\alpha_2 \approx 90°$, $R_2 = \infty$, $R_v = \infty$. In this case, from the last condition the equation $(\mu n_a/R_1)(N_p/n_a - \mu)^2 = 0$ follows. This says that when the coupler focuses the freely propagating beam at a finite distance and perpendicularly to the waveguide, the focusing is stigmatic if $\mu = N_p/n_a$.

In general, four holographic recording parameters $\alpha_1$, $\alpha_2$, $R_1$, $R_2$ must be adjusted in order for the focusing grating coupler to be optimized. However, off-axis aberrations of the full beam can not be fully compensated. For more complete compensation, methods of copying primary holographic gratings or techniques utilizing pre-aberrated recording waves can be used.

## 3. SUBSTRATE-MODE HOLOGRAPHIC INTERCONNECTS

Holographic grating couplers which intercouple freely propagating beams and substrate-modes of the backplane interconnection [13] encounter problems similar to those described above. The difference consists in the contrasting nature of the guided wave and the substrate-mode, which is nothing but a freely-propagating wave traveling at a small angle with respect to the boundary. In this case one cannot assume the angle of propagation of the wave in the backplane to be 90°. At the same time, the cross-sectional intensity distribution of the outcoupled beam is simply given by the distribution in the incident beam, under the assumption of uniform grating modulation.

The simplest scheme for back-plane interconnection consists of a thick dielectric plate with two focusing grating couplers on its lower side (Fig. 16). Inside the plate, zigzag beam propagation takes place with total internal reflections at its boundaries. One of the couplers serves as input, the other couples the light beam and focuses it on a detector. As the two gratings are in series with each other, their dispersion properties can either add or subtract. In the usual case, where the optical backplane interconnection makes use of gratings with parallel straight lines, the dispersion of the gratings subtracts and chromatic aberration can be compensated.[14] Differentiation of the grating equation, under the assumption of coupling between air and a backplane with refractive index $n$, gives for the minus first order an angular dispersion in the form

$$\frac{d\theta_d}{d\lambda} = -\frac{1}{n\Lambda\cos\theta_d}, \tag{15}$$

where $\theta_d$ is the output angle. It can be seen that the dispersion increases with the angle $\theta_d$.

Optical interconnects which distribute signals into many channels need elements to fan-out one beam into a 1-D or 2-D array of beams having the same light power. For this purpose gratings can also be used. It is well known that the repeating feature (period) of the grating is responsible for a discrete set of diffraction orders, while the shape of the single feature determines the distribution of the incident power into the individual orders. For example, a sinusoidal groove profile distributes the incoming beam into diffraction orders according to the square of Bessel functions of the corresponding orders. If the groove acts as a focusing lens with a very short focal length, it can distribute the incident power into many orders very uniformly. This is the case with the parabolic surface-relief groove profile.

## 4. CONCLUSION

Diffractive components recorded holographically can be used to a considerable extent for beam coupling between 3-D and 2-D optics. The principles of optical elements of this kind were described above. The discussion offered general knowledge about surface relief and volume gratings, and their use for coupling. Special attention was paid to focusing grating couplers.

## REFERENCES

1. M.L. Dakss, L. Kuhn, P.F. Heidrich, and B.A. Scott, Grating coupler for efficient excitation of optical guided waves in thin film, *Appl. Phys. Lett.* 16: 523 (1970).
2. E.K. Popov, L.V. Tsonev, and E.G. Loewen, Scalar theory of transmission relief gratings, *Opt. Comm.* 80: 307 (1991).
3. H. Kogelnik, Coupled wave theory for thick hologram gratings, *Bell Syst. Tech. J.* 48: 2909 (1969).
4. T. Tamir and S. Peng, Analysis and design of grating couplers, *Appl. Phys.* 14: 235 (1977).
5. S. Miyanaga and T. Asakura, Intensity profile of outgoing beams from uniform and linearly tapered grating couplers, *Appl. Opt.* 20: 688 (1981).
6. H. Kogelnik and T.P. Sosnowski, Holographic thin film coupler, *Bell Syst. Tech. J.* 49: 1602 (1970).
7. M. Li and S.J. Sheard, Wave-guide couplers using parallelogramic-shaped blazed gratings, *Opt. Comm.* 109: 239 (1994).
8. T. Aoyagi, Y. Aoyagi, and S. Namba, High-efficiency blazed grating couplers, *Appl. Phys. Lett.* 29: 303(1976).
9. A. Katzir, A.C. Livanos, and A. Yariv, Chirped grating output couplers in dielectric waveguides, *Appl. Phys. Lett.* 39: 225 (1977); S.I. Bozhevolnyi, E.M. Zolotov, V.A. Kiselev, A.M. Prokhorov, and E.A. Shcherbakov, Focusing diffraction gratings for integrated optics, *Soviet J. Pisma v Zh. Tekh. Fiz.* 3: 746 (1977); M. Miler, Chirped grating couplers: a holographic approach, *Opt. and Quant. Electronics* 11: 359 (1979).
10. M. Miler and M. Skalský, Stigmatically focusing grating coupler, *Electron. Lett.* 15: 275 (1979); M. Miler and M. Skalský, Grating element for coupling and focusing light into a planar waveguide, *Soviet J. Pisma v Zh. Tekh. Fiz.* 5: 1447 (1979).
11. M. Miler and M. Skalský, Chirped and curved grating coupler focusing both outgoing beam and guided wave, *Opt. Comm.* 13: 13 (1980); M. Miler, C.W. Slinger, and J.M. Heaton, Off-axis holographic zone plates recorded and reconstructed by cylindrical wavefronts, *Opt. Acta* 31: 745 (1984).
12. D. Heitmann and C. Ortiz, Calculation and experimental verification of two-dimensional focusing grating couplers, *IEEE J.Quant. Electron.* 17: 1257 (1981); M. Miler, Holographic focusing grating couplers, *Proc. SPIE* 1183: 659 (1989); G.N. Lawrence, K.E. Moore, and P.J. Cronkite, Rotationally symmetric construction optics for a waveguide focusing grating, *Appl. Opt.* 29: 2315 (1990).
13. J.W. Goodman, F.J. Leonberger, S.Y. Kung, and R.A. Athale, Optical interconnections for VLSI systems, *Proc. IEEE* 72: 850 (1984); F. Sauer, Fabrication of diffractive-reflective optical interconnects for infrared operation based on total internal reflection, *Appl. Opt.* 28: 386 (1989).
14. J. Schwider, Achromatic design of holographic optical interconnects, *Opt. Eng.* 35: 826 (1996).

# A 16 LEVEL $CO_2$ LASER BEAM SHAPER: DESIGN AND FABRICATION

P. Antuofermo, A. Cacucci, A. M. Losacco, and O. De Pascale

Centro Laser
S.P. per Casamassima km 3, 70010 Valenzano (BA), Italy

## 1. INTRODUCTION

In different high power laser applications, beam shaping is required to perform processes such as surface hardening and alloying, texturing, and microdrilling.

Conventional methods of producing intensity distribution suitable for surface treatment, such as multi-segmented mirrors and scanning techniques, are generally not able to produce anything more than an approximation of the required distribution, thereby limiting their use and compromising the performance of the treated component.

Alternately Computer Generated Holograms (CGH) can be designed to implement specific optical functions, by phase transformation, with high accuracy and good efficiency.

The design and fabrication of a reflective diffractive optic which transforms the gaussian spatial profile of a $CO_2$ laser beam into a rectangular top-hat profile (Fig. 1) are described.

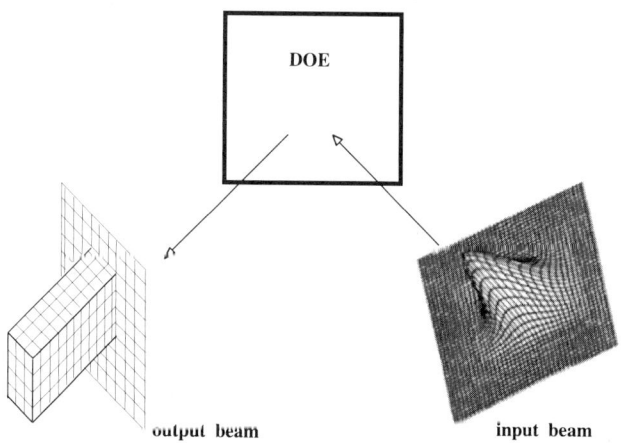

Fig. 1 DOE working conditions.

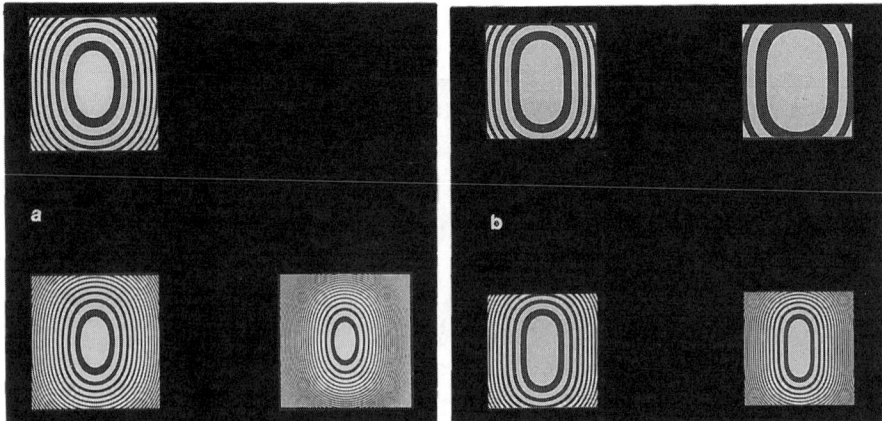

Fig. 2 Micrographs of the masks used: a) 8 level DOE; b) 16 level DOE.

$CO_2$ high power laser applications require the fabrication of relatively large optics with high diffraction efficiency, thus the laser beam lithography and the standard VLSI-technology should be used.

Multilevel phase profiles are achieved by repeated writing and etching to obtain high efficiency.

## 2. DESIGN

Several techniques have been proposed to design beam formers,[1] in this work we calculate the CGHs by means of the scalar diffraction theory[2] with the stationary phase approximation.[3-5]

Two different multilevel binary CGHs were designed, for a $CO_2$ laser ($\lambda$=10.6 µm), whose parameters are reported in Table 1. The first one has 8 levels and the second one has 16 levels.

The designed optics work in reflective mode, so the angle of incidence is an additional parameter to be considered. To limit the shadow effect, a 15° incident angle was chosen; then the maximum depth of surface relief is given by the :

$h_{max} = (\lambda/2)\cos 15°$.

## 3. FABRICATION

Two different materials were tested as substrate: copper and silicon. Both materials are widely used for high power laser conventional optics because they provide good heat transfer from the surface to the bulk.

The multistep process was used to obtain n levels. As is known, $\log_2 n$ masks are necessary. The large area patterns, usually required by high power laser beams, could be written by laser beam lithography with good resolution. Fig.s 2a and 2b show the set of 3 and 4 masks used respectively fabricate the 8 and 16 level optics, whose characteristics are reported in Table I.

Table I. Designed parameters.

|  | 8 LEVEL OPTIC | 16 LEVEL OPTIC |
|---|---|---|
| INPUT DIAMETER | 12 mm | 10 mm |
| OUTPUT SIZES | 3x6 mm$^2$ | 3x6 mm$^2$ |
| FOCAL DISTANCE | 200 mm | 400 mm |
| NUMBER OF PIXELS | 1500x1500 | 1250x1250 |
| PIXEL SIZE | 8 µm | 8 µm |

Pattern transfer on the substrates was performed by repeated contact printing and etching. Both wet and dry etching techniques were tested in order to obtain the best results in terms of optical quality and fabrication costs.

### 3.1. Copper Etching

Different mixtures of acids are commonly used as copper wet etchants. Among these, an aqueous solution of $HNO_3$, $CH_3COOH$ and $H_2SO_4$ was tested to give a rate of 20 nm/s with quite a good control of the etching depth. A photoresist mask was used since it is sufficiently etchant resistant.

Alternately, dry etching in argon plasma (P = 250 W, p = 20 mtorr, T = 50 °C) was tested to give a 9 nm/min rate with a good control of the etching depth, provided a thick enough photoresist mask was used.

### 3.2. Silicon Etching

Silicon etching in aqueous alkali or acid solutions is mainly used for micromachining. KOH/water solutions were shown to etch the different crystallographic planes at different rates. On the other hand, acid solutions etch isotropically but were tested for give a poor control of the etching depth. Both the wet etchants required a silicon nitride or silicon dioxide mask, because the photoresist etching rate was higher than that of silicon.

Reactive ion etching in $CF_4$ plasma (P = 100 W, p = 30 mtorr, T = 20 °C) with a rate of 18 nm/min was found to give better results, even though a thicker photoresist layer had to be used.

A gold thin film was finally evaporated on the multilevel structure to ensure the best reflectivity at the $CO_2$ laser wavelength.

## 4. RESULTS

To evaluate the quality of the etched surfaces and the pattern accuracy, the fabricated optics were inspected by a microprofilometer. The measured profiles are reported in Fig.s 3 and 4 for both the 8 and 16 level structures, on copper and silicon.

As shown in Fig.s 3a and 3b, wet etched copper profiles are characterized by high roughness and low pattern accuracy: this mainly happens for the 16 level optics, which require a lower feature size and an additional etching step.

The high roughness is due to a preferential attack at the boundary grains of the metal structure, both for wet and dry etching, as shown in Fig.s 3a-3d. Even though the roughness of 16 level dry etched copper is lower than that of wet etched copper, it was of the same order of magnitude as the minimum step height. Lower roughness was obtained on silicon substrates as shown in Fig.s 4a and 4b.

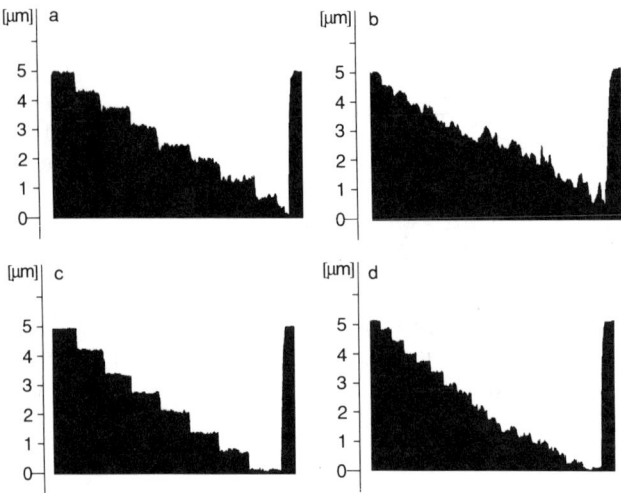

Fig. 3 Microprofilometer graphs of the copper relief structures: a) 8 level wet etched; b) 16 level wet etched; c) 8 level dry etched; d) 16 level dry etched.

These results were also confirmed by SEM analysis as reported in Fig.s 5 and 6: the 16 level structure on silicon shows a very smooth surface as composed with the 8 level copper one.

Optical characterization was performed for the 16 level dry etched silicon substrate. A plexiglass block is positioned to be marked by the transformed $CO_2$ laser beam in order to qualitatively investigate the output beam profile. In Fig. 7 the top view mark is reported and the background millimeter paper confirms the designed beam size (6 x 3 $mm^2$). Side and front view are shown in Fig.s 8 and 9, respectively. They show quite a good profile, confirming the desired pattern. The sligth imperfections in the profile are due to design approximations, fabrication errors and an incident laser beam that is not perfect gaussian.

As is well known, the theoretical diffraction efficiency of 16 level optics is 98.8%; in Table II the measured diffracted power is reported for two input beams and the related diffraction efficiency has been calculated.

## 5. CONCLUSIONS

A 16 level $CO_2$ laser beam shaper was designed and fabricated on copper and silicon substrates in order to support high power density laser beams. It works in reflection mode.

The best performance was obtained using silicon substrates and dry etching fabrication techniques. Diffraction efficiencies of about 72% were measured.

Table II. Optical characterization data.

| INPUT BEAM POWER [W] | POWER IN THE BEAM SIZE [W] | DIFFRACTION EFFICIENCY [%] |
|---|---|---|
| 10 | 7.2 | 72 |
| 25 | 17.9 | 71.6 |

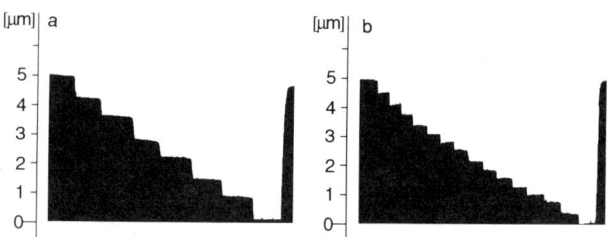

Fig. 4 Microprofilometer graphs of the silicon relief structures: a) 8 level dry etched; b) 16 level dry etched.

Fig. 5 SEM photograph of 8 level dry etched copper relief surface.

Fig. 6 SEM photograph of 16 level dry etched silicon relief surface.

Fig. 7 Top view mark of the output beam on millimeter paper.

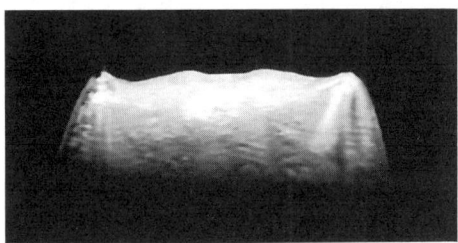

Fig. 8 Side view mark of the output beam.

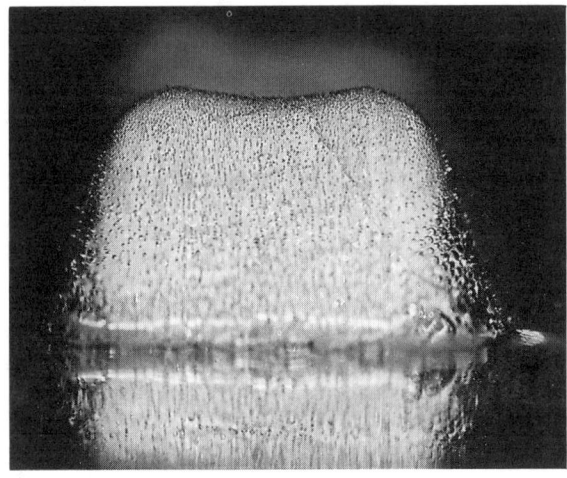

Fig. 9 Front view mark of the output beam.

Better performance can be obtained increasing the number of levels and optimizing fabrication processes.

ACKNOWLEDGMENTS. The authors are very grateful to R. Lassandro and L. Morandi for their technical support.

## REFERENCES

1. M.T. Eismann, A.M. Tai, and J.N. Cederquist, "Iterative design of a holographic beamformer", *Appl. Opt.* 28:2641 (1989).
2. J.W. Goodman, *Introduction to Fourier Optics*, McGraw-Hill, New York (1968).
3. O. Bryngdahl, "Geometrical tranformations in Optics", *J. Opt. Soc. Am.* 64:1092 (1974).
4. N.C. Roberts, "Beam shaping by holographic filters", *Appl. Opt.* 37:31 (1989).
5. N.C. Roberts, "Multilevel computer-generated holograms with separable phase functions for beam shaping", *Appl. Opt.* 31:3198 (1992).

# MICROOPTICS FOR CHROMATIC CONTROL OF EXTENDED POLYCHROMATIC SOURCES: DESIGN METHODOLOGIES AND TECHNOLOGIES TO IMPLEMENT THEM OVER LARGE AREAS

P.Perlo, C.Bigliati, V.Lambertini, P.M.Repetto, and S.Sinesi

FIAT Research Centre
Strada Torino 50, 10043 Orbassano, Italy

## 1. INTRODUCTION

Advanced design packages for diffractive and microoptics based systems are available in several research laboratories and industries. Extensive work on the development of iterative algorithms in both the scalar and more recently in rigorous electromagnetic regimes has lead to a very high degree of design sophistication. Advanced commercial packages are available as well, and complex designs with their use have been reported in many conferences. But in spite of the effort made, the analysis of diffractive and microoptics systems in the presence of extended polychromatic sources remains limited or at the best partial.

The Chapter addresses the design and the fabrication of beam shapers for polycrhomatic light. Design criteria are proposed to correct both the color of the zero order, typical of binary structures, and to correct rainbow effects often visible at the border of the shaped beam. A simple case based on binary zone plates is presented with a simplified explanation of the more general procedure. Large binary and continous profile elements developed directly by a high speed laser writing system and by a mass transfer process are shown.

The design criteria consist of the hypothesis that a polychromatic beam is an incoherent superposition of monochromatic spatially coherent beams [1] and that a diffractive element is composed of many independent sub-elements designed so that the final result is a well defined shaped beam with negligible chromatic aberration (or vice versa, with maximized color separation). In Section 4 we will prove first theoretically and then with experimental results that the assumptions made are satisfactory for the majority of polychromatic light beam shapers. The monochromatic analysis can then be applied iterative for different sub-areas until the superposition of the beams corrects the visible chromatism. With this plan in mind, Section 2 will be dedicated to the design of monochromatic light beam shapers, concentrating on those aspects that determine the speed and the accuracy of the calculation to obtain complex shaped patterns.

The problems related to the manufacturing of large diffractive or microrefractive elements with nonrepeated clusters are discussed in Section 5. A diffractive element which is actually in production for the automotive industry, is presented for the first time to our knowledge.

## 2. MONOCHROMATIC DESIGN OF DIFFRACTIVE OPTICAL ELEMENTS

Phase diffractive optical elements (DOEs) for focusing into a desired area with the required intensity distribution are important in applications such as laser beam shapers and homogenizers.

Multi-order phase diffraction gratings (PDGs) represent a particular case of DOEs generating the intensity distribution in the form of a 1D or 2D array of light spots. With only one period, the PDG can produce a continuous intensity distribution inside a rectangular area; this allows us to consider the phase function of the single period of PDG as the phase function of a DOE, aimed to focus a plane beam with uniform intensity into a rectangle. The use of PDGs has been reported in many applications such as multiple imaging arrays, coherent addition of laser beams, fiber optic star couplers, free-space optical interconnects, optical processing, optical computing and optical communications. [2-7] Many approaches have been used to design multiorder binary as well as multilevel and continuous-relief PDGs. Among these are: stochastic techniques (direct binary search,[8] iterative discrete on-axis,[9] simulated annealing algorithms[10-13]), Gerchberg-Saxton phase retrieval algorithms,[14-17] genetic algorithms,[18] downhill simplex,[19] and gradient methods.[20-21]

The fabrication of DOEs by photolithography requires phase quantization; the phase of the DOE can have only M discrete values. Binary DOEs (M=2) are of particular interest because they are easier and less costly to manufacture; they are also easier to replicate with high accuracy. Besides, it must be noticed that in the case of symmetric distributions, even when they very complex, there is no point in designing other than binary elements.

Recently we have proposed two novel algorithms for the direct calculation of DOEs, [22, 28] the first method has been used for the optimisation of one-dimensional binary PDGs, the second one for 1D or 2D binary as well as quantized phase DOEs.

### 2.1. One-Dimensional Binary PDGs

The solution of a system of N non linear equations is required to compute the binary grating generating $M = 2N+1$ orders[22]. As is well known, the phase profile of a binary grating period consists of K grooves of equal depth but different width (Fig. 1). The coordinates $x_1,...,x_{2K}$ of the grooves represent phase transitions with amplitude $\varphi$. The values of the coordinates and the

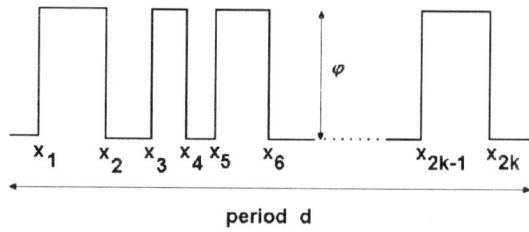

Fig. 1 Phase profile of a binary grating period.

depth of the grooves are the parameters that influence the intensity of the diffracted orders. For asymmetrical grating profiles finding a solution requires more effort than in the symmetric case, on the other hand, solutions with larger minimum feature size and higher efficiency can be found.[23]

It has been demonstrated that in general the error of the scalar theory is significant (> ± 5 %) when the minimum feature size is < 14 λ ,[24] in spite of that, at the optimal depth, even for feature sizes ~ 2 λ the error [24] is minimized when the fill factor approaches 50%. For binary structures the scalar accuracy is not very sensitive to changes of the incidence angle.[24] without any restriction we assume normal incidence in the calculation.

For applications such as laser beam homogenizers and shapers as well as fiber optic networks it is of real value to extend the binary grating design to a large number of orders; when the number of orders increases, the random search of the binary grating profile reguires a long time for calculation and leads to unstable convergence; the solution of the nonlinear equation system is numerically prohibitive, and gradient procedures with random starting points lead to local minima of the merit function which do not represent the solution of the problem.

To calculate multi-order asymmetrical binary gratings we use an initial approximation of the grating profile based on the properties of a one dimensional element focusing into a line; a gradient search algorithm is then used for the optimization. In general, the main problem associated with gradient procedures is the choice of an initial profile to obtain stable convergence. As a rule, with random generation of the initial profile, the algorithm stagnates with an RMS deviation of the final order intensities from the designed ones in the range of 70-80%. Under the initial analytical approximation we proposed we obtained quick and stable convergence with high efficiency and low RMS error using different gradient procedures such as the steepest descent and conjugate gradient methods; we have described [25] the Polak-Ribiere formulation [26] of the latter. We presented [25] the results of binary gratings calculated for different numbers M = 2N+1 of equal intensity diffracted orders; in addition, we used the same procedure for the calculation of gratings with any intensity distribution in the diffracted orders. In Fig. 2 is shown one period of a binary PDG which generates a linear intensity distribution in 101 orders; the relative intensity distribution is shown in Fig. 3.

This method, applied to the calculation of gratings with any number of order up to 281, gave energy efficiencies of 78 - 84% and RMS error of 1 – 5 % within a few seconds on a 586 computer. The high energy efficiency E and the low RMS δ obtained for any grating with order up to 281 is a confirmation of the validity of the approach adopted. It should be noted that there are no restrictions for the method to compute gratings with 1001 orders or more, still remaining within a few minutes of computation: we limited [25] the presentation of the results to closely match the gratings we manufactured. For the asymmetrical gratings under study we noticed that the minimum groove width $\Delta_{min}$ of the initial binary profile remains approximately the same size after the iteration correction. For example, for a 101 order grating we obtained a ratio $\Delta_{min}$ / period ~ 0.01. Furthermore, we noticed that the initial approximation we proposed gives a starting grating profile with fill factor of nearly to 0.5; moreover, the iteration correction does not change the fill factor in a significant way. In the optimization algorithm we constrain the fill factor within ± 10 % of 0.5; this allows a minimum feature size down to 2λ with an acceptable scalar accuracy (close to 5%). We also generalized the initial approximation to the two-dimensional case: the starting point is now given by the binary distortion applied to the geometrical optical focusator in a rectangular area.[25]

## 2.2 Direct Calculation of Two-Dimensional Quantized Phase DOEs

Two dimensional crossed binary gratings have been fabricated as well,[25] combining two 1D gratings with an overall efficiency of about 65%; however better efficiency, over 75%, is possible by direct optimization of the 2D structure.[27]

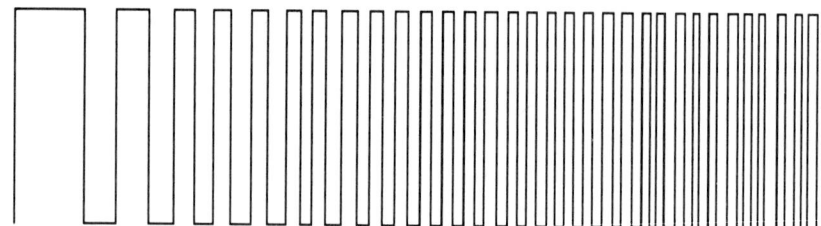

Fig. 2 One period fo a binary PDG.

Recent we have proposed [27] a new iterative method for the calculation of quantized phase DOEs. The method consists of: - the approximation of the quantized DOE complex transmission function (CTF) by a non-quantized CTF - a gradient-search algorithm for the optimisation of the non-quantized CTF - the inverse replacement of the non-quantized CTF with a quantized CTF. The relation between the quantized CTF (first term) and the non-quantized one (second term) is given by:

$$e^{i\varphi_M(x,y)} = e^{-i\frac{\pi}{M}} \sum_{k=-\infty}^{+\infty} (-1)^k \operatorname{sinc}(k+\frac{1}{M}) e^{-i(Mk+1)\varphi(x,y)} \qquad (1)$$

where M is the number of discrete levels and $\varphi(x,y)$ is a continuous phase function.

The starting $\varphi(x,y)$ function can result either from a Gerchberg-Saxton iterative algorithm or from analytical ray-tracing considerations. It must be pointed out that the use of a suitable initial approximation usually enhances both the diffraction efficiency and the convergence speed; furthermore, if the initial approximation is calculated by means of geometrical optics calculations, the result is a fringe-like phase pattern which will be only partially destroyed by the optimisation procedure. Thanks to that the final structure will be easy to manufacture and less sensitive to the fabrication errors. On the other hand, when

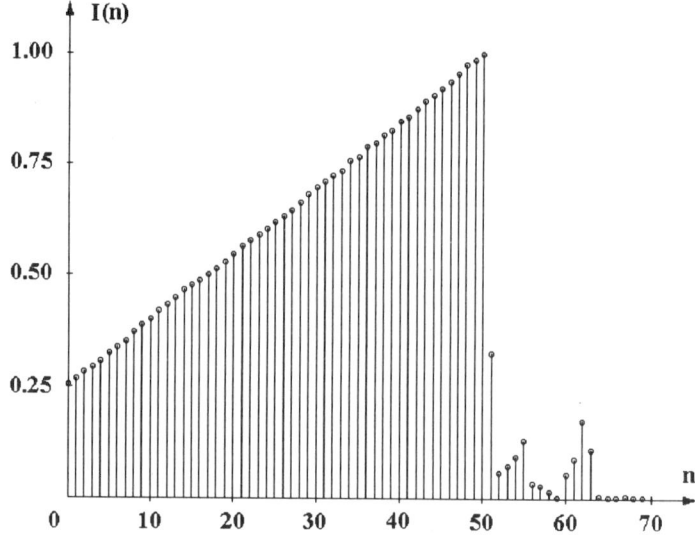

Fig. 3 Intensity distribution relative to the binary PDG of Fig.2.

analytical considerations are not possible due to the complexity of the intensity distributions, the proposed method gives efficient and fast solutions even using a random starting point.

Fig. 4 clearly shows the capabilities offered by a suitable ray-tracing analytical approximation (a,b) optimized in a continuous relief design (c,d), as compared to a purely numerical one (e,f). Each pair of diagram shows corresponding the element and the far-field pattern.

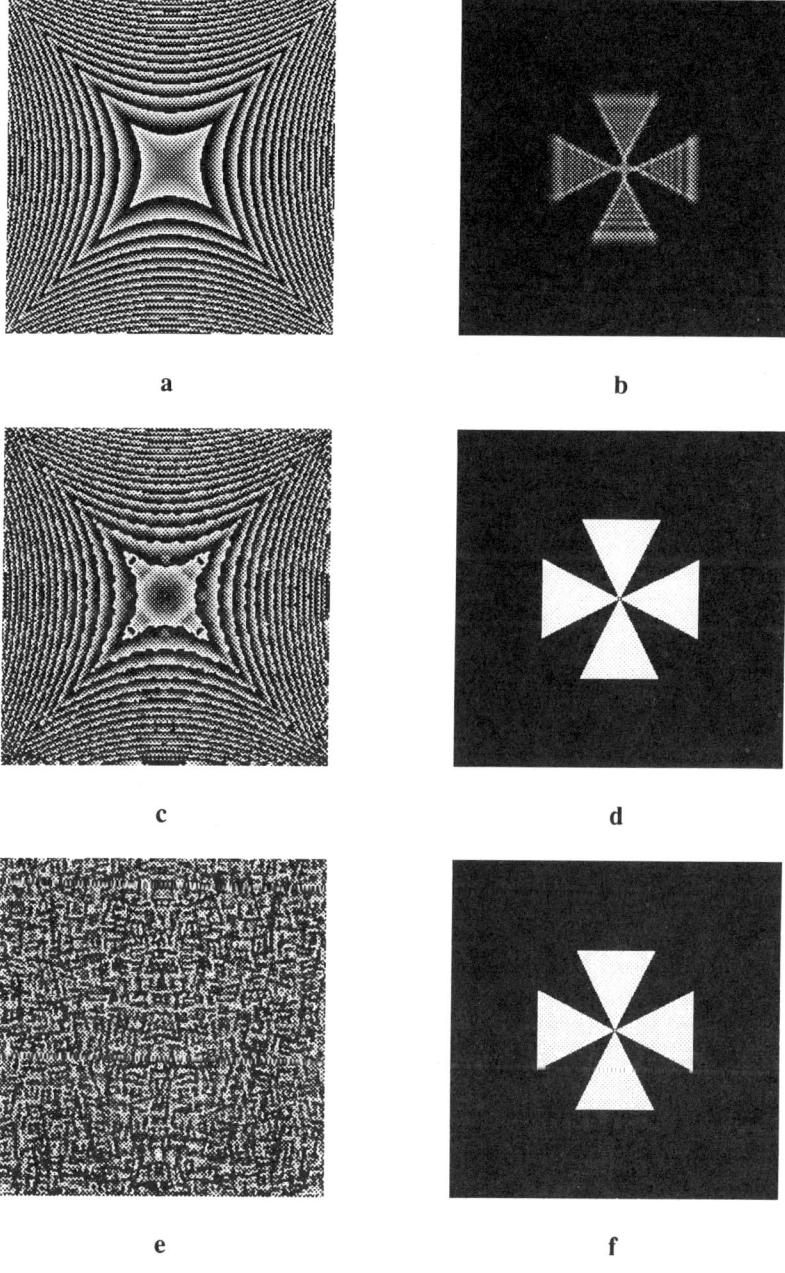

Fig. 4 (a,b) Efficiency 91.2% r.m.s. 42.2% ;(c,d) Efficiency 92.9% r.m.s. 0.2%; (e,f) Efficiency 89.4% r.m.s. 0.0%

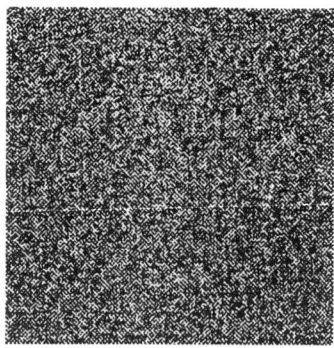 

Fig. 5 Efficiency 91.0 % r.m.s. 0.1 %: a) continuous relief hologram; b) very complex image pattern.

At the same time Fig. 5 shows the capabilities of a pure numerical optimisation for a continuous relief hologram (Fig. 5a), in the case of a very complex image pattern (Fig. 5b).

In comparison with the familiar methods [2÷21, 28÷32] oriented to the design of diffraction gratings, the proposed method has some advantages: it considers the intensity distribution of the illuminating beam, and it is directly two-dimensional, which is a substantial advantage for the non factorable areas of focusing and for obtaining high energy efficiency. The method needs two Fourier transforms per iteration and is equivalent to the GS algorithm in its complexity. The employment of discrete algorithms of the FFT in the course of calculation produces a pixel-like binary DOE which allows one to take into consideration discrete-type effects upon the outputting of a photomask of the DOE on a raster photoplotter.

When the method is applied both to the calculation of beam shapers converting a gaussian beam into a uniform one, and to the calculation of binary gratings of any order up to 51x51, it gives an energy efficiency of 71-75% and a root-mean-square error (R.M.S.) of 3-5% within a 15 minutes on a 586 computer. In Table I we present the results [27] for binary gratings calculated for different numbers of equal intensity diffraction orders. The energy efficiency E of 73-75% and r.m.s. error of 3-5% obtained for any grating with the number of orders up to 51x51 is a confirmation of the validity of the method developed. The optimisation of the 2D binary gratings considered typically took 5-10 minutes for small arrays (up to 11x11 orders) and approximately half an hour for a 51x51 order grating on an IBM PC computer equipped with an Intel 586 processor.

Table I. The results of calculation of binary diffraction gratings.

| Number of orders | Energy efficiency (%) | R.M.S. error (%) |
|---|---|---|
| 5x5 | 74.4 | 3.1 |
| 7x7 | 73.7 | 3.1 |
| 9x9 | 72.6 | 2.9 |
| 11x11 | 73.8 | 3.6 |
| 15x15 | 74.7 | 3.4 |
| 21x21 | 73.8 | 3.7 |
| 33x33 | 75.1 | 4.8 |
| 51x51 | 74.2 | 4.9 |

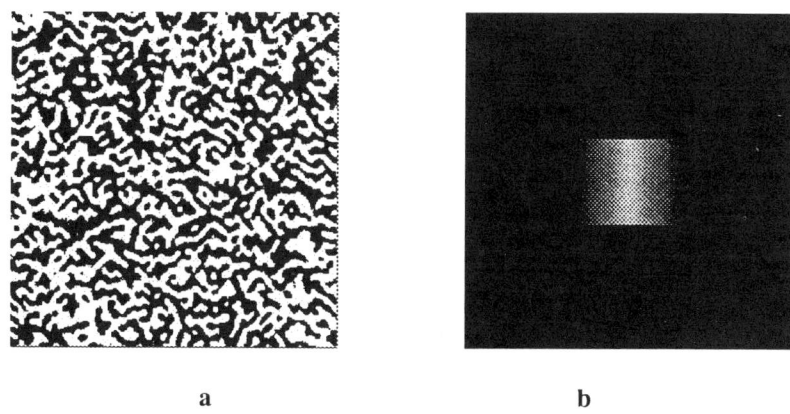

Fig. 6. Binary PDG (a), and the far-field intensity distribution (b).

The binary PDG shown in Fig. 6a has been optimised in a computational time of 10 minutes; the array size is 256x256 pixels, the number of illuminated orders is 91x61, the final efficiency is 78% and final r.m.s. deviation is 6%; the far-field intensity distribution generated by the PDG is shown in Fig. 6b.

## 3. CHROMATIC ABERRATION OF PERIODIC STRUCTURES

To analyze chromatic effects due to non-monochromatic illumination of a periodic phase structure, we can consider a general surface-relief array with period $d$; let be $D$ the size of the input beam and

$$t(\overline{x},\lambda) = e^{i\varphi(\overline{x},\lambda)} = e^{i\frac{2\pi}{\lambda}[n(\lambda)-1]h(\overline{x})} \qquad (2)$$

where $t(\overline{x};\lambda)$ t (is the transmission function inside a single period, $\lambda$ is the input wavelength, $n(\lambda)$ is the refractive index of the dispersive material and $h(\overline{x};\lambda)$ the surface height.

Within the regime of validity of the scalar theory, it is well known from Fourier optics that the angular separation between two adjacent diffracted orders is given by $\lambda/d$, while the width of each order is $\lambda/D$, which means that if the size of the impinging beam is coincident with the period it is impossible to distinguish one diffraction order from another at any input wavelength. The considerations so far are valid for diffractive, refractive and harmonic structures; harmonic structures can be obtained by a modulo – m $2\pi$ distortion of a continuous phase function. To simplify the description we now assume that the structure is one-dimensional and symmetric; the intensity distribution $I_k$ between the orders will be also symmetric and determined by the $h(\overline{x})$ of the relief. Let be N a higher order with non-zero intensity; both N and $I_k$ are strongly wavelength dependent.

If $\lambda_1$ is the design wavelength and $\lambda_2$ the reconstruction wavelength, it is possible to write the transmission function $t(\overline{x};\lambda_2)$ for a general modulo – m $2\pi$ harmonic structure as a non-linear distortion of $t(\overline{x},\lambda_1)$:

$$t(\overline{x},\lambda_2) = \sum_n e^{i\pi[mq(\lambda_1,\lambda_2)-n]} \operatorname{sinc}[mq(\lambda_1,\lambda_2)-n] e^{i\frac{n}{m}\varphi(\overline{x},\lambda_1)} \qquad (3)$$

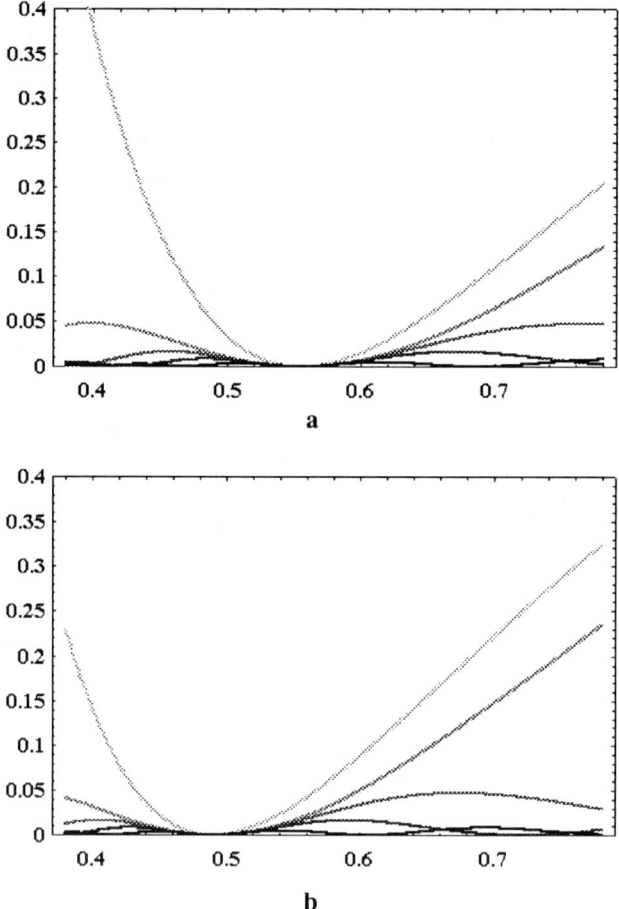

Fig. 7 0-order efficiency *versus* wavelength in µm.

where

$$q(\lambda_1, \lambda_2) = \frac{\lambda_1}{\lambda_2} \cdot \frac{n(\lambda_2) - 1}{n(\lambda_1) - 1} \qquad (4)$$

The transmission function for diffractive and refractive structures can be derived from the above expression in the limits m = 1 and m → ∞.

When the periodic structure is illuminated with a polychromatic field two main effects arise due to chromatic aberration; we call them the Zero-Order Color (ZOC) and the BOrder Rainbow (BOR); the former is related to the 0-order efficiency $I_0$ which depends in turn on the reconstruction wavelength, the latter is related to the $\lambda$ dependence of $N$ and its visibility grows faster as $I_k$ decreases through the orders close to $N$. Fig. 7a,b shows the 0-order efficiency versus wavelength for $\lambda_1$ = 555 nm and $\lambda_1$ = 490 nm (BK7 Schott dispersion); in both cases the differences between binary diffractive (lines respectively from light grey to dark grey), continuous relief diffractive and harmonic structures (M = 2,3,5) can be seen. One can easily see that ZOC aberration is only visible in diffractive or small *m* harmonic structures.

The curves in Fig. 4 are calculated from the equations

$$I_0 = 1 - \sin^2(\frac{\pi}{2} \cdot q) \qquad \text{for the binary case and} \qquad (5)$$

$$I_0 = sinc^2(m \cdot q) \qquad \text{for the other cases.} \qquad (6)$$

The angular width of the BOR aberration can be expressed as

$$\Delta\psi(\lambda_1, \lambda_2) = \frac{N(\lambda_2) \cdot \lambda_2 - N(\lambda_1) \cdot \lambda_1}{d} \qquad (7)$$

The above expression is valid for any periodic structure; the difference between diffractive, harmonic and refractive behaviours is included in the $\lambda$ dependence of $N$. Taking into account only the most efficient term of Eq. (4), for a modulo $-$ m $2\pi$ harmonic structure we can define

$$N(\lambda_2) = N(\lambda_1) \cdot \frac{m_{int}(\lambda_1, \lambda_2)}{m} \qquad (8)$$

$m_{int}(\lambda_1, \lambda_2)$ represents the integer number nearest to $mq(\lambda_1, \lambda_2)$, being the value of $n$ which maximises the function $sinc[mq(\lambda_1, \lambda_2)-n]$ of Eq. (4). The terms of Eq. (4) corresponding to values of n other than $m_{int}$ introduce secondary spectra which arise as additional rainbows. The efficiency of each wavelength inside the main rainbow is given by

$$\eta = sinc\left[mq(\lambda_1, \lambda_2) - n\right] \qquad (9)$$

For diffractive and refractive structures Eq.(9) yields:

$$N_{dif}(\lambda_2) = N_{dif}(\lambda_1) \qquad (10)$$

and

$$N_{ref}(\lambda_2) = N_{ref}(\lambda_1) \cdot q(\lambda_1, \lambda_2) \qquad (11)$$

From Eq.(8) can be derived

$$\Delta\psi_{dif}(\lambda_1, \lambda_2) = \frac{N_{dif}(\lambda_1)}{d}(\lambda_2 - \lambda_1) \qquad (12)$$

$$\Delta\psi_{ref}(\lambda_1, \lambda_2) = \frac{\lambda_1 N_{ref}(\lambda_1)}{d}\left[\frac{n(\lambda_2) - n(\lambda_1)}{n(\lambda_1) - 1}\right] \qquad (13)$$

Fig. 8 a,b shows $\Delta\psi_{ref}(\lambda_1, \lambda_2)$ (expressed in radians) versus $M$ for two different design wavelengths $\lambda_1 = 490$ nm and $\lambda_1 = 555$ nm (BK7 Schott dispersion, ratio $N(\lambda_1)/d$ fixed to 1.); in both cases the curves for $\lambda_2 = 400$ nm (grey line) and $\lambda_2 = 700$ nm (black line) are plotted. The BOR effect gets smaller and smaller as m increases toward the refractive limit

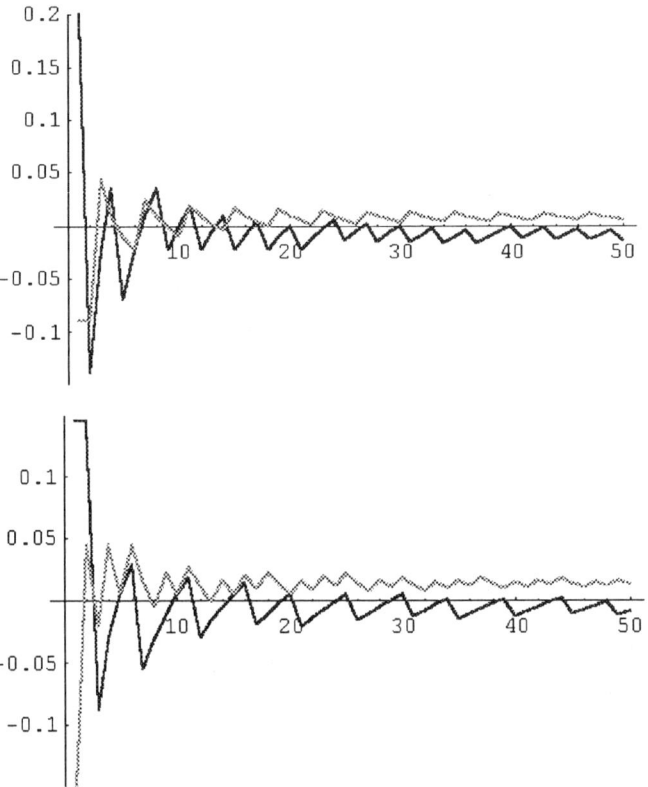

Fig. 8 Rainbow angular width *versus* m. a) $\lambda_1 = 490$ nm; b) $\lambda_1 = 555$ nm.

for $m \to \infty$; it is quite remarkable that the passage from the diffraction to the refraction regime is not gradual but oscillating.

The function $\Delta\psi_{ref}(\lambda_1, \lambda_2)$ only represents the angular width of the main rainbow and does not include any information about the intensity distribution inside the rainbow; this distribution can be easily calculated by Eq. (4).

The description of BOR effects still remains valid for binary structures, as a particular case of diffractive structures. In spite of that the additional effect of secondary rainbows becomes more evident for binary elements.

The Section is devoted to an efficient method to reduce both ZOC and BOR effects for a matrix of binary zone plates, used as a light beam shaper.

As will be described in detail, the ZOC effect can be minimized, taking account of both the spectral emission curve of the impinging polychromatic light and the human-eye sensitivity curve, by an appropriate choice of the design wavelength.

To reduce the BOR effect in binary structures we take into consideration several possibilities. As we showed above, the goal is to reduce the slope of $I_k$ decrease in the orders $k$ close to $\pm N$; this can be done either by a modification of the binary phase structure in order to produce multi-focal effects [33] or by properly arranging inside a single cluster sub-elements with different focal lengths. The reason why we will use the second method is its relative simplicity both in concept and manufacturing. In order to understand our procedure some basic assumptions should be kept in mind.

We consider a white light quasi-collimated beam; such a beam has no temporal coherence on account of its very large frequency bandwidth. On the contrary, due to its collimating properties we can assume a sort of spatial coherence; the temporal incoherence allows us to treat it as an incoherent superposition of monochromatic beams. Each of them is considered spatially coherent, and propagates according to the laws of scalar diffraction theory; the last assumption seems to be quite suitable for the collimating properties of the beam. At the same time, the only partial spatial coherence gives us the possibility to use binary zone plates in the far field without regard to the interference effects which typically arise when a continuous symmetric phase function is turned into a binary one.

## 4. A SIMPLIFIED PRACTICAL METHODOLOGY FOR CHROMATIC CORRECTION WITH EXTENDED POLYCHROMATIC SOURCES

To prove the hypothesis we consider the case of a matrix of binary zone plates. To simplify the example as much as possible we suppose that the impinging polychromatic beam is collimated. This is never true in practice, but the concepts can be easily extended for real designs.

The binary zone plate or Wood-Rayleigh [34, 35] lens, since its origin, has always been known to be subject to strong chromatic effects. Matrices of binary zone plates have been fabricated since microlithography has been applied to binary optics and typically represent the first design and the first fabricated diffractive structure by most of those starting to work in diffractive optics.

### 4.1. BOrder Rainbow Correction (BORC)

For our simplified example let us take a cluster of four different adjacent zone plates having the same size, such as that shown in Fig. 9. The focal lengths are chosen so that, at the same designed wavelength, the borders of the four diffracted polychromatic patterns superpose, minimizing the visible chromatism. Remaining within the scalar regime, at the first diffracted order the angular spread is given by

$$\sin\theta = \frac{\lambda}{\Delta_{min}} \qquad (14)$$

where $\Delta_{min}$ is the minimum zone width and $\lambda$ the designed wavelength.

We suppose that each lens is larger than the spatial coherence of the monochromatic beam impinging on it. The transmitted fields from the four lenses can then be added incoherently. Let $V = 550$ nm be the design wavelength; to the four lenses there correspond four different square patterns having different angular spreads. In a single pattern the three wavelengths $B = 480$ nm, $V = 550$ nm and $R = 640$ nm, which here represent the polychromatic beam, have different angular deviations. The final image produced by the single lens is then a distinguishable three color pattern. The four focal lengths are easily found so that the incoherent superposition of the overall images eliminates the rainbow at the border.

### 4.2. Zero Order Color Correction (ZOCC)

To choose the proper design wavelength so that the zero order beam has indistinguishable chromatism of the projected beam on a white screen, let us consider the equations that relate the Chromaticity Coordinates of the screen and those of the white chosen as a reference:

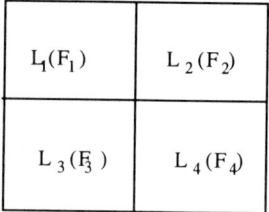

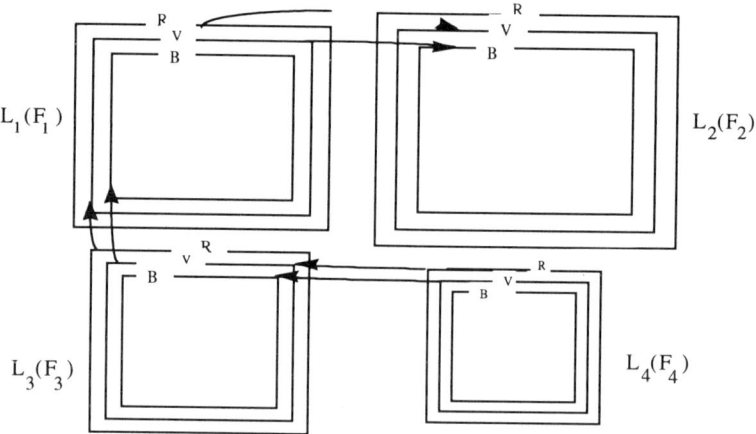

Fig. 9 Four different adjacent zone plates.

$$W(\lambda_{prj}) = A \cdot [x_n - x(\lambda_{prj})] + B \cdot [y_n - y(\lambda_{prj})] + Y$$
$$T(\lambda_{prj}) = C \cdot [x_n - x(\lambda_{prj})] + D \cdot [y_n - y(\lambda_{prj})]$$ (15)

where W and T are Whiteness and Tint of the screen. The coefficients A, B, C, D depend on the source, $(x_n, y_n)$ are the Chromaticity Coordinates of the reference white, (x,y) represents the Chromaticity Coordinates of the screen and Y is the tristimulus Coordinate in the color space (X, Y, Z). Let be the screen a white sheet of paper located in the far field of the cluster of lenses.

To determine the optimal wavelength $\lambda_{prj}$ we constrain the absolute value of the T function to be close to 0 and the W function to be close to 100.[36] The tristimulus Coordinate Y is proportional to the luminance of the screen. The diffractive efficiency of the m-order, defined as the portion of the impinging energy deflected in a defined angle, can be here represented as the intensity in a determined direction if we suppose a point model where the subtended solid angle tends to 0. The screen can be photometrically described in terms of intensity values instead of the luminance ones if the r.m.s. spot size corresponding to the geometrical ray impinging on the screen tends to zero. Instead of the tristimulus value Y we can then use the diffractive efficiency. With the constraint Y=100 and T=0 we have to solve the following system of equations:

$$\begin{cases} I_o = 1 + \varepsilon \\ |T| \prec \varepsilon \end{cases}$$ (16)

where $I_0$ represents the addition of the zero order diffractive efficiencies of the wavelengths composing the polychromatic beam. $\varepsilon$ is a positive quantity as small as desired, introduced to find solutions to the above system of equations.

The Chromaticity Coordinates due to the superposition of different colors are obtained by weighting each wavelength with the photopic (or scotopic) sensitivity curve, the spectral curve of the source and the diffractive efficiency at that wavelength.

## 5. FABRICATION OF A LARGE ARRAY OF NON REPEATED CLUSTER

One of the problems in the fabrication of large nonrepeated clusters of diffractive or refractive microoptics is the ability to generate, with reasonable time and cost, topological structures with minimal feature size below one micron and accuracy of 0.1-0.2 microns at the boundaries of each cell in the cluster. The idea that in most cases the time to fabricate a master is not relevant, because typically a master is reproduced through fast and low cost processes such as stamping or embossing, cannot be accepted as a general rule and more specifically in our case. Keeping in mind the constrains in terms of time, cost and especially a minimum number of fabrication steps, we present two technologies in use at FIAT Research Center. The first is a process based on the use of a high speed laser writer on chrome or color center photoglass; the second consists of exposure through a gray level mask of photopolymeric composition, which by a mass transfer process creates the desired topology.

Even though the literature is rather rich inpublications describing binary multilevel elements, in this paper we focus our considerations primarily on purely diffractive binary structures or continuous microrefractive ones. We fabricate the majority of symmetric polychromatic beam shapers using the former, while with the latter we have a higher degree of freedom in designing the desired pattern and have reduced sensitivity to chromatism.

### 5.1. High Speed Laser Writing System

The system we describe (Fig. 10) is a modified version of the CLWS-300 polar laser writer[37,38] developed by the Technological Design System Institute of Novosibirsk. The original system works at an effective data rate around 4 Mpixel/s, generates minimum

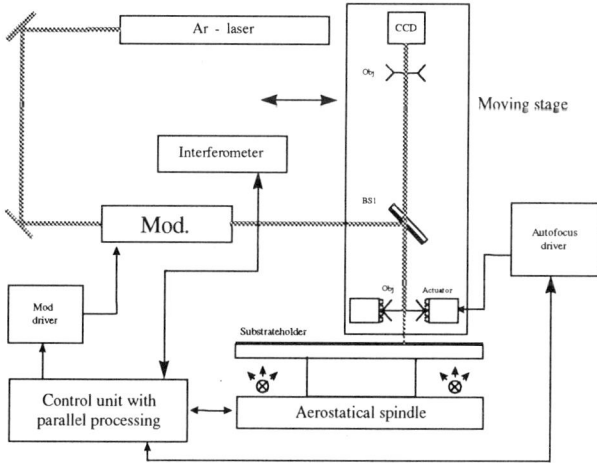

Fig. 10 CL WS-300 polar laser writer (modified).

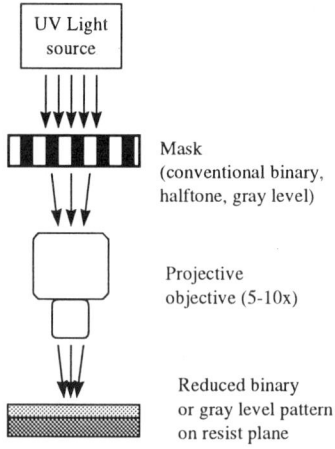

Fig. 11 Writing on resist process.

feature sizes well below 1 micron and has fabrication accuracy better than 0.1 micron at the boundaries; the beam can be modulated at several levels of intensity. In spite of general good performance, the time for writing and handling of the large data files is unacceptable when the system is used to generate large nonrepeated clusters or, more generally large numerical cells. This can be considered a general problem no matter whether the system is a laser writer or an e-beam, and it is certainly a great limit in the development of microstructures. The modified version as developed at FIAT Research Center has an efficiency of beam transfer close to 70% throughout the UV and visible spectrum of the Argon laser. Parallel processing and transfer of the data allows a continuous data transfer close to 80 M pixel/s with modulation at 256 intensity levels, while the resolution and accuracy of the original system remains the same.

The high efficiency of beam transfer allows direct writing on chrome plates, on resist, on photoglass with color centers, on a-silicon films, on dry photopolymer compositions, and on polyimide. Depending on the recording material and on the availability of a dry etchers and or a photoreducer with stepper, several one step or multistep procedures are then possible.

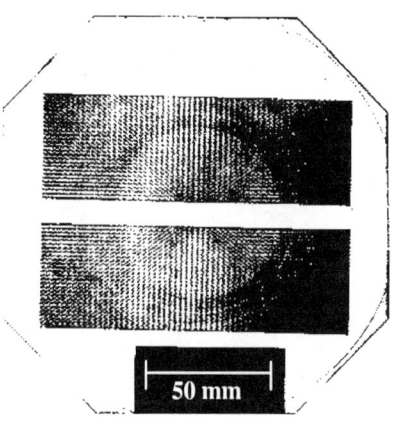

Fig. 12 Laser written 120x120 mm binary element.

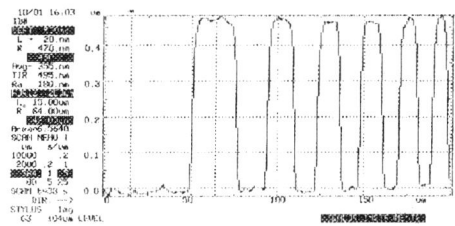

Fig. 13 Profile of the matrix of Fig.12.

Although rotationally symmetric patterns may seem to be more easily produced, we are routinely making matrices of a nonrepeated microlens type pattern in three ways binary amplitude on chrome, binary on resist, and gray scale on color center photoglass. When writing directly on chrome, as originally proposed by Poleshchuk,[37-38] the binary amplitude chrome mask with enlarged size is in a second step photoreduced to produce either a binary or a modulated image on a resist plane (Fig. 11) . After developing the resist, the remaining relief can be easily replicated or transferred into the substrate by dry etching. This procedure is convenient when a high resolution photoreducer with stepper is available, and for large array is a must if the speed of writing is only a few M pixel/s.

An example of a 120x120 mm binary element directly written with the laser writer is shown in Fig. 12. The profile of one element of the matrix is shown in Fig. 13.

The use of color center photoglass such as the Canyon Material Inc. glass is a particularly complex problem in a polar system, as the total energy deposited in the single pixel by the focused laser beam depends on the position of the laser with respect to the center of rotation, on the velocity of the spindle and on the laser power. On the other hand, once the gray level masks are produced, by a simple contact UV exposure of resist or of dry photopolymers high order kinoforms or continuous high quality refractive microlens arrays can be easily produced (Fig. 14).

### 5.2. The Use of Photopolymers

This technology was first proposed by Solovjev [39] for the production of infrared laser beam shapers. In the system developed at FIAT Research Center shown in Fig. 15 a gray level (or binary amplitude) mask is illuminated by an incoherent UV source. A HeNe laser beam is used to monitor the state of polymerization.

Before the polymerization is complete, the UV exposure is stopped and a mass transfer process originates the relief structure. Mass transport or monomer diffusion into oligomeric lattice takes place between adjacent zones with different concentrations of an polymerized monomer. The process can be described to sufficient approximation by Fick's law. [40-41]

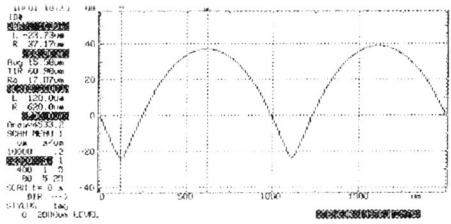

Fig. 14 Microlens array profile.

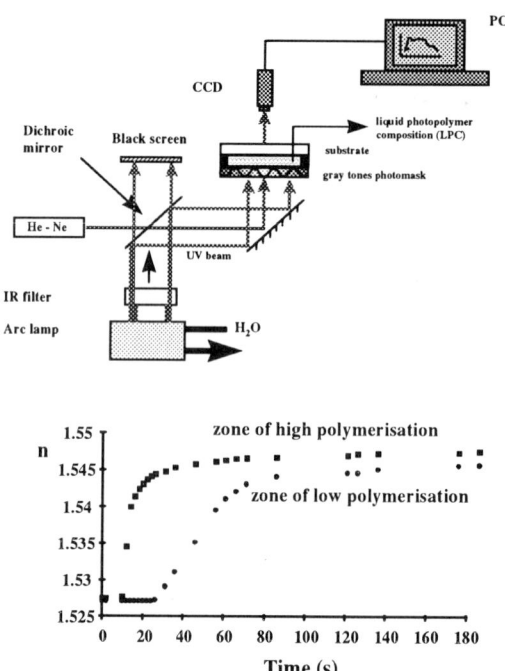

Fig. 15 Photopolimer composition system developed at FIAT Research Center.

$$\frac{\partial C}{\partial t} = D(\eta)\frac{\partial^2 C}{\partial x^2} \qquad (17)$$

where $D(\eta)$ is the diffusion coefficient which changes with thermodynamical parameters and degree of polymerization $\eta$, x is the direction of mass transfer, and C is the monomer concentration.

Relief structures from a few microns up to 150 microns in height commonly fabricated for use as polychromatic beam shapers. Among these structures are higher order kinoform, triangular profiles, multiorder gratings, microrefractive lens arrays and sinusoidal lenses; some measured profiles are shown in Fig. 16.

An 260mm-length element developed with binary microlenses is shown in Fig. 17. The so called third stop light (CHMSL, or center high-mounted stop light) that uses that element is now in production for FIAT cars.

## 6. CONCLUSION

Simple design criteria are proposed for the design of polychromatic light beam shapers. The fast advanced diffractive design packages developed for monochromatic beams can be effectively used with minor modifications to correct for chromatic effects. As examples of monochromatic designs we have reported hybrid analytical-numerical techniques which minimize time of calculations, give very high efficiency and allow the fabrication of large structures minimizing errors and costs. We have given a general treatment for color

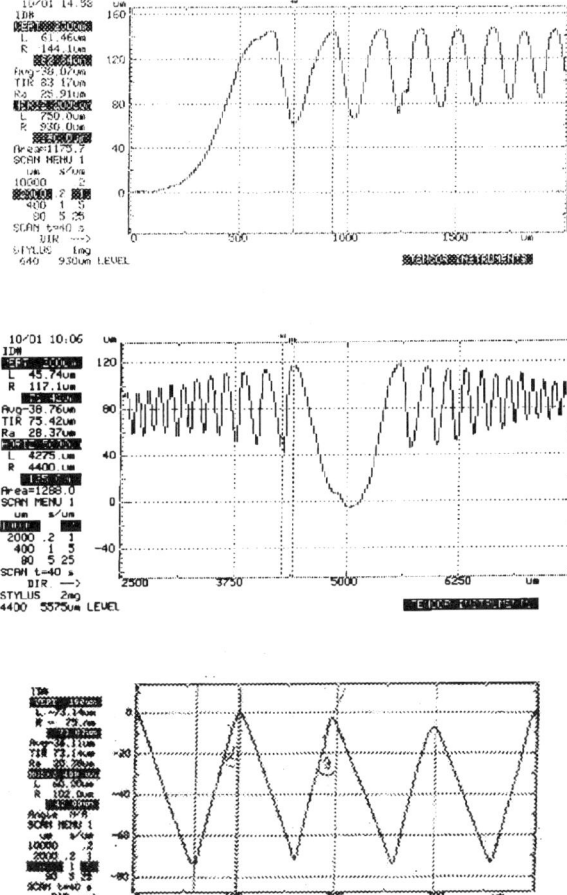

Fig. 16 Profile of 5÷150 μm relief structures.

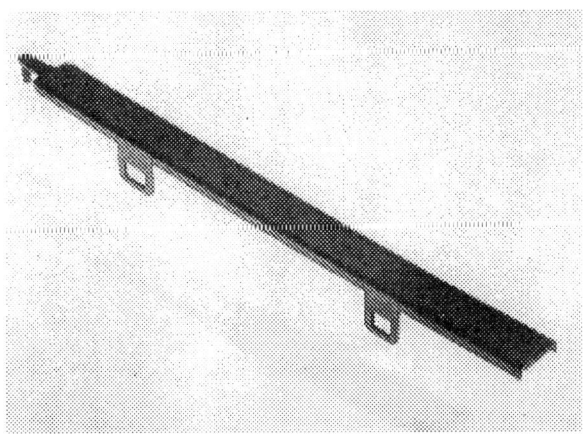

Fig. 17 260 mm-length element developed with binary microlenses.

separation from a polychromatic extended source, and examples based on zone plates have been described to explain the applicable procedure in the simplest possible case of a binary beam shaper. The concepts of Zero Order Color Correction (ZOCC) and of Border Rainbow Correction (BORC) have been introduced. Two fast methodologies for the fabrication of large nonrepeated clusters have been described, and several elements developed for both monochromatic and polychromatic sources have been shown. For the first time to our knowledge, real mass production of diffractive components for automotive applications, based on the above considerations, has been described.

## REFERENCES

1. J.W. Goodman, *Introduction to Fourier Optics*, McGraw Hill Book Co., San Francisco (1968)
2. H. Damman, and E. Klotz, Coherent optical generation and inspection of two-dimensional periodic structures, *Opt. Acta* 24:505 (1977)
3. J. Mait, and K. Brenner, Optical symbolic substitution: system design using phase only holograms, *Appl. Opt.* 27:1692 (1988)
4. W. Veldkamp, J. Leger and G. Swanson, Coherent summation of laser beams using binary phase gratings, *Opt. Lett.* 11:303 (1986)
5. R.L. Morrison and S.L. Walker, Progress in diffractive phase gratings used for spot array generators, *Optic Computing 1991 Technical Digest* 6:144 (1991)
6. U. Killat, G. Rabe and W. Rave, Binary phase gratings for star couplers with high splitting ratio, *Fiber Integ. Opt.* 4:159 (1982)
7. J.L. Brubaker, F.B. McCormick, F.A.P. Tooley, J.M. Sasian, T.J. Cloonan, A. Lentine, S.J. Hinterlong and M.J. Herron, Optomechanics of a free space photonic switch: the components, *Proc. SPIE* Vol. 1533 (1991)
8. M.A. Seldowitz, J.P. Allebach and D.W. Sweeney, Synthesis of digital holograms by direct binary search, *Appl. Opt.* 26:2788 (1987)
9. M.R. Feldman and C.C. Guest, Iterative encoding of high-efficiency holograms for generation of spot arrays, *Opt. Lett.* 14:479 (1989)
10. E. Sidick, A. Knoesen and J.N. Mait, Design and rigorous analysis of high efficiency array generators, *Appl. Opt.* 32:2599 (1993)
11. R.L. Morrison, S.L. Walker and T.J. Cloonan, Beam array generation and holographic interconnections in a free-space optical switching network, *Appl. Opt.* 32:2512 (1993)
12. J. Westerholm, J. Turunen, A. Vasara and A. Salin, Stripe-geometry two-dimensional Damman gratings, *Opt. Comm.* 74:245 (1989)
13. A. Vasara, M.R. Taghizadeh, J. Turunen, J. Westerholm, E. Noponen, H. Ichikawa, J.M. Miller, T. Jaakkola and S. Kuisma, Binary surface-relief gratings for array illuminators in digital optics, *Appl. Opt.* 31:3320 (1992)
14. R.W. Gerchberg and W.O. Saxton, A practical algorithm for the determination of phase from image and diffraction plane pictures, *Optik* 35:237 (1972)
15. J.R. Fienup, Iterative method applied to image reconstruction and to computer-generated holograms, *Opt. Eng.* 19:297 (1980)
16. F. Wyrowski and O. Bryngdahl, Digital holography as part of diffractive optics, *Rep. Prog. Phys.* 54:1481 (1991)
17. V.V. Wong and G.J. Swanson, Binary optic interconnects: design, fabrication and limits on implementation, *Appl. Opt.* 32:2502 (1993)
18. E. Johnson, M.A. Abushagar and A. Kathman, *Optical Design for Photonics Technical Digest*, OSA, Washington (1993)
19. M.T. Gale, M. Rossi, H. Schtz, P. Ehbets, H.P. Herzig and D. Prongué, Continuous-relief diffractive optical elements for two-dimensional array generation, *Appl. Opt.* 32:2526 (1993)
20. H. Damman and K. Gortler, High-efficiency in-line multiple imaging by means of multiple phase holograms, *Opt. Comm.* 3:312 (1971)
21. U. Krackhardt, N. Streibl, Design of Damman-gratings for array-generation, *Opt. Comm.* 74:31 (1989)
22. L. Doskolovich, V. Soifer, M. Shinkarev, A method for stochastically synthesizing binary diffraction grating, *Avtometria* 3:104 (1992)
23. S. Bobrov, B. Kotletsov and Y. Turkevich, New diffractive optical elements, *Proc. SPIE* 1751:154 (1992)
24. A.D. Pommet, M.G. Moharam and E.B. Grann, Artificial uniaxial and biaxial dielectrics with use of two-

dimensional subwavelength binary gratings, *JOSA (A)* 11-2695 (1994)
25. L.L. Doskolovich, V.A. Soifer, G. Alessandretti, P. Perlo and P. Repetto, Analytical initial approximation for multiorders binary grating design , *JEOS* 3:921 (1994)
26. W.H. Press, B.P. Flannery, S.A. Teukolsky and W.T. Vetterling, *Numerical Recipes: the Art of Scientific Computing*, Cambridge University Press, Cambridge (1990)
27. L.L. Doskolovich, N.L. Kazanskiy, P. Perlo, P. Repetto and V.A.Soifer, Direct two-dimensionalcalculation of binary DOEs using a non-binary series expresssion approach, *International Journal of Optoelectronics* 10:4 (1996)
28. J. Jahns, M.M. Downs, M.E. Prise, N. Streibl and S.J. Walker, Dammann gratings for laser beam shaping, *Opt.Eng.* 28:1267 (1988)
29. H. Stark, W.C. Catino and J.L. LoCicero, Design of phase gratings by generalized projections, *JOSA(A)* 8:155 (1991)
30. M.R. Feldman and C.C. Guest, Iterative encoding of high-efficiency holograms for generation of spot arrays, *Opt. Lett.* 14:479 (1989)
31. J. Turunen, A. Vasara and J. Westerholm, Kinoform phase relief synthesis, *Opt. Eng.*28:5 (1989)
32. J.R. Fienup, Phase retrieval algorithms: a comparison, *Appl.Opt.* 21:15 (1982)
33. V.A. Soifer, L.L. Doskolovich, M.A. Golub, N.L. Kazanskiy, S.I. Kharitonov and P. Perlo Multi-focal and combined diffractive elements., *Proc. SPIE* 1993:226 (1992)
34. R.W. Wood, Diffraction gratings with controlled groove form and abnormal distribution of intensity, *Philos. Mag.* 6 23:310 (1912)
35. Lord Rayleigh, *Wave Theory of Light (in Encyclopedia Britannica)*, 9ed, Charles Scribner's Sons, New York, (1888)
36. R.W.G. Hunt, *Measuring Colour*, 2ed, Fountain Press, England (1991)
37. V. Bedernikov, V. Koronkevich and A. Poleshchuk, Precise plotter for synthesizing optical elements, *Opt. Instr. And Data Processing* 3 (1981)
38. V. Koronkevich et al., Production of kinoform optical elements, *Optik*, 67:259 (1984)
39. Y.B. Boiko, V.M. Granchak, I. Dilung, V.S. Solovjev, I.N. Sisakian and V.A. Soifer, Relief holograms recording on liquid photopolymerisable layers, *Proc. SPIE* 1238: 253 (1992)
40. J. Crank, *The Mathematics of Diffusion*, 2ed, Clarendon Press, Oxford (1975)
41. J. Crank, *Diffusion in Polymer*, 2ed, Academic Press, London and New York (1975)

# AN UNCONVENTIONAL OPTICAL ELEMENT FOR SPLITTING AND FOCUSING HIGH POWER LASER BEAMS

V. Russo[*], G. De Angelis[§], and A. Scaglione[§]

[*] Polytechnic of Milan, Department of Physics
Piazza Leonardo Da Vinci 32, 20133 Milan, Italy

[§] University of Salerno
Department of Information Engineering and Electrical Engineering
Via Ponte Don Melillo, 84084 Fisciano (Sa), Italy

## 1. INTRODUCTION

In industrial, medical and research applications of both medium and high power laser radiation, it is frequently necessary that a single source be used by multiple users, who may be physically far apart. Multiple connections are then required to simultaneously couple the laser beam into two or more optical fibers. In order to obtain satisafctory system flexibility such a function should be carried out by a single compact device; this component should not be damaged by the high power laser beam, and should be usable with optical fibers with different core diameters.

One commonly adopted solution is to use dielectric beam-splitters, arranged in a cascade configuration along the axis of the laser beam, with launching optics for each of the fibers to be coupled. Although this approach is efficient in terms of the number of users that can be served, it is made up of a large number of single components and it is sensitive to beam polarization. It provides a fixed (uniform or not) distribution of the incident beam power among the users, which can be changed only by removing or substituting some splitters. In other systems used to excite several optical fibers, the input ends of the fibers are arranged in a corolla configuration on a revolving disk, placed orthogonally to the laser beam.The fibers can be sequentially excited by rotating the disk; obviously, this system cannot excite more than one fiber at time. More recently, diffractive optics have been proposed to split a laser beam into a number of lower power clones.[1] In this case, to excite several optical fibers, either separate launching optics are needed for each of this clones, or a single focussing lens has to be positioned in front of the diffractive element.

In this Chapter, we investigate the possibility of using a monolithic refractive element (multi-spot lens), which simultaneously fulfils the functions of beam splitter and focussing

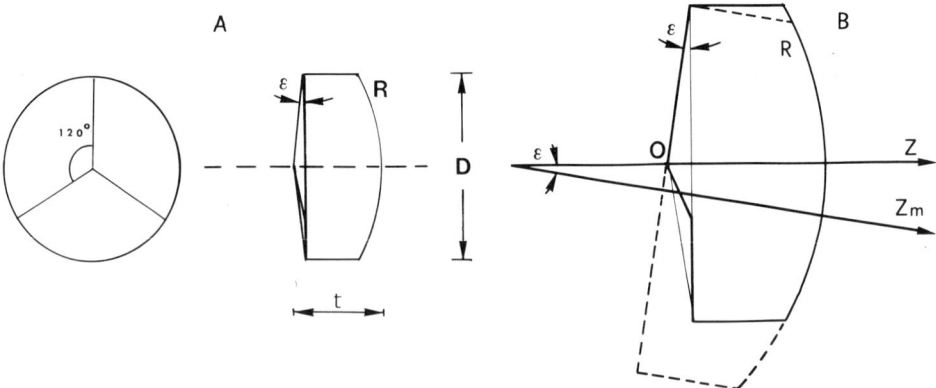

Fig. 1 A: geometry of the multi-spot lens; B: misaligned plano-convex lens modelling each of the three lens's sectors.

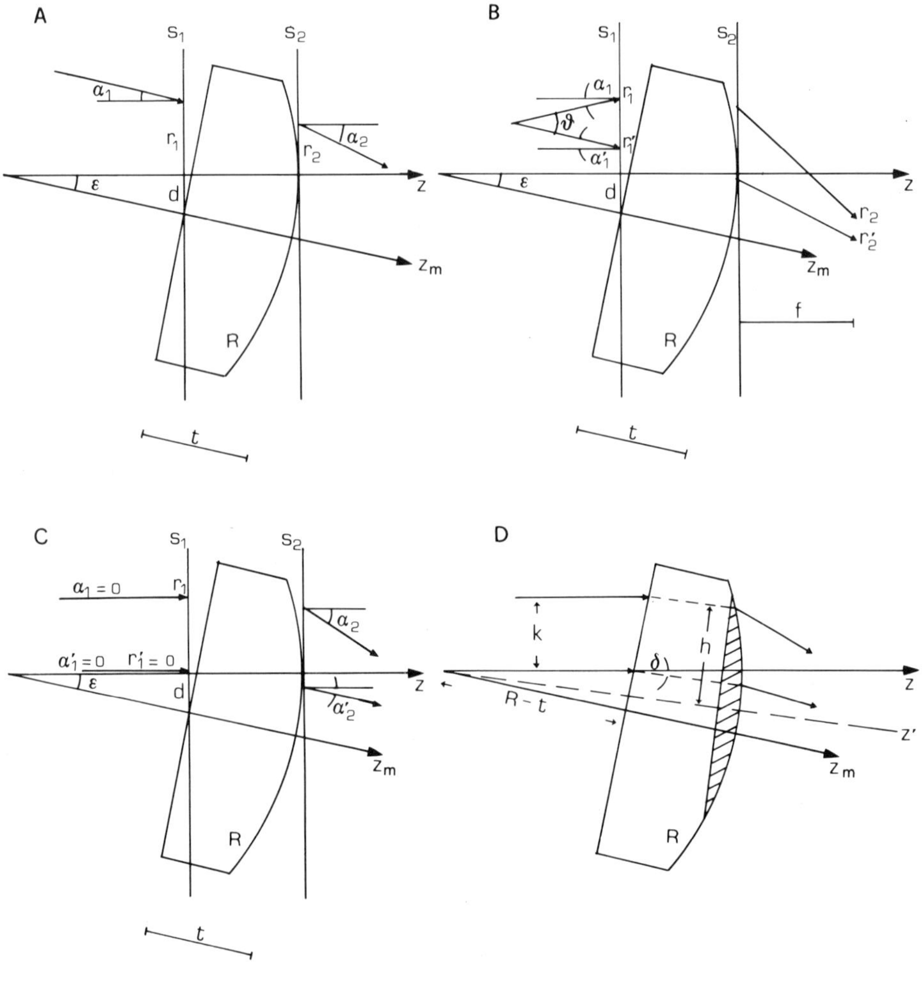

Fig. 2 Misaligned plano-convex lens modelling each of the three multi-spot lens sectors and ray coordinates involved in the calculation of the lens parameters.

optics, to direct the laser beam power into a number of different spots lying in a plane orthogonal to the lens axis. The element consists of an asymmetrical refractive lens, one face of which has been conventionally rounded in a convex shape, while the other face has a prismatic shape in order to split the laser beam. The lens design parameters have been evaluated using simple relations obtained from a matrix formulation of the image-forming characteristics of a misaligned element.

An experimental characterization at $\lambda = 514.5$ nm was carried out on a lens generating three focal spots; this characterization has substantially confirmed the results of the theoretical analysis and it has shown that the lens can be conveniently used to couple three large-core optical fibers (core diameter > 200 μm) to a laser beam. Moreover, it was verified that by translating the lens in a direction orthogonal to the laser beam axis a different amount of power can be addressed into each of the focal spots. The lens was also tested as coupling device between a single large-core optical fiber and three receiving fibers. The proposed component is thus suitable to be used as the basic element of a coupling device in a multi-user beam delivery system providing variable power distribution among the users.

## 2. LENS GEOMETRY AND OPTICAL CHARACTERISTICS

In order to simplify the analysis, a positive lens was fabricated that simultaneously splits and focusses the incident laser beam into three different points on a plane orthogonal to the beam axis. The three focal spots are disposed at the vertices of an equilateral triangle at a relative distance large enough to allow the input sections of three large-core optical fibers to be easily positioned. Fig. 1.A shows the lens geometry. The rear face of the lens is spherical with radius $R$; the front face has a prismatic shape in order to split the laser beam. It therefore consists of three equal circular plane sectors, each of which is inclinated by an angle $\varepsilon$ with respect to the plane orthogonal to the lens axis. The particular geometry enables us to consider each of the three sectors into which the lens can be decomposed as part of a plano-convex lens with optical axis making an angle $\varepsilon$ with the axis of the incoming laser beam (Fig. 1.B). These sectors are identical, therefore, the optical parameters of the multi-spot lens can be evaluated by referring to only one of them. The problem therefore, can be reduced to the study of a simple misaligned plano-convex lens for which, in the framework of the paraxial approximation, the matrix optics method [2-4] can be applied.

Fig. 2.A shows a misaligned plano-convex lens whose optical axis $z_m$ is tilted by an angle $\varepsilon$ compared to the beam axis z. On the two reference planes $S_1$ and $S_2$, the coordinates (inclination angle $\alpha$ and distance $r$ from the z-axis) of the incoming and outgoing ray are $(r_1, \alpha_1)$ and $(r_2, \alpha_2)$, respectively. By introducing the ray augmented vectors $\underline{r}_1=(r_1,\alpha_1,1,1)$ and $\underline{r}_2=(r_2,\alpha_2,1,1)$, one can define the lens transfer augmented matrix $M_D$ of the misaligned lens; such a matrix relates the ray vectors on the planes $S_1$ and $S_2$:

$$M_D = \begin{bmatrix} 1 & \dfrac{t}{n} & 0 & \left(1-\dfrac{1}{n}\right)t\varepsilon \\ -\dfrac{n-1}{R} & 1+\left(1-\dfrac{1}{n}\right)\dfrac{t}{R} & \dfrac{n-1}{R}d & \left(1-\dfrac{1}{n}\right)\dfrac{t}{R}\varepsilon \\ 0 & 0 & 1 & 0 \\ 0 & 0 & 0 & 1 \end{bmatrix}, \quad (1)$$

where $t$ is the lens thickness, $n$ is the refractive index, $R$ is the radius of the spherical surface and $d$ is the height from the z-axis of the «front vertex» of the misaligned lens.

From the matrix $M_D$ it is possible to construct the more general ray transformation augmented matrix $M_T$ describing the ray vector transformation between a plane orthogonal to the z-axis at distance $u$ to the left of the reference plane $S_1$ and a plane at distance $v$ to the right of the reference plane $S_2$ for a system in air:

$$M_T = \begin{bmatrix} 1+\dfrac{n-1}{R}v & u+\dfrac{t}{n}-v\left(-\dfrac{n-1}{R}u+1+\left(\dfrac{1}{n}-1\right)\dfrac{t}{R}\right) & -\dfrac{n-1}{R}vd & \left(1-\dfrac{1}{n}\right)\left(1-\dfrac{v}{R}\right)t\varepsilon \\ -\dfrac{n-1}{R} & -\dfrac{n-1}{R}+u+1+\left(\dfrac{1}{n}-1\right)\dfrac{t}{R} & \dfrac{n-1}{R}d & \left(1-\dfrac{1}{n}\right)\dfrac{t}{R}\varepsilon \\ 0 & 0 & 1 & 0 \\ 0 & 0 & 0 & 1 \end{bmatrix} \quad (2)$$

Finally, by imposing $M_{T12} = 0$ (*abcd* law), the matrix (2) gives the lens image-forming augmented matrix $M_i$ of the misaligned element. For an object at the infinity ($u = \infty$) this matrix takes the simple form:

$$M_i = \begin{bmatrix} 0 & 0 & d & t\varepsilon \\ -\dfrac{n-1}{R} & \infty & \dfrac{n-1}{R}d & \left(1-\dfrac{1}{n}\right)\dfrac{t}{R}\varepsilon \\ 0 & 0 & 1 & 0 \\ 0 & 0 & 0 & 1 \end{bmatrix} \quad (3)$$

In the following, both geometrical considerations and matrices (1)-(3) will be used to evaluate the principal optical parameters of the multi-spot lens.

### 2.1. Focal Length

The focal length $f$ along the propagation axis (the z-axis in Fig. 2.A) of a misaligned plano-convex lens can be obtained from the *abcd* law; for an object at infinity ($u = \infty$) $v = f$ and then:

$$f = -\frac{R}{n-1} \quad (4)$$

in the paraxial approximation, the focal length of each sector thus coincides with that of the misaligned element evaluated along the $z_m$-axis. Such a length can be assumed as the focal length of the multi-spot lens.

### 2.2. Distance between the Focal Spots

For a collimated ($\alpha_1 = 0$) incident beam the relation

$$\underline{r_2} = M_i \underline{r_1} \quad (5)$$

gives the coordinates of one of the beam's rays, in the focal plane. Here the distance of the outgoing rays from the z-axis is given by the element $r_2$ of the coordinate vector $\underline{r_2}$:

$$r_2 = M_{i13} + M_{i14} = d + t\varepsilon \quad (6)$$

Using a first order approximation we can put $d = (R-t)\varepsilon$ and therefore:

$$r_2 = R\varepsilon \; ; \tag{7}$$

consequently the reciprocal distance $L$ between the three spots is :

$$L = \sqrt{3}R\varepsilon \; . \tag{8}$$

It should be noted that $L$ does not depend on the lens's refractive index. It can therefore be concluded that the same material can be employed to design lenses with different focal lengths but the same distances between the spots. Similarly, by varying the beam wavelength the distance between the spots does not change whereas the focal plane is shifted ($n=n(\lambda)$).

## 2.3. Focal Spot Size for a Diverging Laser Beam

In the symmetry plane of each sector into which the lens can be decomposed, the size of the focal spot originating from an incoming beam with full divergence $\vartheta$ can be evaluated by determining the distance between the points in which the lens focusses two tilted collimated beams (tilt angle = $\pm \vartheta/2$). With reference to Fig. 2.B, by applying to each incoming ray the coordinate transformation described by the matrix (2) it can be found that ($u = 0$ and $v = f$ in this case):

$$r_2 = -f\alpha_1 + d + t\varepsilon \tag{9 a}$$
$$r_2' = -f\alpha_1' + d + t\varepsilon \tag{9 b}$$

where $\alpha_1 = \vartheta/2$ and $\alpha_1' = -\vartheta/2$. The size of the focal spot can be evaluated as the difference between $r_2$ and $r_2'$.

$$\Delta r = |r_2' - r_2| = 2f\alpha_1 = f\vartheta \tag{10}$$

## 2.4. Angular Aperture of the Outgoing Beam

In the symmetry plane of each sector, the transformation described by the lens transfer matrix $M_D$ makes it possible to evaluate the angular aperture $\gamma$ of the outgoing beam. With reference to Fig. 2.C and considering the marginal rays of a collimated ($\alpha_1 = 0$) incident beam (coordinate vectors $\underline{r}_1 = (r_1, 0, 1, 1)$ and $\underline{r}_1' = (0, 0, 1, 1)$ respectively), it is possible to evaluate the inclination angles $\alpha_2$ and $\alpha_2'$ of the respective outgoing rays; the difference between them gives the desired angular aperture. From matrix (2) the inclination angles of the two outgoing marginal rays are:

$$\alpha_2 = M_{D21}r_1 + M_{D23} + M_{D24} \tag{11 a}$$

$$\alpha_2' = M_{D23} + M_{D24} \tag{11 b}$$

respectively. The angular aperture of the outgoing beam is thus

$$\gamma = |\alpha_2 - \alpha_2'| \; . \tag{12}$$

## 2.5. Transverse Spherical Aberration

The effect of spherical aberration can be evaluated using a procedure that entails calculating the aberration of a thin lens. In Fig. 2.D an incident collimated beam is refracted, with angle $\delta$, by the plane front surface. The portion of spherical surface subtended by the refracted beam can be regarded as part of the rear surface of a plano-convex thin lens (dashed in Fig.) with optical axis z'. For this lens, $h$ is the height of the marginal refracted ray from the z' axis:

$$h \cong k + (R-t)\delta \tag{13}$$

hence, the transverse spherical aberration satisfies:[5]

$$r_t = h f G \tag{14}$$

where $f$ is the lens focal length (coinciding with the one of the misaligned plano-convex lens), and $G$ is:

$$G = \frac{h^2}{8f^3} \frac{1}{n(n-1)} \left[ \frac{n+2}{n-1} + 4(n-1) + (3n+2) + \frac{n^3}{n-1} \right] . \tag{15}$$

## 3. LENS DESIGN

The relations obtained allow us to evaluate the lens parameters so that the lens can meet the given requirements. In particular, if the lens is to be used to excite three equal large-core optical fibers, the size of the focal spots and the angular aperture of the single outgoing beams must be smaller than the core diameter and the angular aperture of the fiber acceptance cones respectively. For fibers with high numerical aperture (N.A.) values the limitation on the angular aperture of the outgoing beams is not particularly restrictive and then does not play an important role in the design procedure. Further requirements concern the technological aspects of the lens (choice of material and simplicity of fabrication) and its use in a sufficiently compact coupling device. This requires the lens to be small in size, with distances between the generated focal spots large enough to allow the input sections of the optical fibers to be housed.

By setting the fiber core diameter $\Phi_{core}$, given the characteristics of the laser beam, (diameter and divergence) and those of the material (refractive index), and given an upper limit to the transverse spherical aberration, it is possible to obtain the limit value for the focal length. In fact, by requiring that:

$$\Delta r + r_t = f\vartheta + hfG < \Phi_{core} \tag{16}$$

and assuming $r_t \leq 0.3 \Delta r$ it follows that

$$f < \frac{\Phi_{core}}{1.3 \vartheta} = \chi . \tag{17}$$

Hence, from Eq. (4) we have:

$$R < \chi(n-1) . \tag{18}$$

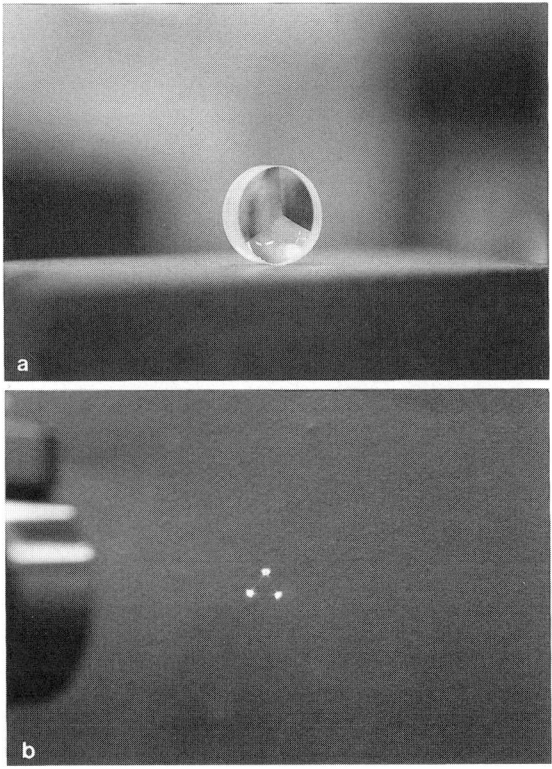

Fig. 3 Photographs of the three-spot lens and of the generated spots.

Assuming a laser beam divergence of 4 mrad (a typical value for high power multimode lasers) and receiving fibers with core diameter of 600 μm, for a lens in BK7 glass ($n = 1.52$ at $\lambda = 514.5$ nm ) Eqs. (17) and (18) give: $f < 115.4$ mm and $R < 60$ mm respectively; practical considerations suggest fixing the distance between focal spots at 3 mm; therefore Eq. (8) provides:

$$\varepsilon > 1.65° \ . \tag{19}$$

On the basis of Eqs. (17)-(19) a multi-spot lens was fabricated in BK7 glass with the following geometrical parameters: $R = 20$ mm, $D = 20$ mm, $\varepsilon = 5°$ and $t = 10$ mm. In Fig. 3 photographs of the lens and of the three generated spots are shown.

## 4. EXPERIMENTAL CHARACTERIZATION

Experimental characterization was performed at $\lambda = 514.5$ nm by using a Coherent Innova 90.6 Argon laser in multimode configuration (beam diameter 3.97 mm, full divergence > 0.5 mrad). Measurements of the intensity distribution, in planes orthogonal to the lens axis set 200 μm apart, were carried out by means of a laser beam analysis system (SPIRICON LBA 100A) and a CCD camera (Cohu 4800); the sensor was positioned on a motorized translator (Microcontrole MT 160) whose translation axis was parallel to the lens axis. The analysis system can make both energy and geometrical evaluations; as far as the

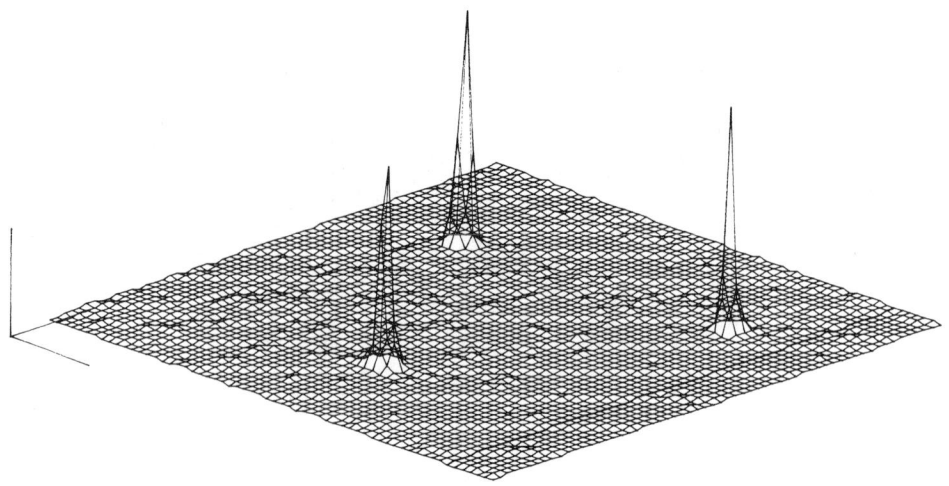

Fig. 4 Intensity distribution on the focal plane of the three-spot lens showing a symmetrical power distribution among the spots (the lens axis coincides with the laser incoming beam one).

latter are concerned, the system has a routine that makes it possible to calculate the diameter, at $1/e^2$ of the peak value, of the relative intensity distribution. The lens's focal length $f$ was measured in the plane in which the three outgoing beams presented the cross section with the smallest diameter. In the same plane the distance between focal spots was measured as the distance between the positions of the peak intensity of each of the three spots.

Fig. 4 shows the intensity distribution in the focal plane of the lens; a qualitative evaluation indicates that the focal spots are almost circular in shape, are equally spaced and have a nearly uniform distribution intensity. Fig. 5 shows, in the same plane as the preceding one, the intensity distribution measured when a small translation (20-30 μm) in the plane orthogonal to the beam axis was given to the lens; in this case an asymmetrical power distribution among the spots was achieved.

In Fig. 6 the diameter of the cross section of one of the three outgoing beams is plotted as a function of the distance from the lens's rear vertex. The trend of the experimental data shows that it is possible to evaluate the position of the plane in which the spot diameter assumes a minimum value. Table I reports the quantitative evaluations regarding the position of this

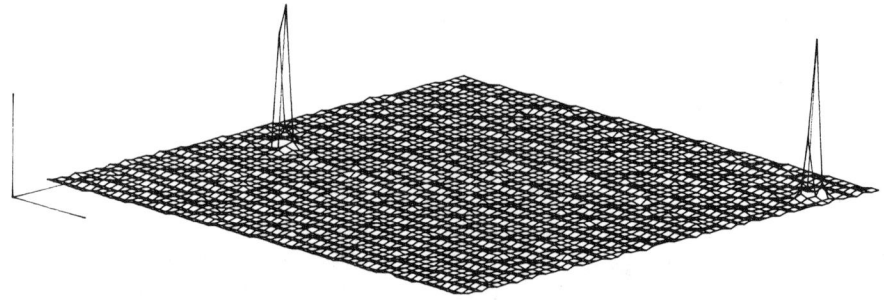

Fig. 5 Intensity distribution on the focal plane of the three-spot lens showing an asymmetrical power distribution among the spots (the lens is translated in the plane orthogonal to the laser incoming beam axis).

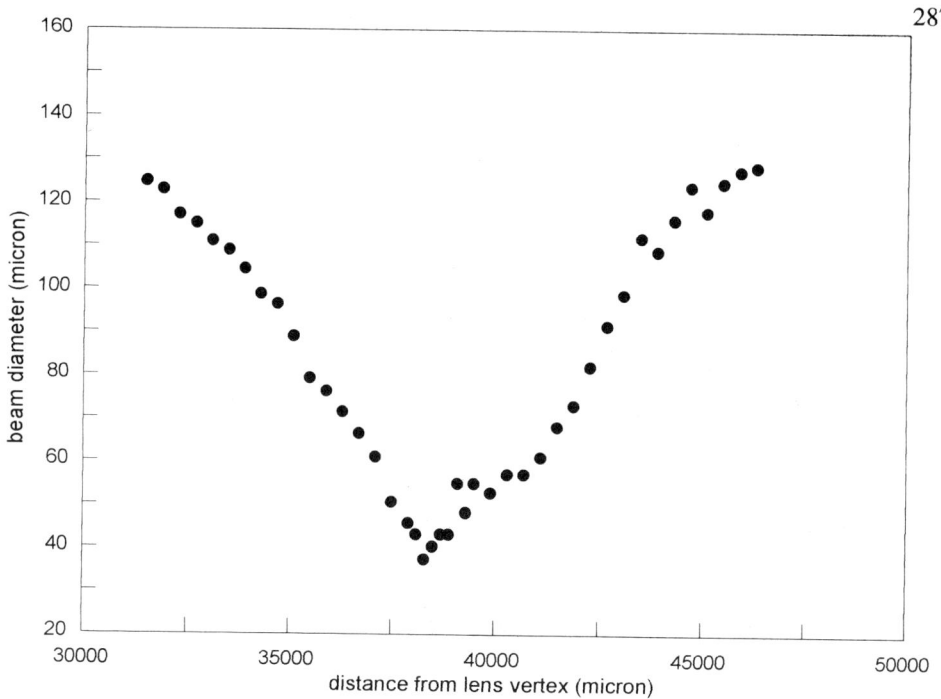

Fig. 6 Beam diameter (at $1/e^2$) of the cross section of one of the lens outgoing beams.

Table I. Characteristic parameters of the three-spot lens in various working geometries.

| beam characteristics | | $f_{th}$ | $f_{meas}$ | $\Delta r_{th}$ | $\Delta r_{meas}$ | $L_{th}$ | $L_{meas}$ |
|---|---|---|---|---|---|---|---|
| diam(mm) | div (mrad) | [mm] | | [µm] | | [mm] | |
| 3.97 | 0.5 | 38.46 | 38.40 | 47.2 | 38 | 3.02 | 2.97÷3.00 |

$f$: focal length; $\Delta r$: spot size; $L$: distance between spots; $th$: theoretical value; $meas$: measured value.

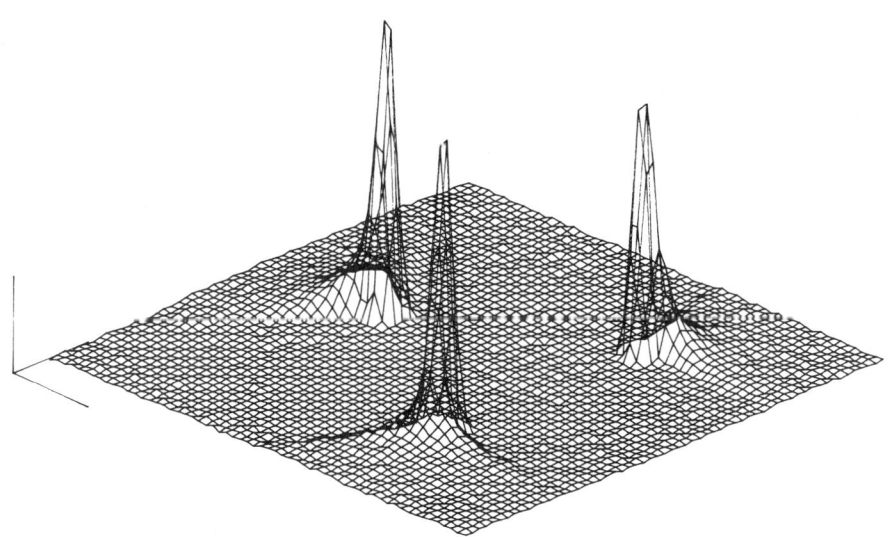

Fig. 7 Intensity distribution on the focal plane of the three-spot lens focussing the output beam from a 400 µm optical fiber (collimated by a 50 mm focal length plano-convex lens).

plane, the spot diameter and the relative distance between focal spots; for a comparison, the theoretical values obtained from matrix optics are also reported.

The data summarized in Table I show that the distance between spots is practically equal (the maximum deviation is - 0.05 mm) to the design value (3.02 mm) while the experimental evaluation of the diameter of the focal spot exhibits a significant difference. This deviation is linked both to the intrinsic approximation of the model used (geometrical optics) and the spatial resolution of the CCD sensor. However, it should be pointed out that the measured diameter is much smaller than the diameter of the coupled fibers (600 μm) for which the lens was designed; therefore, as far as launching efficiency in a coupling device is concerned, the differences encountered are practically negligible. As far the lens focal length is concerned, a negligible deviation (- 0.060 mm) from the theoretical value (38.46 mm) was measured. Fig. 6 points out that the diameter of the beam cross section maintains a value of less than 130 μm over a large range (about 7 mm) on either side of the theoretical focal plane. Consequently, positioning the fibers on the focal plane does not play a particularly critical role in ensuring efficient coupling.

As regards the use of the lens in a laser-fiber coupler device, launching efficiency measurements were carried out by coupling three 5 m long optical fibers (core diameter 600 μm, N.A. 0.30). It was verified that the 91% of the lens outgoing laser power can be uniformly coupled into the three fibers. By translating the lens in a direction orthogonal to the laser beam axis, a different amount of power can be coupled into each of the three fibers. In particular, a condition can be easily reached in which most of the outgoing laser power from the lens (70%) is addressed into a single fiber.

The lens was also tested as coupling device from a single large-core optical fiber to the three 600 μm optical fibers. In this case, a plastic-clad launching fiber (3M FT-400-LAT) with a 400 μm core diameter and a numerical aperture of 0.38 was used. The fiber was excited with a 20x microscope objective (N.A. = 0.40). Stable and well defined operating conditions were achieved by using a launching fiber of a given length (5 m) arranged on a 35 cm diameter cylindrical support; this made it possible to minimize, and hence control, modal coupling along the fiber. A 50 mm focal length plano-convex lens was used for collimation of the fiber's outgoing beam. Fig. 7 shows the intensity distribution on the plane on which the spots have the smallest diameter (~ 340 μm). The distance between spots ranges from 2.89 mm up to 2.92 mm. It was verified that 50 % of the power from the launching fiber can be uniformly addressed into the three spots.

## 5. CONCLUSIONS

On the basis of the experimental characterization performed, it can be concluded that the multi-spot lens can represent a solution to the problem of simultaneous excitation of several large diameter optical fibers. In particular, the small size and the characteristics of the outgoing beams (focal length and distance between spots) allow the lens to be installed in a compact coupling device housing the input fiber sections too; this device can also be provided with mechanical equipment to enable the relative movement of the various elements and thus allows both symmetrical and asymmetrical distributions of the incident beam's power. The lens design data can be obtained, with good approximation, by using simple relations deduced from an analysis based on matrix optics. It is worth nothing that, in order to ensure a correct evaluation of the lens characteristics, it is necessary to have sufficient knowledge of the incident beam's parameters (diameter and divergence). If the device is being used to couple a single optical fiber to more fibers, then a collimating lens must be placed between the launching fiber and the multi-spot lens. In this case, the experimental results showed that the design data still guarantees good launching efficiency in the coupled fibers.

# REFERENCES

1. F. Wryowski and R. Zuidema, Diffractive interconnection between a high-power Nd:YAG laser and a fiber bundle, *Applied Optics* 33:6732 (1994).
2. W. Shaoming and L.Ronchi, Array Optics, in *Atti della fondazione Giorgio Ronchi*, Firenze (1985).
3. W. Shaoming and L.Ronchi, Principles and Design of Optical Array, in *Progress in Optics* XXV, E. Wolf Ed., Elsevier Science Publishers B.V.(1988).
4. A. N. Matveev, *Optics*, Mir Publishers, Moscow (1988).
5. H.E. White and F.A. Jenkins, *Fundamentals of Optics*, Mc Graw-Hill, New York (1976)

**FIBER SENSORS**

# FIBER BRAGG GRATINGS AS TEMPERATURE AND STRAIN SENSORS

R. Falciai[*], R. Fontana[§], A. Schena[*], and A.M. Scheggi[*]

[*] Optoelectronics and Photonics Department (IROE CNR)
"Nello Carrara" Electromagnetic Waves Research Institute
Via Panciatichi 64, 50127 Florence, Italy
[§] Departmento of Electronic Engineering, University of Florence
Via S. Marta 3, 50139 Florence, Italy

## 1. INTRODUCTION

A fiber Bragg grating (FBG) consists of a periodic modification in the refractive core index along a short section of a germanosilicate optical fiber. Photorefractive intracore fiber Bragg gratings are attracting intense interest because of their wide field of application: they can be used in telecommunication systems [1-4] for wavelength division multiplexing (WDM) devices, wavelength-selective couplers, switches, integrated laser cavities (where a pair of gratings acts as partial or total narrow-band reflecting mirrors), dispersion compensators, as well as in sensing systems,[5] either as single element or in a quasi-distributed array. Due to their compatibility with the transmission medium and the relative ease of their fabrication, the use of FBG as sensors, together with existing fiber technology, opens up a whole range of new opportunities for novel applications in areas such as smart structures and materials.

## 2. FIBER BRAGG GRATINGS: WORKING PRINCIPLE AND FABRICATION TECHNIQUES

An advantage of using gratings, as opposed to many other types of interferometric sensors which require sophisticated processing, is that Bragg sensors may be relatively easily processed. For example, it is possible to realize a WDM grating sensor network in which each grating has an individually assigned wavelength.[6] Thus, the status of the grating, i.e. the sensor, can be interrogated by using either a broadband source or a tunable laser source. This type of sensor network is highly adaptable, and can be distributed over a short length or a large area, as in smart skin applications.

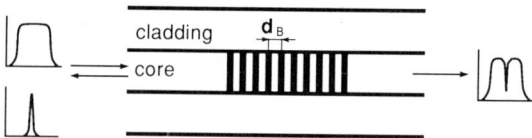

Fig. 1 Behaviour of a typical FBG in reflection and in transmission when illuminated with a broadband source.

## 2.1. Working principle

In order to write a fiber Bragg grating, it is necessary to modify the refractive core index, typically along 5 to 10 mm of the fiber: the modulation is induced by the photorefractive effect by means of an UV interference pattern that impinges laterally on the fiber. The wavelength of the light employed must range from 240 to 255 nm, because an absorption band due to germanium is present in that region.[7] Under such conditions, the core of the fiber undergoes a chemical-physical modification with subsequent permanent variation of the refractive index. This phenomenon is linked to the presence of Ge-defects in the silica vitreous matrix, and is more pronounced as the germanium oxide concentration in the core increases.[8]

According to the first-order Bragg condition, an FBG reflects a narrow linewidth light signal at a wavelength $\lambda_B$ determined by the grating period $\Lambda_B$ and the effective core index $n_{eff}$ of the guided mode in the fiber, according to the relation

$$\lambda_B = 2 n_{eff} \Lambda_B . \tag{1}$$

This device acts as a band rejection or notch filter in reflection and transmission, respectively. An FBG typical geometry, illuminated with a broadband source, is shown in Fig. 1, together with the reflection and transmission spectrum.

A complete description of the interaction between light propagating in the fiber and such a periodic structure can be provided by coupled mode theory.[9] The analysis involves finding the ideal eigenmodes in a uniform fiber and subsequently using them to represent the perturbed fields in a filter. Assuming a periodic modulation of the core along the axis $z$ of the fiber, the index profile can be expressed by

$$n^2(z) = n_0^2 + \delta n^2 \cos(\theta z) \tag{2}$$

where $n_0$ is the index of a uniform fiber, $\delta n$ the induced index perturbation, $\theta = 2\pi/\Lambda_B$ and $\Lambda_B$ the perturbation period.

The reflectivity $R(L, \lambda)$ as a function of wavelength $\lambda$ and filter length $L$ is then [10]

$$R(L,\lambda) = \frac{\kappa^2 \sinh^2(SL)}{\Delta\beta^2 \sinh^2(SL) + S^2 \cosh^2(SL)}, \tag{3}$$

where $\Delta\beta = \beta - (\pi/\Lambda)$, $\beta$ is the eigen propagation constant, $S = \sqrt{\kappa^2 - \Delta\beta^2}$ and $\kappa$ is the coupling coefficient. The latter is dependent on the index perturbation $\delta n$ induced in the

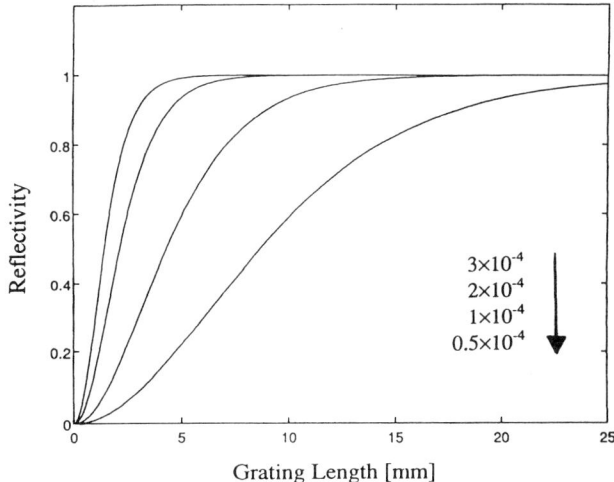

Fig. 2 Theoretical reflectivity as a function of the filter length at Bragg wavelength $\lambda_B$ for different values of induced index perturbation. The curves, from top to bottom, correspond to $\delta n$ of $3 \cdot 10^{-4}$, $2 \cdot 10^{-4}$, $1 \cdot 10^{-4}$ and $0.5 \cdot 10^{-4}$.

optical fiber filter, which in turn varies with the writing power, the fraction of guided optical power in the core and the characteristic parameters of the optical fiber employed (numerical aperture, core diameter).

By imposing the mismatch condition $\Delta\beta = 0$, equivalent in the previous formula to keeping $\lambda$ constant and equal to $\lambda_B$ defined by (1), Eq. (3) can be used to plot the reflectivity as a function of the filter length $R(L)$

$$R(L) = \tanh^2 \frac{\pi \delta n L \eta(V)}{\lambda_B}, \qquad (4)$$

where $\eta(V)$ is a function of the fiber $V$ parameter and represents the fraction of the integrated mode intensity contained in the core. Fig. 2 shows $R(L)$ for four different values of induced index perturbation and for the parameters relevant to the commercial fiber 3M that we used to produce gratings. The growing behaviour of peak reflectivity with increasing filter length is evident. By defining the effective cut-off length $L_{cutoff}$ of the filter as the length at which the reflectivity is equal to 0.9, it can be seen that $L_{cutoff}$ is a decreasing function of $\delta n$. If, on the other hand, $L$ is kept constant and equal to $L_{cutoff}$, Eq. (3) can be used to plot the spectral reflectivity of the filter and the spectral bandwidth can be determined.

The bandwidth of Bragg reflection depends on two things: the number $N$ of grating planes and the strength of the index modulation $\delta n$. A certain portion of the incident light is reflected at each grating plane: if the Bragg condition is not satisfied, then the wavelets reflected at each subsequent plane become progressively more and more out of phase (dephased). If, however, the grating strength is sufficient, a substantial portion of the incident power can be reflected before dephasing becomes important. On the contrary, very little power is reflected before dephasing sets in. The balance between the dephasing length, the physical length of the grating and the length (at exact phase-matching) needed for substantial reflection determines the bandwidth of the Bragg reflection. A general expression for the approximate full-width-half-maximum (FWHM) bandwidth of a grating is

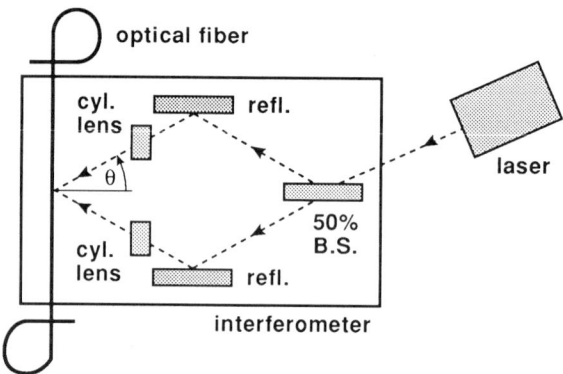

Fig. 3 Schematic of the experimental setup used for the external writing of FBG using the holographic technique.

$$\frac{\Delta\lambda_B}{\lambda_B} = s\sqrt{\left(\frac{\delta n}{2n_0}\right)^2 + \left(\frac{1}{N}\right)^2} \qquad (5)$$

where parameter $s$ equals $\approx 1$ for strong gratings (with near 100% reflection) and $\approx 0.5$ for weak gratings. Multiple reflection to and fro between opposite ends of the grating region causes side-lobes to appear off-resonance; this is particularly noticeable for strong gratings.

### 2.2. Fabrication techniques

There are two methods for inducing a modulation of the refractive index in the core of an optical fiber: *internal* and *external writing*.

*Internal writing*[11] is an holographic process in which the propagation modes to be coupled are launched in the fiber as bond coherent modes; they modify the core refractive index of the fiber by means of a two-photon absorption process.

The basic process for the formation of the gratings is the interference between the incident light from an Ar single-mode laser and counterpropagating light due to the Fresnel reflection from the end of the fiber, at the silica/air interface. The presence of these counterdirectional waves produces in the fiber a stationary pattern that photoinduces an index core variation in the germanosilicate fiber.

Grating formation has been observed at both the emission lines of the Ar laser, i.e. at 514.5 nm and 488 nm. The wavelength resonant with the grating written in the fiber (i.e. the wavelength of the fiber Bragg grating) is the same as that of the writing laser and the grating is distributed over the whole length of the fiber. With fiber lengths of about 30 cm, Hill et al. obtained typical bandwidths for the Bragg peak of about 200 MHz, according to the theoretical predictions.

The power analysis of the bandwidth, length and photoinduced variation of the refractive index of the Bragg peak were extensively studied by Lam and Garside.[11] In particular, the quadratic behaviour of $\delta n$ with writing power suggested a two-photon absorption process. As for all holographic experiments, this method requires high stability for the beam and the measurement apparatus: screening from thermal and vibrational effects is necessary so as to avoid the relative motion of the two fiber ends.

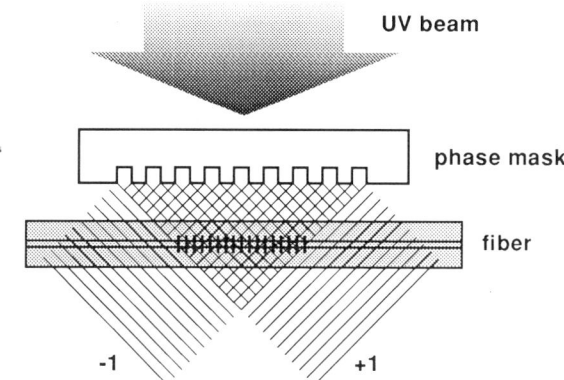

Fig. 4 Writing scheme for the fabrication of FBGs by means of the photolithographic technique.

*External writing* consists of an external projection of an interference pattern laterally on the fiber. This can be achieved by means of the holographic interference of two coherent beams, using the photolithographic method, or through a point-by-point exposure.

The *holographic technique* [12] involves an intensity-division interferometer (Fig. 3): the light from a UV source is split by a 50% beamsplitter and recombined by two reflecting mirrors, so as to interfere at a pre-set angle. Two cylindrical lenses form a focal line along the fiber in such a way that the interference pattern is perpendicular to the fiber axis. The typical size of the illuminated region is 5 to 10 mm by 125 µm. The angle $\theta$ between the two interfering beams ($\lambda_{UV}$ is the writing wavelength) determines the grating pitch according to the relation

$$\Lambda_B = \frac{\lambda_{UV}}{2 \sin \theta} \tag{6}$$

and also the operation wavelength of the filter (see Eq. 1). This is a versatile characteristic that makes this method particularly suitable for large scale production. On the contrary, high mechanical stability is required as well as a source with good spatial and temporal coherence. To bypass the latter problem, a line-narrowing filter is useful when a KrF laser or a doubled-dye laser is employed. The writing times necessary to photoinduce gratings with a given peak reflectivity using the holographic technique can be extremely short, as the upper power limit is given by the fiber breakage threshold: gratings have been written [13] with only one shot, thus avoiding stability problems.

The *photolithographic technique* [14] uses a UV-transparent silica glass phase mask, placed in close contact with the optical fiber (Fig. 4). A phase mask consists of a one-dimensional surface-relief structure whose shape approximates a square wave profile. The amplitude of the periodic pattern is chosen to modulate, by $\pi$ radians, the phase of a UV-laser beam, i.e.

$$\frac{4\pi(n_{silica} - 1)A}{\lambda_{UV}} = \pi \tag{7}$$

where $A$ is the amplitude of the surface-relief pattern. This choice of the surface-relief amplitude results in a diffraction pattern for the designed wavelength with nulled zero-order

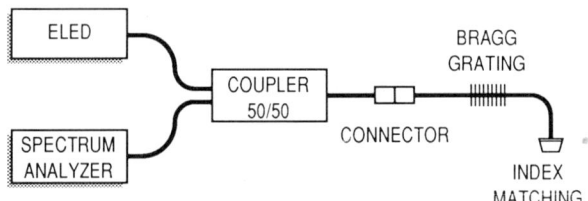

Fig. 5 Experimental set-up used at IROE for FBGs characterization.

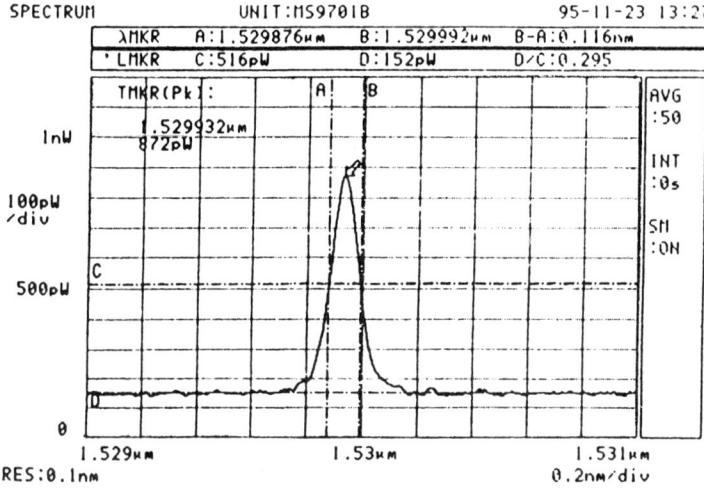

Fig. 6 Spectral behaviour of an FBG in reflection, as seen with the spectrum analyzer in use at IROE.

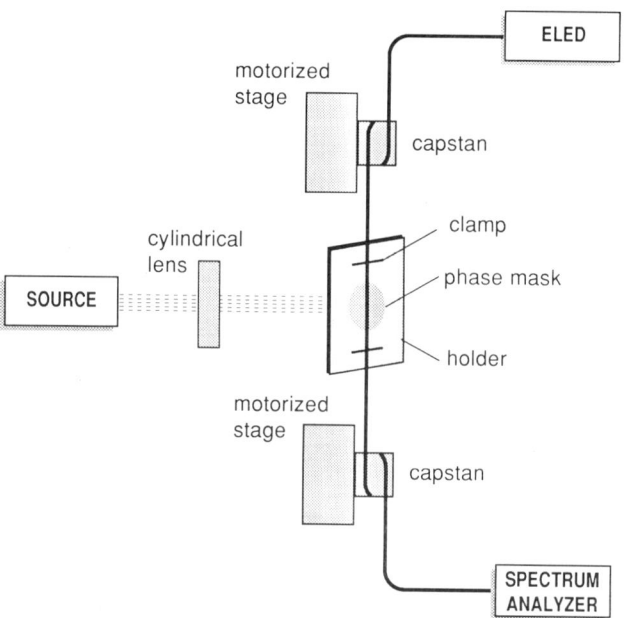

Fig. 7 Fabrication set-up employed for the stretch-and-write technique.

diffracted beam. In practice the zero-order can be suppressed to less than 5% of the light diffracted by the mask. The principal beams leaving the mask are the diverging plus-one and minus-one orders, each of which typically contains more than 35% of the diffracted light. The mask does not need to be shaped to a square wave. Goodman [15] discusses zero-order nulled surface-relief phase masks shaped to a sinusoid that would be equally useful in grating applications. From simple geometrical considerations, the Bragg grating pitch $\Lambda_B$ is the half of the phase mask pitch $\Lambda_{mask}$:

$$\Lambda_B = \frac{1}{2}\Lambda_{mask} .\qquad(8)$$

The principal period of the mask's diffraction pattern $\Lambda_B$ is independent of wavelength: therefore, in principle, it is possible to write a Bragg grating through the mask with a collimated broad-band source.

This technique has the advantages of giving good reproducibility of results: it makes use of a simple writing apparatus, and does not requires coherent sources. On the other hand, to vary the operation wavelength of the Bragg gratings, it is necessary to change the phase mask, the latter being an expensive optical component.

The fiber Bragg gratings produced by means of one of the external writing methods are much more commercial then the internally-written ones: these filters can operate at any wavelength whatsoever, in one of the three optical windows, and can be written on fibers compatible with standard communications fibers. Another advantage of the external over the internal writing method consists of the dramatic reduction in writing power necessary in order to photoinduce a grating by means of a one-photon absorption process: the one-photon absorption cross-section is about five orders of magnitude greater than that for a two-photon absorption process.

In recent years, a new technique has been proposed, the so-called *point-by-point writing technique*.[16] It consists of a selective exposure, by means of a slit, of successive regions of the core of an optical fiber. By moving the optical fiber with respect to the slit, a grating of the desired length and pitch is realized. This technique is only applicable to long-period gratings, due to practical focusing limits and mechanical stability.

In order to write gratings at IROE, the photolithographic technique has been chosen, because in this way it is possible to fabricate gratings even utilizing a low-coherence source and to reduce problems in terms of sensitivity to mechanical vibrations. An excimer KrF Lambda Physic laser (kindly supplied by RTM S.p.A., Italy) that emits at 248 nm and three commercial phase masks, with grating periods of 1060, 1070 and 1078 nm, were used.

Fiber gratings were written on three different kinds of optical fiber. These were a standard telecommunications fiber ($2r = 9$ µm, NA = 0.11), a Pirelli coupler type fiber ($2r = 4.5$ µm, NA ≈ 0.16) having a high germanium concentration, and a 3M optical fiber designed for fiber grating technology ($2r = 7.5$ µm, NA = 0.16). The fibers were exposed over regions of about 5 mm and 10 mm to a fluence level of ≈ 300 mJ/cm$^2$ per pulse, with a repetition rate of 100 Hz for different exposure times.

FBGs were spectrally characterized both in transmission and in reflection: the experimental set-up is shown in Fig. 5. An ELED was used in order to have a wide-band source (FWHM = 45 nm at 1550 nm). The output signal was detected by an optical spectrum analyser with a readout resolution of 4 pm. For the standard telecommunications fiber, a peak reflectivity only up to 20% was reached, due to the low germanium concentration in the fiber core, while different peak reflectivities up to 99% were obtained with several gratings on Pirelli and 3M fibers. Because of its high Ge-doping, the Pirelli fiber was much more photosensitive then the other fibers: the exposure times necessary for imprinting were lower by a factor 5 for the same writing conditions

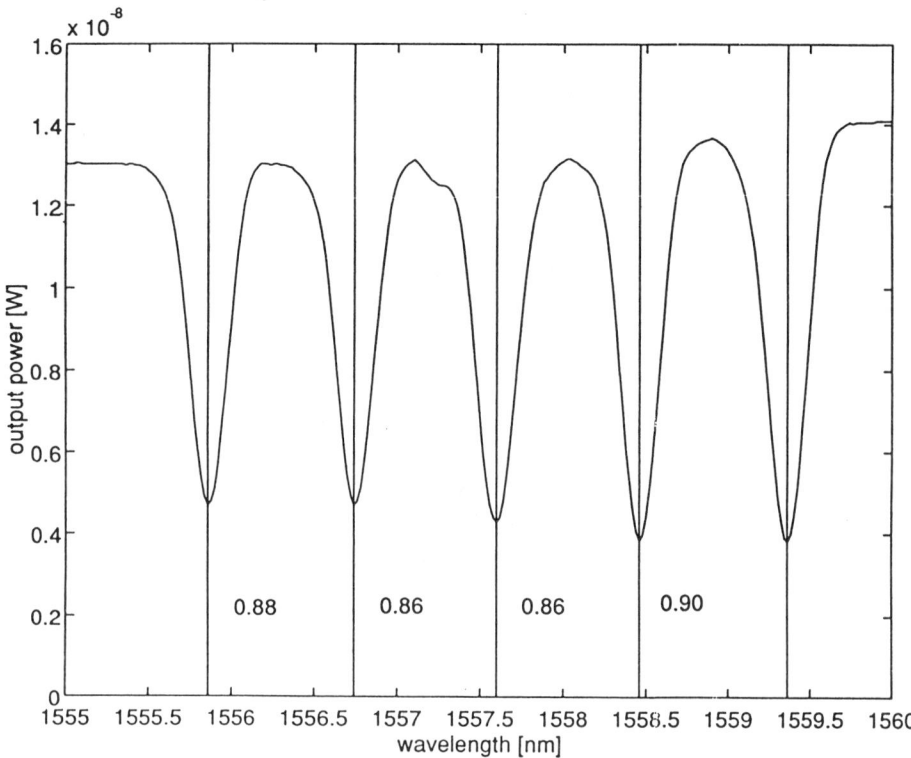

Fig. 8 Transmission spectrum of five FBGs at different wavelengths written on the same fiber in different positions by using the same phase mask.

and desired peak reflectivity. In all cases, a reflection bandwidth of about 0.12÷0.35 nm was measured. A typical reflection spectrum, obtained with the spectrum analyser, is shown in Fig. 6.

In order to realize a quasi-distributed sensor network system, we wrote gratings with different operating wavelengths using an automated tension system with the same phase mask and two motorized stages (see Fig. 7): by setting the displacement of the stages it was possible to stretch the fiber to a desired strain value. The writing wavelength was always the same, as given by Eq. 8; but when the fiber was relaxed, the peak moved towards shorter $\lambda$, due to the strain dependence.

Arrays of five gratings at different wavelengths were written on the same fiber in different positions. In Fig. 8 is shown the transmission spectrum of the five FBGs, with a separation of $\approx 0.88$ nm. The difference in the spacing separation is due to an experimental set-up error of about $\pm 20\ \mu m$.

In principle, it is possible to write Bragg gratings over a 8 nm range of wavelength: this upper limit is due to the maximum strain that the used fiber can stand before breakage. The strain calibration curve shown in Fig. 9 makes possible the setting of the spacing between the adjacent peaks.

## 3. BRAGG GRATINGS AS STRAIN AND TEMPERATURE SENSORS

It is clear that, if temperature and strain can be used to control the properties of gratings and related devices, it is equally true that any external process which alters the optical a)

b)

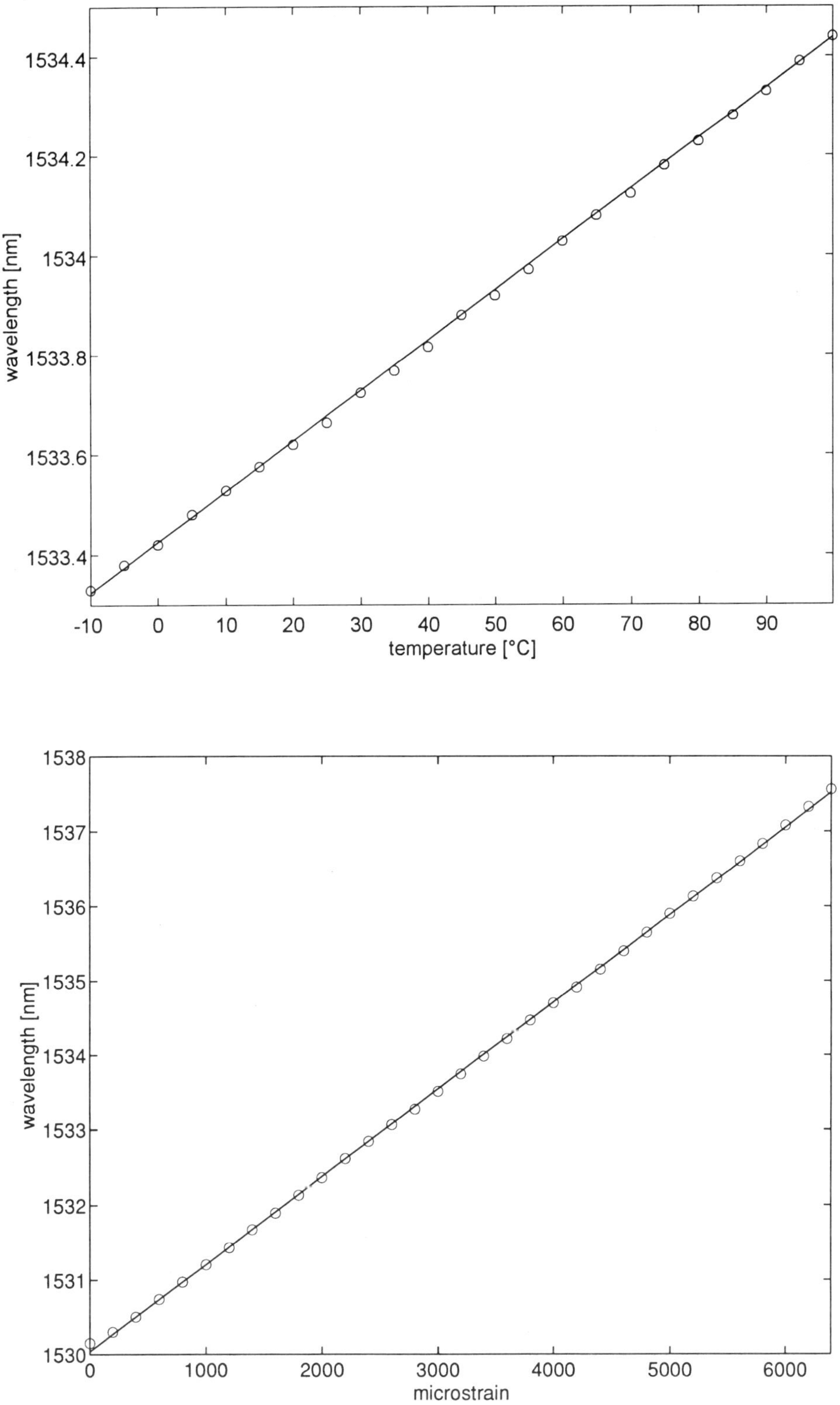

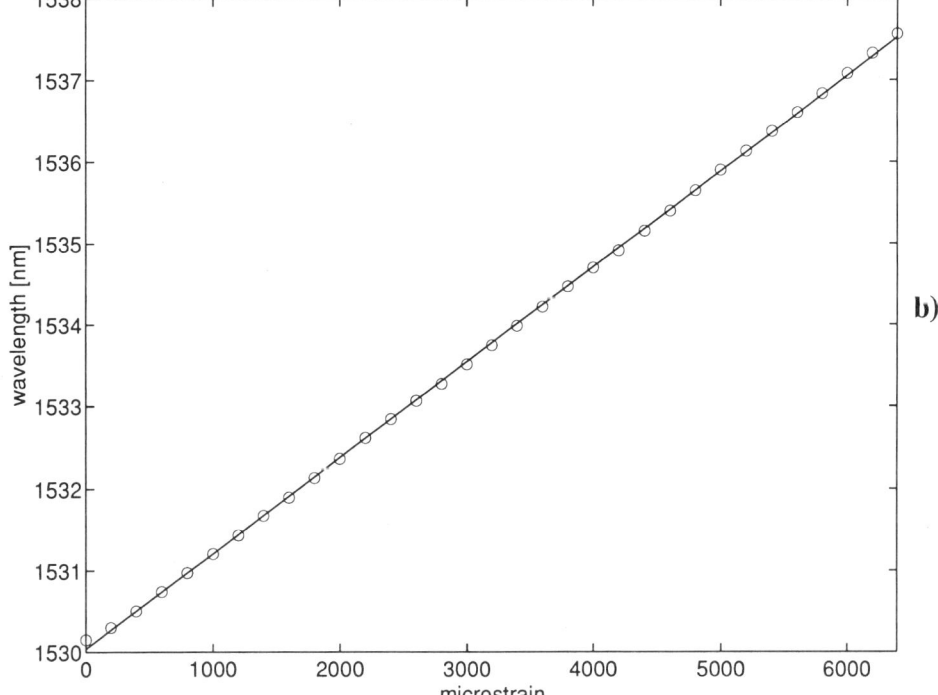

Fig. 9 Temperature (a) and strain (b) response of a FBG sensor.

constant of the fiber can be measured by monitoring certain characteristics of the grating. Thus, the Bragg grating resonance may be altered by temperature and the environmental stress.

When the fiber is influenced by a longitudinal strain $\Delta\varepsilon$, the change in the Bragg wavelength, due to a variation of the grating pitch and to a change in the refractive index induced by the photoelastic effect, is then given by [17]

$$\Delta\lambda_B = \lambda_B[1-P_e]\Delta\varepsilon ,\qquad(9)$$

where $P_e$ is an effective photoelastic constant that depends on the Poisson ratio $v$ and the components $P_{11}$ and $P_{12}$ of the strain-optic tensor of the fiber, according to the relation

$$P_e = \frac{n_{core}^2}{2}[P_{12} - v(P_{11} + P_{12})] .\qquad(10)$$

The cross sensitivity to a change in temperature, due to thermal expansion that causes a variation in the grating pitch and to the refractive index variation with temperature, is expected to be

$$\Delta\lambda_B = \lambda_B(\alpha + \xi)\Delta T ,\qquad(11)$$

where $\alpha$ is the thermal expansion coefficient and equals $0.55\times10^{-6}$ °C$^{-1}$ for silica, and $\xi$ represents the thermooptic coefficient and is approximately equal to $8.3\times10^{-6}$ °C$^{-1}$ for the germania-doped silica core. As $\alpha$ is smaller than $\xi$ by a factor of ~20, the index change is by far the predominant effect.

A single measurement of the wavelength shift of the Bragg peak cannot distinguish between the effects of temperature and strain. Assuming that the strain and thermal response are essentially independent, i.e. the related strain-temperature cross-term $\Delta\varepsilon\cdot\Delta T$ is negligible, the change in Bragg wavelength of the fiber grating can be expressed as

$$\Delta\lambda_B(\varepsilon,T) = K_\varepsilon\Delta\varepsilon + K_T\Delta T ,\qquad(12)$$

where $K_\varepsilon$ and $K_T$ are the said strain and temperature coefficients, respectively, and are defined by Eq.s (10) and (11). Such behaviour has already been found to apply well for small perturbations.

For sensing applications it is very important to distinguish between strain and temperature effects. Several discrimination techniques have been proposed in recent years: these include the use of a second grating element encapsulated in a different material and placed in-line with the first grating, and the use of two superimposed gratings of two different Bragg wavelengths.[18,19] Both schemes require the writing of two gratings, and have limitations in case an array of many sensors has to be interrogated at the same time.

An alternative technique has been proposed [20] that uses both the primary- and second-order diffraction from a single, conventionally-written FBG, when illuminated with light that coincides with the Bragg reflecting wavelengths, so as to decouple the effects of temperature and strain. It is important to note that high reflectivities at both primary and secondary wavelengths are possible, the relative magnitudes of the reflectivities being attributed to the degree of distortion introduced in the writing process. The method requires that the source is matched to the second order reflectivity wavelength of the grating; therefore, careful selection of the source emission characteristic is essential. This technique uses a matrix inversion approach similar to that described by Xu et al. for the two superimposed gratings.[18] A grating illuminated with two distinct wavelengths and exposed to the same level of temperature and strain perturbation will produce different wavelength shifts. A matrix equation may be constructed relating the temperature and strain coefficients at the two different wavelengths of illumination

$$\begin{bmatrix} \Delta\lambda_1 \\ \Delta\lambda_2 \end{bmatrix} = \begin{bmatrix} K_{\varepsilon_1} & K_{T_1} \\ K_{\varepsilon_2} & K_{T_2} \end{bmatrix} \begin{bmatrix} \Delta\varepsilon \\ \Delta T \end{bmatrix},  \quad (13)$$

where suffixes 1 and 2 refer to the two different wavelengths. Once the elements of the matrix are known, via the independent measurement of the Bragg wavelength change with strain and temperature at the wavelengths of interest, inversion of the matrix, under appropriate conditions, is used to determine the absolute temperature and strain information for single measurements at wavelengths 1 and 2.

Recently a novel strain sensor that is intrinsically insensitive to temperature effects has been presented:[21] it consists of a Bragg grating which is written on an optical fiber tapered by chemical etching. The taper profile can be realized in such a way that a tension applied to the fiber results in a linear chirping in the grating, due to the strain gradient that grows along the fiber itself. The sensor response to an external strain results in a broadening of the Bragg peak that does not depend on temperature variations.

The behaviour of FBGs realized in our laboratories as temperature and strain sensors was verified. Temperature measurements in dynamic conditions were carried out in the −10 to 100 °C range, recording simultaneously the Bragg peak position and the temperature every 5 °C. Strain measurements were performed by means of a micrometric translation stage up to the fiber breakage limit. In Fig. 9, the temperature and strain response of a Bragg grating inscribed on a 3M fiber are reported: the corresponding sensitivities were $1.1 \times 10^{-2}$ nm/°C and $1.2 \times 10^{-3}$ nm/µε, respectively. The sensors behaved in linear manner over wide ranges of strain and temperature without showing any hysteresis.

For a Pirelli fiber, we measured the Bragg peak displacement due to fiber tensioning in two different ways: by means of a micrometric translation stage and by means of the application of known weights, in the 0 to 1800 µε and 0 to 400 g ranges, respectively. From the stress-strain relation we obtained the Young modulus for the Pirelli Ge-doped fiber, i.e. $E = (6.671 \pm 0.014) \times 10^{10}$ N/m$^2$, that is smaller than the value for pure silica.

## 4. APPLICATIONS TO MATERIALS AND WIDE STRUCTURES

When surface-attached to or embedded within a structural material, an array of FBGs can provide detailed strain mapping without significantly modifying the structural properties. They constitute a powerful sensing tool for the analysis of structural loading and deformation, shape analysis, the characterization of vibrational modes, monitoring of residual stress, and damage detection.

The geometry of optical fibers makes the Bragg sensor a device that is intrinsically compatible with materials such as composites. FBGs are also suitable for applications to metals, because of their ability to tolerate high temperatures, as well as for ceramics and carbon-carbon.

Recent studies have shown that, in order to optimize fiber embedding within composites, fibers must be placed parallel to the material reinforcing structure, and the fiber jacket must be as similar as possible to the host matrix.

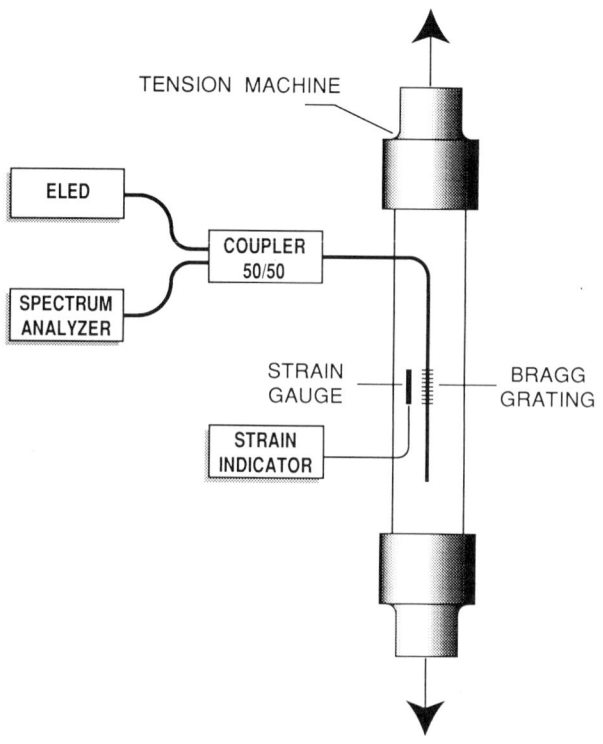

Fig. 10 Experimental set-up used for testing a specimen of steel.

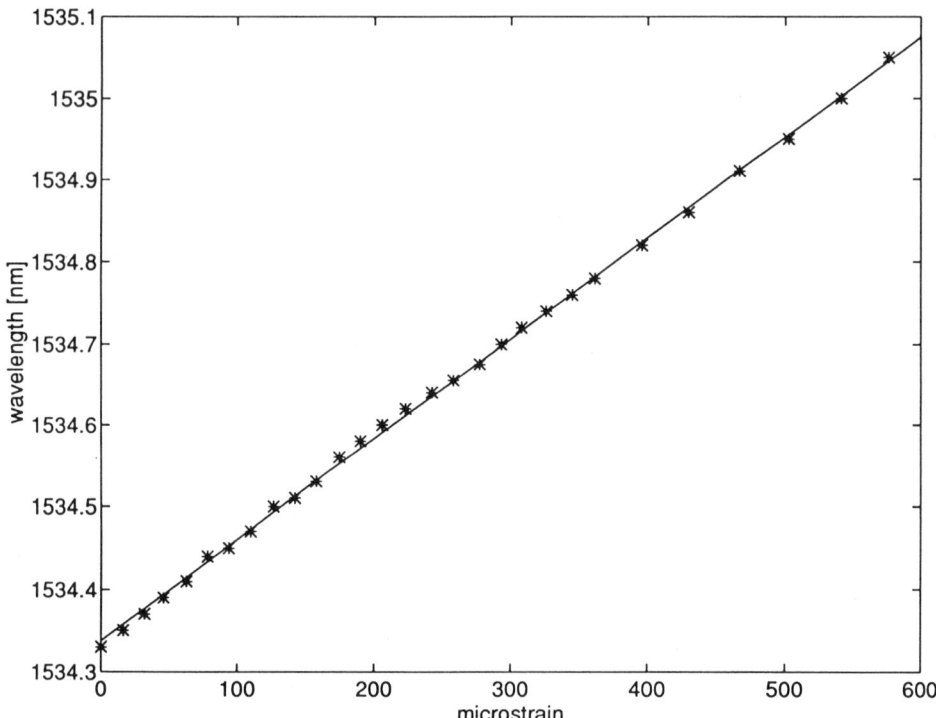

Fig. 11 FBGS shift related to strain gauge for a standard plate specimen.

## 4.1. Materials

A standard plate specimen of steel FEP04 with a 1 mm thickness was tested in the 0 to 600 µε range in our laboratories. A fiber Bragg grating was surface-attached to one side of the sample and a conventional strain gauge was attached on the other side. The experimental setup is shown in Fig. 12. The specimen was stretched by means of a tension machine, and the results of the two sensors were simultaneously recorded. The two methods showed good agreement over the whole range investigated (Fig. 13), and have comparable measurement resolutions (about 10 µε). Measurements on various samples such as glass and composites are in progress, along with strain mapping of car chassis and windscreens.

It is possible to find many examples in the literature that show various applications of fiber Bragg grating sensors: here in what follows a few significant examples are reviewed.

Among the materials of technological interest can certainly be included composites that, depending on the reinforcing fiber, have different fields of application. Fiber-resin composites are finding increasing numbers of military and civil applications because of their high stiffness-to-weight ratio, high strength, and immunity to deterioration from chemical exposure. These include airplane wing panels and fuselages, submersible hulls, bridges, decks, trusses for civil structures, and spacecraft.

Friebele et al.[22] describe the fabrication of multielement FBG arrays, and report the embedding of these arrays in glass/vinyl-ester composite products during the CRTM$^{TM}$ process (Continuous Resin Transfer Molding). Distributed strain measurements of the products during bending tests using the embedded FBG strain sensors are compared with calculated strains and measurements made using conventional electrical gauges. A 10 ply glass-vinyl ester composite panel ≈15 cm wide was used for the initial demonstration of FBG embedding during CRTM$^{TM}$. The optical fiber sensor array (7 gratings, working at 820 nm, separated in wavelength by 2 nm and in length by 0.4 m, plus two reference gratings) was inserted between the 9$^{th}$ and the 10$^{th}$ ply so as to distance them as far as possible from the neutral axis. The composite panel was placed in a 3-point bending geometry, with the gratings located on the concave side of the board: the spectra of the embedded FBG array in both flat and loaded panels are shown in Fig. 12. A and B were the two reference gratings,

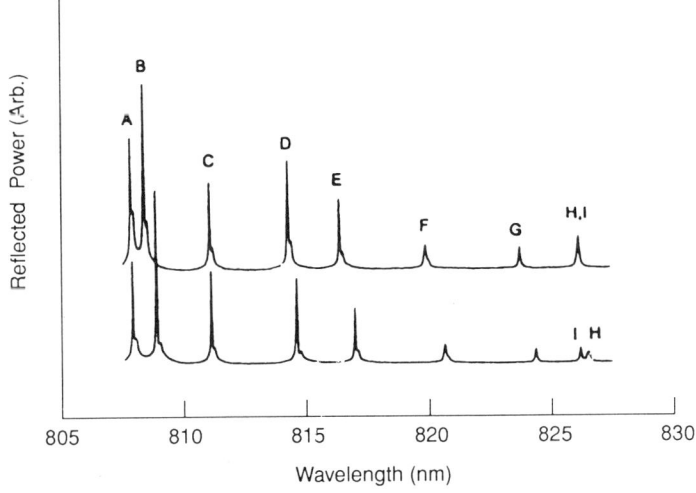

Fig. 12 Reflectance spectra of FBG arrays embedded in a CRTM$^{TM}$ composite panel measured flat an under 3-point bending with 1.35 kg load at its mid-point.

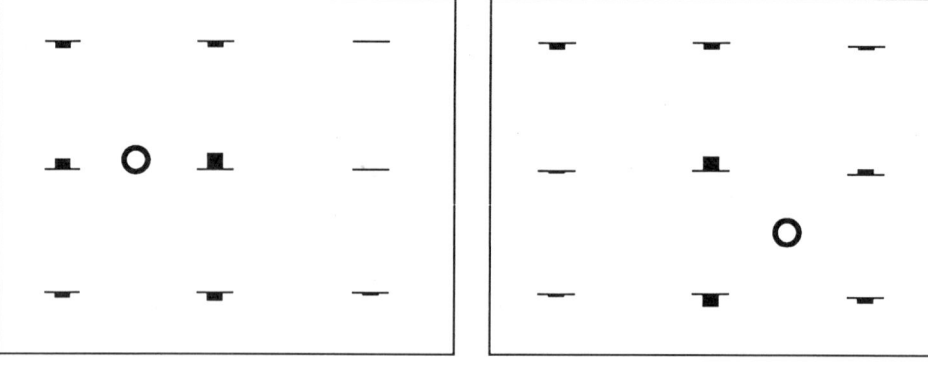

Fig. 13 Screen plot of he mapped strain infomation from the multiplexed 9 element FBG array for 2 diferent plate deflcting forcs (indicated by the open circle). The marker indicates the amplitude corisponding to a 500 μstrain level.

and the panel was supported on pivots located approximately under gratings C and I; as expected, their Bragg wavelengths did not shift significantly, indicating minimal additional stress. In contrast, each of the sensors between the pivots showed a decrease in Bragg wavelength due to the compressive stress induced in the gratings. The strains were derived from the wavelength shifts of each of the FBGS. The experimental results were compared with the results of two calculations; the beam midpoint strains were calculated analytically from the measured deflected shape of the beam, and non-linear finite element analysis.

An example of strain mapping over a plate which is subjected to deflection forces is given by Davis et al.[23] They describe a multiplexed nine-element Bragg grating array attached to a thick aluminium plate, which was firmly clamped on all edges. This plate was divided into 9 cells (i.e. a 3 by 3 matrix form), and each FBG element was attached approximately in the center of each cell, perpendicular by oriented to the closest boundary of the plate. The system was subjected to four different deflecting forces that were applied to the plate in different sites: the resulting strain mapping is shown in Fig. 13 for two different loadings. The strain distribution indicates the highest strain reading at a FBG when the force is close to that element. The response was found to be highly localized, indicating that the stress concentrations are close to the point of the applied force.

## 4.2. Wide structures

The corrosion of steel within large concrete structures, such as bridges, is prompting consideration of carbon fiber-based composite material replacements for steel. In cold-climate countries, this replacement is deemed to be even more important and urgent, due to the greatly accelerated deterioration of steel-reinforced concrete structures by the use de-icing salts. Since composite materials are unproven in their replacement for steel in concrete structures, there is considerable motivation to instrument-test structures that incorporate these materials in order to gain a better understanding of how well composite materials serve in this new challenging role.

In 1993, the city of Calgary commissioned the first highway bridge in the world to use carbon fiber composite prestressing tendons. Measures et al.[24] were asked to instrument the precast concrete girders with fiber optic Bragg grating sensors in order to monitor the changes in their internal strain over an extended period of time, and to compare the performance of

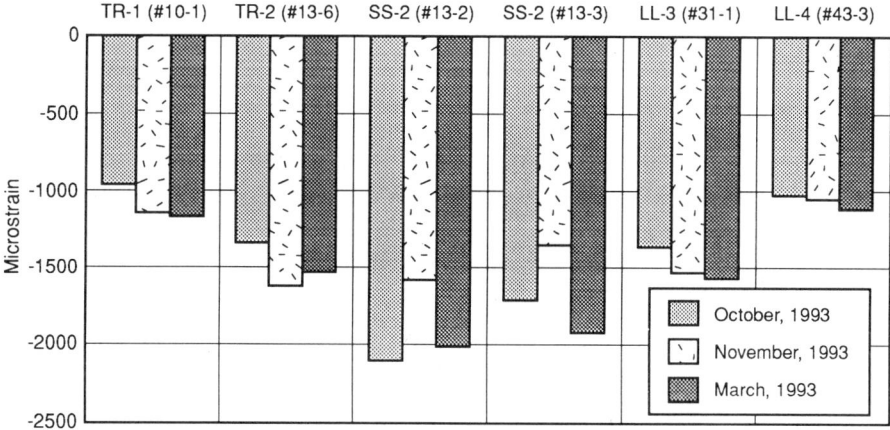

Fig. 14 Preliminary internal strain measurement from the bridge of Calgary revealing the stress relaxation in the carbon and steel tendons from the combined effects of: destressing, concrete shrinkage, dead loading of the bridge deck, and the post-tensioning applied across the two spans. Sensor code: TR-Tokyo Rope, LL-Leadline and SS-Steel tendons.

different prestressing tendons (steel and two types of carbon fiber composite). The authors realized arrays of 18 gratings that were embedded during the construction of the concrete girders: only 15 of them survived and functioned correctly.

The long-term characteristics and relaxation behaviour are shown in Fig. 14: the first set of strain data were obtained after placement of the deck and immediately following post-tensioning of the girders. These measurements represent the stress relief in the tendons from the combined effects of destressing, concrete shrinkage, creep, dead loading of the bridge deck, and post-tensioning of the two spans. The second measurement was taken approximately one month after the bridge opening, and shows some further relaxation of prestress for the carbon fiber composite prestressing tendons. However, the steel tendons show the opposite effect when proper allowance is made for the apparent thermal strain

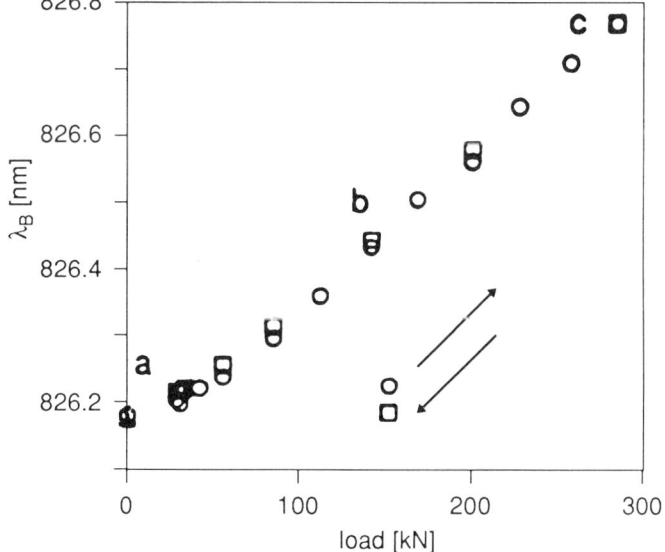

Fig. 15 Bragg wavelength of grating no. 3 versus load.

commensurate with the lower temperatures in November. Monitoring of traffic, extreme load events and load history have also been performed: such information is useful in bridge maintenance procedures and scheduling, in assessing bridge designs, and in comparing actual loading to the design loads. The authors have been able to demonstrate that structurally-integrated fiber optic sensors can measure the changes in the internal strain within deck girders that arise from both static and dynamic loading of the bridge with a large truck.

Nellen et al.[25] report on civil-engineering applications of wavelength multiplexed optical fiber Bragg grating arrays produced directly on the draw tower for testing and surveying advanced structures and materials such as carbon-fiber reinforced-concrete elements and prestressing tendons. They equipped a concrete T-shape cantilever beam reinforced with carbon-fiber lamellas over the whole area of its bottom side, with an array of 7 FBG sensors. At two points, forces of up to 283.4 kN were introduced statically and dynamically at low frequencies (four-point-bending). Fig. 15 shows the Bragg wavelength displacement of grating 3 in one cycle, from zero up to the maximum load and back.

## 5. CONCLUSIONS

The working principle of fiber Bragg gratings and their behaviour as temperature and strain sensors were described. Several examples of applications to smart structures and materials found in the literature were presented, together with a description of the activity at IROE. The intrinsic properties of the fibers and the high strain sensitivity of FBG sensors make them very attractive for monitoring various materials and wide structures of industrial and civil interest. There is growing attention to new signal demodulation techniques to make FBG sensor networks competitive as compared with to traditional sensor networks.

## REFERENCES

1. R. Kashyap, Optical fibre Bragg gratings for applications in telecommunications, *Proc. 21th Eur. Conf. on Opt. Comm. (ECOC '95 - Brussels)*, 23 (1995)
2. P A.L. Mason, J.A.J. Fells, R.V. Penty and I.H. White, Optical communication system performance using fibre Bragg grating dispersion compensators, *Proc. 20th Eur. Conf. on Opt. Comm. (ECOC '94 - Firenze)*, 435 (1994)
3. F. Ouellette, P.A. Krug, T. Stephens, G. Dhosi and B. Eggleton, Broadband and WDM dispersion compensation using chirped sampled fibre Bragg gratings, *Electron. Lett.*, 31, 11, 899 (1995)
4. P.A. Morton, V. Mizrahi, S.G. Kosinski, L.F. Mollenauer, T. Tanbun-Ek, R.A. Logan, D.L. Coblentz, A.M. Sergent and K.W. Wetch., Hibrid soliton pulse source with fibre external cavity and Bragg reflector, *Electron. Lett.*, 28, 6, 561 (1992)
5. G.P. Brady, S. Hope, A.B. Lobo Ribeiro, D.J. Webb and L. Reekie, Bragg grating temterature and strain sensors, *10th Int. Conference on Optical Fibre Sensors - Glasgow*, 510 (1994)
6. A.D. Kersey, T.A. Berkoff and W.W. Morey, Multiplexed fiber Bragg grating strain-sensor system with a fiber Fabry-Perot wavelength filter, *Opt. Lett.*, 18, 16, 1370 (1993)
7. M. Josephine Yuen, Ultraviolet absorption studies of germanium silicate glasses, *Appl. Optics*, Vol. 21, No.1, 136 (1982)
8. T.E. Tsai, C.G. Askins and E.J. Friebele, Photoinduced grating and intensity dependence of defect generation in Ge-doped silica optical fiber, *Appl. Phys. Lett.*, 61, 4, 390 (1992)
9. D Marcuse, *Theory of dielectric optical waveguide* (Academic Press, New York) (1974)
10. P.St J. Russel, J-L. Archambault and L. Reekie, Fibre gratings, *Physics World*, 41 (1993)
11. D.K.W. Lam and B.K. Garside, Characterization of single-mode optical fiber filters, *Appl. Optics*, 20, 3, 440 (1981)
12. W.W. Morey, G. Meltz and W.H. Glenn, High temperature capabilities and limitations of fiber grating sensors, *Proc. SPIE*, Vol. 1169, Fiber Optic and Laser Sensors VII, 98 (1989)
13. C.G. Askin, M.A. Putnam, G.M. Williams and E.J. Friebele, Stepped-wavelength optical-fiber Bragg grating arrays fabricated in line on a draw tower, *Optics Lett.*, 19, 2, 147 (1994)

14. K.O. Hill, B. Malo, F. Bilodeau, D.C. Johnson and J. Albert, Bragg gratings fabricated in monomode photosensitive optical fiber by UV exposure through a phase mask, *Appl. Phys. Lett.*, 62, 10, 1035 (1993)
15. G.W. Goodman, *"Introduction to Fourier Optics,* (Mc Graw-Hill, San Francisco) (1968)
16. R.J. Campbell and R. Kashyap, The properties and applications of photosensitive germanosilicate fibre, *Int. J. of Optoelectronics*, 9, 1, 33 (1994)
17. M.G. Xu, L. Reekie, Y.T. Chow and J.P. Dakin, Optical in-fibre grating high pressure sensor, *Electron. Lett.*, 29, 398 (1993)
18. M.G. Xu, J.L. Archambault, L. Reekie and J.P. Dakin, Simultaneous measurement of strain and temperature using fibre grating sensors, *10th Int Conference on Optical Fibre Sensors - Glasgow*, 191 (1994)
19. D.A. Flavin, R. McBride and J.D.C. Jones, Temperature-insensitive interferometric measurement of strain using grating-coupled LED sources, *10th Int Conference on Optical Fibre Sensors - Glasgow*, 195 (1994)
20. K. Kalli, G. Brady, D.J. Webb, L. Reekie, J.L. Archambault and D.A. Jackson, Possible approach for the simultaneous measurement of temperature and strain via first and second order diffraction from Bragg grating sensors, *10th Int Conference on Optical Fibre Sensors - Glasgow* (1994)
21. M.G. Xu, L. Dong, L. Reekie, J. A. Tucknott and J.L. Cruz, Temperature-independent strain sensor using a chirped Bragg grating in a tapered optical fibre, *Electron. Lett.*, 31, 10, 823 (1994)
22. E.J. Friebele, C.G. Askins, M.A. Putnam, J. Florio, A.A. Fosha, R.P. Donti and C.D. Mosley, Distributed strain sensing with fiber Bragg grating arrays embedded in CRTM composites, *2nd European Conf. on Smart Structures and Materials - Glasgow*, 338 (1994)
23. M.A. Davis, D.G. Bellemore and A.D. Kersey, Structural strain mapping using a wavelength/time division addressed fiber Bragg grating array, *2nd European Conf. on Smart Structures and Materials - Glasgow*, 342 (1994)
24. R.M. Measures, T. Alavie, R. Maaskant, S. Huang and M. LeBlanc, Bragg grating fiber optic sensing for bridges and other structures, *2nd European Conf. on Smart Structures and Materials - Glasgow*, 162 (1994)
25. P.M. Nellen, R. Bronnimann and U. Sennhauser, Applications of distributed fiber Bragg grating sensors in civil engineering, *SPIE* Vol. 2507, 14 (1995)

# RADIALLY GRADIENT-INDEX LENSES: APPLICATIONS TO FIBER OPTIC SENSORS

A. G. Mignani, A. Mencaglia, M. Brenci, and A. Scheggi

Optoelectronics and Photonics Department (IROE CNR)
"Nello Carrara" Electromagnetic Waves Research Institute
Via Panciatichi 64, 50127 Florence, Italy

## 1. INTRODUCTION

Monomode and multimode optical fibers are characterized by an intrinsic numerical aperture. Consequently, when designing fiber optic probes for sensing applications, some optics must be used to convert the fibers' divergent beam into a collimated one, in order to minimize insertion losses. The radially-graded refractive-index (GRIN) rod-lenses are particularly suitable for use in fiber optic sensors. In fact, the cylindrical shape of the GRIN lens, similar to the fiber shape, makes possible a compact, stable and rugged fiber-to-GRIN connection. GRIN-rod lenses have planar input and output faces, enabling the fiber to be glued directly onto the lens without optical paths in air, thus reducing cross-talk effects and pollution problems.

This Chapter reviews some relevant examples of fiber optic probes that make use of GRIN-rod lenses which have been designed for fiber optic sensors based on the intensity modulation technique. All the probes are simple, low in cost and rugged, and have been implemented not only as demonstrators but also as industrial prototypes. In particular, probes for proximity, vibration, temperature and gas-absorption measurements are presented, as well as probes for fluid-suspended particle characterization by means of static and dynamic light scattering techniques.

## 2. FIBER-GRIN COUPLING

A GRIN-rod lens is a lens that has a cylindrical shape, planar input and output faces, characterized by a refractive-index distribution with a maximum at the rod axis, then radially decreasing with a parabolic law described by the equation

$$N(r) = N_0 \left(1 - \frac{A}{2} r^2\right) \tag{1}$$

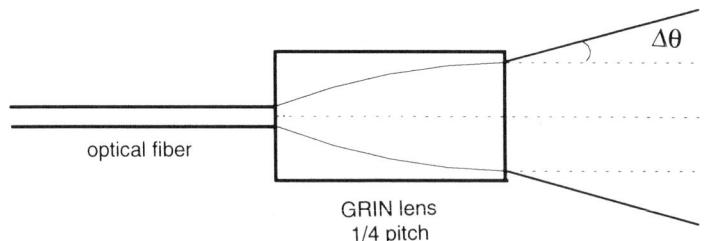

Fig. 1 Optical fiber butt-coupled to a GRIN lens.

where $N_0$ is the refractive index on the rod axis, $r$ is the radial position, and $A$ is the quadratic gradient constant which is the coefficient shaping the parabolic function.[1-6] Meridional rays trace sinusoidal paths having the same period, also called pitch, $P$. The lens length, $Z$ is usually expressed as a function of $P$ as

$$Z = \frac{2\pi P}{\sqrt{A}} \quad (2)$$

Quarter-pitch GRIN lenses, which are characterized by $P=1/4$, are those most used in fiber sensing applications, since they convert the fibers' divergent beam into a nearly-collimated one. As shown in Fig. 1, an optical fiber butt-coupled to a 1/4-$P$ GRIN lens provides a slightly-divergent beam, the divergence of which, $\Delta\theta$, is given by

$$\Delta\theta = N_0 \sqrt{AR_f} \quad (3)$$

that is, it depends on the GRIN optical constants $\sqrt{A}$ and $N_0$, as well as on the fiber radius $R_f$. Table I summarizes the reduction in the fiber's beam divergence, provided by the commercially-available GRIN lenses made by Nippon Sheet Glass Co. (NSG) under the trademark SELFOC.[7]

Table I Beam divergence given by an optical fiber butt-coupled to a GRIN lens

| 1/4-$P$ GRIN lens @ $\lambda$ = 0.83 µm | Fiber characteristics | | |
|---|---|---|---|
| | Singlemode 6/125 (NA=0.12 or 6.8°) | Multimode GI 50/125 (NA=0.22 or 12.7°) | Multimode HCS 200/230 (NA=0.38 or 22.3°) |
| SLS 1.0 | 0.13° | 1.09° | 4.36° |
| SLS 2.0 | 0.06° | 0.55° | 2.17° |
| SLW 1.0 | 0.16° | 1.38° | 5.5° |
| SLW 1.8 | 0.09° | 0.76° | 3.04° |
| SLW 2.0 | 0.08° | 0.68° | 2.73° |
| SLW 3.0 | 0.05° | 0.46° | 1.85° |

Fiber-GRIN centering and aligning can be easily achieved by inserting the fiber in a precise capillary (ferrule or sleeve) with an outer diameter equal to the GRIN diameter, then butt-joining capillary and GRIN by means of a V-groove aligner or another precise cylinder to form an assembled unit. In addition to GRIN lenses, NSG also provides lens and fiber housings, made of steel, suitable for several standard fiber and lens types. Should a completely dielectric assembly be required, glass ferrules with extremely precise mechanical tolerances can be purchased from Nippon Electric Glass (NEG), which also offers non-standard capillaries to the customer's design.[8]

## 3. FIBER OPTIC SENSORS: BASICS

A fiber optic sensor (FOS) is a device in which the light guided by the fiber undergoes a modulation in response to an external perturbation.[9] Broadly speaking, FOSs are classified according to the optical property of the guided light that is modulated, i.e., intensity, phase, wavelength and state of polarization. Fiber optic sensors based on the intensity modulation technique will be considered here as follows. Depending on the sensing mechanism, these sensors can be additionally classified as spectral, intrinsic, and extrinsic:

- *Spectral* FOSs are *direct sensors*, performing direct analyses without the necessity for actually taking a sample. These sensors are a sort of spectrophotometers, operating at suitable wavelengths, in which the optical part has been modified with a fiber optic link for highly-localized measurements. Absorbance and fluorescence analyses need a specific measuring cell at the fiber-end, so as to establish the optical path and to provide backtransmission. Reflectance analyses, on the contrary, are performed using bare fibers, the numerical aperture of which determines the view angle inspected.

- *Intrinsic* FOSs use the fiber itself as the sensing element, and are generally *indirect sensors* in which modulation is induced through a modified portion of core or cladding. The natural cladding of the fiber can be chemically treated, or stripped away and replaced with a layer, so that the core-cladding interface acts as the sensing element and evanescent-wave spectroscopy is performed. The core can be suitably doped, so that a modulation of absorption or fluorescence occurs.

- *Extrinsic* FOSs are *indirect sensors*, making use of an additional sensing element, fixed at the fiber end, which causes optical modulation of the guided light. Many FOSs are of this type, since it is relatively easy to find materials that can change their own optical characteristics (i.e. refractive index, reflectivity, absorbance, fluorescence) by interacting with the parameter being tested.

## 4. PROXIMITY SENSORS AND APPLICATIONS

A mirror or a diffusive target moving in front of an optical fiber is the transducer for an optical fiber proximity sensor. It can be extended to the monitoring of position, displacement, pressure, temperature and vibration, or any other parameter influencing the target position with respect to the fiber end-face.[10] A GRIN lens interposed between the fiber and the target allows for obtaining a wider sensitive range, together with a custom distance between the optical head and the target.

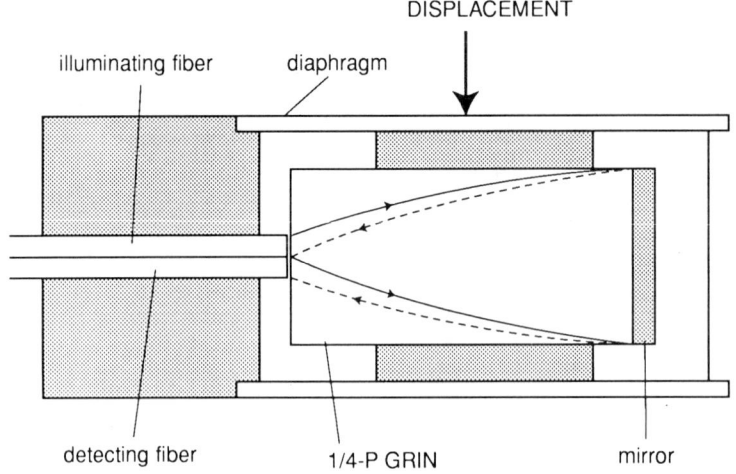

Fig. 2 General Purpose proximity sensor.

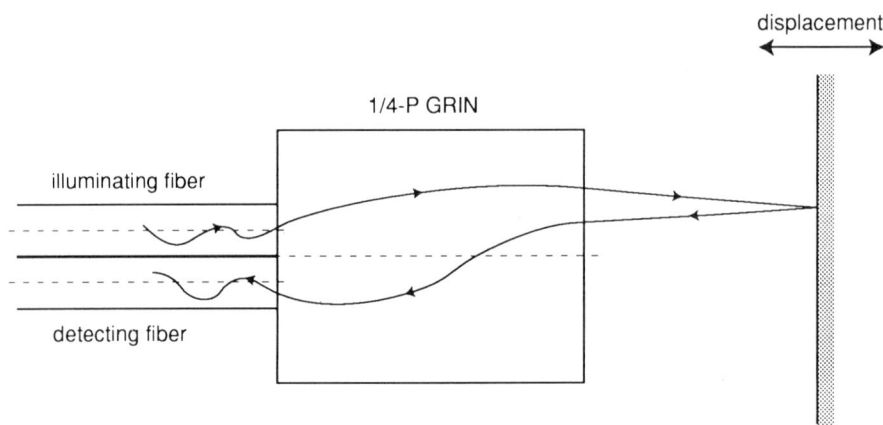

Fig. 3 Pressure sensor.

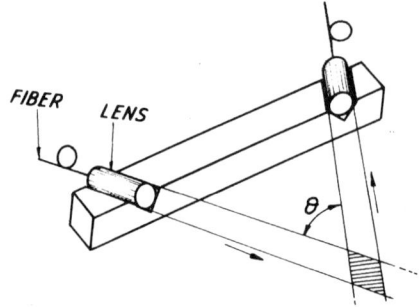

Fig. 4 Proximity sensor with two fibers and two GRIN-rod lenses.

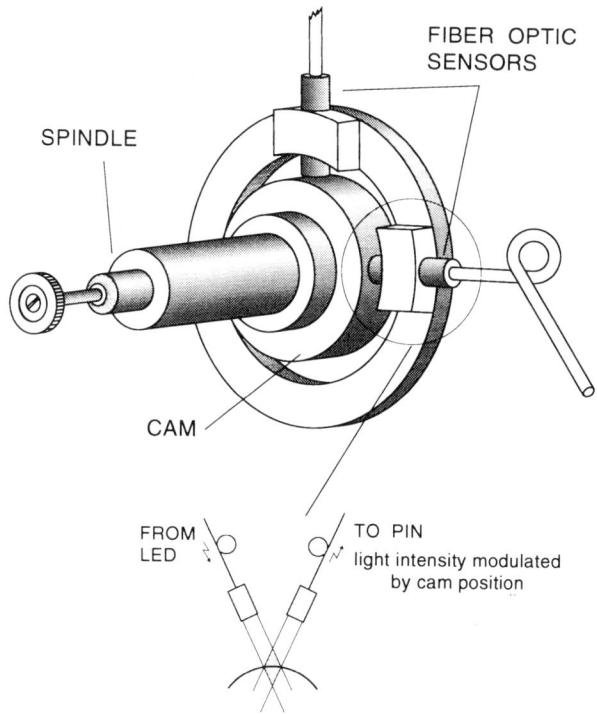

Fig. 5 Robotics applications: fiber optic proximity sensors to check the spindle-stop position.

Fig. 2 shows a typical example of a GRIN-based proximity FOS making use of 50/125 graded-index fibers and of a SLW-2.0-0.25-P GRIN lens which exhibits a linear response for mirror displacements in the 5-8 mm range, with a micrometric sensitivity.[11] Another sensor developed for displacement measurements, applied to pressure measurements, exploits the relative motion of a GRIN positioned in front of the fiber link, as shown in Fig. 3. The fibers are kept in a fixed position and are symmetrically located with respect to the 1/4-P GRIN with a reflective end-face that is fixed to a mobile holder.[12] The light from the illumination fiber undergoes a folded half-pitch path and is refocused in the receiving fiber, as the light intensity is modulated by the relative transverse displacement between the GRIN and the fibers. A sensitivity of 2.5 nm and a dynamic range of 25 µm have been obtained using 50/125 fibers and a SLS-2.0 GRIN; a better sensitivity is foreseen with the use of singlemode fibers.

A different approach, illustrated in Fig. 4, allows for better shaping the distance and the dimension of the sensing range. It uses two GRIN-ended fibers: the first one provides the illumination light cone, while the second one identifies a detection view-cone. When a target is placed in the illumination-detection overlapping region, the scattered and/or reflected light is detected and target motion induces an intensity modulation. The optical characteristics of both GRINs and fibers, together with the mutual geometrical position of the GRINs, determine the position and the shape of the sensing region.[13] A geometrical arrangement with θ = 60° together with HCS fibers and SLW-1.8 GRINs made it possible to obtain a sensing zone located at 10 mm distance from the optical head. A linear response was obtained in a 1.6 mm range with a sensitivity of 35 µm.

The optical head of Fig. 5 with a proper geometrical arrangement of the GRINs has been used in robot applications to check the correct stop-position of a fast spindle. A

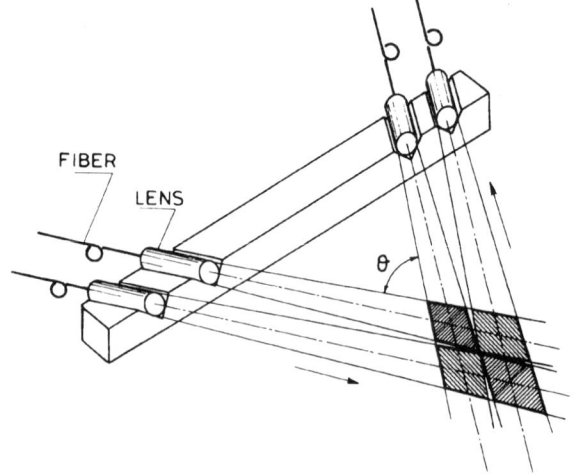

Fig. 6 Two proximity sensors giving four slightly-overlapped sensing-zones.

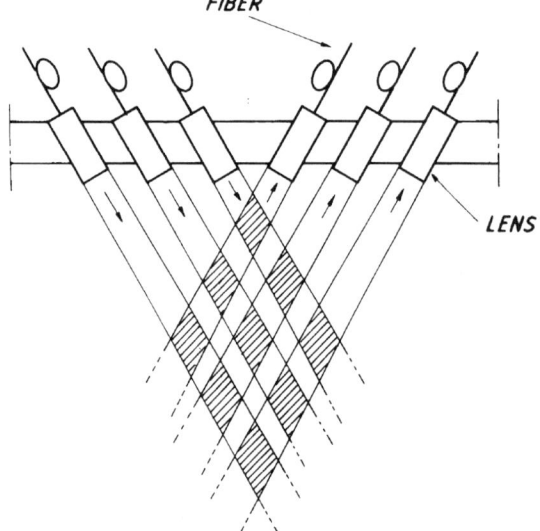

Fig. 7 Array of proximity sensors.

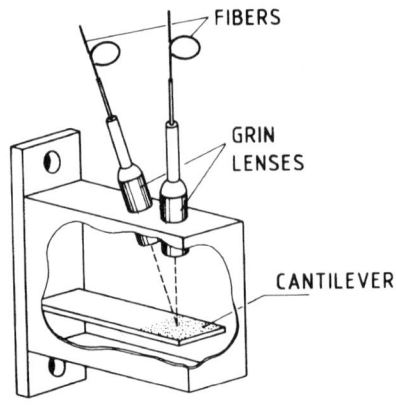

Fig. 8 Fiber optic vibration sensor.

circular cam, eccentrically fitted to the axis of the spindle, is used to monitor the angular position of the spindle. The cam motion is controlled by two fiber-optic proximity sensors, of the Fig. 4-type, angularly positioned with respect to the cam in order to generate two signals with 90° relative phase shift. Cam eccentricity produces displacement of the cam edge in the range of the sensors' linear response. For each sensor, the intensity of the light scattered by the cam edge is collected by the fiber connected to the PIN whose electric output represents the sensor output. The correct spindle stop position is achieved by processing the sensor outputs and closing the loop with the feedback on the spindle drive motor. With repetition of the startup/stop procedure, the preset spindle stop position was attained in less than 5 seconds, with a stop position repeatability of better than 0.5°.

More extensive spatial information than that given by the sensor of Fig. 4 can be obtained by means of an array of suitably-positioned GRIN-ended fibers. Possible arrangements of the sensing head are illustrated in Figs. 6 and 7. Optical heads with θ =60° have been utilized with different spacings between the lenses, giving separated or slightly-overlapping sensing zones which were located at various distances from the optical head, from 45 mm to 80 mm. These sensors made possible proximity measurements with different amplitudes of the sensing range and sensitivities from 25 to 60 µm.

In addition to the robotics applications discussed above, the optical head of Fig. 4 is the basis for a wide variety of FOSs, since there can be many transducers positioned inside the sensing zone, as shown in Fig. 8. A diffusive-surface elastic-cantilever oscillating inside the sensing zone made possible the development of a FOS for vibration or vibration-and-temperature monitoring at the fixed frequency of 100 Hz, the vibrations of the stator end-windings in an electric power station. The optical head used SLW-1.8 GRINs and HCS fibers positioned at a distance between their centers of 2.5 mm, and an angle between their axes of 13°, so as to locate the sensing zone at a distance of 10 mm from the optical head. By exploiting cantilever motion in the range of linear response, only vibration amplitude was measured in the 0-250 µm peak-amplitude range with a 0.1 µm resolution.[14] The additional temperature monitoring can be achieved by exploiting the temperature-induced sensor-case expansion and the cantilever motion in the parabolic range of response, with a reduction however in the vibration sensitivity.[15]

## 5. SENSORS BASED ON STATIC AND DYNAMIC LIGHT-SCATTERING MEASUREMENTS

Being able to characterize fluid-suspended particles from fractions of a micron to a hundred microns is becoming increasingly important in numerous fields, including environmental, industrial, biochemical and medical applications. A particulate can be characterized by light scattering techniques, since its characteristics (size, shape, particle mobility and molecular interaction) determine those of the light that it scatters (intensity, angular distribution and fluctuations of the intensity, polarization, and frequency displacements). The particulate's characteristics are thus determined using inversion techniques to analyze the scattered light. The intensity of the light incident at an observation point, scattered by a certain number of particles, can be expressed as the sum of a constant term $I$, and a term that fluctuates in time as a result of particle mobility, $I(t)$,[16]

$$I(t) = I + I(t) . \qquad (4)$$

The term $I(t)$ averages zero, so that the mean intensity of the particle-scattered light coincides with the first constant term $I$. Static light scattering (SLS) techniques are based on the measurement of the light's mean intensity, by which characteristics such as size and

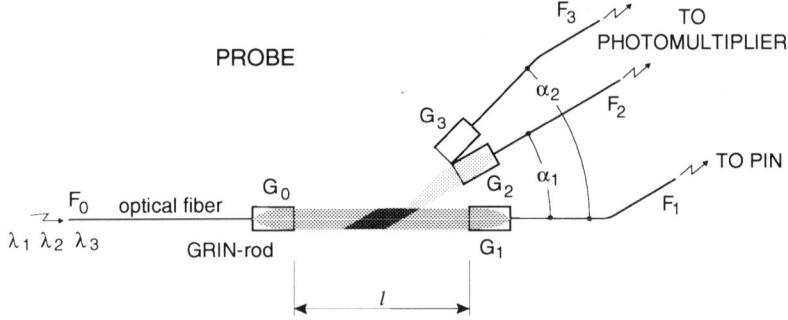

Fig. 9. SLS-based fiber optic sensor for combined measurement of particle size, density and feractive ondex.

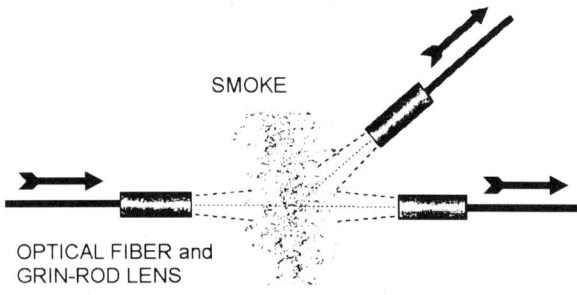

Fig. 10. SLS- based fiber optic smoke sensor.

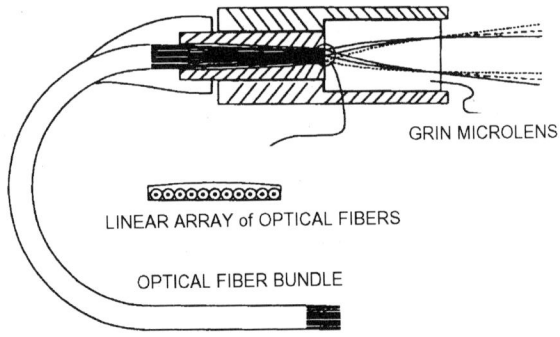

Fig. 11. Scattering volume created by means of optical fibers and ¼-$P$ GRIN lens.

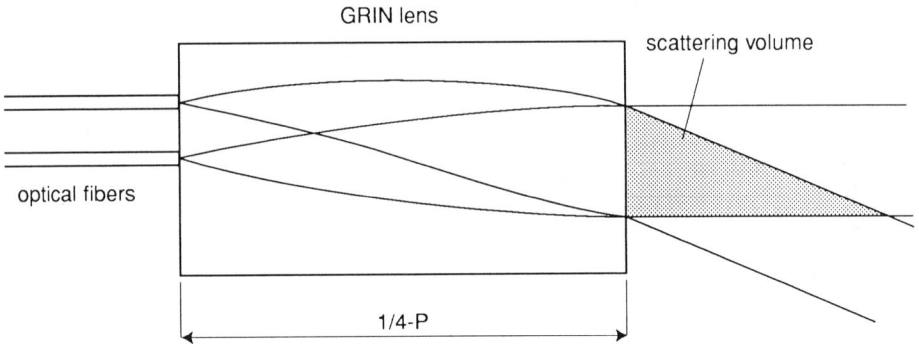

Fig. 12. DLS probe using a linear array of fibers.

molecular weight in particle equilibrium conditions can be determined. Dynamic light scattering (DLS) techniques are based on the measurement of intensity fluctuations caused by particle motion; characteristics such as size and velocity distribution can be determined in relation to the particle mobility.

The geometric versatility of optical fibers enormously facilitates measurements in highly localized, hazardous, hard-to-access, and/or electromagnetically-disturbed sites. In addition, their low attenuation enables remote detection and real-time signal processing, while reducing alignment problems. Most of the FOSs based on SLS or DLS measurements make use of GRIN lenses, since these lenses are the most suitable, compact and easily connectable optics for fiber-beam modification. In general, since light-scattering measurements are performed by means of collimated beams, 1/4-$P$ lenses are used.

The SLS technique, using the optical head illustrated in Fig. 9, has made possible the combined measurement of particle size, density and refractive index of a monodisperse particulate. Four identical components were used, $F_0G_0$, $F_1G_1$, $F_2G_2$, $F_3G_3$, each comprising a 50/125 optical fiber coupled to a SLW-1.8-0.25-$P$ GRIN lens. $F_0G_0$ and $F_1G_1$ were placed along the same axis, while $F_2G_2$ and $F_3G_3$ were positioned with respect to $F_0G_0$ to form angles between their relative axes of $\alpha_1 = 30°$ and $\alpha_2 = 45°$, respectively. Fiber $F_0$ provided sequential illumination at three wavelengths $\lambda_1 = 543$ nm, $\lambda_2 = 670$ nm and $\lambda_3 = 830$ nm, respectively. In practice, scattering and transmission measurements at multiple wavelengths and multiple detection angles were performed. The use of GRIN-ended optical fibers was of fundamental importance in order to optimize the dimension of the illumination and detection beams. Calibration of the sensor by means of standard polystyrene microspheres showed the possibility of measuring particle density in the $10^6$-$10^9$ pp/cm$^3$ range, particle dimension in the 0.3-1.0 µm range, and refractive index of the fluid where particles were suspended in the 1.33-1.43 range.[17,18]

A much simpler FOS, utilizing a single wavelength and the simplified optical head of Fig. 10, was tested for smoke detection with a sensitivity of $10^8$ part/cm$^3$. The presence of the smoke in the sensing zone provided light scattering, the detection of which could be used for safety purposes to trigger an alarm at a certain threshold value. This optical head made possible not only smoke detection, but also the control of the correct sensor operation. In fact, in addition to the GRIN-ended fiber for light-scattering detection, another GRIN-ended fiber was used which detected possible failures in the light-delivery system.[19]

As far as DLS measurements are concerned, it must be noted that the use of monomode optical fibers is the fundamental factor, because these are able to provide small numerical apertures and dimensions, hence an extremely small angle of coherence. Consequently, the spatial coherence factor is close to the ideal value of 1, which would be difficult to obtain with a bulk optics system.[20] Moreover, the use of a bundle of optical fibers coupled to a 1/4-$P$ GRIN lens enhances the versatility of the DLS measurements, as shown in Fig. 11. One fiber serves to illuminate the sample, while another fiber collects the scattered light. In addition to collimating the light beam from the fiber connected to the source, the 1/4-$P$ GRIN lens tilts the beam by an angle related to the fiber's radial distance on the lens surface. The same thing occurs in the collecting fiber beam, and each receiving fiber collects the scattered light at a characteristic angle, according to the position of the fiber on the lens surface. This creates a scattering volume so that each collecting fiber receives only the light scattered by the particles inside a certain volume. Therefore, by properly selecting pairs of transmission/collection fibers, the light scattered can be measured close to backscattering, which is the most convenient situation. In addition, the proper pair selection makes it possible to fit the scattering volume to be anlyzed, thus enhancing the precision of the measurement.

Fig. 12 shows a probe designed for the real-time monitoring of polymerization processes by means of a linear array of eleven single-mode fibers coupled to a GRIN

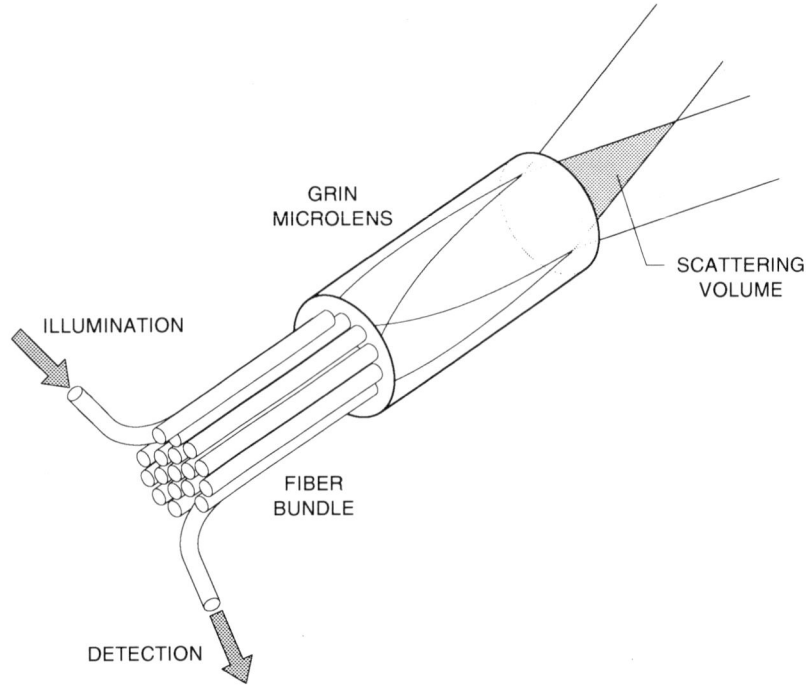

Fig. 13. DLS probe using a circular array of fibers.

microlens.[21] The central optical fiber provides the illumination beam, while the lateral fibers detect the light scattered in the range 150°-167°. Another DLS probe using a circular array of singlemode fibers is illustrated in Fig. 13 .[22] Two different fiber-arrays were tested, both giving satisfactory results. The first array consisted of 19 fibers with a central fiber and two contiguous rings of 6 and 12 fibers, making possible measurements in the scattering-angle range 172°-178°. The other array had 18 fibers with two central fibers and an external 16-fiber ring, making possible measurements in the much wider scattering-angle range 159°-176°. The DLS probes making use of the circular array presented some novel features, the most relevant of which was the demonstrated capability of performing measurements in small drops of colloidal suspensions. Both DLS probes, based on either planar or circular fiber-arrays, demonstrated a coherence factor higher than 0.8, very close to the ideal value.

## 6. ABSORBANCE MEASUREMENTS BY MEANS OF OPTRODES

There are countless extrinsic FOSs making use of a chemical reaction region positioned at the distal end of an optical fiber which provides the intensity modulation of the guided-light as a function of the physical or chemical parameter to be measured. Such probes are called *optrodes*, from a combination of the terms *opt*ical and elect*rode*. Many optrodes are based on absorption measurements performed on solid- or liquid-phase chemistry. Fiber optic absorption measurements can be approached by means of two different schemes of fiber-probe connections:

- a twin-fiber connection, using one fiber for lighting, the other fiber for detection, and the transducer in between;

- a single-fiber connection, using a single fiber for both lighting and detection, together with a coupler for separating the two channels. Solid transducers are generally glued to the fiber end-face, while liquid transducers are confined in membranes or suitable casings wrapping at the end of the fiber. In both cases, a reflecting surface is necessary to provide the backtransmission of the intensity-modulated light.

Due to the thickness of the transducer and because of the fiber's numerical aperture, a light-intensity loss is introduced by the presence of the transducer, independently of its own absorption properties. The intrinsic loss of an empty probe is defined as:

$$L = -10 \, Log \frac{I_{out}}{I_{in}} \tag{5}$$

where $I_{in}$ is the light power illuminating the probe and $I_{out}$ is the light power guided to the detector after probe transmission. Such a loss represents a critical parameter when designing FOSs based on optrodes, especially those using a single-fiber connection. In fact, because of the presence of the coupler, such a scheme is characterized by a critical sensor efficiency which dramatically decreases as the probe loss increases. The following discussion will take into account only probes designed for the single-fiber scheme and using multimode fibers, representing a frequent sensing condition.

An optrode consisting of a transducer butt-coupled to the fiber is characterized by a high intrinsic loss, and provides very poor sensor efficiency. Independently of transducer absorption, a transducer having a thickness of 0.5 mm produces an intrinsic loss of between 8 and more than 15 dB, according to whether large or small core fibers are used. When a transducer with a thickness of 5 mm is used, the dramatic loss of more than 20 dB occurs. The insertion of a 1/4-P GRIN lens between the fiber and the transducer, as illustrated in Fig. 14, enormously reduces probe intrinsic-loss. Also, for a transducer having a thickness of 5 mm, probe loss can be less than 3 dB. This kind of probe has been used also as a general-purpose cell for absorption measurements, leaving the interspace between the GRIN and the mirror empty.[23] Unfortunately, this probe scheme has a problem which cannot be overlooked: even a slight tilting of the reflecting surface causes an additional and noticeable loss.[24] For example, by using SLS-2.0 GRINs, a mirror tilting of 0.5° with respect to the optical fiber axis produces an additional loss in the 1-3 dB range, depending on the fiber dimension.

Fig. 15 shows a probe for FOSs which is able to minimize the intrinsic loss even in the presence of slight tilting or misalignment of the optical components. Two facing 1/4-P GRIN lenses confine the transducer: the first GRIN is cemented to the optical fiber, while the other has a reflecting end-face. By considering the paraxial ray approximation, the position and the orientation of an output meridional ray, $r_{output}$ and $\theta_{output}$, are related to the input parameters, $r_{input}$ and $\theta_{input}$ by the expressions

$$\theta_{output} \approx \theta_{input} - 2h\sqrt{A} - 2Ar_{input} \, DN_0/n_t \qquad r_{output} \sim r_{input} \tag{6}$$

where $D$ and $n_t$ are the transducer thickness and refractive index, respectively; $h$ is the axial offset between the GRINs, and the approximations

$$h\,tg\alpha \ll D \qquad r_{input} \ll 1/\sqrt{A} \tag{7}$$

have been assumed, where $\alpha$ is the GRIN's angular misalignment. It turns out that output and input positions are nearly the same, independently of the angular offset between the

322

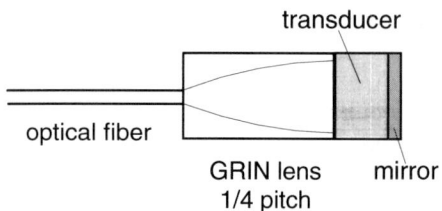

Fig. 14 Optrode for FOSs minimizing intrinsic loss.

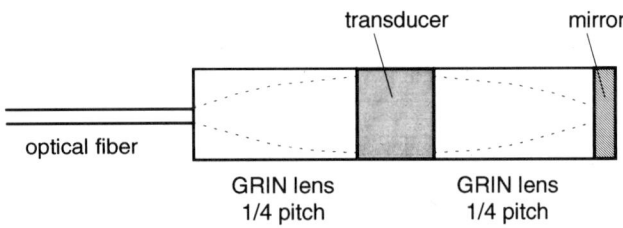

Fig. 15 Probe for Foss based on an autofocusing scheme.

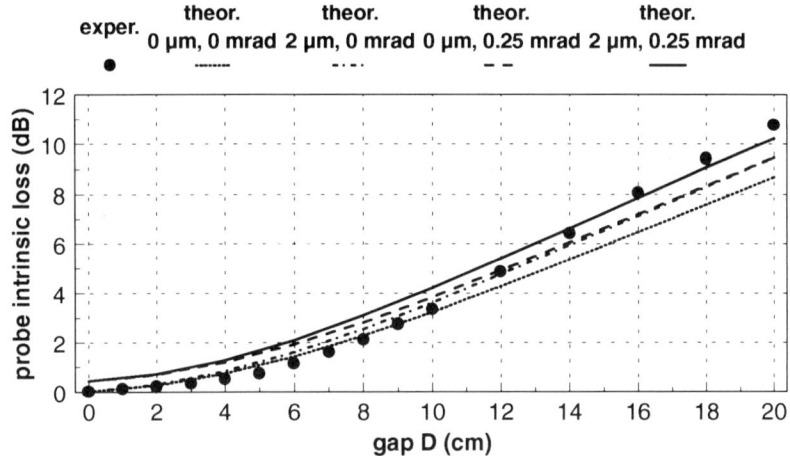

Fig. 16 Intrinsic loss of the reflective-type breadbord probe for gas-absorbance measurements: experimental results compared with theoretical expectations.

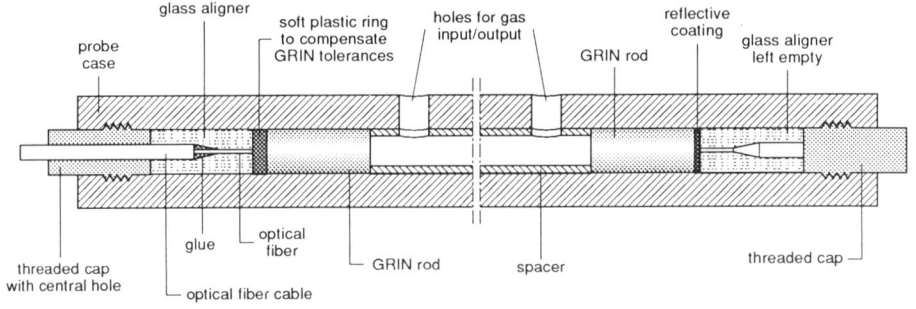

Fig. 17 Microoptics gas-cell for gas absorption measurements.

GRINs; the ray orientations are only slightly different, so that the probe can be considered nearly autofocussing.[25, 26]

The nearly-autofocussing probe has been used for the implementation of a fiber optic temperature sensor which utilizes a thermochromic solution as transducer having a thickness of 5 mm and a refractive index of 1.377. Lenses and transducer were confined in a precise-hole glass microcapillary, and an air bubble was left at the probe bottom so as to compensate for solution volume change due to temperature variations. By using 100/140 silica fibers and SLS-1.0-0.63-0.24-$P$ lens the intrinsic probe loss was $\approx$ 2 dB, with an experimental sensor efficiency of 63% (theoretical 82%). For comparison, a similar probe with the transducer butt-coupled to the fiber should have an intrinsic loss of more than 24 dB and a very poor theoretical sensor efficiency of 0.5%.

## 7. MICROOPTICS PROBE FOR GAS-ABSORBANCE MEASUREMENTS

Taking its inspiration from the design of the self-focussing probe shown previously, a microoptic probe for gas absorption-measurement has been implemented, that makes use of singlemode optical fibers and GRIN lenses. Such a probe can be used for the remote monitoring of explosive and toxic vapours, a pressing need in many industrial and mining complexes.[27, 28] The current availability of reliable and powerful semiconductor laser diodes having emission in the 0.7-2.0 µm spectral range makes it possible to detect many gases by direct absorption-spectroscopy, since combination or overtone bands of the fundamental molecular vibrations are present in the near infrared spectral range (NIR).[29, 30] The combined use of fiber optics and laser diodes amplifies the potentialities of both, making possible the real-time, continuous and localized analysis of many gases: for example, CO, $CO_2$, $SO_2$, HCl, HI, HCN, $NH_3$, $NO_x$, $CH_4$, $C_2H_4$, $C_2H_2$, $C_6H_6$, and others. The all-optical approach presents relevant advantages such as a high degree of safety especially in hazardous conditions, together with the possibility of implementing all-optical sensor-networks.

Two kinds of probes were modelled, the autofocussing one and another scheme based on two facing 1/4-$P$ GRINs, each lens coupled to the singlemode fiber so as to perform transmission measurements. The intrinsic loss for each kind of probe was theoretically evaluated as a function of gap $D$, by considering the GRIN's axial and angular misalignments. The theoretical evaluation was obtained by means of an optical design software (Solstis-Optis, Gaussian and Photonics modules) that assumes the following conditions: illumination wavelength $\lambda$ =1.3 µm, singlemode optical fibers @ 1.3 µm providing an input Gaussian beam with a waist of 10 µm, SLW-3.0-0.25-$P$ GRIN-lenses, 100%-reflectivity mirror.

The theoretical probe losses evaluated for the transmissive and reflective probes, assuming 0, 2, 4 and 6 µm fiber-GRIN axial offsets, showed that the reflective probe is much more efficient than the transmissive one, except in the case of gaps greater than 10 cm and offsets larger than 4 µm. Also, by assuming 0, 0.25 and 0.5 mrad GRINs' angular offsets, the reflective probe was more efficient, especially for 0.5 mrad tilt. In addition to low-losses, the reflective probe offered a much more compact design with respect to the transmissive one, necessitating, however, a high-quality fiber coupler or an isolator for reducing the crosstalk signal from the illumination fiber. A breadboard probe was used to experimentally verify the calculated intrinsic-loss of the reflective probe as a function of gap $D$.

Fig. 16 shows the experimental results in comparison with the theoretical curves that best fit the experimental curve. It is notable that a probe with a gap of 5 cm, making possible absorption measurements over an optical path-length of 10 cm, was characterized by an intrinsic loss of 0.75 dB, and probe with a gap of 10 cm still presented a low loss, that is, 3.3 dB.

In addition to the breadboard assembly, a probe was implemented as demonstrator, manufactured according to the design illustrated in Fig. 17. The probe case housing all the optical components was a brass tube with threaded inner edges and an external diameter of 10.5 mm and an internal diameter of 3.0 mm. A thin steel sheath was positioned at the center of the probe in order to keep the GRIN lenses spaced at the desired distances. The probe case and the spacer had holes on their side surfaces, so as to provide input and output channels for gas venting. The optical elements were butt-coupled by compression, without the use of glue, by means of two plastic caps (Delrin or PVC) which were screwed to the inner edges of the probe case.

Two probes were manufactured, intended for field-measurements of methane at 1.3 µm. Since only alarm measurements were supposed, capable of measuring 2% LEL, short pathlengths were considered. In particular, gaps between the lenses of 1.25 cm and 2.5 cm were tested, thus making possible absorption-measurements with optical paths of 2.5 cm and 5.0 cm, respectively. The intrinsic losses were 0.6-1.0 dB and 0.7-1.1 dB for the two probes, in fairly good agreement with the theoretical expectations.

## 7. CONCLUSIONS

The overview presented here has demonstated that many different FOSs gain benefit from the use of GRIN-rod lenses. Compactness, lightness, easy connection and also relatively low-cost make GRIN an irreplaceable tool when designing a FOS. All the sensors presented made use of commercially-available standard GRIN, especially the 1/4-*P* type; in case specific working wavelengths or lens-lengths are necessary, however, it is possible to purchase custom components. In addition, by changing the fiber position with respect to the GRIN, as well as the lens length, it is possible to model the divergence of the lens emerging beam, the size, and, consequently, the dimension of the Gaussian image-planes.[31]

## REFERENCES

1. E.W. Marchand, *Gradient Index Optics,* 1978, Academic Press, New York
2. K. Iga, Theory for graded-index imaging, *Appl. Opt.* 19:1039 (1980).
3. A. Sharma, D.V. Kumar, and A.K. Ghatak, Tracing rays through graded-index media: a new method, *Appl. Opt.* 21:984 (1982).
4. A. Sharma, Computing optical path length in gradient-index media: a fast and accurate method, *Appl. Opt.* 24:4367 (1985).
5. C. Gómez-Reino, E. Larrea, M.V. Pérez, and J.M. Cuadrado, Imaging, transforming, and modal propagation parabolic gradient-index rod, *Appl. Opt.* 24:4379 (1985).
6. J.R. Flores, C. Gómez-Reino, E. Acosta, and J. Liñares, Geometrical optics of gradient index lenses, *Opt. Eng.* 28:1173 (1989).
7. Nippon Sheet Glass Co. Ltd., Fiber Optics Division, Sumitomo Fudosan Shiba Bldg., 1-11-11, Shiba, Minato-ku, Tokyo 105, Japan.
8. Nippon Electric Glass Co. Ltd., Miyahara 4-chome, Yodogawa-ku, Osaka 532, Japan.
9. B. Culshaw, and J. Dakin Eds., *Optical Fiber Sensors - vol. 1: Principles and Components - vol. 2: Systems and Applications*, 1988, Artech House Pbl
10. See Ref. 9, volume 2, chapter 17 *Physical and Chemical Sensors for Process Control,* pp. 653-699.
11. S.D. Cusworth, J.M. Senior, A reflective optical sensing technique employing a GRIN rod lens, *J. Phys. E: Sci. Instrum.* 20:102 (1987).
12. P.J. Murphy, and T.P. Coursolle, Fiber optic displacement sensor employing a graded index lens, *Appl. Opt.* 29:544 (1990).

13. M. Brenci, G. Conforti, A. Mencaglia, A.G. Mignani, and A.M. Scheggi, Fibre-optic position sensor array, *Int. J. Optoelectr.* 3:473 (1988).
14. G. Conforti, M. Brenci, A. Mencaglia, and A.G. Mignani, Fiber optic vibration sensor for remote monitoring in high power electric machines, *Appl. Opt.* 28:5158 (1989).
15. M. Brenci, A. Mencaglia, and A.G. Mignani, Fiber-optic sensor for simultaneous and independent measurement of vibration and temperature in electric generators, *Appl. Opt.* 30:2947 (1991).
16. B. Chu, *Laser Light Scattering*, 1991, Academic Press, New York.
17. M. Brenci, D. Guzzi, A. Mencaglia, A.G. Mignani, and M. Pieraccini, Quasi-monodisperse particulate characterization with optical fibers and a three-wavelength scattering technique, *Sens. Act. B* 29:115 (1995).
18. M. Brenci, D. Guzzi, A. Mencaglia, A.G. Mignani, and M. Pieraccini, An optical-fiber sensor for the measurement of the size and density of monodisperse particulates, *Sens. Act. A* 48:23 (1995).
19. M. Brenci, D. Guzzi, A. Mencaglia, and A.G. Mignani, Fibre-optic smoke sensor, *Sens. Act. B* 7:780 (1992).
20. A.J. Macfayden, and B.R. Jennings, Fibre-optic systems for dynamic light scattering-a review, *Opt. Las. Tech.* 22:175 (1990).
21. H.S. Dhadwal, and R.R. Ansari, Multiple fiber optic probe for several sensing applications, *SPIE Proc. Fiber Optic and Laser Sensors IX*, 1584:262 (1991).
22. M. Brenci, A. Mencaglia, A.G. Mignani, and M. Pieraccini, "A circular-array optical fiber probe for backscattering photon correlation spectroscopy measurements", *Appl. Opt.* 35:6775 (1996).
23. D.A. Landis, C.J. Seliskar, and W.R. Heineman, Fiber-optic-graded-indes-lens absorbance sensor with wavelength-scanning capability, *Appl. Opt.* 33:3432 (1994).
24. G. Conforti and M. Brenci, Power loss in optical-fiber graded-index-rod components, *Opt. Lett.* 13:59 (1988).
25. G. Conforti, M. Brenci, A. Mencaglia, and A.G. Mignani, Fiber-optic thermometric probe utilizing GRIN lenses, *Appl. Opt.* 28:577 (1989).
26. A.G. Mignani, M. Brenci, and G. Conforti, Nearly autofocusing cavity for fiberoptic sensor probes, *SPIE Proc. Micro-optics*, 1014:42 (1988).
27. S.O. Ryding, *Environmental Management Handbook*, 1992, IOS Press.
28. C.P. Straub, *Practical Handbook of Environmental Control*, 1990, CRC Press.
29. D.E. Cooper, and R.U. Martinelli, Near-infrared diode lasers monitor molecular species', *Laser Focus World* 28:133 (1992).
30. R.U. Martinelli, R.J. Menna, D.E. Cooper, C.B. Carlisle, and H. Riris, Near-infrared InGaAs/InP distributed-feedback lasers for spectroscopic applications, *SPIE Proc. Laser Diode Technology and Applications* VI, 2148:292 (1994).
31. D.A. Landis, and C.J. Seliskar, Fiber-optic/GRIN lens coules for use in chemical spectroscopy, *Appl. Spectr.* 49:547 (1995).

# INTEGRATED OPTICAL INSTRUMENTATION FOR FIBER GRATING SENSORS

M. Varasi

ALENIA Research Department
Rome, Via Tiburtina km 12,4 , 00131 Italy

## 1. INTRODUCTION

The objective of this Chapter is the development of fiber compatible, portable, rugged and low cost instrumentation for the processing of the optical signals from Fiber Bragg Grating (FBG) sensors embedded in composite materials in order to monitor local static and dynamic stresses and strains of the structure. The FBG sensor is a grating realised by UV holography[1] in an optical fiber that is monomodal at 1300 nm, and reflects radiation whose wavelength satisfies the Bragg relation: $\lambda_B = 2n\Lambda$ (n is the effective refractive index of the fiber, $\Lambda$ is the period of the grating)(Fig. 1). The fiber is positioned between two plies of the composite material structure and embedded by the curing process (Fig. 2), with input-output in the external surface of the structure. In this configuration the fiber tracks the stress status of the structure in which it is embedded, and compressive or tensile stresses parallel to the fiber axis at the location of the FBG induce changes of $\lambda_B$ in accordance with the relation:

$$\Delta\lambda_B / \lambda_B \approx (1-p_e)\varepsilon \qquad (1)$$

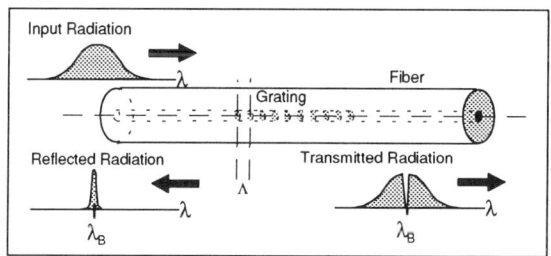

Fig. 1: Fiber Bragg Grating (FBG) sensor operating mode

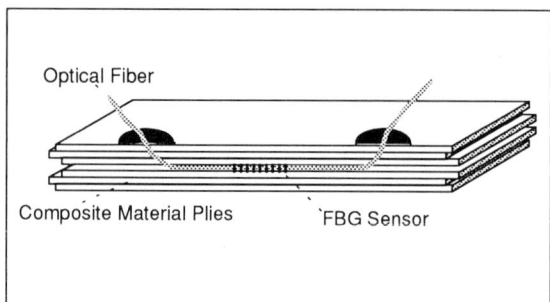

Fig. 2 Sensorised optical fiber in the composite material

where $p_e$ is the effective photoelastic constant of the fiber and ε is the deformation measured in μstrains. The measurement of this wavelength shift enables the FBG sensor to measure the structures static and dynamic stresses and strains at the location of the sensor.

In the generic configuration of the optical measurement system, optical radiation from a broadband source (Super Luminescent Diode) is coupled into the fiber and sent to the FBG sensor. The reflected optical signal is sent to the instrumentation through an "x" coupler. The instrumentation is used to measure the value of $\lambda_B$ and its change when the composite structure is loaded.

The conventional instrumentation includes an Optical Spectrum Analyser (OSA) based on an air path grating monocromator. This instrument dramatically limits the industrial application of the system because of its weight, its volume, its sensitivity to vibrations and its low scanning speed. The objective of this work is the development of instrumentation able to overcome these limitations, possibly fully integrated with the optical fiber. The result is an instrument based on an Acoustic Tunable Optical Filter (ATOF) integrated in a planar optical circuit on LiNbO$_3$ and named the Integrated Optical Spectrometer (IOS).[3]

The IOS device is based on an Acousto Optical Tunable Filter [4] realised in X cut LiNbO$_3$. The fabrication processes and the improvements obtained in resolution, sidelobe ratio, and polarisation insensitivity [5] are discussed in detail.

IOS based instrumentation will be also described in detail adopting three main operating modes of the IOS:[6] as spectrometer, as discriminator and as tracking filter.

The results obtained fully comply with the industrial application requirements in terms of speed and sensitivity of the IOS, allowing IOS based instrumentation to exceed the requirements of aeronautical applications and allow the monitoring of low level high frequency vibrations.

## 2. INTEGRATED OPTICS AOTF TECHNOLOGY

The IOS device is based on an Acousto Optical Tunable Filter realised in X cut LiNbO$_3$, in which the monomode optical waveguide, the acoustical waveguide, the SAW (Surface Acoustical Wave) transducers, the input-output optical polarisers and the interfaces with the fibers are integrated.

The AOTF circuit exploits a collinear interaction between the guided optical radiation and the collinear surface acoustical wave, both propagating parallel to the Y crystal axis. The interaction causes the transfer of optical power between the two polarizations of the guided optical radiation, when the following phase matching relationship is satisfied:

$$\vec{K}_{TM} - \vec{K}_{TE} = \vec{K}_A \qquad (2)$$

Eq. (2) is the vectorial description (momentum conservation) of the collision between the input photon ($\vec{K}_{TE}$ or $\vec{K}_{TM}$) and phonon ($\vec{K}_A$) to produce the output photon ($\vec{K}_{TM}$ or

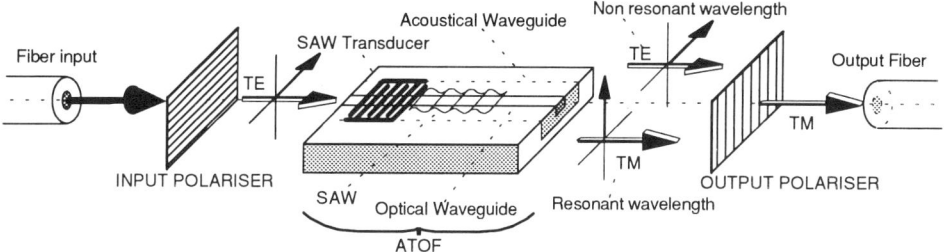

Fig. 3 Schematic view of the components integrated in the IOS planar circuit

$\vec{K}_{TE}$). Because of the collinear nature of the collision it is possible to obtain from the scalar version of the Eq. (2) the relation:

$$\lambda_p = V_{ac}(n_{TM} - n_{TE}) / f \qquad (3)$$

where $\lambda_p$, the peak wavelength in which the interaction has the maximum efficiency, is related to the frequency f of the signal exciting the acoustical wave, $n_{TM}$ and $n_{TE}$ being the effective refractive indexes of the two waveguided optical polarizations TM and TE. When the ATOF is included between two crossed polarisers, as shown in Fig. 3, the transmission function of the system is a sinc $^2$ function centred at $\lambda_p$ and the device acts as an optical filter tunable by frequency exciting the SAW with a FWHM given by the following expression:

$$\Delta\lambda_p \approx 0.8 \, \lambda_p^2 / L\Delta n \qquad (4)$$

where L is the length of the acousto-optical interaction region and $\Delta n = n_{TM} - n_{TE}$.

The corresponding integrated optical circuit realised in X cut LiNbO$_3$ is schematically shown in Fig. 4. The optical waveguide was realised by diffusion at 1050°C for 8 hours, of a 75 nm thick Titanium strip whose width is 6 µm, this allowing a monomode regime at 1300 nm for both the polarised modes. The Titanium strip was evaporated by an electron gun source and defined by a lift-off process. The acoustical waveguide has been included in order to minimise the diffraction dispersion of the acoustical radiation and thus reduce power consuption and increase the acousto-optical interaction length; this allows obtaining high filter resolution. The acoustical waveguide was realised by increasing the acoustic propagation velocity ($\Delta v / v \approx 0.01$) in the areas around the guiding channel (width = 100 µm),

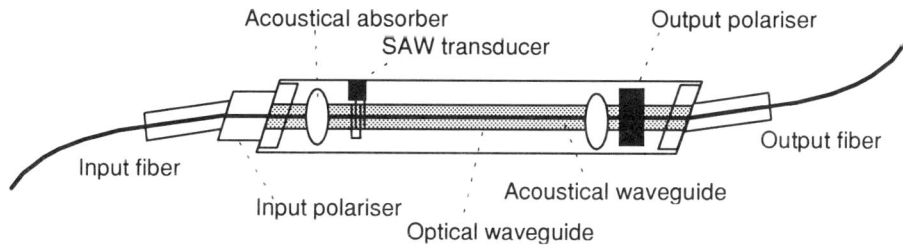

Fig. 4 Schematic view of the LiNbO$_3$ integrated optical circuit

by high temperature diffusion (1050°C, 30 hours) of a thin (120 nm) layer of Titanium. The interdigital SAW transducer was realised with a thin (100 nm) Al film and a period of 16 μm, corresponding to a central frequency of 210 MHz. The acousto-optical interaction region was delimited by two acoustical absorbers, in order to eliminate undesiderable interferences by reflected acoustical waves. A thin metallic layer acts as TM polarisation filter at the output of the interaction region, and a short TAPE (Thermal Annealed Proton Exchange) [7] optical waveguide, whose plane is normal to the plane of the IOS circuit, acts as TE polarisation filter at the input of the IOS, being butt coupled between the output of the AOTF and the input of the output fiber.

## 3. HIGH RESOLUTION IOS

The application of IOS based instrumentation to the processing of radiation from FBG sensors requires a narrow band IOS filter transmission function in order to obtain high sensitivity in stress and strain measurement. Eq. 4 indicates that the only practical way to increase filter resolution is to increase the interaction length as much as possible.

Working in the 1300 nm wavelength range an interaction length of 40 mm is necessary to obtain a FWHM in the range .4-.45 nm . In these conditions great uniformity is required in the fabrication processes, in order to ensure the necessary uniformity of acousto-optic coupling conditions along the interaction region.[8]

In spite of efforts to improve process uniformity, results were still not satisfactory and a residual inhomogeneity causes asymmetric degradation of sidelobes in the filter transmission function. In Fig. 5 an experimental result is shown that was obtained by an high uniformity standard processed high resolution IOS (L = 40 mm). The IOS has been characterised using a narrow linewidth (10 KHz) solid state ring laser at 1318.7 nm. Becouse the laser line is narrower than the IOS transmission function, the filter function is fully characterised by scanning the IOS control frequency. The asymmetry of the sidelobe and the - 4.5 dB sidelobe ratio have to be compared with the symmetrical sidelobes distribution and the - 9.6 dB sidelobe ratio expected in a sinc $^2$ condition. Since it is not possible to obtain acceptable results solely by improvements in process uniformity, the adoption of some trick in the device configuration is necessary.

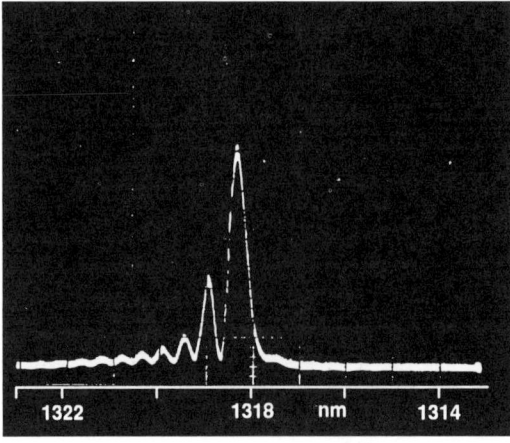

Fig. 5 Experimental IOS transmission function with evidence of degradated sidelobe ratio.

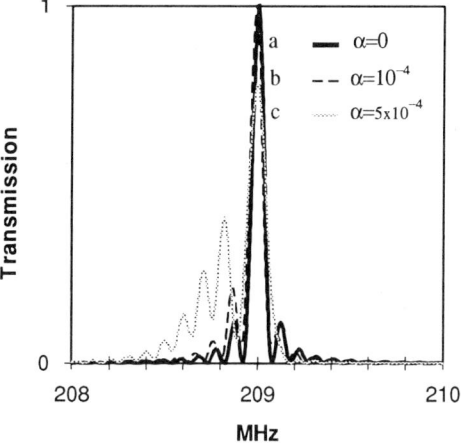

Fig. 6 Simulated IOS transmission function including parabolic Δn inhomogeinity.

In order to analyse the problem a simplified mathematical model has been adopted for the collinear acousto-optical interaction, in which the evolution of the complex optical mode field amplitudes $E_{TE}$ and $E_{TM}$ are given by the solution of the following system of equations system:

$$\frac{dE_{TE}}{dy} = kE_{TM} \exp\{-iy\beta\} \tag{5.1}$$

$$\frac{dE_{TM}}{dy} = -kE_{TE} \exp\{iy\beta\}, \tag{5.2}$$

where $k$ is the coupling coefficient between the two modes, $y$ is the spatial coordinate along the interaction region, and $\beta = \Delta n / \lambda - f / V_{ac}$ is the phase matching parameter.

Adopting this model it is possible to correlate the observed asymmetry with a parabolic variation of the parameter $\Delta n$ along the Y direction, centered at $L/2$, as expressed by the Eq. (6):

$$\Delta n(y) = \Delta n_0 \left\{ 1 - \alpha \left[ 1 - \frac{2y}{L} \right]^2 \right\}, \tag{6}$$

where $\Delta n_0$ is the unperturbed difference between the effective refractive indexes of the TE and TM guided modes, and $\alpha$ is the parabolic perturbation coefficient.

Including this perturbation in the model, it is possible to obtain a simulation in accordance with the experimental results (Fig. 6). From this point, the research has been oriented to look for a design solution ehich is able to compensate the effect. This was attained simply by observing that a spatial variation of the acoustic velocity $V_{ac}(y)$ can compensate the variation of the effective refractive index difference $\Delta n(y)$ in the expression of $\beta$:

$$V_{ac}(y) = V_{ac} \left\{ 1 - \xi \left[ 1 - \frac{2y}{L} \right]^2 \right\}. \tag{7}$$

If a variation of the SAW velocity as expressed in Eq. (7) is considered, it is possible to have $\beta$ constant along the interaction length, and thus obtain the targetted interaction uniformity, when Eq. (8) is satisfied:

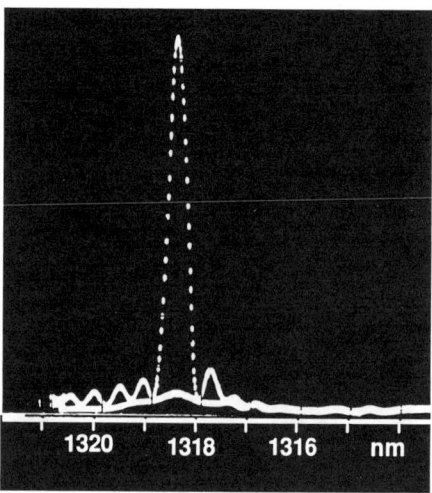

Fig. 7 Experimental Δn inhomogeinity compensated IOS transmission function

$$\xi = -\left(\frac{\Delta n_0 V_{ac}}{f\lambda}\right)\alpha . \tag{8}$$

A modified acoustic waveguide was then adopted in which the width of the acoustic channel follows a parabolic function of the y coordinate (centered at L/2), satisfing Eq. (8).[9]

The adoption of this solution allowed us to solve the problem, and the repeatibility of the results was validated by a large number of production runs. A typical experimental result is shown in Fig. 7, in which the FWHM is 0.43 and the sidelobe ratio is lower than 9 dB, in good agreement with the expected value when working in the 1300 nm wavelength range.
The normalised efficiency of this class of IOS is shown in Fig. 8 vs RF control signal power. 90% efficiency is obtained with only 8 mw of RF power. In this condition the overall optical insertion loss (TM output/TE input) is less than 5 dB.

## 4. POLARISATION INSENSITIVE IOS

The practical utilisation of the device discussed above is limited by its sensitivity to the polarisation status of the radiation received from the FBG sensor. The FBG is insensitive to the polarisation of the radiation, so it is necessary to develop an IOS able to process the signal independently of its polarisation status. Two solution have been given to this problem, by including two IOS in the same optical circuit and by adopting an heterodyne detection scheme in the single channel IOS.

### 4.1. Double Channel Optical Circuit

This is the most widely adopted solution, in which the two interaction regions are included between two polarisation splitters/combiners (Figs. 9a and 9b).[10] The first polarisation splitter separates the two polarisation to be processed in the two AOTF, and the splitters at the output of the interaction regions mix the results of the interactions and filter out the residual radiation. In our solution the key device, the polarisation splitter/combiner, is realised by adiabatic coupling between the Titanium diffused waveguide and a TAPE

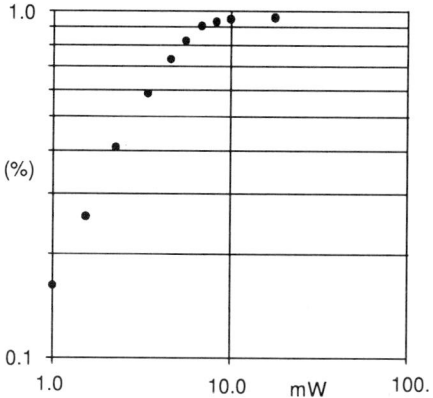

Fig. 8 Normalised IOS efficiency vs RF control signal power.

waveguide. This solution allows us to avoid the thermal instabilities of polarisation splitters based on interference between modes. The results obtained allowed us to produce two classes of devices, the first characterised by the inclusion of both the optical waveguides in the same acoustical waveguide (Fig. 9a), and the second in which the two IOS are completely separated in two different acoustical waveguides (Fig. 9b). The first type of device is usually adopted as the IOS in the processing of radiation in FBG sensor systems. The second configuration has been developed in order to obtain the same frequency shift of the optical radiation in both the IOS's in order to comply with the requirement for a frequency shifter in a soliton transmission system.

### 4.2. Heterodyne Detection

This device exploits the possibility of obtaining a heterodyne signal detecting the optical radiation at the output of the ATOF. As consequence of conservation of energy, the frequency of the output optical radiation is shifted by the frequency of the SAW.

Consider for example the case in which the input optical radiation is TE polarised and its field is expressed by:

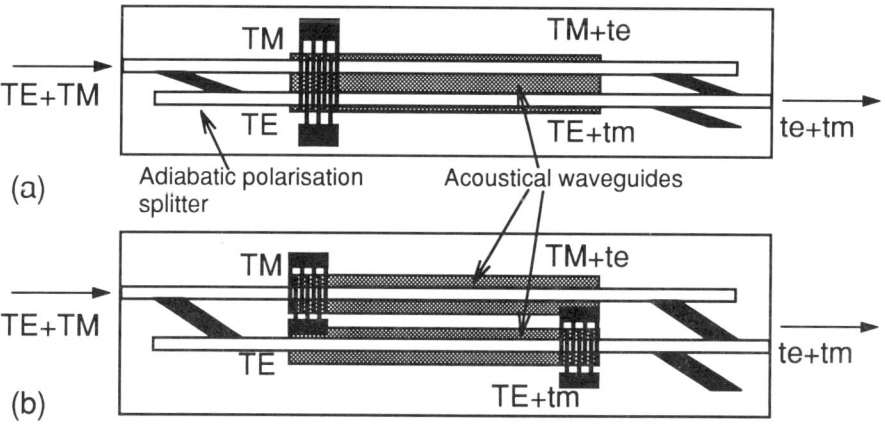

Fig. 9 Schematic view of the double channel optical circuit for polarisation insensitive IOS operation.

$$E_{TE} = A_{TE} \exp\{j[\omega_0 t - \Phi_{0,TE}]\}. \tag{9}$$

The optical fields at the output of the acousto-optical interaction are then given by the following expressions (10.1) for residual TE mode and (10.2) for TM mode product:

$$E_{TE,TE} = A_{TE}(1-\alpha_{TE})\exp\{j[\omega_0 - \Phi_{0,TE} - \Phi_{TE,TE}]\} \tag{10.1}$$

$$E_{TE,TM} = A_{TE}(\alpha_{TE})\exp\{j[\omega_0 + 2\pi f - \Phi_{0,TE} - \Phi_{TE,TM}]\}. \tag{10.2}$$

These two orthogonally polarised modes coherently beat in the optical detector when the output polariser is crossed at 45°, with respect to the waveguide polarisation axis (input polarisation axis), and the corresponding photocurrent is given by the expression (11):

$$i_{ph} = \text{const.}[A_{TE}^2(1-\alpha_{TE})^2 + A_{TE}^2(\alpha_{TE})^2 + 2A_{TE}^2(1-\alpha_{TE})\alpha_{TE}\cos(2\pi f + \Phi_{TE,TE} - \Phi_{TE,TM})] \tag{11}$$

where the heterodyne signal appears at frequency f. The monitoring of this heterodyne signal is an alternative to the monitoring of the TM output DC photocurrent, given by $E_{TE,TM}$, in order to measure the efficiency function of the ATOF. If the noise bandwidths are the same, the ratio between the corresponding peak signals is given by Eq. (12):

$$i_{ph(TM-DC)}/i_{ph(f)} = \alpha_{TE}^2/(1-\alpha_{TE})\alpha_{TE}. \tag{12}$$

The heterodyne signal at frequency f offers an additional and decisive advantage for realising a Polarisation Independent IOS (PIIOS). In fact, the polariser placed at the output of the ATOF, crossed at 45° with the polarisation axis of the planar optical device, allows not only the beat between the TM product of the acousto-optical interaction with the corresponding TE input, but also the TE product to beat with the corresponding TM input.

This allows one to obtain the heterodyne signal when the phase matching Eq. (2) is satisfied with any polarisation status of the input optical radiation.

A short TAPE optical waveguide (5 mm), in which only TE mode is guided, acts as output 45° polariser, being coaxially butt coupled at the output of the ATOF circuit with an angle of 45° between the planes of the two X cut crystal surfaces. No polarisation filter is included at the input of the AOTF. The resulting device is schematically shown in Fig. 10.

In order to characterise the device, the same ATOF circuit has been used in both polarisation sensitive IOS and PIIOS configurations, with the objective to compare the two approaches.

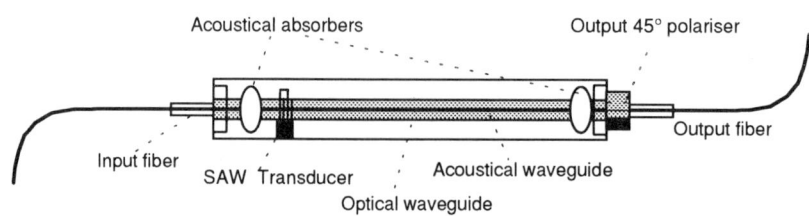

Fig 10. Schematic view of the heterodyne detection PIIOS

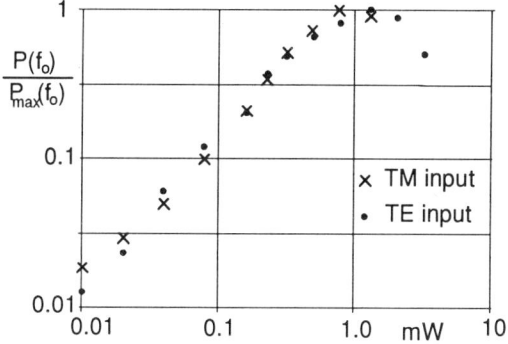

Fig. 11 Normalised PIIOS heterodyne signal output vs RF control signal power.

The RF power of the output heterodyne signal in the PIIOS configuration has been characterised first vs control signal power with $f=f_o$, with TE and TM polarised optical input (Fig. 11).

The maximum efficiency is observed when the power of the control signal is 1.0-2.0 mw for both input polarisations. This condition corresponds to $TE(\omega_0)$ and $TM(\omega_0 + \omega_{ac})$ output optical fields roughly at the same level ($\alpha_{TE} \approx 0.5$). The transmission function of the device has then been characterised vs the frequency of the control signal (input power = 0.2 mw) around $f_o$. The experimental results are shown in Fig. 12, where they are compared with the ATOF transmission function, previously characterised as IOS, obtaining acceptable agreement.

## 5. IOS BASED INSTRUMENTATION

The IOS, integrated with the VCO, allowing one to control the tuning of the filter by simply varying a DC voltage, is the key element of a very effective instrument enabling three main operating modes: spectrometer, discriminator and tracking filter.

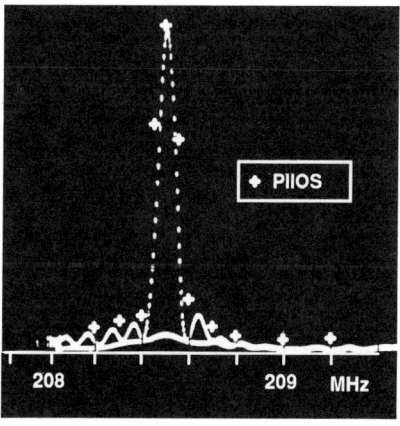

Fig. 12 PIIOS transmission function compared with the original AOTF transmission function.

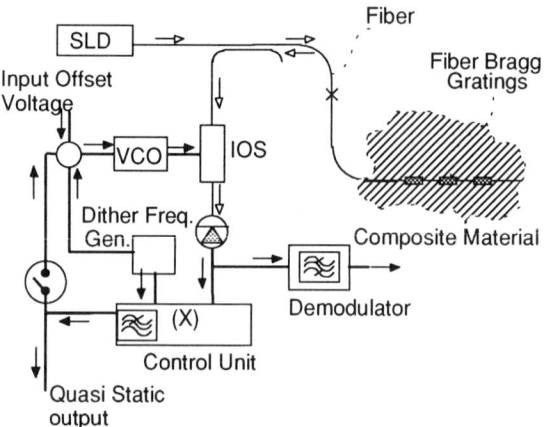

Fig. 13 IOS based instrumentation scheme for tracking the filter operating mode.

### 5.1. IOS as Spectrometer

In the simplest operating mode, as a spectrometer, the IOS acts as a scanning filter. The transmission function of the filter is tuned in a repetitive fashion over the entire region of optical wavelengths of interest, in which the reflection of the grating is contained. When the scanned filter function coincides with the reflected optical signal from the remote fiber grating sensor, the optical detector responds with the largest electrical response, thereby uniquely identifying the wavelength of the optical signal reflected by the sensor. This operating mode is clearly useful for measuring the static strains applied to fiber grating sensors with a sensitivity of 400 µstrains, limited by the resolution of the IOS.

### 5.2. IOS as Discriminator

Working as a discriminator, the opto-acoustic filter is tuned such that the sensor signal is generally positioned on the linear portion of the slope of the IOS transmission function. The opto-acoustic filter is temporarily maintained in this position, and high speed dynamic changes imposed on the fiber grating sensors can be detected. In this manner, dynamic strains with very high frequency content can be measured with very high sensitivity (<10 µstrains).

### 5.3. IOS as Tracking Filter

The previously described operating modes are generally optimum for either static strains (Spectrometer) or for dynamic strains (Discriminator). In some applications the measurement of both static and dynamic strains is required, and in such a case a third mode of operation may be beneficial. The third operating mode configures the opto-acoustic filter with a tracking controller so that the tunable filter can be automatically tuned to stay centered on the sensor signal of interest even while the sensor is caused to shift due to large static perturbation. In this manner, the filter can continue to perform low level dynamic strain detection, regardless of the static load and thermal state. The system configuration for the tracking mode is provided in Fig. 13. In this arrangement, the tracking controller output signal is simply proportional to the displacement of the average wavelength $\lambda_B$ of the opto- acoustic filter with respect to the initial condition, allowing one to measure the static strain level. Furthermore, dynamic strain signals at frequencies greater

Table I. Experimental comparison of the three IOS based instrumentation.

| IOS Operating Mode | Strain Sensitivity | $\delta\lambda$ (nm) |
|---|---|---|
| Spectrometer | 400 μstrains | 0.4 |
| Discriminator | 5 - 10 μstrains | 0.005 - 0.010 |
| Tracking | 0.1 μstrains | 0.0001 |

than the tracking controller bandwidth are not compensated, and they appear on the controller input modulated by the dither frequency. This signal could be used to monitor the vibrations of the structure and could be sent to suitable actuators for a closed loop vibration damping system. Experimental results indicate a sensitivity of this instrumentation in the range of 0.1 μstrains for static deformations and 0.0023 μstrains/Hz$^{1/2}$ for dynamic strains.

## 6. CONCLUSIONS

ALENIA Research Department's IOS has been realised, characterised and its industrialisation is proceeding. A significant effort has been expended to realise high resolution AOTF devices (FWHM = 0.45 nm) avoiding the sidelobe ratio degradation caused by fabrication process non uniformities and polarisation insensitive devices. An innovative approach has been investigated and demonstrated in order to obtain a polarisation insensitive IOS exploiting an heterodyne detection aproach. The IOS device has been used to realise portable, high performance instrumentation for interrogation of FBG sensors embedded in composite material, to monitor both static and dynamic strains of the structure. Three operating modes have been demonstrated and the results are listed in Table I. The comparison with conventional instrumentation (OSA based) is shown in Table II. This IOS based instrumentation is rugged, low cost and adequate for industrial applications, in particular for aeronautical structures. The speed and sensitivity of the IOS allow IOS based instrumentation to exceed the requirements of aeronautical applications and allow the monitoring of low level high frequency vibrations.

ACKNOWLEDGEMENTS. The author wishes to acknowledge Dr. Jim Dunphy for fruitful discussion and concepts for of the instrumentation system, Dr. Sabato Inserra (ALENIA Composite Materials, Foggia - Italy) for his support in the fabrication of the composite material test samples, and his co-workers in ALENIA Research Department A.Vannucci, M.Signorazzi, A.Evangelisti and M.Ricci.

Table II. Comparison between IOS based instrumentation and conventional Optical Spectrum Analyser (OSA).

| Parameter | IOS | OSA | Units |
|---|---|---|---|
| Resolution ($\delta\lambda$) | 0.4 | 0.1 | nm |
| Range ($\Delta\lambda$) | 300 - 600 | 500 | nm |
| Scanning Speed | 35000 | 0.5 | nm/sec |
| Optical Insertion Loss | 6 | 6 - 10 | dB |
| Volume | 0.5 | 30 | lt |

## REFERENCES

1. G.Meltz, W.W.Morey and W.H.Glenn, Formation of Bragg gratings in optical fibers by a transverse holographic method, *Optics Letters*, Vol. 14, pag. 823-825 (1989).
2. W.W.Morey, G.Meltz and W.H.Glenn, Bragg grating temperature and strain sensors, *Springer Proceedings in Physics Vol.44 Optical Fiber Sensors*, Paris (France), 18-20 September 1989, pag. 526-531.
3. M.Varasi, M.Signorazzi, A.Vannucci and J.Dunphy, A high resolution integrated optical spectrometer with applications to fiber sensor signal processing, *Meas.Sc.Technol.*, Vol. 7, pag. 173-178, (1996)
4. Y.Ohmachi and J.Joda, $LiNbO_3$ TE-TM mode converter using collinear acoustooptic interaction, *IEEE Journal of Quantum Electronics*, Vol. QE-13, n°2, pag.43-46 (1977).
5. M.Varasi, M.Signorazzi and A.Vannucci, Polarisation independent integrated optical spectrometer, Optical Fiber Sensors - OFS 10, 11-13 October 1994, Glasgow-Scotland, *SPIE Proceedings n°2360*, pag.273-276.
6. J.Dunphy, G.Ball, F. D'Amato, P.Ferraro, S.Inserra, A.Vannucci and M.Varasi, Instrumentation development in support of fiber grating sensor arrays, Distributed and multiplexed fiber optic sensors III, 8-9 September 1993, Boston-Massachusetts, *SPIE Proceedings n°2071*, pag. 2-11.
7. M.Varasi, A.Vannucci and M.Signorazzi, Lithium Niobate proton exchange technology for phase-amplitude modulators, Integrated optical circuits, 3-4 September 1991, Boston - USA, *SPIE Proceedings n°1583*, pag.165-169.
8. D.A.Smith et al., Source of sidelobe asymmetry in integrated acousto-optic filters, *Applied Physics Letters*, Vol. 62, pag. 814-816 (1993)
9. G.Coussot, Rayleigh wave guidance on $LiNbO_3$, *Applied Physics Letters*, Vol. 22, pag.432-433 (1973)
10. D.A.Smith et al., Polarisation independent acoustically tunable optical filter, *Applied Physics Letters*, Vol. 56, pag. 209-211 (1990)

**MEASUREMENTS**

# MICROMECHANICS:
# NEW CHALLENGES FOR OPTICAL MEASUREMENTS

K. Patorski and M. Kujawinska

Warsaw University of Technology, Faculty of Mechatronics
Institute of Micromechanics and Photonics
8 Chodkiewicza St., 02-525 Warsaw, Poland

## 1. INTRODUCTION

Competitive trends to miniaturize Micro Electro Mechanical Systems (MEMS) introduce unprecedented requirements on their designs. To satisfy these challenges new materials and structural designs are being employed. Complex structures of modern micromechanical assemblies and electronic packages must withstand wide ranges of applied or generated mechanical, thermal or electrical loads. To ensure their optimum design, functionality and reliability, hybrid experimental/numerical methods of analysis called ACES (Analytical, Computational and Experimental Solution) are being implemented.[1] These include:
- forming theoretical models of structure behaviour and implementing them into FEM (Finite Elements Method) analysis; and,
- incorporating knowledge about the accurate geometry of the object being modeled, including local small scale irregularities, which due to unique processing procedures, high stress levels and small feature sizes may differ from bulk material properties.[2]

Considering experimental methodologies which may be unified with FEM for microscale component analysis, it can be easily recognized that the conventional procedures involving strain gauges, photoelasticity, mechanical probing, etc., are generally not applicable to these measurements. An alternative to the conventional methods can be provided by full-field optical methods including the most popular: electronic holography (EOH), shearography and grating interferometry supported by optical and scanning electron microscopy. Recent advances in semiconductor coherent light sources, CCD detectors and fibre optics technology enable us to realize compact, remote control, automatic measuring systems which provide data about static and dynamic behaviour of microelements with nanometer sensitivity and micrometer spatial resolution. These experimental methods are merged with FEM in the general scheme of a hybrid approach to the design of microstructures,[2] Fig.1.

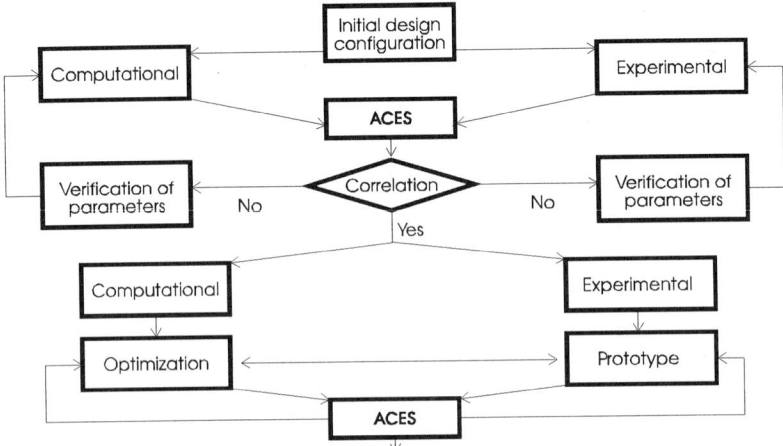

Fig. 1 The scheme of hybrid approach to design and optimization of microstructures based on ACES (Analytical, Computational and Experimental Solution) methodologies [3].

## 2. APPLICATION AREAS

Today, micromeasurements of displacement and strain are greatly required in three basic situations.[3]
- local approach to experimental mechanics, i.e., analysing micro regions in bulk samples (analysis of contact and fracture mechanics and fatigue problems in regions with ambiguous or unknown analytical solution).
- local approach to material engineering, i.e., analysing micro regions in bulk samples made of various polycrystalline,[4,5] composite [5,6] or "smart" materials.[6] The main aim relates to material structure studies.
- analysis of microobjects. This includes two basic groups:
  - MEMS which are micron sized mechanisms (a few µm to a few hundred µm), often manufactured using VLSI techniques adapted from the microelectronics industry and used in sensor and actuator applications in biomedicine, microrobotics and nanometrology.
  - Electronic Components and Assemblies with respect to electronic packaging (heat transfer phenomena and vibrational analysis).

Micromeasurements of displacement/strain as mentioned above are often supported by the knowledge of the element surface: surface roughness, waviness, shape, etc. These parameters are critical to product quality and to the manufacturing process.

## 3. EXPERIMENTAL TOOLS

The experimental methodology considered will be limited to automated interferometric methods providing convenient tools for microelement analysis under static and/or dynamic conditions,[3] Fig. 2.

The main advantages of these methods are: noninvasive character, high sensitivity and accuracy (typical resolution of displacements 10-20 nm, strain $10^{-4}$, surface microstructure height range from less than 0.1 nm to several hundreds of micrometers), high resolution of

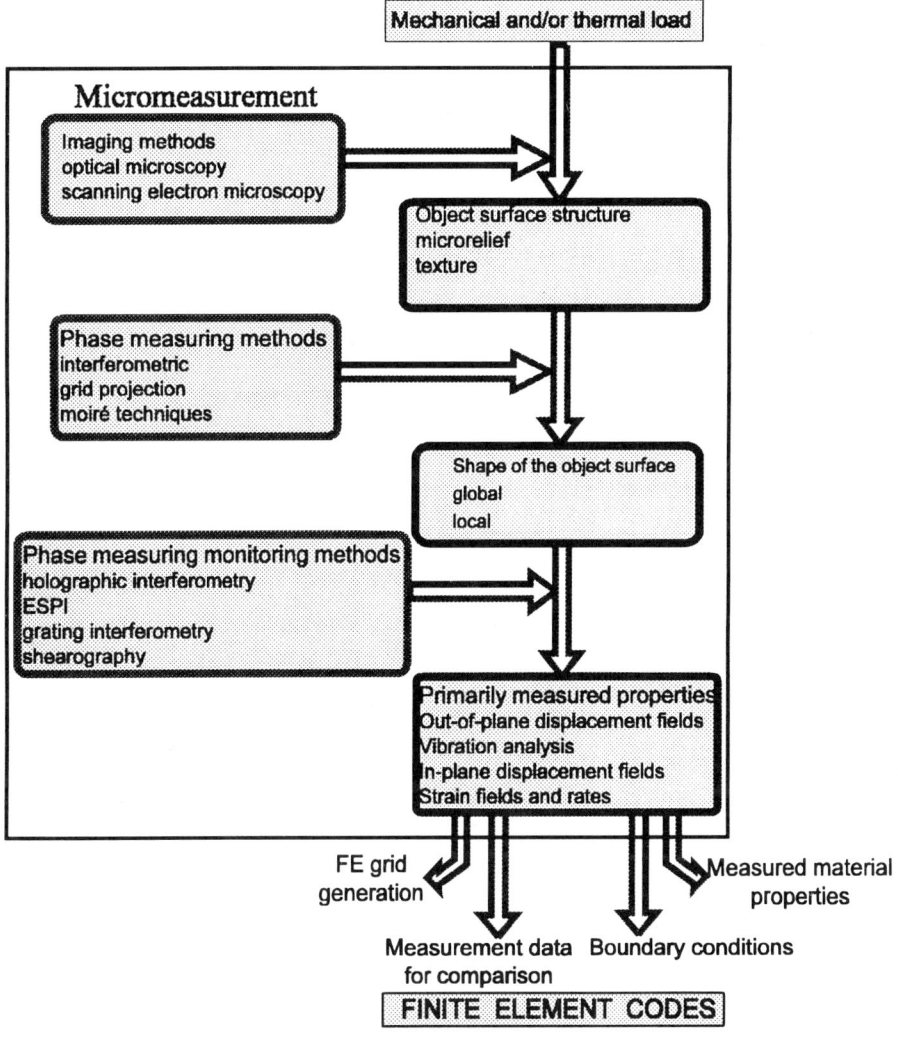

Fig. 2 The measurement scheme required for microobjects.[3]

data points (512x512 points over sub-mm field of view), long term monitoring ability, adaptivity for macro- and microscale measurements, automatic analysis of the results and data preprocessing to meet FEM requirements.

Commercial systems for object microstructure 3-D measurements such as scanning electron and atomic force microscopes as well as interference microscope systems with phase extraction algorithms [7] are available and well established. However, there is a lack of suitable commercial equipment for in-plane displacement/strain measurement.

This lecture will be focused on three basic measurement systems:
- electronic holography (EOH) for static and transient event studies including in-plane and out-of-plane displacements and vibrations,
- grating (moiré) interferometry used for static and time dependent measurements of in-plane parameters and determination of material properties,

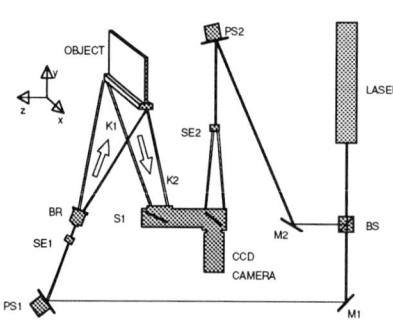
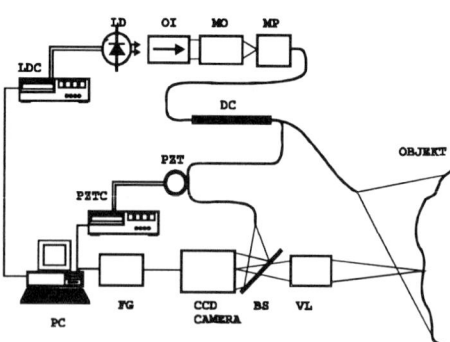

Fig. 3 Optical configurations of the EOH system a) bulk optic arrangement (BS - beamsplitter; M1, M2 - mirrors; PS1, PS2 - phase steppers; SE1, SE2 - spatial filter beam expander assemblies; BR - object beam rotator; SI - speckle interferometer); b) fibre-optics arrangement (LD - laser diode; OI - optical isolator; MO - microscope objective; MP - micropositioner; DC - directional coupler; PZT - piezoceramic transducer phase shifter; VL - video lens; FG - frame-grabber).

- computerized interference microscopy for surface roughness measurements of a wide variety of materials within a greatly extended vertical range up to 500 μm, with subnanometer resolution.

### 3.1. Electronic Holography (EOH)

The EOH system has been described in several papers. The diagrams of conventional and fibre optics configurations shown in Figs. 3a and 3b are taken from.[8,9] The systems include several devices which enable the analysis of objects under static and dynamic loads.

### 3.2. Grating Interferometry (GI)

The variety of opto-mechanical configurations implementing grating (moiré) interferometry including laboratory, workshop and sensor type systems can be found, for example, in Refs. 10-12. Figures 4a and 4b show the basic three-mirror solution (with the Twyman-Green interferometer added for out-of-plane displacement measurement) and the fibre optics version, respectively.

### 3.3. Computerized Interference Microscopy

A typical computerized interference microscope consists of an interference microscope having a CCD array interfaced to a Pentium computer running Microsoft Windows for data analysis and display. The WYKO RST Plus surface measurement system can serve as an example. The particular type of interferometer depends on the magnification.[7,13] The configuration shown in Fig. 5 utilizes a two-beam Mirau interferometer at the microscope objective.

Two modes of operation are available. For smooth surfaces, the phase shifting mode is used since it gives sub-nanometer height resolution. For rougher surfaces, up to greater than 500 μm surface height variations, a vertical scanning coherence sensing technique [14] is used that gives nm height resolution.

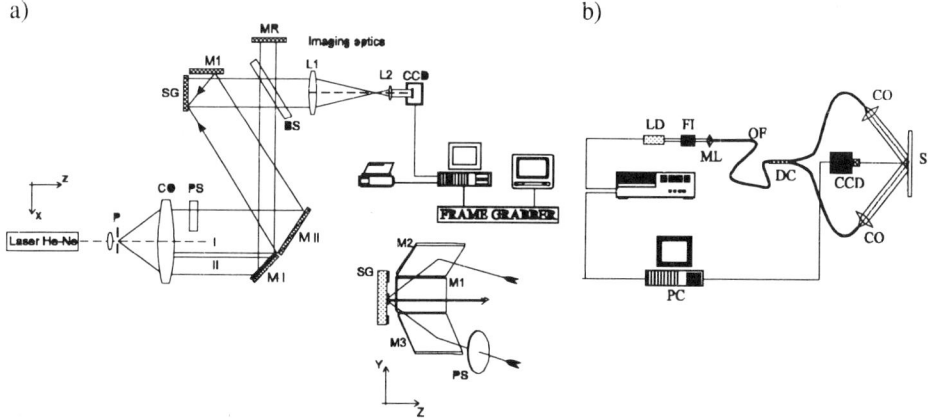

Fig. 4 Grating interferometry arrangements: a) three-mirror grating interferometer for u, v and w displacement measurement (CO - collimating objective; M1, M2, M3, M I, M II, MR - mirrors; L1, L2 - imaging optics; SG - specimen grating; PS - parallel plate phase shifter); b) fibre optics grating interferometer (LD - laser diode; FI - Faraday isolator; ML - microscopic objective; OF - optical fibre; DC - directional coupler).

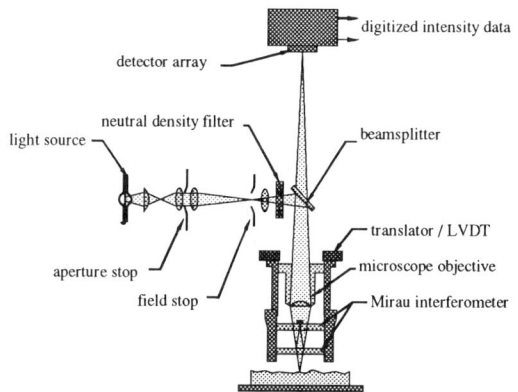

Fig. 5 Optical schematic of computerized WYKO RST Plus interference microscope.

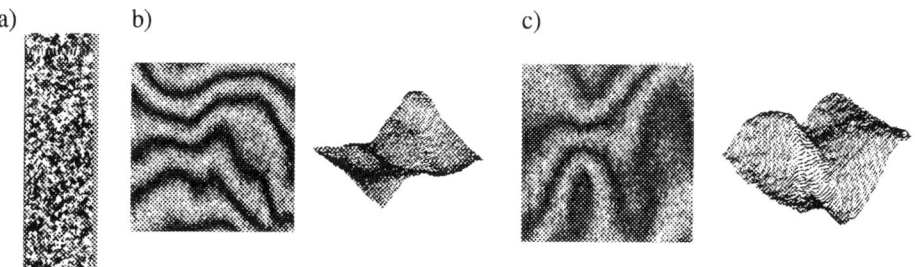

Fig. 6 The local analysis of: a) polycrystalline aluminium sample, b) local u - interferogram and u - displacement 3D map, c) local v - inteferogram and v - displacement 3D map.[4]

346

## 4. SELECTED APPLICATIONS

### 4.1. Local Approach to Materials Engineering

Recent technologies enable production of various types of materials including polycrystalline materials, various alloys, composites, and smart structures [5] with enhanced mechanical and environmental properties. These properties, especially in local regions, depend on the chemical and geometrical composition of material and the history of its manufacturing. Here we focus on the problems connected with polycrystalline materials. [3, 4, 5] The published data are based on the measurements of bulk samples with hundreds or

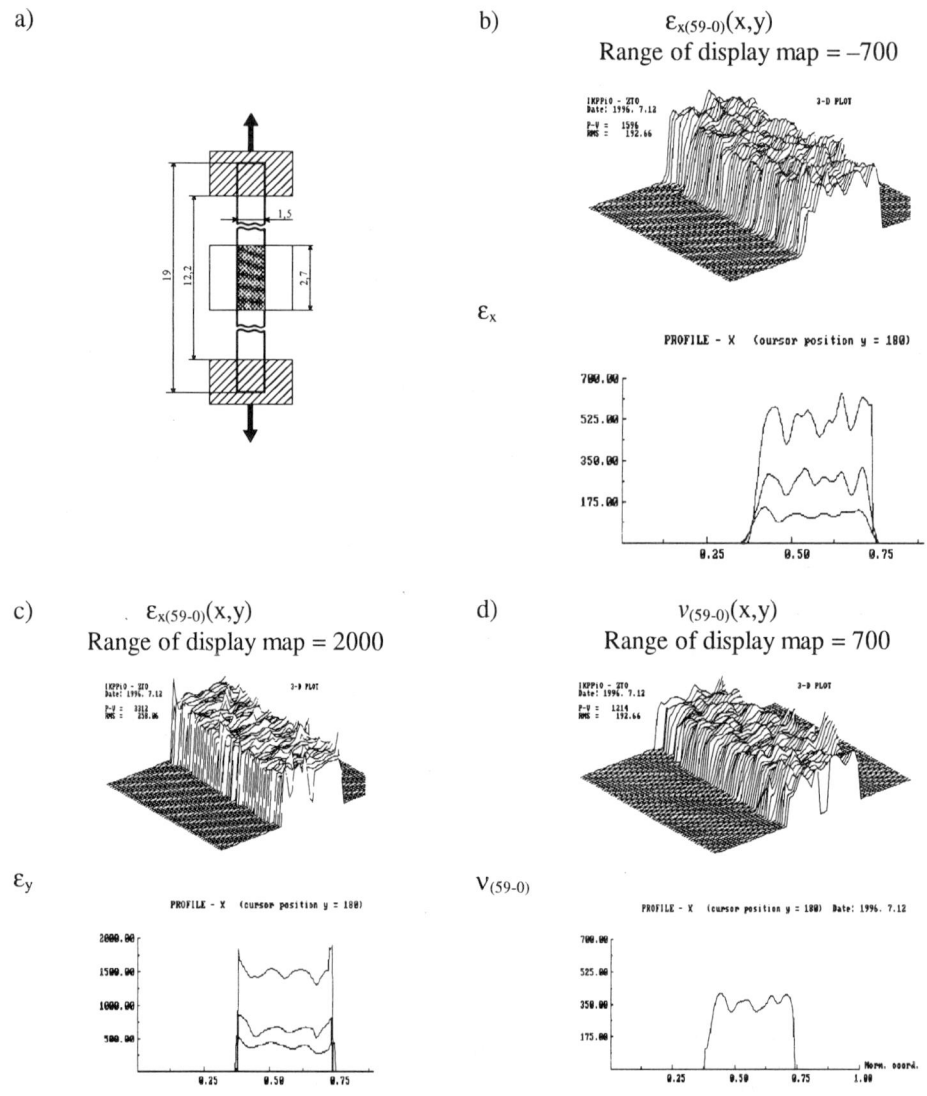

Fig. 7 Strain analysis of a brass microbeam under tensile load: a) specimen geometry, b) 3D plot of $\varepsilon_x$ distribution and $\varepsilon_x$ x-profiles for sequential loads, c) 3D plot of $\varepsilon_y$ distribution and relevant $\varepsilon_y$ x-profiles, d) 3D plot of Poisson ratio distribution and its x-profile.[4]

thousands of grains. In the case of microelements it is necessary to assume, in general, that the polycrystalline materials, due to crystallographic grain anisotropy, have directional properties. Additionally, this anisotropy may be enhanced by different type of mechanical (e.g., cold work) and thermal (e.g., annealing) processes.

Knowledge about this anisotropy and possible non-homogeneous plastic strains is very important for proper application of the materials. Here high sensitivity automatic grating interferometry is applied to indicate several unexpected phenomena which occur in microscale material engineering.

The local analysis of a polycrystalline aluminium sample [4] shown in Fig. 6 clearly indicates the displacement/strain inhomogeneity correlated with grain boundaries and mutual rotation of grains during loading. This behaviour may occur with single grains or with grain complexes which were formed due to the sample processing, e.g., hot rolling. The variations in the $\varepsilon_x$, $\varepsilon_y$ and $\gamma_{xy}$ strains may introduce significant changes in material constants such as the Poisson ratio or Young's modulus, which should be taken into consideration during FEM analysis. Such an effect was noticed during tensile loading of a microbeam which was manufactured of 0.25 mm thick brass sheet. Fig. 7 shows examples of three-dimensional $\varepsilon_x$ and $\varepsilon_y$ and Poisson ratio maps obtained for the load corresponding to a pusher displacement of 59 μm.

## 4.2. Hybrid Analysis of Stress Relaxation in Microelements

Stress relaxation is a process leading to a decrease of stress due to rearrangements in the material. A typical stress relaxation curve [9] is shown in Fig.8. Following application of the load, t = 0, the internal structure of the material changes, resulting in a decrease of stress while the strain is held constant. Determination of the time rate of change of the stress is important for predicting the long term behaviour of the material. Usually the time for full relaxation is of the order of hundreds of hours and depends on the environmental conditions.

The required lifetime of a microconnector at many levels of electronic packaging is several years or more. Therefore the determination of stress relaxation of materials which are used to manufacture microelectronic connectors is essential for the design of reliable electrical interconnections. Besides taking into account the electrical performance, the designer has to deliver an adequate contact force and must maintain this force during the life of the microconnector. The analysis of the mechanical design may be facilitated by FEM, if experimental knowledge about the time-dependent stress characteristic of the element is available.

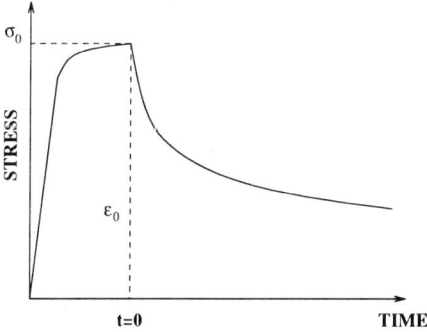

Fig. 8 A typical stress relaxation curve. [9]

348

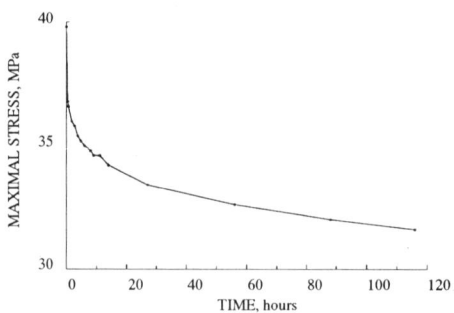

Fig. 9 Maximum of stress in the cantilever beam FEM model based on the EOH data.[9]

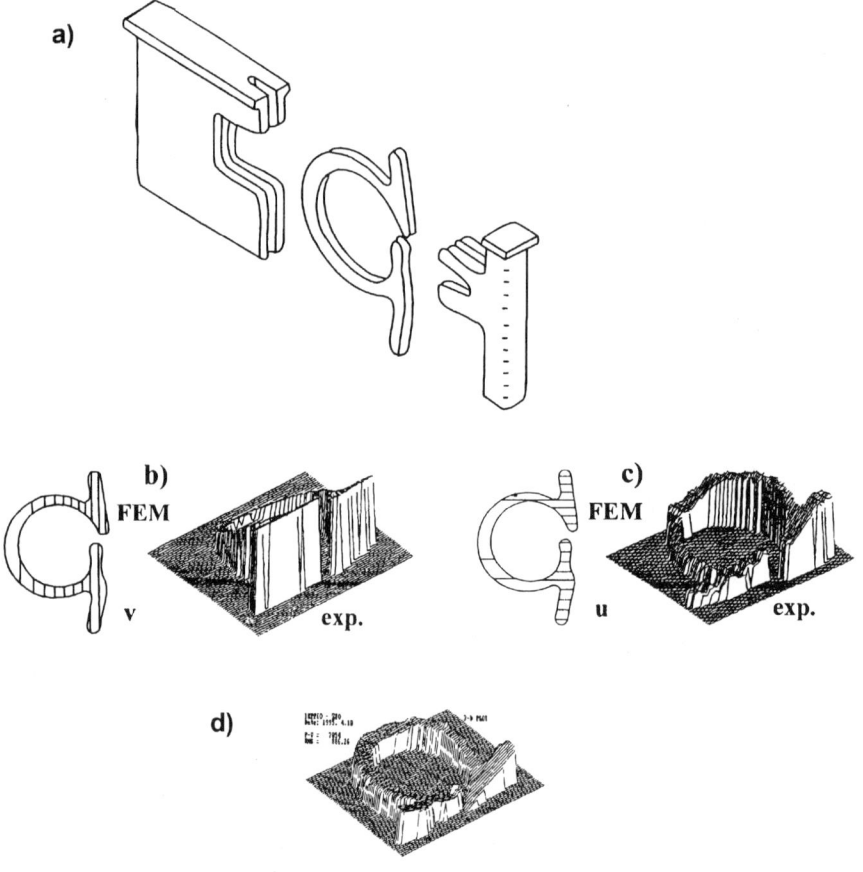

Fig. 10 The performance of the MicroInterposer under load: a) expanded view of the contact and components of the cartridge; b) 3D plot of u – displacement and its FEM representation; c) 3D plot of v - displacement and its FEM representation; d) 3D plot of out – of – plane - displacement.[15]

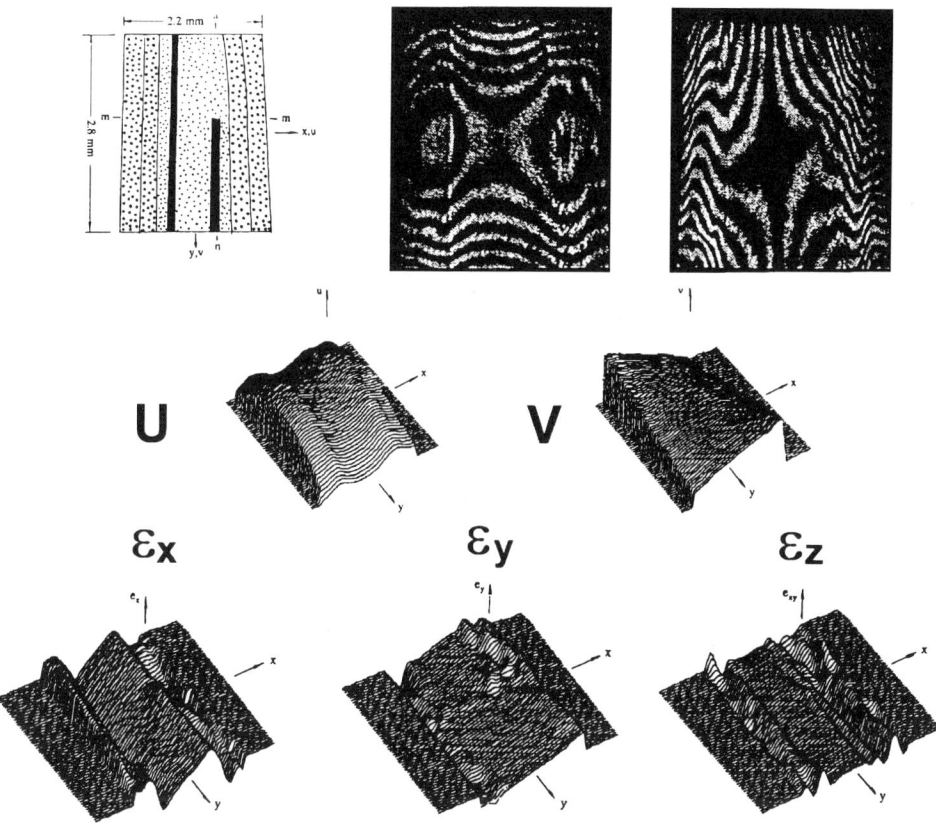

Fig. 11 Composite specimen under four-point bending: a) laminate lay-up geometry; b) u- displacement fringe pattern; c) v- displacement fringe pattern; d) u map, p – v = 2818 nm; e) v maps, p – v = 8712 nm; f) $\varepsilon_x$ map, p – v = 4012 μs; g) $\varepsilon_y$ map, p – v = 9606 μs; h) $\varepsilon_{xy}$ map, p – v = 12450 μs (where μs = microstrains).[6]

The quantitative analysis of the relaxation problem for a cantilever beam made of brass was performed using a hybrid FEM-EOH approach.[8] The sample was monitored in the arrangement shown in Fig.3. The beam was rigidly fixed at the lower end and loaded with a displacement at the upper end. The cantilever beam was illuminated by a parallel light beam normal to the surface to ensure sensitivity to motion only in out-of-plane direction. Immediately after loading, the reference image was taken and stored in the computer. A consecutive series of images was taken at specific time intervals to calculate the time-dependent deformation of the cantilever beam due to the relaxation process. Based on the EOH data, stress distribution in the FEM model was recalculated and the time plot of the maximum stress induced in the cantilever beam is shown in Fig.9.

Our next example describes interconnector testing for a new geometry of a microconnector which is specifically designed to be arranged in rows staggered with respect to each other. It resembles a Greek letter the Ω and is shown together with its cartridge in Fig.10a. The performance of Ω-like Interposer was tested using the grating interferometry approach [15] in which the in-plane and out-of-plane displacements were measured for sequential loads (see Fig.4a). Exemplary u, v and w displacement maps are shown in Fig.10 b-d, respectively. Along with the u and v experimental maps are shown their FEM predictions.

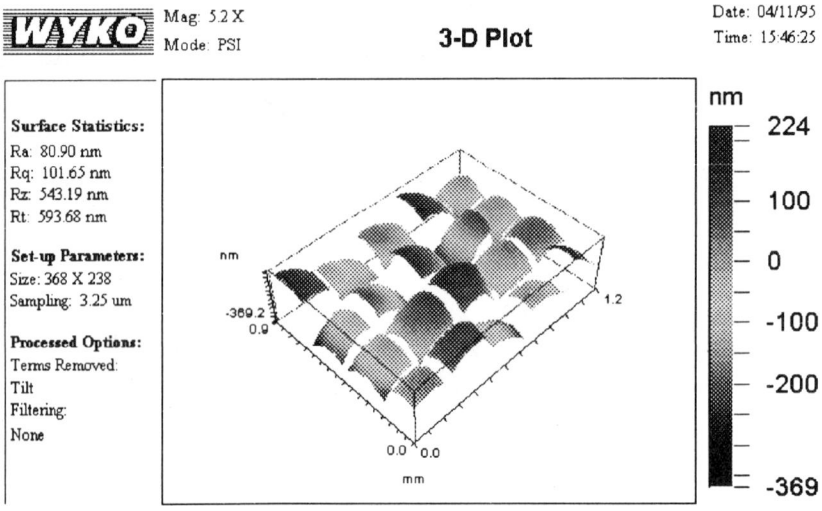

Fig. 12 Phase-shifting measurement of a deformable mirror. The shape of the mirror can be changed by applying voltages to the piezoelectric crystals (courtesy of WYKO Corp.).

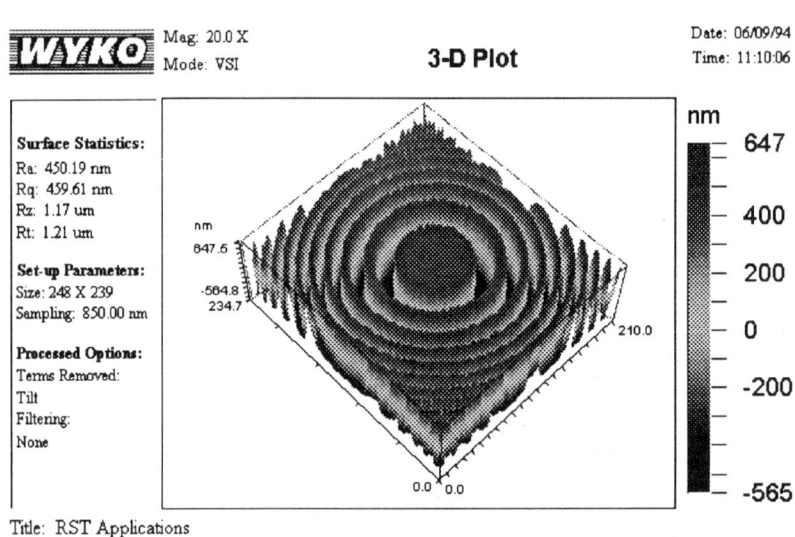

Fig. 13 Vertical scanning coherence peak sensing measurement of binary lens having $R_q = 459.62$ nm, $R_a = 450.19$ nm, $R_t = 1.21$ μm (courtesy of WYKO Corp.).

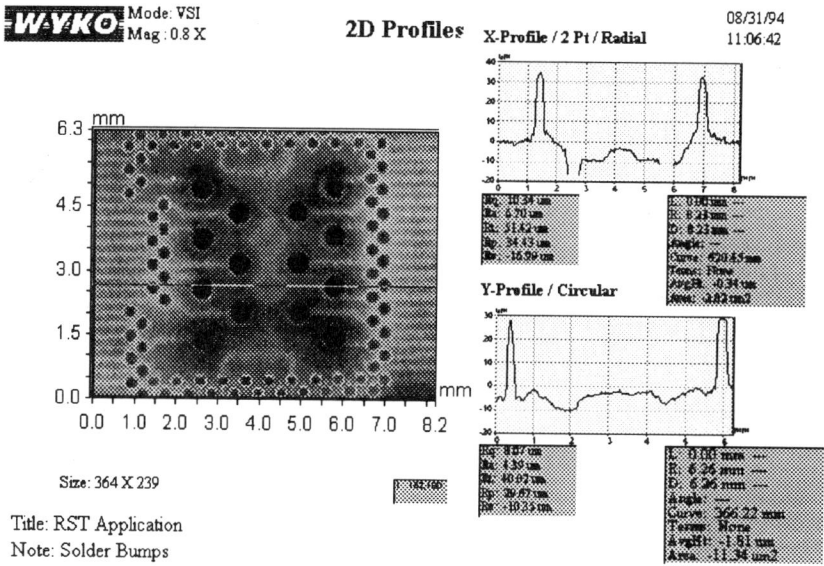

Fig. 14 Vertical scanning coherence peak sensing measurement of solder bumps. Left: Colour contour map (here, reproduced in black and white); right: - 2D profiles (courtesy of WYKO Corp.).

### 4.3. Testing of Complex Composite Materials

Grating interferometry offers significant help in determining the displacement/strain distribution in complex composite materials.[5,6] Figure 11a presents an exemplary carbon fibre/epoxy laminate lay-up geometry. One ply terminates in the middle to form a tapered panel.[6] The specimen was subjected to four-point bending and u and v displacement fields were monitored (Fig.11b-e). Differentiation of displacement maps yielded the strain distributions $\varepsilon_x$, $\varepsilon_y$ and $\varepsilon_{xy}$ shown in Fig.11f-h. The experimental knowledge of these maps enables checking of the existing FEM models, or forming new FEM models of the local behaviour of composite materials.

### 4.4. Surface Microstructure Profiling

The applications of computerized inteference microscopy for the measurement of fine structure for quality control, manufacturing process development and research are diverse,[7,13] they range from optical surfaces and elements, magnetic media and thin film read/write heads, machined surfaces, print rollers, heart valves, knee implants and hip joint replacements for medicine, to almost any surface imaginable. Some examples of measurements of various surfaces are shown in Figs.12-14 using three-dimensional plotted format, cross-sections and colour contour maps.

## 5. SUMMARY

The designs of modern Micro Electro Mechanical Systems (MEMS) are characterized by the increasing miniaturization of the components. This miniaturization requires a detailed knowledge of a micromechanical behaviour of the materials used and a design validation of the final products. For this purpose computational and preferably full field, noninvasive experimental methods are used, resulting in hybrid approaches to the design and optimization of microstructures. This lecture illustrated design and testing problems in the case of three optical methods: electronic holography (EOH), grating interferometry (GI) and interferometric microscopy. They can be fully computerised and are very useful for determining local material properties, stress distributions, relaxation of microelements and dynamic characteristics [3, 16, 17] of MEMS.

Further progress in experimental arrangements of EOH and GI, based on fibre optics, laser diodes and smart fringe processing systems, as well as improver loading devices would enable convenient and reliable support for the analysis of spatial and temporal material characteristics and dynamic (vibration) studies.

ACKNOWLEDGEMENT. The authors would like to thank Professor J.C.Wyant and Dr. J.Schmit of WYKO Corp. and Professor R.J.Pryputniewicz of Worcester Polytechnic Institute for providing the materials used in Chapter.

The preparation of this work was sponsored in part under KBN (Polish State Committee for Scientific Research) grants No 7T07D01108 and 8 T10C 009 08.

## REFERENCES

1. R.J.Pryputniewicz, A hybrid approach to deformation analysis, *Proc.SPIE* 2342: 282-296 (1994).
2. R.J.Pryputniewicz, and D.G.Grabbe, Developments in micromechanics through analysis and experimentation, *Proc.11th Int. Invitational 4ACEM Symp.Soc.Exp.Mech.*, Bethel, 506-532 (1993).
3. M.Kujawinska, and J.R.Pryputniewicz, Micromeasurement: a challenge for photomechanics, *Proc. SPIE*, v. 2782, 15-24 (1996).
4. M.Kujawinska, L.Sabut, and G.Dymny, Polycrystalline material studies by automatic grating interferometry, *Proc.SPIE*, 2782 (1996), in press.
5. L.Sabut, and M.Kujawinska, Novel material studies by automatic grating interferometry, *Proc.SPIE*, 2861: 212-219 (1996).
6. C.Y.Poon, and M.Kujawinska, C.Ruiz, Strain measurement of composite using an automated moiré interferometry, *Measurement*, 11: 45-57 (1993).
7. J.C.Wyant, Computerized interferometric measurement of surface microstructure, *Proc.SPIE*, 2576: 122-130 (1995).
8. R.J.Pryputniewicz, Hologram interferometry from silver halide to silicon and ..... beyond, *Proc.SPIE*, 2545: 405-427 (1995).
9. A.Olszak, and R.J.Pryputniewicz, ESPI-FEM hybrid for studies of time-dependent stress characteristics in small components, *Proc.SPIE*, 2445: 43-53 (1995).
10. D.Post, B.Han, and P. Ifju, High Sensitivity Moiré - Experimental analysis for mechanics and materials, Springer-Verlag, New York (1994).
11. M.Kujawinska, L.Sabut, Recent development in instrumentation of automated grating interferometry", *Optica Applicata*, v.25, 211-232 (1995).
12. G.Dymny, and M.Kujawinska, "Optoelectronic / Image Processing Module for enhanced fringe pattern acquisition and analysis, *Proc.SPIE*, 2784 (1996), in press.
13. J.C.Wyant, Computerized interferometric measurement of surface microstructure, *Optics & Photonics News*, 6: 40-42 (1995).
14. I.J.Caber, Interferometric profiler for rough surfaces, *Applied Optics*, 32: 3438-3441 (1993).
15. M.Kujawiska, T.Tkaczyk, and R.J.Pryputniewicz, Computational and experimental study of deformation in a microelectronic connector", Proc.SPIE, v.2545, 54-70 (1995).
16. G.C.Brown, and R.J.Pryputniewicz, "Experimental and computational determination of dynamic characteristic of microbeam sensors, *Proc.SPIE*, 2545: 108-119 (1995).
17. C.E.Leak, and R.J.Pryputniewicz, Quantitative EOH and FEM hybrid study of vibration characteristics of avionics, *Proc.SPIE*, 2545: 286-299 (1995).

# WAVEGUIDES IN LiNbO3 FOR OPTICAL SENSORS: CHARACTERISATION BY CERENKOV EFFECT

R. Ramponi

Polytechnic of Milan, Department of Physics
CNR Quantum Electronics and Electronic Instrumentation Centre
Piazza Leonardo da Vinci 32, 20133 Milan, Italy

## 1. INTRODUCTION

Sensors and microsystems are likely to become a major application for integrated optic technologies, second only to communications applications. As compared with the latter, this field covers much more diverse needs and it may involve mass production of consumer goods. Despite their great potential, and although a large number of devices have been studied and tested, only a few of them have led to commercially available products. Actually, except for very specific cases, integrated optic sensors have to compete with well developed and qualified non-optical (e.g. microelectronic) devices. Thus, novel optical systems have to meet very high standards in terms of cost-to-performance ratio and reliability.

The needs of optical fibre communication systems gave a strong impulse to the development of integrated optic technologies that today have reached a high degree of maturity. This is particularly true for waveguide technology. Hybridisation and packaging techniques are slowly but constantly improving, addressing the need to combine electronic and optical devices on a single chip or in microsystems, and the prospect of developing all-optical devices. Much work has been performed in recent years to develop sensors in optical fibres.[1] More recently, attention has moved to optical sensors realised in waveguides [2,3] due to the possibility of integrating them into optoelectronic circuits. Applications are in many fields: [2,4] chemistry, biology, environmental sciences, etc. In this paper the basic working principle of integrated optic sensors will be discussed and some examples presented.

Depending on the characteristics required for the sensor, the guiding structures are realised on different substrates (e.g. silicon, lithium niobate, glass, polymers) with different technologies. The advantage of integrated optic sensors realised on silicon is evident since electronics may be integrated; polymers exhibit high potential for chemical, biological and medical sensors; lithium niobate is polarisation dependent and second order non-linearities may be exploited.

Our laboratory has already acquired some experience in the design and characterisation of planar lithium niobate waveguides obtained both by titanium indiffusion and by proton exchange. In particular, second harmonic generation in the Cerenkov configuration allows determination of the effective refractive index of the fundamental mode coupled into the waveguide through the measurement of the output angle of the second harmonic radiation mode, and thus it allows full characterisation without having to outcouple the fundamental mode itself. Second harmonic generation in the Cerenkov configuration from both proton exchange and titanium-indiffused waveguides will be shown and measurements reported. Moreover, a model will be proposed for the realisation of titanium-indiffused lithium-niobate waveguides with predetermined optical characteristics. The potential of second-order non-linear effects in integrated optic sensors will also be described.

## 2. INTEGRATED OPTIC SENSORS: AN OVERVIEW

It is impossible to give a complete overview of integrated optic sensors. Only the general working principles and the main types will be considered.

Indeed, optics has always offered high potential for the field of sensors thanks to its insensitivity to electromagnetic noise and the possibility for no-contact and remote sensing. However, size and weight of components, alignment problems and expensive packaging prevented extensive application of traditional optics in this field. A lot of work has been performed in recent years to develop sensors in optical fibres that are indeed rather easy to handle.[1] More recently, sensors in optical waveguides were studied, [2,3] with considerable advantages: integrated optics allows miniaturisation, integration with other components and thus batch fabrication of fully packaged devices; moreover, the integrated optic sensing head can be combined on a chip with electronics for signal processing, thus allowing the realisation of smart sensors.

Several types of integrated optic sensors have been proposed, in particular, sensors for displacement, vibration, velocity, humidity and temperature, gyrometers, chemical sensors, optomechanical devices, etc. Correspondingly, the number of application fields is also large, ranging from dimensional metrology to chemistry, biology, environmental sciences, gyrometry, and opto-mechanics.

The working principle of integrated optic sensors is the same for the different types: the measurand affects the boundary conditions of the waveguide, thus modifying the propagation characteristics of the guided mode. From these modifications, information can be obtained on the measurand. In particular, in the case of displacement sensors, the position of the measurand affects the optical path. Despite their common working principle, integrated optic sensors can be divided into two main categories in terms of measurement techniques. In fact, depending on the structure of the sensor, both direct measurements of intensity, phase, effective refractive index, or polarisation of the guided mode and interferometric measurements in different configurations (Fabry-Perot, Mach-Zender, Michelson) are possible.

An example of direct measurement is represented by intensity encoded sensors, where, as shown in Fig. 1, the measurand absorbs part of the evanescent field, thus producing an attenuation of the power of the guided light. This kind of sensor is generally used for concentration measurements.[5] Other sensors exploit the effective refractive index changes of the guided mode induced by changes in the external conditions that modify the coupling (or out-coupling) angle. Fig. 2 shows an example of an input grating sensor, where $\alpha$ is the coupling angle. The measurement of the change in $\alpha$ allows one to evaluate the variation of the external parameter. These sensors have been succesfully used as chemical, biochemical and gas sensors.[6] Also, polarisation-encoded sensors have been designed, where the

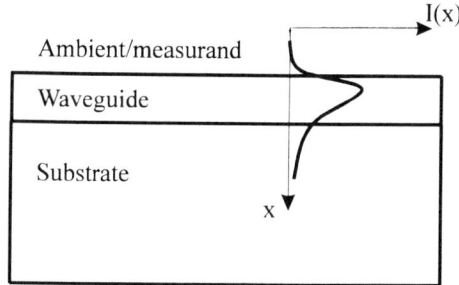

Fig. 1 Intensity distribution of the fundamental mode propagating in a waveguide.

polarisation of the guided mode changes under the influence of the measurand, for example via the change of the birifringence of the guiding film, as happens in the humidity sensor realised by Reuter and Franke.[7, 8]
Another type of sensor, the evanescent fluorescence sensor, performs immunosensing by measuring the fluorescence of the sample excited by the evanescent field in the patterned surface region of ion-exchanged buried waveguides, as shown by Zhou et al.[8]

The second category, as previously mentioned, is represented by interferometric sensors that have been realised in different configurations (Fabry-Perot, Mach-Zehnder, Michelson).[2] Interferometric sensors measure phase differences between the guided mode propagating in the reference branch and that propagating in the sensing branch, which is affected by the sample. As it is well known from conventional optics, interferometric measurements are very sensitive, precise and accurate and thus find many applications in different fields. Since a well defined light phase is necessary, the integrated optic interferometers are generally based on monomode waveguides. All these devices have quite complex guiding structures, and these are not always easy to realise with the necessary precision. In some cases the detection system is very complex.

Various technologies have been developed for fabricating integrated optic waveguides depending on the required characteristics (typically refractive index profile, difference between the refractive index of the surface and that of the substrate, optical depth and width) and on the substrate used. When designing a sensor, the substrate, the waveguide characteristics and the structure geometry must be suited to the application and to the integration needs, and optimised for sensitivity.

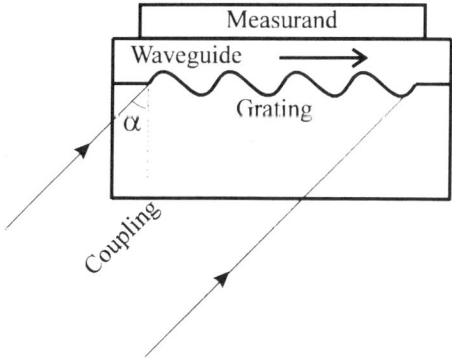

Fig. 2 An input grating coupler for sensor application.

Depending on the desired characteristics, different materials can be used.

Glass waveguides realised by the ion exchange technique (using $K^+$, $Na^+$, $Ag^+$), have been widely used thanks to their low losses and low fabrication cost. These waveguides allow the implementation of purely passive components and are well suited for high sensitivity interferometric sensors (such as refractometers, displacement and biochemical sensors) and absorption sensors. The high sensitivity of these sensors is due to the low refractive index of the substrate.[4]

In chemical and biochemical applications another type of glass waveguides has been successfully used for grating-coupler sensors. Indeed, planar $SiO_2$-$TiO_2$ thin films produced by a dip-coating process on a glass substrate result in step-index waveguides with a large refractive index difference between the film and the substrate, thus allowing one to measure film refractive index variations of about $5\times10^{-5}$, as shown by Clerc and Lukosz.[10]

Waveguides realised on silicon substrates are very attractive for integrated optic sensors thanks to the possibilities of integrating detectors and electronic circuitry and of combining micromechanical structures on the same chip. Waveguides are generally fabricated by depositing materials such as silicon nitride, silicon oxide (doped with phosphorus or boron) and silicon oxy-nitride on buffer layers of $SiO_2$. The main applications of these devices are interferometric pressure sensors[11] and displacement sensors.[12]

Polymeric materials are very promising thanks to their large piezo-electric and electro-optic coefficients. Polymer waveguides are fabricated by deposition from liquids, spin coating and dip coating techniques. They exhibit high potential for chemical, biological and medical sensors due to measurand-specific optical changes.[8]

III-V compounds, thanks to their electro-optic properties and to their possible integration with light sources, detectors and electronic circuitry on the same substrate, are very promising for the realisation of compact and handy sensors. The waveguides are realised by epitaxy. There is great interest in the applications of these materials in the near future.[13]

Finally, lithium niobate ($LiNbO_3$) is a particularly interesting material due to its transparency, as well as its piezo-electric, thermo-optic and electro-optic properties. It is possible to achieve phase modulation by the application of an electric field. Due to the crystal birefringence, its behaviour is polarisation dependent. Moreover, the fabrication techniques of waveguides in $LiNbO_3$ by Ti-indiffusion and by proton exchange (PE) are well standardised. As a consequence, $LiNbO_3$ waveguides are well suited for the realisation of electric field sensors,[14] pressure and temperature sensors[15] and for interferometric displacement sensors.[16] In the latter type of sensors, the electro-optic coefficient of $LiNbO_3$ is generally exploited to obtain a better signal-to-noise ratio by phase-modulating the input signal.[17]

When dealing with sensor applications, sensitivity is a major point, and the sensitivity of a sensor depends on the magnitude of variations of the measured parameter (in the case of integrated optic sensors, typically phase, intensity, etc. of the guided mode), compared with variations in the measurand characteristics. For all sensors exploiting the evanescent field, maximum sensitivity is achieved using thin waveguiding films with refractive index high difference between the surface and the substrate. When variations of the film refractive index are monitored, for a given refractive index difference between the surface and the substrate the sensitivity increases with the guide depth.

Although the working principle of integrated optic sensors is rather simple from a theoretical point of view, they are sophisticated guiding structures, thus requiring high precision and accuracy in the fabrication process. It is therefore very important to develop models that allow the realisation of the waveguides within small tolerances and to be able to characterise them with high precision.

All the sensors described above are based on changes of the optical parameters of the

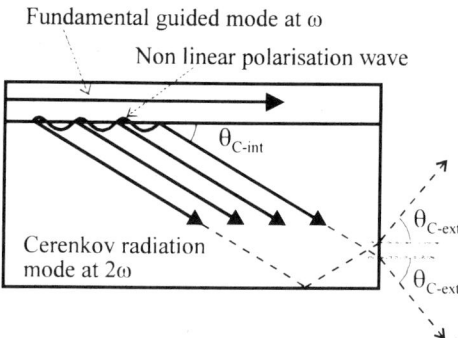

Fig. 3 Schematic diagram of second harmonic generation of Cerenkov radiation at 2ω from the fundamental guided mode at ω.

waveguide operating in the linear regime. However, the nonlinearities of suitable waveguides can also be exploited for sensor applications, as shown by Frlan et al.,[18] who realised a nonlinear waveguide optical spectrometer and deformation sensor. Indeed, nonlinear phenomena are in general very sensitive to any change in the boundary conditions that affect the propagation characteristics, and thus in principle well-suited for sensing. Yet this makes waveguide design and characterisation even more critical.

## 3. CHARACTERISING AND MODELLING LiNbO$_3$ WAVEGUIDES

Our laboratory has already acquired some experience in the design and characterisation of planar lithium niobate waveguides obtained both by titanium indiffusion (graded-index waveguides) and by proton exchange (step-index waveguides).

As previously mentioned, lithium niobate is a well-known material and waveguide fabrication techniques related to it are standardised. It exhibits high second-order nonlinear coefficients, and its guiding characteristics depend on the cut and on the fabrication process.

Second harmonic generation can be obtained in such waveguides, not only as a guided mode but also as a radiation mode in the so-called Cerenkov configuration.[19] A schematic diagram of Cerenkov second harmonic generation is shown in Fig. 3. The phase matching condition between the two interacting modes, $\beta(2\omega) = 2\beta(\omega)$ (where $\beta$ is the propagation constant of the mode), is automatically satisfied by a self-adjustment of the Cerenkov output angle provided that the $n_{eff}$ of the fundamental guided mode is smaller than the substrate refractive index "seen" by the second harmonic radiation mode ($n_{eff}(\omega) < n_{sub}(2\omega)$). Owing to the dispersion phenomenon, this condition is fulfilled for almost all modes, the only possible exception being the lower order modes in multimode step index waveguides with a large difference between the refractive indices of the guide and of the substrate.

Second harmonic generation in the Cerenkov configuration makes it possible to evaluate the effective refractive index of the fundamental mode coupled into the waveguide through the measurement of the output angle of the second harmonic radiation mode, and thus it allows a full characterisation without having to outcouple the fundamental mode itself, as shown by Sanford and Robinson.[20] This property can be usefully exploited in all second-order nonlinear waveguides where conventional characterisation by prism coupling is difficult to achieve (e.g., channel waveguides, or waveguides realised on soft materials where the pressure of the coupling prism can damage the film surface).

We verified the possibility of utilising Cerenkov second harmonic generation as a tool for waveguide characterisation in step index planar Z-cut LiNbO$_3$ waveguides fabricated by

Fig. 4 Far field of the Cerenkov second harmonic radiation (direct and reflected beam).

the proton exchange process in benzoic acid. This kind of waveguide was chosen due to its good conversion efficiency. The parameters of the two waveguides (PE#1 and PE#2) used in the experiment reported below are:

$d_1 = 0.40$ µm $\quad \Delta n_1(\lambda = 0.6328$ µm$) = 0.1179$
$d_2 = 1.38$ µm $\quad \Delta n_2(\lambda = 0.6328$ µm$) = 0.1179$

where d is the optical depth and $\Delta n$ is the difference between the surface and the substrate refractive index. By coupling the beam of a cw Nd:YAG laser ($\lambda = 1.064$ µm) into the two waveguides, TM guided modes were obtained that generated a second harmonic Cerenkov radiation mode (at $\lambda = 0.532$ µm). Figure 4 shows the far field pattern of the Cerenkov radiation: the two spots that can be seen on the lower and upper part of the screen correspond respectively to the radiation emitted directly through the substrate and the part of it reflected from the lower surface of the substrate. Comparative characterisation measurements were performed in the linear regime by the traditional prism coupling technique.

Table I shows the comparison betweeen the $n_{eff}$ obtained from the standard technique and that calculated from the Cerenkov angle measurements. The two sets of values differ from each other only in the fourth decimal place, corresponding to the resolution of the Cerenkov angle measurement method.

The results obtained are indeed encouraging in terms of resolution, and confirm the potential of Cerenkov second harmonic generation as a tool for second order nonlinear waveguide characterisation. Waveguide characterisation by exploiting Cerenkov second harmonic generation offers several advantages as compared with the conventional prism coupling technique. First of all, the $n_{eff}$ of the fundamental mode as calculated from the output angle of the Cerenkov radiation is somehow an average value over the whole waveguide, since second harmonic generation takes place all along the waveguide. On the contrary, the $n_{eff}$ obtained by the prism coupling angle depends on the prism position along the waveguide and is therefore affected by possible inhomogeneities of the guide itself. Moreover, in the prism coupling technique, the value of $n_{eff}$ is affected by the coupling conditions (dimensions of the coupling area and pressure of the prism on the waveguide). Last, but not least, the calculation of $n_{eff}$ in the case of the prism coupling technique requires knowledge of the prism characteristics (refractive index and angles) with very high

Table I. Determination of $n_{eff}$ at 1.064 µm for the modes guided in the two PE:LiNbO$_3$ waveguides described in the text, by the standard prism coupling technique and by measurement of the Cerenkov angle.

| Guide | Mode | $n_{eff}$ (prism coupling) | $n_{eff}$ (Cerenkov) |
|---|---|---|---|
| PE#1 | $TM_0$ | 2.1572 | 2.1576 |
| PE#2 | $TM_0$ | 2.2260 | 2.2263 |
|      | $TM_1$ | 2.1665 | 2.1668 |

precision, which is not always easy to achieve. This problem is not present for Cerenkov characterisation; however, in this case, the calculation of $n_{eff}$ requires the knowledge of the substrate refractive indices which is also rather difficult to achieve with the necessary precision, but which is at least of general interest for a full characterisation of the waveguide.

We also performed some preliminary tests on Ti:indiffused waveguides. In this case, the conversion efficiency is much lower, since it is proportional to the amplitude of the refractive index increase and is higher, the more discontinuous is the index profile.[21] Hence, a gradient index waveguide with a slight index increase, such as a Ti:indiffused waveguide, is indeed not the most suitable for Cerenkov second harmonic generation. As an example, we report the results obtained in a Y-cut, X-propagation Ti:LiNbO$_3$ waveguide with the following optical parameters:

ordinary optical depth = 4.24 µm $\qquad \Delta n(\lambda = 0.6328 \text{ µm}) = 0.0056$

extraordinary optical depth = 3.53 µm $\qquad \Delta n(\lambda = 0.6328 \text{ µm}) = 0.0076$.

The prism coupling characterisation gives an $n_{eff}$ of 2.2341 for the $TM_0$ mode at 1.064 µm (against an expected value of 2.2337 according to fabrication parameters), whereas Cerenkov characterisation gives a value of 2.234. Only three decimal digits are reported since the precision of the measurement was lower due to the lower efficiency, as compared with PE waveguides, and to the fact that the waveguide exhibited phase matching for Cerenkov second harmonic generation from $TM_0$ at $\omega$ to a TE radiation mode at $2\omega$, hence a very small output angle, rather difficult to measure. We expect to achieve better results in the characterisation of Ti:LiNbO$_3$ by means of the Cerenkov effect (that is to obtain a precision at least on the fourth decimal digit of the effective refractive index) by using a pulsed high-power laser source to increase the process efficiency. Since the conversion efficiency also increases with mode confinement,[22] good results should more easily be obtained in the case of Ti:LiNbO$_3$ channel waveguides.

The sensitivity of the output angle of Cerenkov second harmonic radiation to variations of the $n_{eff}$ of the corresponding fundamental guided mode suggests the possibility of exploiting the phenomenon for sensor applications. Preliminary measurements recently performed in our laboratory gave encouraging results.

Characterisation of fabricated waveguides is not the only critical issue for practical applications. Indeed, especially in the case of sensors, it is very important to achieve good control of the optical characteristics of the waveguides as related to the fabrication parameters, so as to realise waveguide sensors with the desired characteristics. Thus, for

Table II. Comparison between experimental and expected $n_{eff}$ of a Ti:LiNbO$_3$ waveguide fabricated according to the model described in the text (the values shown in boldface correspond to the phase matching condition for guided second harmonic generation).

| Wavelength (µm) | Mode | $n_{eff}$ (experimental) | $n_{eff}$ (expected) |
|---|---|---|---|
| 0.633 | TE$_0$ | 2.2100 | 2.2102 |
|  | TE$_1$ | 2.2060 | 2.2060 |
|  | TE$_2$ | 2.2034 | 2.2036 |
|  | TM$_0$ | 2.2908 | 2.2908 |
|  | TM$_1$ | 2.2883 | 2.2884 |
|  | TM$_2$ | 2.2867 | 2.2867 |
| 0.532 | TE$_0$ | 2.2429 | 2.2427 |
|  | TE$_1$ | 2.2390 | 2.2390 |
|  | TE$_2$ | 2.2359 | 2.2362 |
|  | **TE$_3$** | **2.2345** | **2.2344** |
| 1.064 | **TM$_0$** | **2.2346** | **2.2344** |

proper design, accurate modelling is needed. In our laboratory we developed a model for titanium indiffused lithium niobate waveguides able to generate guided second harmonic radiation at fixed operating wavelength and temperature ($\lambda = 1.064$ µm and T = 25°C).[23] These waveguides were actually designed for applications in fields different from sensing, but the model developed, in terms of the relationship between optical and fabrication parameters of the waveguide, can be exploited to design waveguides for different applications simply by choosing different optical parameters. The results obtained demonstrate that it is possible to achieve a fine control on the realisation of the waveguides even in an application as delicate as guided second harmonic generation. Our investigation was performed on planar Y-cut X-propagation waveguides. To satisfy the phase matching condition between the interacting guided modes (fundamental and second harmonic), the birefringence of the material was exploited. Due to the cut and the propagation direction chosen for the waveguides, the fundamental radiation was TM polarised whereas the second harmonic was a TE mode. As the starting point for our model, we assumed the relationships given by Fouchet et al.[24] between the optical characteristics of the waveguide (refractive index increase and optical depth) and the fabrication parameters (diffusion temperature T, indiffusion time t and titanium thickness $\tau$). The diffusion temperature was kept at 1000°C, where we know from our previous experience the titanium diffusion coefficient in lithium niobate. We then chose fabrication conditions that allowed the phase matching condition to be achieved with low sensitivity with respect to fluctuations in the titanium layer thickness and indiffusion time, within our typical tolerances. The analysis led to the following fabrication parameters: t = 18 h 20 min and $\tau$ = 480 Å, corresponding to phase matching between the TM$_0$ mode at $\lambda = 1.064$ µm and the TE$_3$ mode at $\lambda = 0.532$ µm.

The waveguide was fabricated according to the above design and characterised by the prism coupling technique at different wavelengths (0.532, 0.633 and 1.064 µm). The results are reported in Table II, together with those calculated from the model. At 0.633 µm all the characterisation data are reported, whereas at the fundamental (1.064 µm) and second harmonic (0.532 µm) wavelengths, only the data of interest for second harmonic generation are shown. The agreement between the values calculated according to our model and the experimental values confirm the accuracy of the proposed model. Moreover, nonlinear testing of the waveguide demonstrated guided second harmonic generation between the expected modes with the expected conversion efficiency.

## 4. CONCLUSIONS

Integrated optic sensors are widely studied today due to their great potential for applications. For the design of such sensors, accurate waveguide modelling is required and very accurate characterisation techniques are needed. Among the possible characterisation techniques, when sensors are realised in second-order nonlinear waveguides, second harmonic generation in the Cerenkov configuration can be exploited.

Since nonlinear effects are in general very sensitive to any change in the boundary conditions and therefore in the propagation characteristics, they are very likely to play an important role in integrated optic sensors in the near future. The use of nonlinear materials might also allow the realisation of smart sensors performing all-optical signal processing.

ACKNOWLEDGMENTS. The author wishes to acknowledge the valuable scientific contributions of V. Russo, S. Bruno, D. Conti, M. Marangoni, R. Osellame (Integrated Optics Laboratory) and of S. De Silvestri, V. Magni, G. Cerullo, M. Zavelani-Rossi (Solid-State Laser Laboratory) of the Department of Physics, Polytechnic of Milan, where the experimental work was performed. Ti:LiNbO$_3$ waveguides were fabricated by L. Palchetti at IROE-CNR, Florence, whereas PE-LiNbO$_3$ waveguides were kindly provided by M. Varasi of Alenia, Rome.

The research work was partially supported by the «Non linear Optical Guiding Structures for All-Optical Signal Processing» Research Project of the CNR Technological Sciences and Innovation Committee.

## REFERENCES

1. A.J. Rogers, Optical-fiber sensors, in: *Sensors - a Comprehensive Survey*, W. Göpel, J. Hesse and J.N. Zemel, ed., vol.6: Optical Sensors, E.Wagner, R. Dändliker and K. Spenner, ed., p. 355, VCH, Weinheim (1992).
2. A. Brandenburg, V. Hinkov and W. Konz, Integrated optic sensors, in: *Sensors - a Comprehensive Survey*, W. Göpel, J. Hesse and J.N. Zemel, ed., vol.6: Optical Sensors, E.Wagner, R. Dändliker and K. Spenner, ed., p. 399, VCH, Weinheim (1992).
3. O. Parriaux, Integrated optics sensors, in: *Advances in Integrated Optics*, S. Martellucci, A.N. Chester and M. Bertolotti, ed., p. 227, Plenum Press, New York (1994).
4. W. Lukosz, Integrated optical chemical and direct biochemical sensors, *Sens. Actuators B* 29:37 (1995).
5. S.-W. Kang, K. Sasaki and H. Minamitami, Sensitivity analysis of a thin-film optical waveguide biochemical sensor using evanescent field absorption, *Applied Optics* 32:3544 (1993).
6. W. Lukosz and K. Tiefenthaler, Sensitivity of integrated optical grating and prism couplers as (bio)chemical sensors, *Sens. Actuators* 15:273 (1988).
7. R. Reuter and H. Franke, Monitoring humidity by polymide lightguides, *Appl. Phys. Lett* 52:778 (1988).
8. H. Franke, D. Wagner, T. Kleckers, R. Reuter, H.V. Rohitkumar and B.A. Blech, Measuring humidity with planar polyimide light guides, *Appl. Opt.* 32:2927 (1993).
9. Y. Zhou, J.V. Magill, M.R. De La Rue, P.J.R. Laybourn and W.Cushlay, Evanescent fluorescence immunoassays performed with a disposable ion-exchanged patterned waveguide, *Sens. Actuators B* 11:245 (1993).
10. D. Clerc and W. Lukosz, Integrated optical output grating coupler as refractometer and (bio-) chemical sensor, *Sens. Actuators B* 11:461 (1993).
11. K. Fischer, D. Zurhelle, R. Hoffmann, F. Wasse and J. Muller, Fully integrated optical force and pressure sensor based on SiON layers, in: *Proc. of ECIO 93*, P. Roth, ed., p. 12-7, CSEM, Neuchâtel, CH, April 18-22 (1993).

12. S. Ura, M. Shinouara, T. Suhara and H. Nishihara, Integrated-optic grating-scale-displacement sensor using linearly focusing grating couplers, *IEEE Photonics Technology Letters* 2:239 (1994).
13. T. Suhara, T. Taniguchi, M. Umegaki, H. Nishihara, T. Hirata, S. Iio and M. Suehiro, Monolithic integrated-optic position/displacement sensor using waveguide gratings and QW-DFB laser, *IEEE Photonics Technology Letters* 10:1195 (1995).
14. C.H. Bulmer, Sensitive, highly linear lithium niobate interferometers for electromagnetic field sensing, *Appl. Phys. Lett.* 53:2368 (1988).
15. S. Cucurachi, A. D'Orazio, M. De Sario, V. Petruzzelli and F. Prudenzano, Design of a Ti:LiNbO$_3$ sensor for the simultaneous measurement of stress and temperature, in: *SPIE, Measurement Technology and Intelligent Instruments*, Vol.2101, p. 340 (1993).
16. A. D'Alessandro, M. De Sario, A. D'Orazio and V. Petruzzelli, Design criteria of an integrated optics microdisplacement sensor, in: *SPIE, Optical Testing and Metrology*, Vol. 1332, p.554 (1990).
17. J. Bauer, E. Dammann and E. Fritsch, Integrated optical 3-beam interferometer for distance measurements, in: *Proc. of ECIO 93*, P. Roth, ed., p. 12-25, CSEM, Neuchâtel, CH, April 18-22 (1993).
18. E. Frlan, J.S. Wight, S. Janz, H. Dai, F. Chatenoud, M. Buchanan and R. Normandin, High resolution surface-emitting spectrometer and deformation sensors with nonlinear waveguides, *Optics Letters* 19:1657 (1994).
19. P.K. Tien, R. Ulrich and R.J. Martin, Optical second harmonic generation in form of coherent Cerenkov radiation from a thin film waveguide, *Appl. Phys. Lett.* 17:447 (1970).
20. N.A. Sanford and W.C. Robinson, Direct measurement of effective indices of guided modes in LiNbO$_3$ waveguides using the Cerenkov second harmonic, *Opt. Lett.* 12:445 (1987).
21. M.J. Li, M. De Micheli, Q. He and D.B. Ostrowsky, Cerenkov configuration second harmonic generation in proton-exchanged lithium niobate guides, *IEEE J. Quantum Electr.* 36:1384 (1990).
22. M.P. De Micheli, Second harmonic generation in Cerenkov configuration, in: *Guided Wave Nonlinear Optics*, D.B Ostrowsky and R. Reinisch, ed., NATO-ASI Series, p. 147, Kluwer Academic Publishers, Boston-London (1992).
23. R.Ramponi, V.Russo, V.Magni, L.Palchetti, R.Osellame, M.Zavelani-Rossi, Waveguides in Ti:LiNbO$_3$ for second harmonic generation: design and experimental tests, in: *SPIE* vol. 2954 (1996) in press.
24. S. Fouchet, A. Carenco, C. Daguet, R. Guglielmi and L. Rivière, Wavelength dispersion of Ti induced refractive index change in LiNbO$_3$ as a function of diffusion parameters, *J. Lightwave Technol.*, LT-5:700 (1987).

# MICROSYSTEMS AND APPLICATIONS

# MICROOPTICAL COMPONENTS AND SYSTEMS FABRICATED BY THE LIGA PROCESS

C. Müller

Albert-Ludwigs-Universität Freiburg, Institut für Mikrosystemtechnik
Universitätsgelände Flugplatz, Gebaude 052, 79085 Freiburg, Germany

## 1. INTRODUCTION

In microsystem technology the LIGA technique is well established for the fabrication of microstructures. By use of the combination of deep-etch X-ray lithography with synchrotron radiation, electroforming and moulding techniques microstructures, with mechanical and optical functions for optical communication technology and fibre optical sensing are produced. In this paper the development and quality of LIGA microoptical components and systems, as well as their applications in industrial products will be discussed.

The LIGA process,[1] developed at the Forschungszentrum Karlsruhe, allows the fabrication of microstructures of any lateral shape, with structural heights of serveral hundred micrometers, lateral dimensions down to one micrometer, and submicron accuracy over the total height of the structure.[2] For optical applications additional qualities like small material attenuation on the order of 0.3 dB/cm in the visible and near infrared region, good material homogeneity, and a mean surface roughness of 40 nm are of importance. Furthermore, the moulding techniques involved in the process permit mass fabrication and the use of a broater variety of materials. Another main advantage of the LIGA technique is the aligned and simultaneous patterning of microoptical components, mounting supports, and fixing elements, e.g. fibre fixing grooves. Thus, compact and stable microoptical set-ups with a minimum of assembly and mounting efforts can be realised.

The results presented in this Chapter were obtained at Forschungszentrum Karlsruhe Institut für Mikrostrukturtechnik.

## 2. FIBRE COUPLING ELEMENT FOR MULTIMODE FIBRES

Fibre coupling elements [3] can be manufactured from miniaturised optical components, such as prisms or mirrors, and fibre fixing grooves, accurately positioned with respect to these components. The optical elements are the size of the fibre core diameter, so that losses due to divergence are minimised. As a consequence of the high precision in

Fig. 1 SEM picture of a fibre coupling element for use with multimode fibres.

Fig. 2 SEM picture of the microoptical bench.

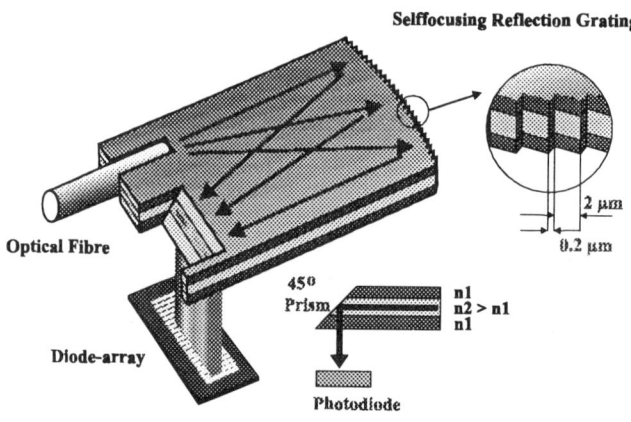

Fig. 3 Principle of the microspectrometer device.

positioning achieved by the lithographic technique, no critical alignment of the fibre is required. In Fig. 1 a coupling element for use in a bi-directional data transfer line is shown.

Three fibres are arranged with respect to each other to form a **T**. Some of the light emitted by fibre 1 is passed though the air gap into fibre 2 and is guided to a fibre optical sensor. The modulated sensor signal is transferred back into fibre 2 and is partly reflected by the prism hypotenuse in fibre 3 which can be detected.

When such a coupling element is used as a part of a pneumatic actuator system it allows on-line monitoring of the actuator motion. The design of the coupling element is flexibly adaptable to any multimode fibres and to any desired coupling ratio. Due to the small size the elements can be easily integrated in microsystems. By this means, various applications in fibre sensing or in LANs with multimode fibres are possible.

## 3. BI-DIRECTIONAL TRANSCEIVER MODULE

For the set-up of a low-cost high-performance bi-directional optical transceiver module[4] for subscriberapplications, a microoptical bench has been fabricated (Fig. 2). The bench enables low-cost and self-aligned lateral positioning of passive optical components with lithographic precision. The module is constructed from a laser diode, a photo diode, a microoptical bench containing two ball lenses and a wavelength filter and single mode fibre. The light of the laser diode, the transmitter, is collimated by the ball lens in front of it, passes trough the wavelength filter and is focused with a second ball lens into the fibre core.

Incoming light from the fibre is collimated by the same ball lens and reflected by the filter into the photo diode. Due to this optical set-up, the duplexer module is relatively insensitive to thermal displacements and dust contamination.

For mounting the system the diodes are laser-welded and the tube is soldered to the housing. Initial experimental results showed that the concept is capable of accomplishing the objectives. Nevertheless, environmental tests between 0°C and 65°C caused changes in the insertion loss of the modules of 1 dB, and call for further optimisation.

## 4. MICROSPECTROMETER

As an example of a light guiding microstructure, the spectrometer[5] set-up shown in Fig. 3 has been developed. The waveguide consists of a polymethylmetacrytale (PMMA) or fully deuterated PMMA-d8 core and a cladding from copolymers of PMMA and tetrafluoropropyl methacrylate (TFPMA). The mean material attenuation in the visible and near infrared spectral range is on the order of 0.3 dB/cm, which is small enough for the optical path length of 2 cm to 3 cm. The thickness of the different layers is matched to the fibre parameters (50/125 µm) multimode fibre, (N.A. = 0.2) by which the light is launched into the waveguide. The position of the fibre end face is given by a fibre fixing groove and is adjusted to the grating on a Rowland circle. The blazed reflection grating is structured with optimised individual positions of each single grating tooth. Thus, imaging losses on the linearised focal line can be neglected.

A high reflectivity is achieved by coating the grating with a 100 nm thick silver layer. The dispersed light is totally reflected by the inclined sidewall of the focal line (45° inclination angle) and can be analysed with a photodiode which is positioned on top of the waveguide. The photodiode used is a HAMAMATSU type S5464 with 512 pixels and a pixel size of 25 µm x 500 µm. The pixel length of 500 µm makes the positioning of the photodiode with respect to the focal line very simple. The dispersed light is focused on the photodiode surface.

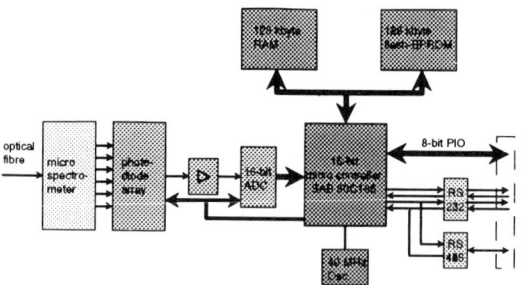

Fig. 4 Block diagram of the evaluation board.

## 5. EVALUATION BOARD

An electronic board using surface mounted device packaging has been developed for the readout of the photodiode array. As a result, the total size of the system is only 60 mm x 70 mm. The evaluation board includes the LIGA microspectrometer, the photodiode array and the complete electronic circuit for data acquisition and evaluation. The system is powered by a 5 V power source and consumes only 800 mW. The diode output signal is processed analytically and fitted to the input level of the A/D-converter. The drive of the diode array is generated by a µ-controller which also controls the A/D-converter, the read in for the spectral data, the pre-processing and evaluation of the data as well as the communication with the host computer. Fig. 4 shows the block diagram of the evaluation board. The communication with the host computer is performed by serial ports (RS232, RS485).

The spectral range from 400 nm to 1100 nm is covered with one grating. Other spectral regions or spectral bandwiths can be obtained by varying the grating parameters. The dynamic range of the system is a function of the spectral distributions of the lightsource, the waveguide, the grating and the photodiode array.

Results achieved in reflection, transmission and flourescence measurements are comparable to results achieved with conventional spectrometers and prove the performance of the microspectrometer system to be sufficient for the envisioned applications, e.g. on-line process control or colour detection. The integrated optical set-up has the potential of cost effective fabrication by moulding techniques. There is potential for further miniaturisation

Fig. 5 Photograph of a mounted microspectrometer system in an alluninum housing.

Table I. Features of the microspectrometer system.

| features | value |
|---|---|
| spectral range | 380 nm - 780 nm, 400 nm - 1100 nm |
| transmission at $\lambda_{blaze}$ | 25 % |
| spectral resolution | 5 nm, 7 nm |
| attenuation of scattered light | 25 dB |
| optical fibre | 50/125 µm, 105/125 µm |
| photodiode array | Hamamatsu S 5464-256Q |
| micro controller | Siemens SAB 80C166 |
| 16 Bit A/D-converter | Burr-Brown ADS 7807U |
| dynamic range | 10.000 - 25.000 :1 |
| measuring time | 40 ms - 2560 ms |
| application | colour measurement, spectral analysis |

and cost reduction by fabrication of highly adapted photoasics, including the whole electronic evaluation board. In this case the moulding step can be performed directly on top of the photoasic, which both decreases the size of the system and assembling efforts.

The input parameters, e.g. integration time, amplification factor and number of the averaged spectra are inquired by the PC and transferred to the system. The pre-processed readout data are transmitted to the PC and are displayed on a monitor. A complete system in a housing is shown in Fig. 5 and Table I lists some features.

Encouraged by the performance of the spectrometer system for the visible wavelength range based on the three layer polymer waveguide, we also fabricated a spectrometer for the NIR, with a design modified to overcome material absorption of the polymer. Thus, the microspectrometer for the infrared wavelength range is based on the principle of light guidance by reflection between plates, with an integrated self-focusing blazed reflection grating coated with evaporated gold. The gating and the fixing structures for the optical fibres to inject and detect the radiation are patterned in PMMA on a quartz or silicon substrate, and positioned with respect to each other in one fabrication step.

Fig. 6 shows the basic layout of the spectrometer system.

Unlike the visible range, no sufficiently broad windows in the absorption spectrum of plastics exist in the IR range to allow a light guiding three layer polymer to be build up. Therefore we decided to fabricate a plate-type wave guide spectrometer in the IR, where radiation is conducted between the plate surfaces though reflection on the metal. Normally, the hollow space is filled with air but it can be filled with other gases as well. The optical

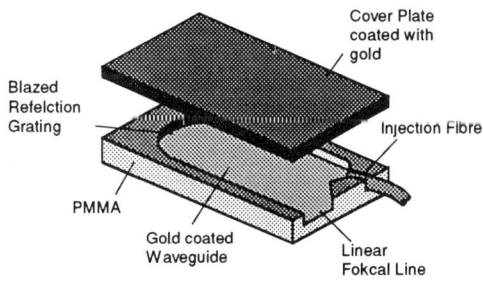

Fig. 6. Basic sketch of the IR spectrometer element.

Table II. Features of two prototypes for the infrared wavelength range.

| features | value | value |
|---|---|---|
| microspectrometer | prototype 1 | prototype 2 |
| spectral range | 1.6 µm - 2.0 µm | 2.7 µm - 3.3 µm |
| transmission at $\lambda_{blaze}$ | 22 % | 24 % |
| spectral resolution | 9.8 nm | 34 nm |
| optical fibre | 90/100 µm quartz | 100/150 µm fluoride |
| application | analysis of plastic reflection spectra | detection of burnable gases |

function elements are integrated in this planar structure. As for the visible range, the unit in its minimal configuration consists solely of the grating and of structures for in and out coupling the radiation. In addition, a mirror [6] is provided which allows increasing the local distance between the coupling in and coupling out structures. Thus, it is also possible to use the unit together with linear detector arrays without any optics between the spectrometer component and the detector array. Features of two prototypes for the infrared wavelength range are shown in Table II.

## 6. CONCLUSION

The examples for both free-space and light guiding applications demonstrate the quality of the LIGA microoptical components and systems. The fabrication of microoptical benches for applications with single mode fibres, demonstrates the advantages of hybrid mounted optical systems.

## REFERENCES

1. P. Bley, J. Göttert, M. Harmening, M. Himmelhaus, W. Menz, J. Mohr, C. Müller, and U. Wallrabe: The LIGA-process for Fabrication of Micromechanical and Microoptical Components, *Micro System Technology*, ICC Berlin, 1991
2. J. Göttert, J. Mohr, and C. Müller: Examples and Potential Applications of LIGA Components in Microoptics, *Teubner-Texte zur Physik*, Bd. 27, B.G. Teubner Verlagsgesellschaft, Stuttgart, Leipzig (1993), S. 219-247
3. J. Göttert, J. Mohr, C. Müller, and H. Sautter: Coupling Elements for Multimode Fibres by the LIGA Process: in H. Reichl (Ed.), *Micro Systems Technologies 92*, VDE-Verlag, Berlin, Offenbach, (1992), S. 297-307
4. A. Müller, J. Göttert, and J. Mohr: LIGA-Microstructures on Top of Micromachined Silicon Wafers Used to Fabricate a Microoptical Switch, *Pro. MME'93, Neuchâtel*, September 1993
5. C. Müller, O. Fromhein, J. Göttert, T. Kühner, and J. Mohr: Microspectrometer System Based on Integrated Optic Components in Polymers as Spectral Detection System for the VIS- and NIR-Range, *Proceedings 7$^{th}$ Eur. Conf. on Integrated Optics (ECIO '95)*, pp. 491-494, Delft, April 1995
6. P. Krippner, J. Mohr, C. Müller, and C. van der Sel: Microspectrometer for the infrared range, *Proceedings European Symposium on Lasers, Optics, and Vision for Productivity in Manufacturing I, Besancon, France*, June 1996

# SENSORS AND MICROSYSTEMS: ELECTRONIC NOSE

C. Di Natale, A. D'Amico, F.M. Davide, and G.Saggio

University of Rome "Tor Vergata", Electronic Engineering Department
Via Ricerca Scientifica s.n.c., 00133 Rome, Italy

## 1. INTRODUCTION

This Chapter has two purposes: The first purpose is to illustrate in some detail a design of an advanced chemical microsystem, also called a micro total analysis system (µTAS) [1] which can use either surface acoustic wave or optical chemical sensors in the test chamber, and which can be fabricated by silicon micromachining technology and semiconductor-glass bonding techniques; this µTAS can acquire data from chemically sensitive devices located in an ambient fluid, transform them in a suitable format for further signal processing and perform internal flow management and control; The second purpose is to present and discuss the Rome Tor Vergata electronic nose (EN), applied as an example in this paper to food analysis, using quartz micro balance (QMB) sensors and metallo-porphyrines as promising chemically interactive materials (CIM's). The current study on QMB sensor matrices is considered and oriented to achieve all the necessary information for the development of the more advanced µTAS introduced above.

## 2. SYSTEM DESCRIPTION AND PRELIMINARY CONSIDERATIONS

Figs.1 and 2 show two approaches to a silicon micromachined EN which have resulted from our more recent studies. These two figures differ from each other in their sensing strategy. The first uses a surface acoustic wave sensor array (SAW) which can be integrated on silicon by ZnO technology; the second uses an optical sensor array. Four blocks represent the functionality of this system: fluid sampling performed by a micropump and other passive micromachined elements; flow control circuits consisting of pressure microsensors, flaps as microactuators, valves and microheaters; an optical or saw sensor array located in the test chamber; signal processing conditioning and control microelectronics. Wafer to wafer and wafer to glass bonds are necessary to achieve the final system configuration.

The delivery of the fluid to the sensor array, which is considered one of the most important problems for the development of ENs, is performed by an electrostatically driven

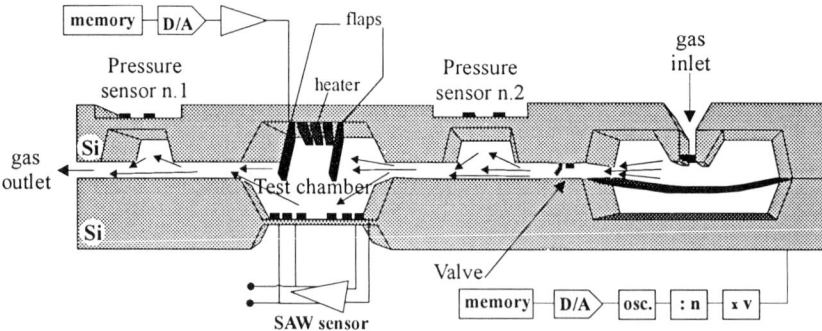

Fig. 1 Schematic of a µTAS containing the SAW sensor matrix.

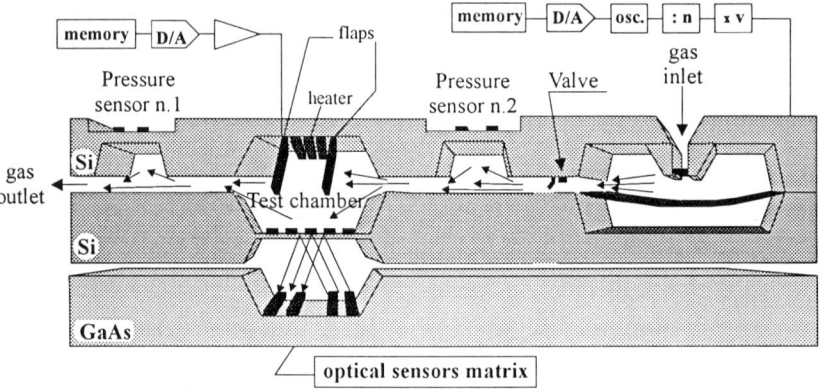

Fig. 2 Schematic of a µTAS containing the optical sensor matrix.

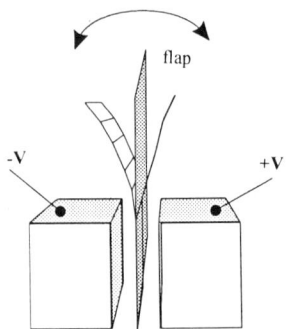

Fig. 3 Schematic of a flap together with the electrostatic driving system.

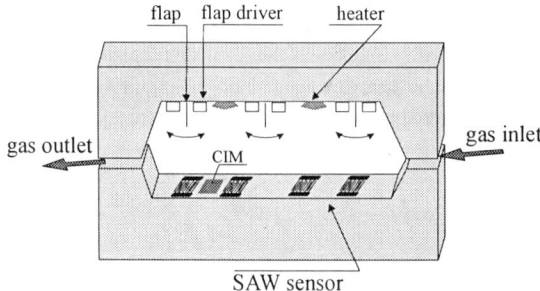

Fig. 4 Particular of the test chamber of Fig. 1.

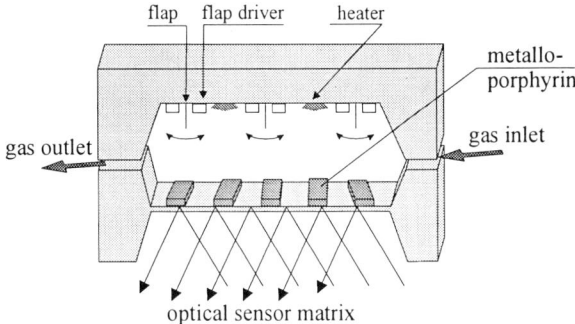

Fig. 5 Particular of the test chamber of Fig. 2.

micropump, which should be highly efficient and consume low power.

Optical gas sensors (either evanescent wave or adsorption type) are generally greatly dependent on the fluidic turbulence occurring at their surface. As a consequence a complete design of the µTAS has to include flow control which can be obtained by suitably driven microheaters, used for the formation of the necessary temperature gradients (turbulence modifiers), and miniflaps for the modulation of the flow field. Two integrated pressure sensors are utilised for the evaluation of the flow-rate.

Microheaters having dimensions in the 100 microns square range should be highly reliable, resisting thermal stress degradation, and be capable of reaching rather high temperature values (up to 400 °C); microflaps should be driven electrostatically as shown in Fig. 3.

The electronics should preferably be fabricated in CMOS technology for low power dissipation, for all the following conditioning circuits: calibration, offset correction and temperature compensation, flaps and micropump drivers.

A µTAS design for nose emulation must overcome a number of difficult problems, representing technological bottlenecks. These are, in order of complexity: fabrication of reliable micropumps, implementation of the sensor matrix, flap construction together with their electromechanical drivers, control of the turbulence in the test chamber, electronic interconnections, and finally the overall packaging.

The sensing part appearing in Fig. 2 is assumed to be optical, with the data reading apparatus initially located outside the µTAS. Based on research results, especially those related to full control of the test chamber environment, this part can also eventually be integrated.

Possible designs of the two test chambers are indicated in Fig. 4, where SAW sensors are considered as elements of the sensor matrix, and in Fig. 5, where this matrix is formed by optical sensors.

The internal top part of the test chamber is similar for the two figures; microheaters, used as heat generators, and flaps, used as heat distributors, are employed for the purpose of creating and controlling the degree of turbulence of the internal gas flow. Turbulence near the sensitive surface of the chemical sensors affects the local velocities and accelerations of the fluid molecules, which can influence the sensitivity. On the other hand, turbulence can also produce unwanted noise, which can affect the resolution. Thus having effective control of the time behaviour of the flow field is critical for the overall performance of the nose. Simulation work is currently being done in order to provide a clearer picture of the fluid phenomena occurring inside test chambers of different geometries containing different numbers of microheaters and microflaps.

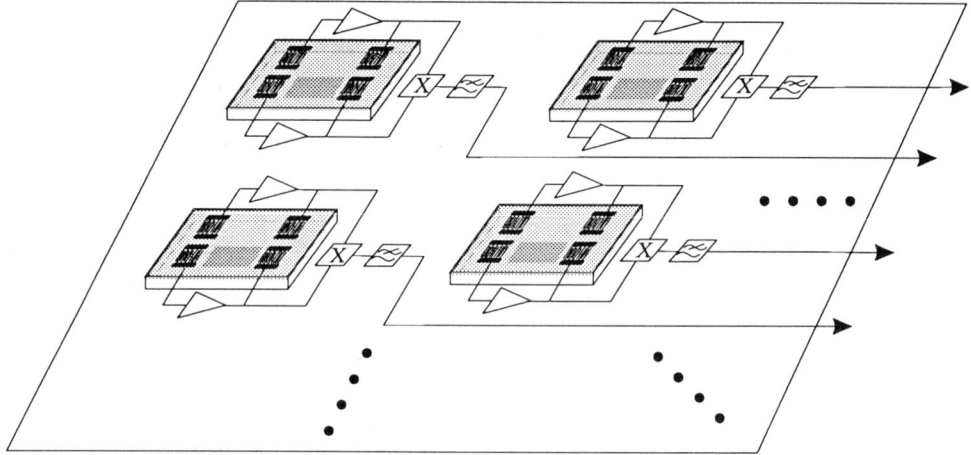

Fig. 6 Schematic of the differential SAW sensor elements. Each element has it's own mixer.

## 3. SENSORS FOR THE NOSE MATRIX

For this nose project two kinds of basic sensors have been anticipated: the first uses surface acoustic wave (SAW) sensors implemented on silicon.[2] The choice of the SAW sensor matrix integrated in silicon follows the idea, considered for future developments, of electronic integration on the same chip. For this purpose the $Si/SiO_2/ZnO$ structure has been thought to be suitable for our purpose due to the availability and reliability of ZnO technology. The ZnO is necessary for electroacoustic transduction from the interdigital transducers (IDT) to the substrate. Fig. 6 shows an example of a part of a matrix where each sensing element is formed by a differential SAW structure with a reference path; each dark region in the SAW path represents the CIM, while Fig. 7 represents five SAW sensors, each with their CIM, and with only one SAW structure used as a reference.

In the first case we have a mixer for each element; in the second case a single more complex mixer serving all 5 SAW sensors can be utilized. The number of SAW elements

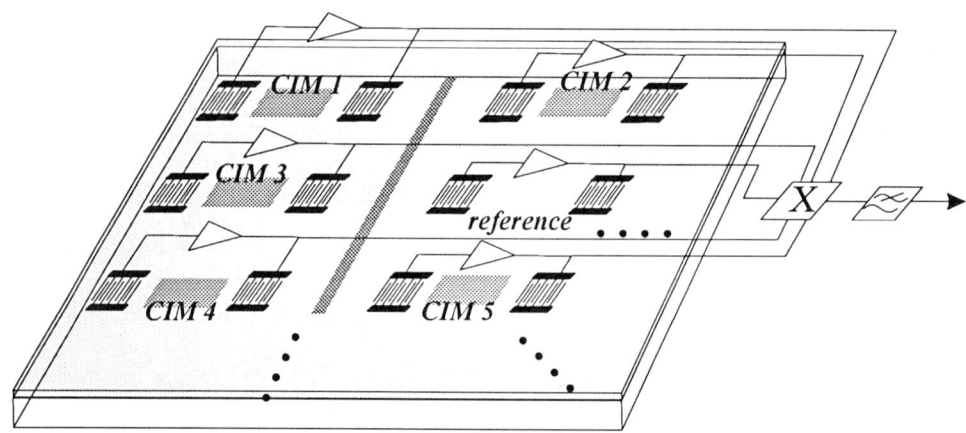

Fig. 7 Schematic of a part of a SAW matrix formed by only ne mixer serving 5 SAW sensor elements.

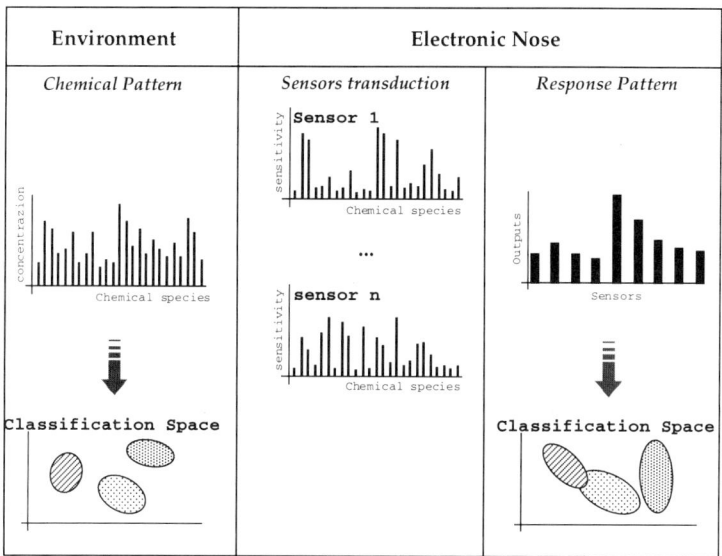

Fig. 8 Electronic nose strategy: from the chemical pattern through sensor transduction a response pattern is obtained for the construction of the final estimated classification space.

constituting a matrix is dependent on a variety of considerations at the design level and certainly it is correlated to the type of environment, its chemical complexity and the degree of total sensitivity needed for the best chemical image presentation. The differential structure is essential for the reduction of undesired common mode error sources such as temperature, pressure, relative humidity.

As far as the optical sensor is concerned; a material called metallo-porphyrin has been widely utilized in the last few years, and is considered satisfactory as a CIM.

The basic principles of electronic noses are illustrated in Fig. 8. Chemical patterns occurring in a certain environment are "translated" by the sensors into a response pattern. This translation is obtained through a deconvolution process between the chemical pattern and the spectral sensitivity of each sensor.

Compared with the original chemical pattern, the response pattern is characterized by a reduced dimensionality. Basically it is a non-linear combination of all the components which form the original pattern.

Optimal performance is achieved when the pattern translation process preserves those features which allow discrimination among the classes which are considered relevant to the particular application. This means that the translation process and the consequent scaling of dimensions does not appreciably change the qualitative properties of the patterns. In fig. 8 the classification space of the EN looks similar to that of the environment even if not identical. Of course, this process is limited by the fact that a reduction of dimensions gives, as a consequence, a reduction of the information content. For this latter reason it is very difficult, at present, to develop an electronic nose for general purpose applications, like natural noses.

On the other hand it is possible today to construct electronic nose prototypes for specific applications where only a subset of the species in the environment are relevant and, with a suitable selection of the sensor selectivities, it is possible to preserve in the response pattern, to some extent, the information relevant to the specific purpose.

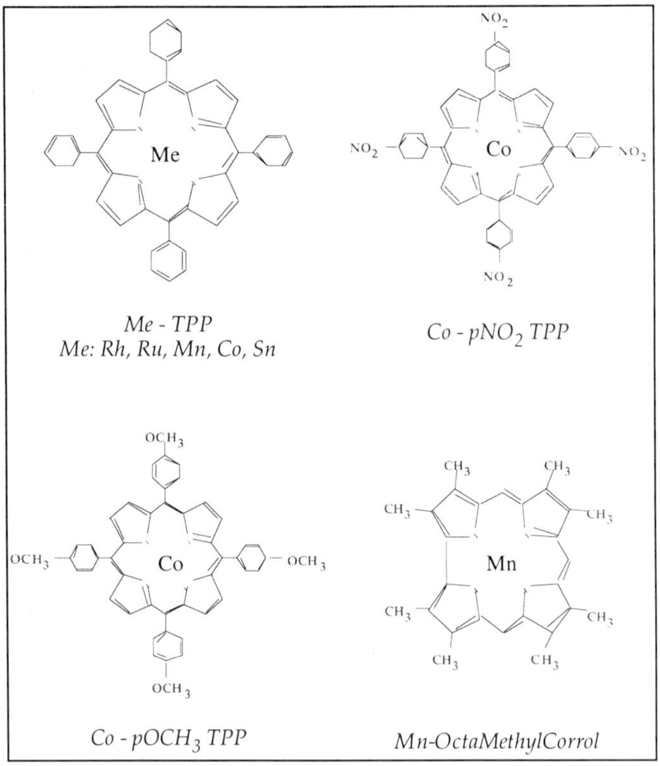

Fig. 9 Structure of some metallo-porphyrines.

## 4. THE ROME "TOR VERGATA" ELECTRONIC NOSE

During the last three years at the University of Rome "Tor Vergata" extensive studies aimed at fabricating reliable and general purpose ENs have been carried out. The project is based on the utilization of metallo-porphyrins and their related compounds as chemically interactive materials for chemical sensing.

Metallo-porphyrins are basically formed by four pyrole rings linked by methenyl groups to form a macrocycle. This basic structure can be simply modified by complexing a metal at the centre of the structure and/or linking some lateral groups around the macrocycle.

One of the first achievements of the project was the fabrication of chemical sensors using metallo-porphyrins as coatings on quartz microbalances (QMB).[3] The main feature of such sensors is the dependence of the selectivity and sensitivity on the nature of the central metal and on the lateral groups. Thus, with a little variation in the synthesis process, it is possible to get sensors which behave differently. This flexibility makes this class of compounds extremely appealing for electronic nose applications, not only in the context of the QMB sensor strategy but also when, we take advantage of their optical frequency dependence of the adsorption coefficient of these CIMs.

Here an electronic nose formed by eight such sensors is considered and briefly described. The sensors are coated with the following compounds:

Ru-*meso*TetraPhenylPorphyrin, Rh-*meso*TetraPhenylPorphyrin,

Mn*meso*TetraPhenylPorphyrin, Co-Tetra*meso*PhenylPorphyrin,
Sn-*meso*TetraPhenylPorphyrin, Co-*meso*TetrapNO2-PhenylPorphyrin,
Co-*meso*Tetra-pOCH3-PhenylPorphyrin and Mn-*meso*OctaMethylCorrole.

Fig. 9 shows the structures of these compounds.

These metals have been selected for their properties. It is well known that they act as catalysts when co-ordinated to porphyrin. For this reason they have been widely used in the past to catalyze, for example, oxidation reactions. As an example of the expected selectivity, manganese-porphyrin sensors, on the basis of the HSAB (hand soft acid basis) principle, are expected to show a great affinity for oxygen ligand donors, whereas Rh-porphyrin sensors should be more efficient for the binding of nitrogenous ligand donors. These hypotheses have been confirmed by previous experiments concerning their sensitivity to alcohols and amines.[3,4,5] It is interesting to note that these compounds play a key role in determining the freshness of food, in particular fish.

Sensors have been prepared by coating AT-cut quartz, trimmed for a fundamental frequency of 5 MHz with metallo-porphyrins. Each sensor is part of an oscillator circuit. As a consequence of a chemical binding on the metallo-porphyrin surface, the mass of coated quartz changes and the frequency of the relative oscillator changes accordingly. A linear law (called the Sauerbrey law) describesthe dependence of the quartz resonance frequency on the mass.

An extensive set of tools for data analysis has been developed at the University of Rome "Tor Vergata". A number of chemometric based methods (Principal Component Analysis, PCA, and Cluster Analysis) and Neural Networks (Feed Forward Back-Propagation Trained Networks, Self Organizing Maps, Adaptive Resonance Theory Based Networks) are utilized to analyse electronic nose data and extract the relevant information.

Basically these techniques can be divided into two groups: the first group is formed by those that can be called *exploratory techniques*, namely those aiming at revealing the intrinsic classification properties of the sensors for a particular application. The second group of techniques can be called *regressive techniques*. These are aimed at establishing a correlation between the sensor array output and the classification scheme which is important for a given application.

Principal Component Analysis, Cluster Analysis and Self Organizing Maps are examples of exploratory techniques; while Feed Forward Back-Propagation Networks and Adaptive Resonance Theory Networks are representative of the regressive techniques group.

The aptitude of the electronic nose to be utilized for the analysis of foods has been tested measuring sensor sensitivities towards several substances whose pressure and metrology are of interest in food analysis. These compounds are representative of various classes of species such as organic acids, alcohols, amines, sulphides, carbonyls.

Fig. 10 shows the sensitivities of the eight sensors, constituting the front end of the EN, to fourteen substances. Different sensitivities are mandatory for efficent behaviour of the EN. An analysis of the sensitivities shows that, as expected, the sensors behave differently one from another. These results confirm the theoretical conjectures and are in agreement with the previous published sensitivities and selectivities.[3,4,5]

It is worth keeping in mind that, due to the specific nature of the transducer, the differences in sensitivities are due not only to the intrinsic selectivity of the molecular sensing mechanism but, since the transducer is sensitive to mass variations, also to the specific mass of the molecules which are bound at the coating surface. For these reasons, for example, the strong difference in sensitivity between long-chain alcohols (such as octanol) and short-chain alcohols (such as methanol or ethanol) is mainly due to the different masses of the molecules.

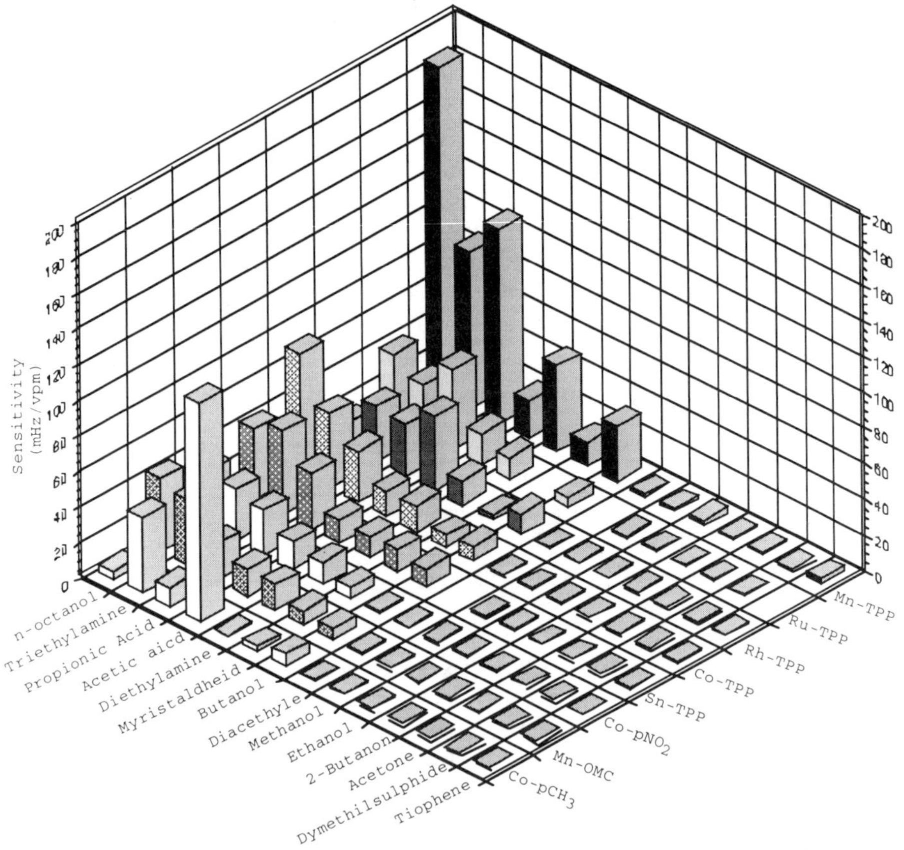

378

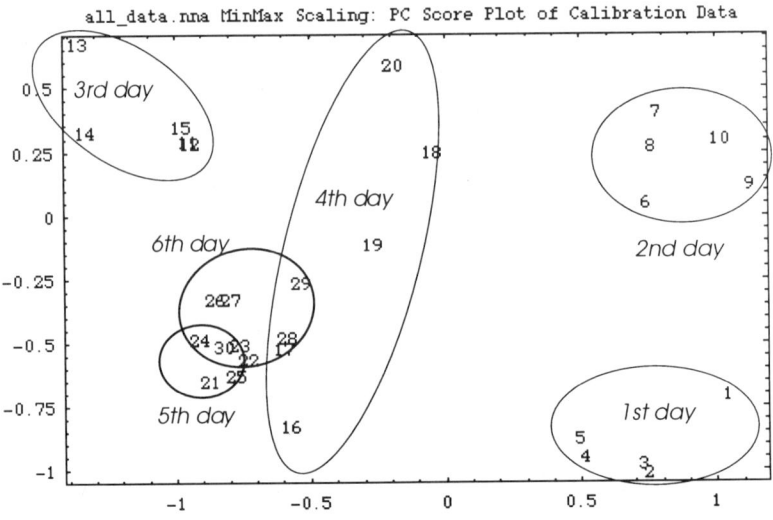

Fig. 10 Three dimensional presentation of the sensitivities versus some substance and versus different types of metallo-porphyrines.

Fig. 11 Classification space obtained by the principal component analysis technique.

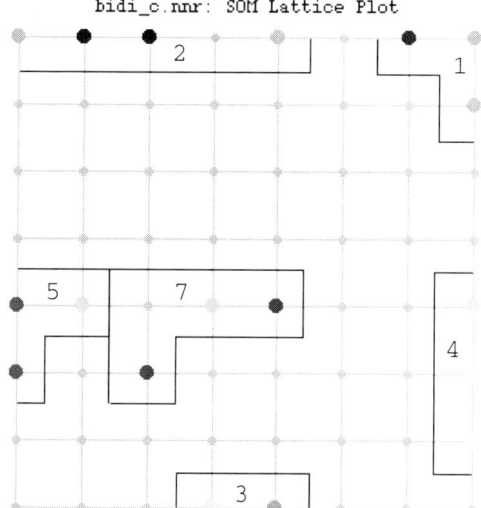

Fig. 12 10x10 SOM lattice showing the subareas relative to 6 days of food-fish ageing.

## 5. EXAMPLE OF APPLICATION TO FOOD ANALYSIS

In this paragraph some results of the application of our EN to gas environments generated by some foods are presented and discussed. These analyses concern codfish and veal meat.

The scope of the experiment was the classification and the determination of the days of storage due to the fact that such a classification is in some way related to the freshness of the products.

The samples were prepared by placing about 20 g. of food into sealed bottles (250 ml). The bottles were kept at a constant temperature of 5°C in a refrigerator. The airspace of each bottle was then sampled daily and injected into the test chamber by a calibrated gas syringe. After each measurement the consumed air was compensated in order to maintain the necessary aerobic conditions; furthermore, it is worth pointing out that this operation avoids gas concentration build-up during the experiment. Data have been collected for some days. The number of storage days were the classes on which the classification scheme was based.

In order to show the intrinsic classification properties of the electronic nose, data analyzed by exploratory techniques are presented.

Fig. 11 shows the PCA results obtained for the fish experiment. Good classification can be observed for the first three days, while the last three days are mixed. The data can be disentangled by the self organizing map (SOM). In addition to its other features, SOM can also be regarded as a data modelling method which provides a bidimensional representation of the data. It has been demonstrated that SOM is a neural implementation of a sort of principal curve analysis.

Fig. 12 shows a 10 by 10 SOM lattice and the area of it which is pertinent to each day. As can be seen, the areas are separated one from another, and an optimum classification is achieved.

Fig. 13 shows the SOM representation of the veal-meat data. In this case also, the sensors provide an intrinsic classification of the data which is in agreement with the storage days of the food.

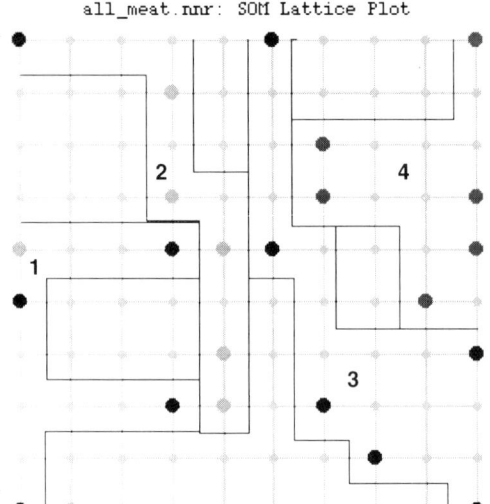

Fig. 13 10x10 SOM with subspaces identifying the classification spaces relative to the meat.

## 6. CONCLUSIONS

This paper illustrates future technological advances, feasibility and problems relative to the fabrication of silicon integrated electronic noses using either SAW or optical sensors, and indicate that sniffing action is one of the most important features for achieving satisfactory overall performance.

To this purpose, the design must incoerporate a fine turbulence control of the flux of the odour under test, through microheaters in combination with flaps as microactuators. The main problem here is relative: the overall calibration of the fluidic microenvironment. Experiments have been performed and utilized as a test vehicle for the development of a µTAS, with an EN made by quartz coated with metallo-porphyrines as a promising chemically interactive material.

The deconvolution procedures adopted have shown a satisfactory pattern recognition capability in the case of food analysis, in particular for codfish and veal meat.

## REFERENCE

1. A.van der Ber and P.Bergueld, *Micro Total Analysis Systems*, Kluwer Publ., Dordrecht ,The Netherlands (1995).
2. A.D'Amico, E.Verona and C.Di Natale, *Handbook of Biosensors and Electronic Noses*, E.Kress-Rogers,ed CRC Press, Boca Raton, USA (1996).
3. J. Brunink, C. Di Natale, F. Bungaro, F. Davide, A. D'Amico, R. Paolesse, T. Boschi, M. Faccio, and G. Ferri, The application of metalloporphyrins as coating material for QMB Based chemical sensor, *Analytica Chimica Acta*, 325 (1996) 53-64
4. C. Di Natale, J. Brunink, F. Bungaro, F. Davide, A. D'Amico, R. Paolesse, T. Boschi, M. Faccio, and G. Ferri, Recognition of fish storage time by a metalloporphyrins coated QMB sensor array, *Measurement Science and Technology*, 7 (1996) 1103-1114
5. C. Di Natale, J. Brunink, F. Bungaro, F. Davide, A. D'Amico, R. Paolesse, T. Boschi, M. Faccio, and G. Ferri; Metallo-porphyrin coated quartz micro balance sensor for amine detection, *Proceedings of the 2nd East Asia Conference on Chemical Sensors, Xi'an (China)*, Oct. 1995; International Academic Publ., Beijing, China (1995).

# ADVANCES IN THE DEVELOPMENT OF OPTICAL MICROSYSTEMS AT DEA-OPTOLAB: INTEGRATED OPTICAL MEASURING DEVICES AND DIFFRACTIVE OPTICAL COMPONENTS FOR INDUSTRIAL APPLICATIONS

F. Docchio and U. Minoni

University of Brescia,
Department of Electronic for Automation
Via Branze 38, 25123 Brescia, Italy

## 1. INTRODUCTION

There is an increasing demand from the manufacturing industry for simple, compact and yet high-performance instruments for dimensional gauging, measurement of profiles in 2 and 3 dimensions, measurement of surface properties, radiometry, polarimetry, etc. So far, the achievement of small size and low cost for optical tools has been hindered by (i) the dimensions of the light sources (mainly lasers) used for the applications, and (ii) the dimensions of optical components (lenses, prisms, etc.).

Recently, considerable progress has been achieved in the development of integrated optics components and diffractive optics lenses/prisms. Integrated optics components form the basis of modern communication technology and optical computing: several active and passive optical components can be integrated into a single chip by modern lithographic techniques, combined with deposition techniques. On the other hand, bulk optical components based on the principle of refraction have found a miniaturized counterpart in optical components based on diffraction.

Both integrated optics and diffractive optics concepts can be successfully applied to the development of optical devices which are building blocks of measuring systems. The integration of optical components in miniaturized optical chips, with interconnections based on the use of integrated optics, can fulfil the needs for small-sized and low-cost electrooptical sensing and measurement instrumentation.

The Laboratory of Optoelectronics of the University of Brescia has been recently involved in a number of projects aimed at the development of innovative optical measurement instrumentation. The projects have mainly been conducted within the framework of national and supranational programmes, in collaboration with leading research centers and industries in Italy, such as IROE, Centro Laser Bari, Pirelli S.p.A., and Microcontrol S.p.A.. The aim of these projects was to provide the Laboratory a background of knowledge concerning the development of miniaturized sensors, instrumentation, and optical microsystems.

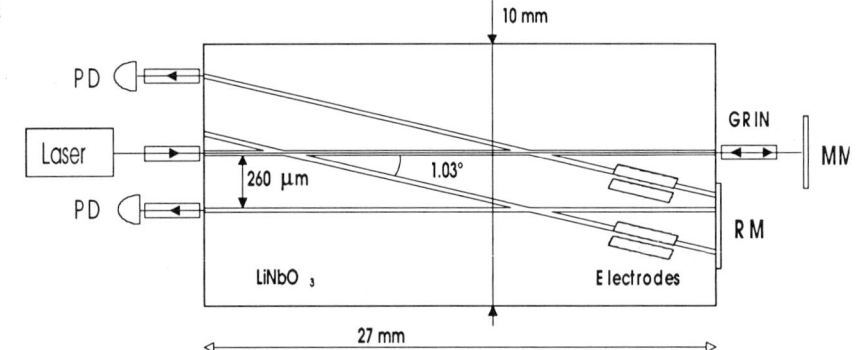

Fig. 1 Optical layout of the integrated double interferometer. MM: measuring mirror; RM: reference mirror. PD: photo-diode.

This Chapter is intended to give a summary of the most recent developments at the Laboratory of Optoelectronics of the University of Brescia, in cooperation with Italian Research Institutions and industries, in the fields of optical sensing and measurement, and of innovative optical components. An interferometric measuring head, based on a double interferometer in integrated optics, has been developed and characterized. A modified version of this setup is being proposed as a high-speed polarimeter for testing various kinds of materials. A laser photolithographic setup for the development of integrated optics circuits has been designed and developed. Finally, the fabrication and characterization of optical diffractive components has been carried out.

## 2. DEVELOPMENT AND CHARACTERIZATION OF AN INTEGRATED OPTICS DOUBLE INTERFEROMETER FOR DISTANCE MEASUREMENTS

The first activity dealt with the design and development of an interferometric sensor in integrated optics. It consists of a double Michelson interferometer, with full phase control on each arm by means of the electrooptical effect, and with control of the splitting ratio of the incoming light between the interferometers. One of the interferometers is used as a reference interferometer, with both distal mirrors fixed. The other interferometer has the mirror of the measuring arm connected to the target.

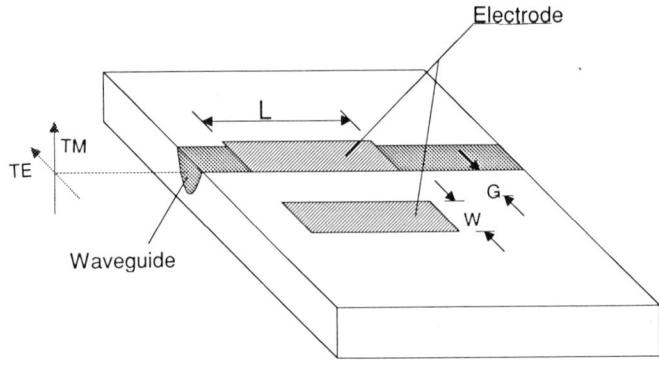

Fig. 2 Layout of the electrooptical modulator section. The parameters L, G, and W were chosen to obtain a halfwave voltage of 5V.

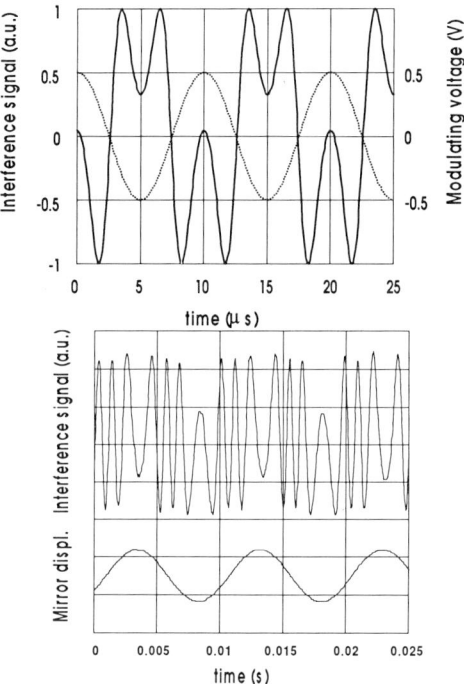

Fig. 3 Examples of performance of the integrated interferometer. a) example of sinusoidal modulation of one of the interferometer modulators; b) example of interferometer readout in the case of sinusoidal displacement of the measurement mirror in the absence of e-o modulation.

Care has been taken in the optimization of the reinjection of the light from the target into the interferometer, to maximize the signal to noise ratio. This development has been carried out in collaboration with IROE and Microcontrol. The envisaged application of the device is dimensional control of manufactured articles, as an alternative to Differential Transformers (LVDT's).

To produce the sensor, we used a z-cut Lithium Niobate crystal in order to optimize the diffusion depth of Titanium. The deposition parameters are: Ti diffusion depth: 20 nm; waveguide width: 6 mm. With these parameters, bimodal waveguides were obtained.

Fig. 2 shows a layout of each of the two electrooptical modulator sections of the interferometer. With the parameters chosen, the half wave voltage of the modulator was determined to be about 5V, with an overall modulation bandwidth of about 1 GHz, mainly limited by electrode capacitance.

The performance of the integrated optics interferometer has been studied. In particular, measurements were performed in relation to (i) the behaviour of the electrooptical modulator, and (ii) the ability of the interferometer to perform distance measurements over a specified distance range. Fig. 3.a) shows, for a fixed path of the interferometer, a typical waveform pattern obtained by modulating at half-wave (the respective voltage has been measured to be about 5 V) the electrooptical modulator of the measuring interfermeter. Fig. 3.b) shows, on the other hand, a typical waveform obtained from the interferometer without modulation, by moving the measuring mirror, placed at xx mm from the end of the interferometer, sinusoidally driven by means of a piezo modulator. From the set of measurements, it was concluded that the interferometer could indeed be used effectively as a position sensor, suitable to be included in a contact head for the measurement of diameters of mechanical workpieces.

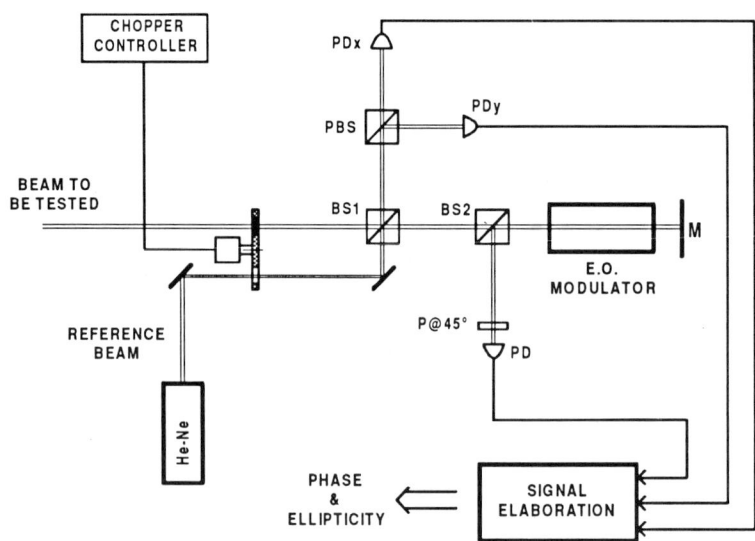

Fig. 4 Optical layout of the polarization measurement system. $PD_x$, $PD_y$, PD: photodiodes; BS1, BS2: non-polarizing beam-splitters; PBS: polarizing beam-splitter; P@45°: polarizer; M retroreflector.

## 3. DEVELOPMENT OF A LASER ELLIPSOMETER FOR POLARIZATION MEASUREMENTS

The integrated circuit used as a displacement measuring device can, with minor changes, offer a solution for the development of a fast ellipsometer for the measurement of the birefringence of films, LCD panels, filters, etc. Theoretical and experimental acivity was therefore initiated at our Laboratory, under the auspices of the National Institute for the Physics of Matter (INFM), to develop an electrooptical ellipsometer in bulk, with the idea that the optical layout could be transposed easily into an integrated optics chip.[1-8] The optical layout of the polarization measurement system is shown in Fig. 4. It is basically composed of a section for the measurement of the orientation of the laser polarization, and of a section for the measurement of the phase delay between the two orthogonal vectors. The former is simply composed of a polarization beam splitter combined with two photodetectors. The latter is composed of an interferometer to which the beam under test is sent. The interferometer is equipped with an electrooptical modulator, which delays one of the polarization components by (ideally) 360° with a sawtooth time function. The two components are recombined by means of a beam analyser composed of a polarizer oriented at 45° with respect to the plane of the diagram. The interference of the two polarizations at the detector, combined with the linearly varying delay in one of the components induced by the electrooptical modulator, contains the information of the phase delay of the two polarizations. Suitable electronics have been developed to analyse the phase delay signal and the orientation signal. Extensive report of the performance of the system is given in.[9] Suffice it here to say that the system has shown excellent linearity and accuracy up to 0.1%.

In the preliminary version shown here, the performance was limited by (i) temperature drifts in the operation of the electrooptical modulator, (ii) the low-frequency operation of the modulator, and (iii) the overall temperature- and mechanical vibration sensitivity of the bulk apparatus. To overcome these limitations, a reference interferometer has been added by means of an additional laser beam with linear polarization sent to the interferometer.

All the limitations of the present configuration will be overcome by using an integrated

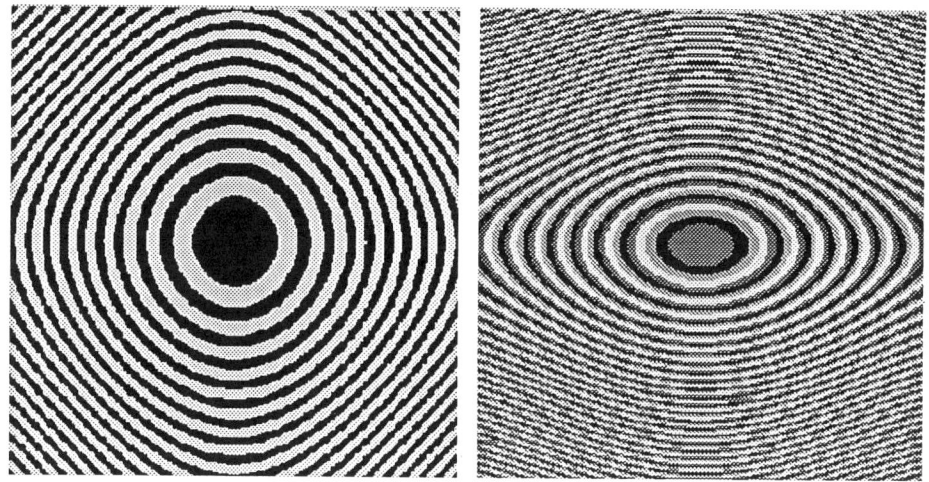

Fig. 5 Examples of DOE patterns, and of far-field beam-forming functions.

optics structure. In fact, in integrated optics, it will be possible to achieve a frequency modulation of up to 1 GHz, thus making it reasonable to perform measurements at a measuring rate of up to 1 MHz, and therefore to measure transient polarization changes.

## 4. DESIGN OF A MICROLITHOGRAPHIC SETUP FOR THE DESIGN OF PROTOTYPES OF INTEGRATED OPTICS CIRCUITS

As an attempt at providing facilities to build prototypes of integrated circuits, a laser microlithographic setup has been designed and realized in collaboration with Pirelli S.p.A. The system is able to produce masks with 1-10 µm thick guides by direct etching of photoresists. Particular care was taken to optimize the channel width through careful control of the beam focusing position with respect to the target.

## 5. DESIGN OF DIFFRACTIVE OPTICAL ELEMENTS (DOE) FOR SHAPING OF LASER BEAMS

Finally, a collaboration of the Laboratory with the Centro Laser Bari led to the design and test of two- and four-level diffractive optical elements (DOEs), in particular beam-shapers that transform the intensity profile of a Gaussian laser beam into suitable intensity distributions. Indeed, as is evident from various contributions to this volume, diffractive optics are key elements for the realization of small, compact, low-cost sensors and optical systems. Thus, it was imperative for our group to acquire a suitable know-how in this field, as well as a pool of possible collaborators for a combined project.

The design of the diffractive elements is carried out by means of appropriate software, developed by Centro Laser Bari and upgraded by our goup. Particular emphasis was given in this work to simulate the performances of the beam at the exit of the DOEs, by means of suitable software. Examples of DOEs developed and characterizad within this project are given in Fig. 5. Two-level and four-level DOEs have been developed and tested, performing different beam forming functions such as flat-top and cross-type patterning.

## 6. CONCLUSIONS

In this Chapter, several projects carried out at the laboratory of Optoelectronics of the University of Brescia have been described. All the developments undertaken in the field of optical microsystems and miniaturized optics have placed the Laboratory in a favourable position for partnerships in international research and development programmes, together with the institutions with which it has collaborated. It should be emphasized that research in the domain of optical microsystems for industrial measurement should be encouraged, due to the vast potential market in the context of quality control and production optimization.

## REFERENCES

1. T.T. Charampopoulus, and B.J. Stagg, High-temperature ellipsometer system to determine the optical properties of materials, *Appl. Opt.* 33:1930 (1994).
2. J.E. Hayden, and S.D. Jacobs, Automated spatially scanning ellipsometer for retardation measurement of transparent materials, *Appl. Opt.* 32:6256 (1993).
3. P.K.A. Wai, and C.R. Menyuk, Polarization decorrelation in optical fibers with randomly varying birefringence, *Opt. Lett.* 19:1517 (1994).
4. I. An, H.V. Nguyen, A.R. Heyd, and R.W. Collins, Simultaneous real-time spectroscopic ellipsometry and reflectance for monitoring thin-film preparation, *Rev. Sci. Instrum.* 65:3489 (1994).
5. C. Ye, M.O. Sundstrom, and K. Remes, Microscopic transmission ellipsometry: measurement of the fibril angle and the relative phase retardation of single intact wood pulp fibers, *Appl. Opt.* 33:6626 (1994).
6. B. Scholl, T. Stein, A. Neues, and K. Mertens, In-line fiber optic polarimeter with a 99% coupler, *Opt. Eng.* 34: 1669 (1994).
7. H. Fu, T. Goodman, S. Sugaya, and J.K. Erwin, M. Mansuripur, Retroreflecting ellipsometer for measuring the birefringence of optical disk substrates, *Appl. Opt.* 34:31 (1995).
8. P. Daveze, H. Sahsah, and J. Monin, A new automatized device for high-precision measurements of optical retarder plates, *Meas. Sci. Technol.* 7:157 (1996).
9. U. Minoni, G. Scotti, and F. Docchio, Real-time polarimetric characterization of laser beams, *Proc. IMEKO XIV, Tampere, Finland*, Vol. II, 79-84 (June 1997).

# MICRO-OPTICS FOR MICRO TOTAL ANALYSIS SYSTEMS

A. E. Bruno[*], B. Krattiger[*], S. Barnard[*], M. Ehrat[*], R. Völkel[§],
Ph. Nussbaum[§], H. P. Herzig[§], and R. Dändliker[§]

[*] Ciba-Geigy Ltd., Bioanalytical Research
K-127.1.76, 4002 Basel, Switzerland
[§] Institute of Microtechnology
Rue A.-L. Breguet 2, 2000 Neuchatel, Switzerland

## 1. MINIATURIZED ANALYTICAL SYSTEMS (µTAS)

The development of miniaturized analytical systems has gained considerable momentum in the last few years. Planar micro-optics and solid state light sources have a great potential for reducing the size and simplifying the architecture of analytical systems[1]. Our investigation is focused on microlens arrays for µTAS. Arrays of refractive microlenses (2 µm to 5 mm lens diameter) have been fabricated by melting resist technology and replicated in polycarbonate and PMMA by embossing and casting techniques. Microlens arrays have been integrated into miniaturized analytical systems.

We have demonstrated several miniaturized detectors featuring diffractive optical elements (DOEs), gradient index lenses (GRINs) and refractive microlens arrays as light focusing elements. LEDs and laser diodes have been used as light sources. Some of these devices are: absorption [2], fluorescence [2-3], thermo-optical absorption [4-5] and refractive index [6-7] detectors for capillary electrophoresis. Emphasis is presently given to the technology aspects related to the fabrication of microlens arrays and to the development of a miniaturized fluorescence µTAS sensor array featuring ultra-bright blue LEDs for the determination of critical gases and electrolytes in blood.

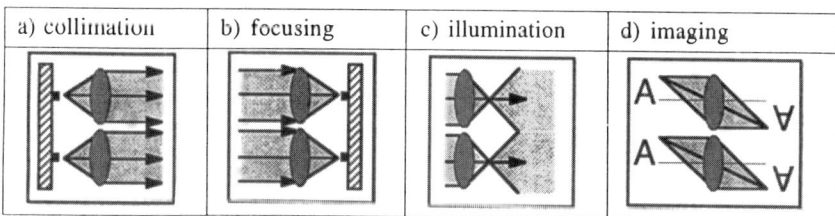

Fig. 1 Basic optical functionality carried out by microlens arrays.

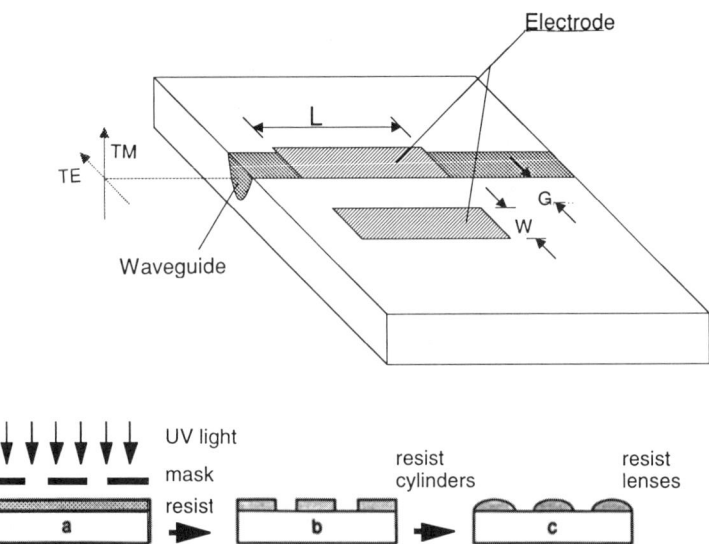

Fig. 2 Melting resist lens fabrication: a) lithography, b) developing and c) melting.

## 2. REFRACTIVE MICROLENS ARRAYS

Fig. 1 shows the basic optical functionality of microlens arrays for µTAS. Microlenses are used for collimation or focusing of light sources (LEDs, lasers or fibers), for illumination, e.g. of fluids or immobilized membranes, and for imaging, e.g. to collect fluorescence light.

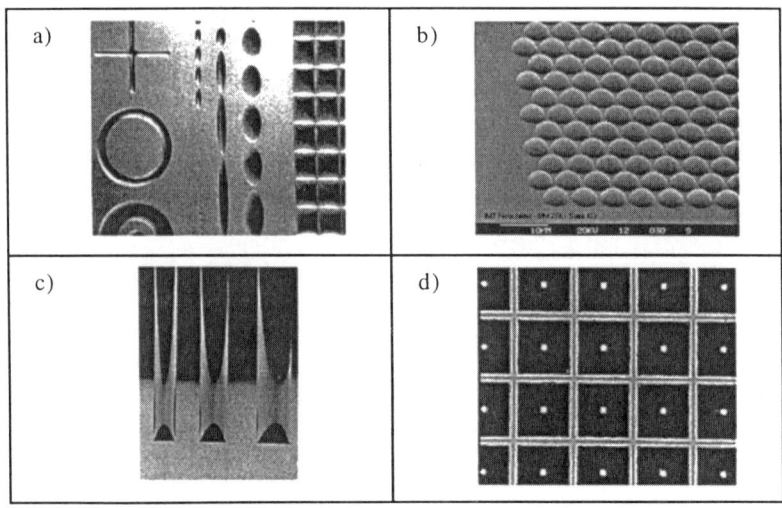

Fig. 3 Photographs of melting resist microlens arrays. a) Elliptical, rectangular and ring lenses, b) SEM picture of very small microlenses ($\phi \sim 5$ pm), c) cylindrical lenses (1.2 mm, 1.4 mm, 1.8 mm width) and d) focal plane of square lenses (400 µm width).

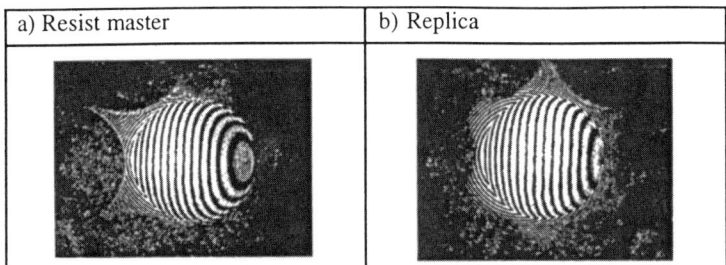

Fig. 4 The wave aberrations of: a) the lens master; and, b) its replica (Mach-Zehnder[12]).

Arrays of refractive microlenses (2 μm to 5 mm diameter) have been fabricated [8-10] by melting resist technology as shown in Fig. 2. Photoresist cylinders are formed by lithography and subsequently melted. Surface tension forms plano-convex spherical lenses.

Resist microlens arrays may serve as a master for replication. The resist master is used to generate a Ni shim by electroforming in a Ni-plating bath. The Ni shim serves as a mold for embossing, casting or CD-injection molding. Fig. 3 shows different microlens arrays fabricated by melting resist technology.

Fig. 4 shows the wave aberrations [11] of a resist master replicated [12] by casting in a PMMA-type organic glass. Wave aberrations of 0.14 λ (rms.) were found for both master and replica. No significant difference between the master and the replicated lens arrays could be observed.

ACKNOWLEDGMENTS. The project was partially supported by the Swiss Priority Program 'Optique'. The authors like to thank M. Gale (Paul Sherrer Institute, Zurich, Switzerland), E. Noordhanus (Philips, Eindhoven, Holland), and S. Haselbeck, M. Eisner and J. Schwider (Erlangen, University, Germany) for their support.

# REFERENCES

1. H. P. Herzig (Ed), in *Micro-Optics: Elements, Syst. and Appl.*, Taylor & Francis, London, (in Press) 1996.
2. A. E. Bruno, F. Maystre, B. Krattiger, P. Nussbaum, P. and E. Gassmann, The pigtailing approach to optical detection in CE, *TrAC.*, 1994, 13, 190-198.
3. A. E. Bruno, S. Barnard, M. Rouilly, A. Wallner, J. Berger and M. Ehrat, All-solid-state miniaturized fluorescence sensor array for the determination of critical gases and electrolytes in blood, *Anal Chem.*, (submitted).
4. B. Krattiger, A. E. Bruno, H. M. Widmer and R. Dändliker, Hologram-based thermooptical absorbance detection in CE: Separation of nucleosides and nucleotides, *Anal Chem.*, 1995, 67(1), 124-130.
5. J. M. Saz, B. Krattiger, A. E. Bruno, J. C. Dez-Masa and H. M. Widmer, Thermooptical absorbance detection of native proteins separated by CE in 10 Am i.d. tubes, *J. of Chromatogr.*, 1995, 699, 315-322.
6. B. Krattiger, G. J. M. Bruin and A. E. Bruno, Hologram-based refractive index detector for CE: Separation of metal ions, *Anal. Chem.*, 1994, 66, 1-8.
7. A. E. Bruno, B. Krattiger, Ch. 11 in *Carbohydrate analysis: HPLC and CE*, El Rassi (Ed), Elsevier, NY, 1995.
8. Z. D. Popovic, R. A. Sprague and G. A. Neville-Connell, Technique for the monolithic fabrication of microlens arrays, *Appl. Opt.* 27, 1281 (1988).
9. Daly, R. F. Stevens, M. C Hutley and N. Davies, The manufacture of microlenses by melting photoresist, *J. Meas. Sci. Techn.* 1 759-766 (1990).
10. Ph. Nussbaum, R. Völkel, H. P. Herzig and R. Dändliker, Micro-optics for sensor applications, European Symposium on Lasers, Optics, and Vision for Productivity in Manufacturing, *Proc. SPIE 2783*, 1996.
11. Fabricated at Philips Components, Eindhoven/Holland
12. Measured at University of Erlangen, Prof. Schwider, Erlangen/Germany

# OPTICAL TWEEZERS: LASER MANIPULATION OF MICROPARTICLES

G.C. Righini

Optoelectronics and Photonics Department (IROE CNR)
«Nello Carrara» Electromagnetic Waves Research Institute
Via Panciatichi 64, 50127 Florence, Italy

## 1. INTRODUCTION

Although the forces of radiation pressure have been well known to physicists,[1] for a long time it has been difficult to study them in the laboratory due to the presence of temperature gradients causing thermal forces which could obscure the radiation pressure. It has only been since 1970, after the advent of suitable laser sources, that practical applications could be investigated, beginning with the experimental demonstration by A. Ashkin of acceleration and trapping of micron-sized particles using the force of radiation pressure from a cw visible laser.[2] Additional steps were made in the following years,[3-6] but a major practical advancement was achieved in 1986 when Ashkin and coworkers demonstrated optical trapping of dielectric particles (in the size range from about 25 nm up to 10 μm) in water solution by a single-beam gradient force radiation-pressure trap.[7] An year later, Ashkin and Dziedzic coined the term «optical tweezers» when reporting the demonstration of the optical trapping and manipulation of individual viruses and bacteria in aqueous solution by laser radiation pressure. They «have used the trap as an *optical tweezers* for moving live single and multiple bacteria while being viewed under a high-resolution optical microscope».[8] During the last decade there has been increased attention to the use of optical tweezers: this occurred mainly in microbiology, but optical tweezers are expected to play an important role in the future development of nanometric-scale technology for manipulation, control and analysis of ultra-small structures.

In this Chapter a short description of the fundamentals of single-beam optical traps will be given, together with an overview of the demonstrated applications and an extended reference to published results.

## 2. FUNDAMENTALS OF LASER TRAPPING

The first experiments on accelerating and trapping particles by the forces of radiation

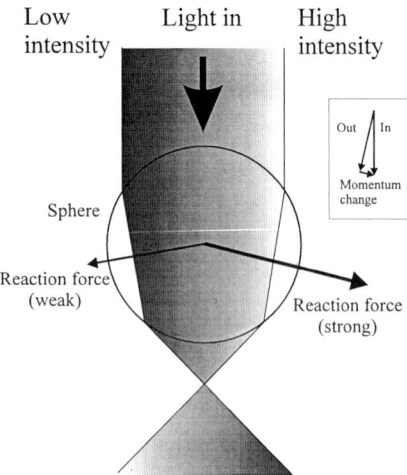

Fig. 1 A light beam with a gradient in intensity produces a reaction force, which pulls the dielectric sphere towards the brighter light. The inset shows the momentum change in ne of the bright rays.

pressure were performed using micrometer-sized transparent latex spheres suspended in water. A true optical potential well (or "optical bottle", as Ashkin also called it) was constructed using two counterpropagating identical TEM$_{00}$ laser Gaussian beams with their beam waists located symmetrically about the particle to be trapped: the particle was seen to be in stable equilibrium at that point.[2]

It was quite later, in 1986, that optical trapping of dielectric particles was demonstrated by using a single laser beam, i.e., in a much more convenient experimental arrangement for practical applications. All the optical traps, of which the single-beam gradient force trap is one of the simplest both conceptually and practically, are based on the basic scattering and gradient forces of radiation pressure. Without entering into a detailed explanation, we can summarize that the scattering force is proportional to the optical intensity and directed as the incident light, while the gradient force is proportional to the gradient of intensity and directed along the intensity gradient. For the sake of simplicity, let us consider the effect of radiation pressure forces on a spherical dielectric particle. The particle is supposed to be lossless and in the Mie size regime, that is having a diameter large compared with the wavelength. In such a simplified situation, ray-optics can be used to describe the scattering and optical momentum transfer to the particle; in this regard, it has been shown that the ray optics model is highly accurate for the prediction of axial forces and reasonably accurate for prediction of transverse forces in the size regime larger than 10 μm.[9]

A typical situation is that of a glass or polystyrene microsphere in water, and in this case the effective refractive index $n_e$, which is equal to the particle index divided by the index of the surrounding medium, is in the range 1.1 to 1.2. Thus, the particle behaves like a weak positive lens. Referring to Fig. 1, let us consider an incident beam of light with a gradient in intensity (in the picture the intensity is increasing in the beam from right to left): this is what happens when one uses a laser beam with gaussian intensity profile and the particle is placed off-axis. Surface reflections can be neglected. The rays are bent inside the particle and the resulting forces due to ray refractions in the weak-lens regime pull the particle towards the higher intensity region (the axis of the gaussian beam).

No net radial force would exist in absence of the light gradient because the two reaction forces would be equal and thus their transverse components would cancel each other. In the case of a particle with $n_e < 1$ (e.g. a dielectric hollow sphere or a water droplet in liquid paraffin[10]) the overall radial force would pull the particle out of the region of high intensity.

A trap may therefore be constructed by producing a three-dimensional light gradient with the highest intensity in the center: this is just the situation which occurs near any focal spot. The sharper the focus, the steeper the gradient: very steep gradients are needed in order to ensure that the gradient forces overcome the scattering forces in the focal region and the particle will not be ejected from the trap along the direction of the incident beam.

Single-beam trapping is also effective for submicrometer Rayleigh size particles, having diameter much smaller than the wavelength of the trapping laser. In this case, obviously, one is in the wave optics regime and accurate calculations of trapping forces become quite complex, except for the simplest geometries. When one has to deal with particles in the intermediate size regime, namely when their size is of the same order as the wavelength, the electromagnetic approach seems to provide better results than the ray-optics model;[9] in such an intermediate regime, however, the agreement between theory and experiment is not always good, and calibration measurements may be necessary, especially for applications in cell biology.[11]

Trapping has been demonstrated for dielectric particles with sizes in the range from about 10 μm down to 26 nm: in these two extreme cases, the required power of the laser beams was of the order of 50 mW and 1.4 W, respectively.[12, 2] The forces exerted are of the order of several picoNewtons, and it is interesting to note that an optical trapping force sensor has been suggested as a unique tool capable of measuring the dynamic properties of a single micrometer particle.[13] Ghislain and coworkers used the transmitted beam of the trapping laser to monitor the deflection of a trapped microsphere within the potential well of the trap: for a polystyrene sphere of 1 μm diameter trapped with a 60 mW laser beam they were able to detect small ($10^{-12}$ N), dynamic ($\approx$ 1 ms) forces with high spatial resolution (10 nm).

A typical experimental arrangement is depicted in Fig. 2, where a high magnification and high numerical aperture microscope objective focuses the collimated laser beam inside the sample cell, a sandwich structure consisting of a microscope slide, a suspending medium, and a cover glass. Spacers between bottom and cover glass slides have typical thickness around 100 μm. In the figure a particle of diameter 1 μm is shown, which is trapped in the focus

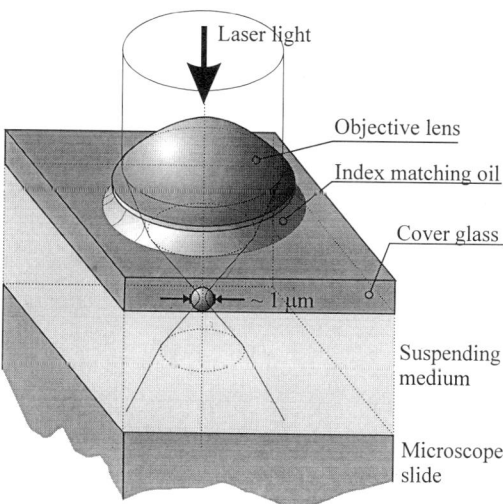

Fig. 2 Optical tweezers. A laser beam is focused by an objective lens onto a micrometre-sized object, placed in a suspending medium. A light gradient is created in the beam waist, and this traps the object.

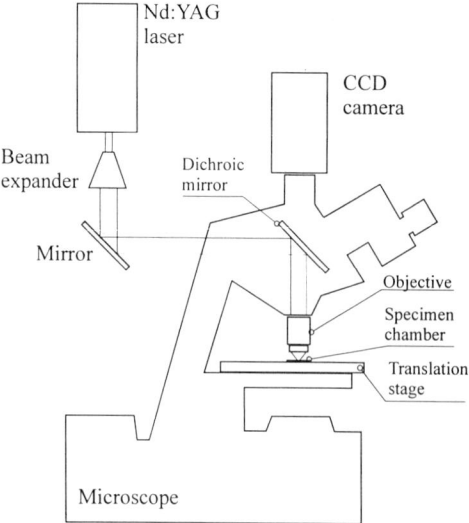

Fig. 3 Sketch of the optical tweezers setup. the dichroic mirror is used to reflect the infrared laser beam into the microscope objective, while transmitting the image to the CCD camera (illuminating rays, coming from below, are not shown). The translation stage allows one to manipulate the trapped particle down or up while it remain in focus.

of the laser; if laser wavelength is around 1 μm, the diffraction-limited beam waist can be expected to be around 0.5 μm.

The choice of the laser itself is quite important, because absorption of light by the particles, besides possibly causing damage to the particles, decreases the force generated by the refraction of the rays. Thus, in many applications lasers operating in the near-infrared region have been used: the most frequently used source has been a cw Nd:YAG laser at 1.064 μm, but laser diodes both at 1.3 μm [14] and at around 800 nm [12, 15, 16] are being increasingly used, which allow the development of compact, reliable and cheaper devices.

## 3. APPLICATIONS IN MICROBIOLOGY AND MICROMECHANICS

The application area for optical tweezers which has grown most rapidly is undoubtly microbiology: traps have been demonstrated to be one of the most effective ways of manipulating biological matter without physical contact and without damage. Especially when working in such a field, it is very convenient to introduce the trapping laser beam into an optical microscope, so that its objective is focusing the beam onto the biological particle (which can be manipulated accordingly) while the particle itself is being viewed under high magnification.

The optical tweezers setup is sketched in Fig. 3: it is the combination of a conventional, high-resolution optical microscope and a laser trap which allows the operator to observe and manipulate micro-biological specimens (from living cells to bacteria or viruses).[8, 17] The use of a laser diode, as already mentioned, makes the equipment more compact, and commercial devices of this type are already available.

Demonstrated applications in microbiology [18 - 22] include sorting and classification of living cells, [23, 24] cell fusion, [25] and intracellular surgery.[26 - 28] Objects within a living cell, such as organelles or filaments of cytoplasm, have been manipulated without damaging the cell wall.[27] The mechanical compliance of bacterial flagella was also measured, e.g. for Escherichia Coli and streptococcus; [29] the technique relies on the measurement of the trapping force while the bacterium is tethered to a glass surface by its flagellum, and it is moved (while still being trapped) against the Stokes drag and the flagellum torque.

New biologically relevant information was obtained by studying dynamic light-scattering of single biological cells; using two laser diodes, a double-beam trapping configuration was designed, which was combined with a separate probe beam in order to perform angle-dependent measurements of light-scattering by single particles, such as polystyrene beads and human lymphocytes.[30] Optical tweezers have also been suggested as a tool to perform *in situ* analyses on marine phytoplankton cells and other marine organisms, in conjunction with UV fluorescence, flow cytometry and microphotometry techniques.[31]

Another interesting area is related to the study of the enzymes, such as kinesin, myosin and dynein, which play a crucial role in the conversion of ATP (adenosine triphosphate) energy into mechanical work.[32-35] These enzymes behave like linear motors which convert chemical energy into mechanical energy and thereby power the transport along specific biologic substrates, such as actin filaments (in the muscles) and microtubules. Kinesin, in particular, couples ATP hydrolysis to movement along microtubules; for the experiments, microscopic silica beads were pre-coated with carrier protein, exposed to low concentrations of kinesin, and individually manipulated by the optical tweezers directly into microtubules.[32] The enzyme's stepping action propels the bead along the substrate, while the optical trap provides the capability of applying controlled loads to the bead and of analyzing forces and displacements with high accuracy. The experiments performed have allowed biologists to conclude that kinesin moves in nanometer steps along well-defined successive sites, which also depend on load conditions and ATP levels.[35]

Other recent work has been focused on the study of the stretching properties[36] and the elastic response[37] of DNA molecules. In the former work the stretching of single, tethered DNA molecules by a flow was directly visualized with fluorescence microscopy; molecules with length ranging from 22 to 84 µm were held stationary against the flow by the optical trapping of a latex microsphere attached to one end of a DNA molecule, while the other end remained free. In this way it was possible to investigate the hydrodynamic interaction between the polymer and the fluid; this is particularly significant because direct observation and controlled deformation of individual DNA molecules provides insights into the previously inaccessible regime of single polymer dynamics. In the latter work, a dual-beam laser trap was used to investigate the elastic properties (bending and twisting rigidity) of double-stranded DNA (dsDNA); these properties, in fact, affect much of the behavior of DNA (e.g. how it bends upon interaction with proteins and loops to connect enhancer and promoter regions). A simple way to test DNA elasticity is to stretch a single macromolecule from both ends, measuring the force as a function of its end-to-end distance: for such a purpose, each end of a single dsDNA molecule was attached to a separate microscopic latex bead. One bead was held by suction with a glass micropipette while the other bead was held in the optical trap, which also operated as a force transducer. In this way it was possible to show that dsDNA molecules in NaCl buffer undergo an abrupt but reversible transition into a stable extended form (about 70% longer than its B-form contour) when the longitudinal stress reaches 65 pN.[37]

Optical tweezers, however, may find important applications not only in microbiology and microchemistry, but also in micromechanics and, more generally, in all the fields where sub-micrometer-scale technology is required to fabricate, analyse and control nanometric structures and devices.[38-40] The use of radiation pressure to move small particles on the surface of a substrate was demonstrated by Kawata and Sugiura in 1992.[41] Polystyrene latex spheres with diameters of 1 to 27 µm and glass spheres with diameters of 2 and 8 µm were moved along the flat surface of a high-index prism in the evanescent field produced by the total reflection of the beam from a Nd:YAG laser inside the prism itself. The velocity of the particles was measured as a function of the laser incidence angle (because the depth of the evanescent field decreases as the incident angle increases), and running speeds up to 20 µm/s were achieved using 150 mW laser power. More recently, the same group demonstrated the possibility of driving the microparticles

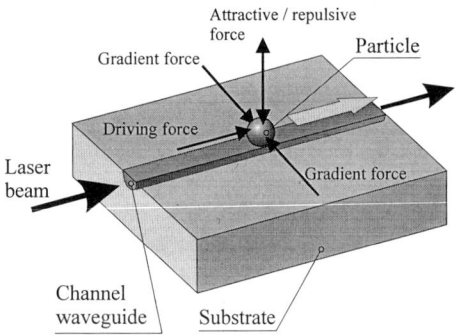

Fig. 4 A dielectric particle can be driven by the evanescent field of a channel waveguide.

in the evanescent field that is generated in the vicinity of a channel waveguide.[42] The experimental configuration is sketched in Fig. 4: the beam from a cw Nd:YLF laser, emitting at 1.047 μm with 2.3 W power, is end-fire coupled into a diffused channel waveguide in a glass substrate. The waveguide is covered with water containing microparticles. The effective laser power incident upon the waveguide was about 80 mW and induced the movement of polystyrene spheres at speeds as large as 14 μm/s. Metallic spheres of gold (0.5 μm diameter) and platinum (1 μm diameter) were also trapped and moved along the waveguide. The potential of this mechanism for the realization of laser-driven linear micromotors has to be further explored.

A demonstrated application in microdynamics concerns optical trapping and directional high-speed rotation of anisotropic microfabricated objects in a liquid.[43] Such microobjects, which were flat on top and bottom, had been fabricated by reactive ion-beam etching of a 10 μm thick silicon dioxide layer deposited by RF-sputtering onto a sacrificial GaAs substrate; typical diameters were in the range of 10 to 25 μm. Their cross section, however, was dissymetric; that is, they did not have bilateral symmetry, while keeping rotational symmetry. This anisotropic geometry produces a torque as a result of the net radiation pressure on their surfaces produced by a symmetrical intensity profile of the laser beam: rotation speeds of 22 rpm were obtained with a laser power of 80 mW. Optically induced rotation, requiring no mechanical contact and no electric wires, can be expected to play a significant role in microsystems such as remotely driven motors and actuators; it can offer some advantages over microfabricated electrostatic motors.

## 4. CONCLUSIONS

Laser trapping is a mature technique, which is being investigated as an important engineering tool which is noncontacting and noninvasive: neglecting here the subjects concerned with the manipulation of atoms (namely laser cooling, optical molasses, etc.), [3, 44-47] most of the applications today are in the microbiology field, but there is increasing interest in micromechanics and nanosystems. The results already obtained indicate that optical tweezers have applicability in the development of novel laser-driven micromotors and microactuators. In this area, exciting opportunities also arise related to the use of integrated optical waveguides [42] and of optical fibers [48, 49] to create light-force traps.

Potential applications in mesoscopic particles and micro-optical devices could also follow from the work of Sato and Inaba, who reported the first observation of second-harmonic generation from $LiNbO_3$ particles induced by a cw Nd:YAG trapping laser [50] and, more recently, second-harmonic and sum-frequency generation from $KTiOPO_4$ (KTP) microscopic particles during the optical trapping of this nonlinear material.[51] In the latter experiment, KTP particles of a few micrometers diameter (both size and shape were not uniform because they were obtained by grinding a KTP crystal into powder with a mortar) were suspended in water and trapped by using a Nd:YAG and a tunable Ti:Al$_2$O$_3$ laser. Besides providing the possibility of new insights for the study of various nonlinear optical processes, these experiments might lead to the development of novel or advanced all-optical micro-devices.

ACKNOWLEDGEMENTS. The author wishes to thank Dr. M.A. Forastiere for his assistance, and Prof. S. Martellucci for his interest and encouragement. The financial support by the CNR Strategic Project on *Materials and Devices for Optoelectronics* is also acknowledged.

# REFERENCES

1. see for instance E.F. Nichols and G.F. Hull, The pressure due to radiation, *Phys. Rev.* 17, 91 (1903).
2. A. Ashkin, Acceleration and trapping of particles by radiation pressure, *Phys. Rev. Lett.* 24, 156 (1970).
3. A. Ashkin, Trapping of atoms by resonance radiation pressure, *Phys. Rev. Lett.* 40, 729 (1978).
4. G. Roosen, A theoretical and experimental study of the stable equilibrium positions of spheres levitated by two horizontal laser beams, *Opt. Commun.* 21, 189 (1977).
5. A. Ashkin and J.M. Dziedzic, Observation of light scattering from nonspherical particles using optical levitation, *Appl. Opt.* 19, 660 (1980).
6. M. Lewittes, S. Arnold and G. Oster, Radiometric levitation of micron sized spheres, *Appl. Phys. Lett.* 40, 455 (1982).
7. A. Ashkin, J.M. Dziedzic, J.E. Bjorkholm and S. Chu, Observation of a single-beam gradient force optical trap for dielectric particles, *Opt. Lett.* 11, 288 (1986).
8. A. Ashkin, J.M. Dziedzic, Optical trapping and manipulation of viruses and bacteria, *Science* 235, 1517 (1987).
9. W.H. Wright, G.J. Sonek, M.W. Berns, Parametric study of the forces on microspheres held by optical tweezers, *Applied Optics* 33, 1735 (1994).
10. H. Misawa, M. Koshioka, K. Sasaki, N. Kitamura and H. Masuhara, Three-dimensional optical trapping and laser ablation of a single polymer latex particle in water, *J. Appl. Phys.* 70, 3829 (1991).
11. H. Felgner, O. Muller, and M. Schliwa, Calibration of light forces in optical tweezers, *Appl. Opt.* 34, 977 (1995).
12. T.C. Bakker Schut, E.F. Schipper, B.G. de Grooth, J. Greve, Optical-trapping micromanipulation using 780-nm diode lasers, *Optics Letters* 18, 447 (1993).
13. L.P. Ghislain, N.A. Switz, and W.W. Webb, Measurement of small forces using an optical trap, *Review of Scientific Instruments* 65, 2762 (1994).
14. S. Sato, M. Ohyumi, H. Shibata, H. Inaba, and Y. Ogawa, Optical trapping of small particles using a 1.3-µm compact InGaAsP diode laser, *Opt. Lett.* 16, 282 (1991).
15. R.S. Afzal and E.B. Treacy, Optical tweezers using a diode laser, *Rev. Sci. Instr.* 63 pt.1, 2157 (1992).
16. G.J. Escandon, Y. Liu, G.J. Sonek, and M.W. Berns, Beam magnification and the efficiency of optical trapping with 790-nm AlGaAs laser diodes, *IEEE Photon. Technol. Lett.* 6, 597 (1994).
17. A. Ashkin, J.M. Dziedzic, and T. Yamane, Optical trapping and manipulation of single cells using infrared laser beams, *Nature* 330, 769 (1987).
18. W.H. Wright, G.J. Sonek, Y. Tadir, and M.W. berns, Laser trapping in cell biology, *IEEE J. Quantum Electron.* 26, 2148 (1990).
19. I.A. Vorobjev, H. Liang, W.H. Wright, and M.W. Berns, Optical trapping for chromosome manipulation: a wavelength dependence of induced chromosome bridges, *Biophys. J.* 64, 533 (1993).
20. G.J. Sonek, W.H. Wright and M.W. Berns, Optical tweezers: getting a handle on the microscopic biological world, *LEOS '93 Annual Mtg. Conf. Proc. (IEEE, New York, 1993)* Cat. No.93CH3297-9, 234.
21. Y. Liu, D.K. Cheng, G.J. Sonek, M.W. Berns, C.F. Chapman and B.J. Tromberg, Evidence for localized cell heating induced by infrared optical tweezers, *Biophys. J.* 68, 2137 (1995).
22. H. Pausewang, Applications and trends of laser technology and photophysics in the field of biological *Sciences*, Laser u. Optoelektr. 28, 63 (1966).
23. T.N. Buican, M.J. Smyth, H.A. Crissman, G.C. Salzman, C.C. Stewart, and J.C. Martin, Automated single-cell manipulation and sorting by light trapping, *Applied Optics* 26, 5311-5316 (1987).
24. H. Tashiro, M. Uchida and M. Sato-Maeda, Three-dimensional cell manipulator by means of optical

trapping for the specification of cell-to-cell adhesion, *Opt. Eng.* 32 p 2812 (1993).
25. R. Steubing, S. Cheng, W.H. Wright, Y. Numajiri, and M.W. Berns, Laser-induced cell fusion in combination with optical tweezers: the laser-cell fusion trap, *Cytometry* 12, 505 (1991).
26. M.W. Berns, W.H. Wright, B.J. Tromberg, G.A. Profeta, J.J. Andrews, and R.J. Walter, Use of a laser-induced optical force trap to study chromosome movement on the mitotic spindle, *Proc. Natl. Acad. Sci. USA* 86, 4539 (1989).
27. A. Ashkin and J.M. Dziedzic, Internal cell manipulation using infrared laser traps, *Proc. Natl. Acad. Sci. USA* 86, 7914 (1989).
28. H. Liang, W.H. Wright, S. Cheng, W. He, and M.W. Berns, Micromanipulation of mitotic chromosomes in $PTK_2$ cells using laser-induced optical forces (optical tweezers), *Exp. Cell Res.* 204, 110 (1993).
29. S.M. Block, D.F. Blair, and H.C. Berg, Compliance of bacterial flagella measured with optical tweezers, *Nature* 338, 515 (1989).
30. R.M.P. Doornbos, M. Schaeffer, A.G. Hoekstra, P.M.A. Sloot, B.G. de Grooth, and J. Greve, Elastic light-scattering measurements of single biological cells in an optical trap, *Appl. Opt.* 35, 729 (1996).
31. G.J. Sonek, Y. Liu, and R.H. Iturriaga, *In situ* microparticle analysis of marine phytoplankton cells with infrared laser-based optical tweezers, *Appl. Opt.* 34, 7731 (1995).
32. S.M. Block, L.S.B. Goldstein and B.J. Schnapp, Bead movement by single kinesin molecules studied with optical tweezers, *Nature* 348, 348 (1990).
33. A. Ashkin, K. Schutze, J.M. Dziedzic, U. Euteneur, and M. Schliwa, Force generation of organelle transport measured *in vivo* by an infrared laser trap, *Nature* 348, 346 (1990).
34. S.C. Kuo and M.P. Sheetz, Force of single kinesin molecules measured with optical tweezers, *Science* 260, 232 (1993).
35. K. Svoboda, C.F. Schmidt, B.J. Schnapp, and S.M. Block, Direct observation of kinesin stepping by optical trapping interferometry, *Nature* 365, 721 (1993).
36. T.T. Perkins, D.E. Smith, R.G. Larson, and S. Chu, Stretching of a single tethered polymer in a uniform flow, *Science* 268, 83 (1995).
37. S.B. Smith, Y. Cui, and C. Bustamante, Overstretching B-DNA: the elastic response of individual double-stranded and single-stranded DNA molecules, *Science* 271, 795 (1996).
38. H. Misawa, K. Sasaki, M. Koshioka, N. Kitamura, and H. Masuhara, Multibeam laser manipulation and fixation of microparticles, *Appl. Phys. Lett.* 60, 310 (1992).
39. M. Miwa, H. Misawa, T. Araki and T. Yoshimura, Laser manipulation technique and its role in study of micromachine, *Proc. $7^{th}$ Intl. Symp. on Microsystems, Intelligent Materials and Robots*, J. Tani and M. Esashi, Eds. (Tohoku Univ, Sendai, 1996) 67.
40. E. Higurashi, O. Ohguchi, H. Ukita and T. Tamamura, Rotational manipulation of artificial micro-objects based on the radiation pressure exerted on their internal sides, *Proc. $7^{th}$ Intl. Symp. on Microsystems, Intelligent Materials and Robots*, J. Tani and M. Esashi, Eds. (Tohoku Univ, Sendai, 1996) 63.
41. S. Kawata and T. Sugiura, Movement of micrometer-sized particles in the evanescent field of a laser beam, *Opt. Lett.* 17, 772 (1992).
42. S. Kawata and T. Tani, Optically driven Mie particles in an evanescent field along a channeled waveguide, *Opt. Lett.* 21, 1768 (1996).
43. E. Higurashi, H. Ukita, H. Tanaka, and O. Ohguchi, Optically induced rotation of anisotropic micro-objects fabricated by surface micromachining, *Appl. Phys. Lett.* 64, 2209 (1994).
44. S. Stenholm, The semiclassical theory of laser cooling, *Rev. Mod. Phys.* 58, 699 (1986).
45. S. Chu, Laser manipulation of atoms and particles, *Science* 253, 861 (1991).
46. L. Moi, S. Gozzini, C. Gabbanini, E. Arimondo and F. Strumia, Eds., *Light induced kinetic effects* (ETS Editrice, Pisa, 1991).
47. M.D. Hoogerland, H.F.P. Debie, H.C.W. Beijerinck, E.J.D. Vredenbregt, K.A.H. Vanleeuwen, P. Vanderstraten and H.J. Metcalf, Force, diffusion, and channeling in sub-Doppler laser cooling, *Physical Review A* 54, 32064 (1996).
48. A. Constable, J. Kim, J. Mervis, F. Zarinetchi, and M. Prentiss, Demonstration of a fiber-optical light-force trap, *Opt. Lett.* 18, 1867 (1993).
49. E.R. Lyons and G.J. Sonek, Confinement and bistability in a tapered hemispherically lensed optical fiber trap, *Appl. Phys. Lett.* 66, 1584 (1995).
50. S. Sato and H. Inaba, Observation of second harmonic generation from optically trapped microscopic $LiNbO_3$ particle using Nd:YAG laser, *Electron. Lett.* 28, 286 (1992).
51. S. Sato and H. Inaba, Second-harmonic and sum-frequency generation from optically trapped $KTiOPO_4$ microscopic particles by use of Nd:YAG and Ti:$Al_2O_3$ lasers, *Opt. Lett.* 19, 927 (1994).

# LASER TIME-OF-FLIGHT VELOCIMETRY: PROPOSALS FOR MINIATURISATION

H. Imam[*], B. Rose[*], S. G. Hanson[§], and L. Lading[§]

[*] Ibsen Micro Structures A/S, CAT
PO Box 30, Frederiksborgvej 399, 4000 Roskilde, Denmark
[§] Optics and Fluid Dynamics Departement, Risø National Laboratory
PO Box 49, Frederiksborgvej 399, 4000 Roskilde, Denmark

## 1. INTRODUCTION

The need for miniaturised sensors in a production environment is rising due to the pressure for on-line quality control. Efficient monitoring of the production process reduces raw material waste, increases productivity and lowers operating costs. Currently, mechanical contact sensors and magneto/electro non-contact sensors are employed. Optoelectronic sensors can provide an attractive alternative, as the optical signal probe is non-contacting, is immune to electric and magnetic fields, and can deliver a more accurate measurement. However, in order for optical sensors to be a viable alternative, accuracy and cost effectiveness is an issue, especially as compared to the savings obtained as a result of the increased accuracy.

In this Chapter, the concept of a miniaturised velocimeter for continuous production processes is developed, based on *Laser Time-of-Flight Velocimetry* (LTV). The Chapter describes a thorough experimental investigation conducted in order to ascertain the critical parameters that affect measurement accuracy as a result of the miniaturisation process. The experimental investigation was conducted with the aid of a bulk optical setup, which allowed analysis of different surfaces and the influence of alignment errors and detector cross-talk. Armed with this knowledge, design proposals are presented for LTV miniaturisation.

## 2. THE LTV PRINCIPLE

The laser time-of-flight velocimeter was introduced by Thompson[1] the measurement in turbulent flows. However, its principle has been extended in order to probe the velocity and length of solid surfaces. Analytical descriptions describing the performance of LTV systems have already been published[2]. It will be shown that these existing models are inadequate in

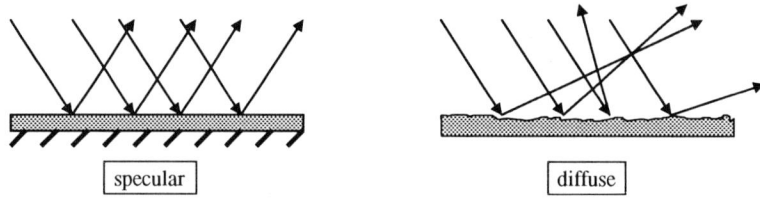

Fig. 1 Difference between specular and diffuse reflection.

predicting the performance of the sensor. It will also be shown how these models may be improved in order to accurately describe the sensor performance.

The sensor relies on the characteristic interaction between laser light and a rough surface, i.e. the *speckle effect*. Fig. 1 illustrates the difference between *specular* and *diffuse* reflection.

Under diffuse reflection, laser light is scattered randomly in all directions from the rough surface. The randomly scattered waves interfere with each other, leading to a random pattern of constructive and destructive interference, otherwise known as speckle. This speckle pattern is unique to the surface that is illuminated. Thus, the same speckle pattern can be only generated if the same area is illuminated on the solid surface. This property is the basis for the LTV system.

Fig. 2 below illustrates the concept of the system. Two identical laser spots are directed to the moving solid surface as shown. The speckle pattern from each spot is detected (the receiver concept is shown in Fig. 3). Each detector is a photodiode, which detects the total intensity as a sum over the detected area. As indicated, the surface is moving in a direction from left to right. The left most laser spot generates a speckle pattern emanating from the area of the moving surface bounded by the laser spot and is detected by its matching detector. After a time $\tau$, this same area will pass under the rightmost spot. As this laser spot is identical to the leftmost spot, an identical speckle pattern will be generated by this area and hence, an identical intensity will be detected by the rightmost detector. A continuous comparison of the detector signals will reveal identical signals in both detectors at a time $\tau$. Thus, a method is required to ascertain this time $\tau$.

This can be achieved by calculating the *cross-correlation* of the two detector signals in time. The peak of this correlation will correspond to the time at which maximum correlation of the two detector signals occur i.e. when the same speckle intensity pattern falls on both detectors. The cross-correlation of the two detector signals is given by

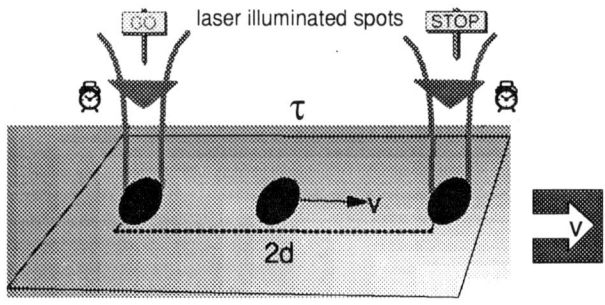

Fig. 2 The LTV Principle.

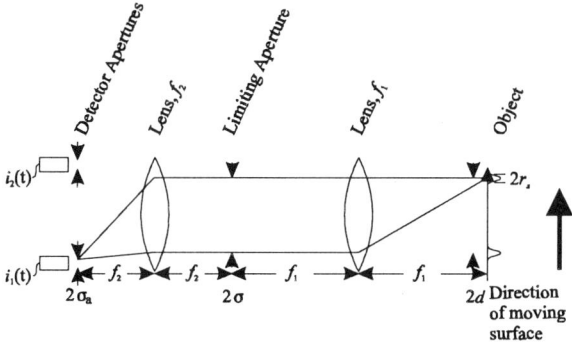

Fig. 3 The LTV receiver scheme.

$$C(\tau) = <\iota_1(t)\, \iota_2(t+\tau)> - <\iota_1(t)><\iota_2(t+\tau)> \qquad (1)$$

where $\iota_1(t)$ is the current given by the detector 1 at time t and $\iota_2(t)$ is the current given by detector 2. The angular brackets denote the ensemble average. The statistics of the speckle optical field are described by a Gaussian normal distribution[3] and it can be shown[2] that the correlation function given in Eq. (1) is also a Gaussian function:

$$C(\tau) = K <\iota>^2 exp\ [-(v\tau - 2d)^2/\sigma^2] \qquad (2)$$

where $K$ is a constant related to the optical system, $v$ is the velocity (in a direction given by the centres of the two spots) of the moving solid surface, $2d$ is the distance between the two laser spots and $\sigma$ is the laser spot radius. Fig. 4 shows a schematic of the correlation function.

## 3. FACTORS AFFECTING CROSS-CORRELATION

Previous mathematical and experimental treatments of the performance of an LTV system have assumed that the moving surface gives rise to an idealised form of speckle described as being *fully developed speckle*. In this case, the surface roughness is such that the height of the scattering structures is greater than the wavelength of the illuminating light and the lateral scale of the structures is of the order of the wavelength. However, when designing the sensor, it is imperative that all possible deviations from this idealised theory are

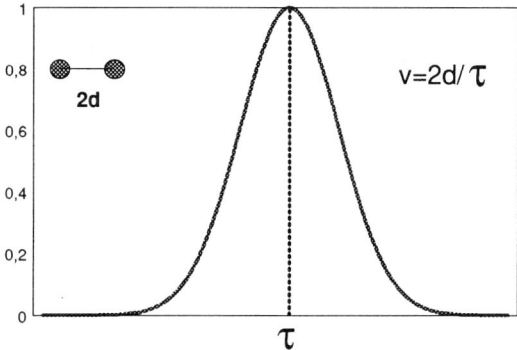

Fig. 4 Schematic of the correlation function.

characterised and that any compromises to sensor performance are investigated.

There are two parts to the sensor. First, an optoelectronic sensor head delivers the twin optical spot probes to the moving surface and detects the resultant speckle patterns. Second, an electronic processing unit is responsible for finding the peak of the cross-covariance and then calculating the instantaneous velocity. This paper will only concern itself with the optical head. Although the design and manufacture of the processing electronics is not a trivial task, standard electronic component strategies are available to perform the required functions and have already been demonstrated at the PCB level [4]. The integration of this design into an ASIC chip is deemed to be straiglitforward.

The task of the sensor design study was to identify the factors, both due to surface characteristics and miniaturisation, that affect the correlation function. Three effects are of concern:

1) *Widening of the correlation peak.* As mentioned previously, the time at which the correlation peaks corresponds to the time lag for the speckle pattern to "move" from one detector to the other detector. The sharpness of this peak will define the accuracy of determining the time lag $\tau$ and hence, define the accuracy of velocity determination.
2) *Shifting of the correlation peak.* A change in the velocity of the moving surface will change the value of the time lag, or in other words, the peak of the correlation will shift to another time lag. However, other circumstances may cause the peak to move, even if the velocity of the moving surface remains constant. Thus, the new time lag obtained will be incorrect, which will lead to an erroneous velocity determination.
3) *Reduction of the correlation peak.* This is known as *decorrelation* and is described as the reduction in the amplitude of the correlation peak. It occurs if the two detectors do not "see" two identical speckle patterns.

In order to design an accurate and robust miniature sensor, these effects have to be quantified and qualified.

## 4. EXPERIMENTAL EVALUATION

It is imperative that the above mentioned phenomena are evaluated both theoretically and experimentally. This will allow tolerance evaluation of the sensor and provide essential information on the behaviour of the sensor under varying circumstances. Thus, it is necessary to build an experimental testbed system. Fig. 5 shows a schematic of the experimental system used. The two diode lasers used were state-of the-art Vertical Cavity Surface Emitting Lasers[5] (VCSELs) emitting in single mode at a wavelength of 850nm. Such

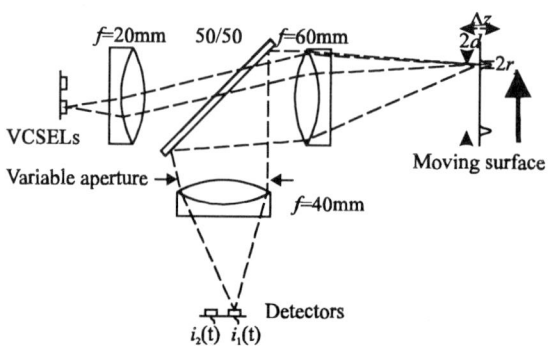

Fig. 5 The experimental LTV testbed.

Fig. 6 Influence of spot size.

lasers were candidates for the light sources in the miniaturised sensor. Light from the two lasers was focused onto a moving surface via two lenses in a telecentric lens configuration. This produces two circular illuminated spots on the moving target surface. The backscattered light travels back through the target lens, is deflected by the beamsplitter and passes through the detector lens (also in a telecentric configuration with the target lens) to the detectors. In other words, the two backscattered spots are imaged onto the two detectors. These signals are then temporally cross-correlated, which produces a Gaussian correlation curve. The peak of this curve corresponds to when the detectors have observed what is perceived to be the *same* speckle pattern. The time at which this peak occurs is the delay at which one detector observes the speckle pattern relative to the other detector (i.e. the time it takes for one scattering element to travel from one spot to the other). The distance between the spots are known, so the velocity of the moving surface can be determined.

A series of experiments has been conducted under differing conditions. These included surfaces of varying roughness, varying illumination spot sizes, varying spot spacings and intentional misalignments in the system. This provided information about the performance of an LTV system and allowed the formulation of a theoretical model describing the LTV system[6]. This information was vital in order to obtain tolerances and specifications that are required to be met when designing the miniaturised velocity sensor.

### 4.1. Accuracy and effects due to different surfaces

In order to evaluate the accuracy of the system, it is necessary to determine those parameters that affect the width of the cross-covariance. Inspection of Eq. 2 indicates that the width of the cross-covariance is dictated by the ratio of the size (diameter) of the laser probe spots to their separation ($2d$). A sharp cross-covariance is obtained by having small spot sizes, however, this is not usually a desirable requirement. Fig. 6 illustrates the problem of having small spot sizes at the same fixed distance $2d$ apart. It is clearly shown that if smaller spots are used, the degree of acceptance of measurement is drastically reduced. This compromises the robustness of the sensor and needs to be avoided. Hence, there is a trade-off between the accuracy of the sensor and the robust performance of the sensor. This affects the design of the optical system and the choice of the laser parameters (in effect, the laser spacing and the divergence). For the system under design a spot radius:spot spacing ratio of 1:10 was used.

A number of different surfaces with varying roughnesses were measured. It was found that the relative width of the correlation peak changed due to the nature of the surface. Fig. 7 shows the change in correlation width with differing spot sizes for a backside silicon wafer.

By changing the position of the moving surface from the focal point of the lens, the spot sizes of the lasers on the surface could be altered by defocusing (at the same spot spacing). The diameter of the spots for each defocus was measured by a beam scanner. By using Eq. (2), the relative width of the cross-covariance is given by:

$$\langle \Delta\tau/\tau_0 \rangle = \sqrt{(\ln 2)} \times r_s / d. \tag{3}$$

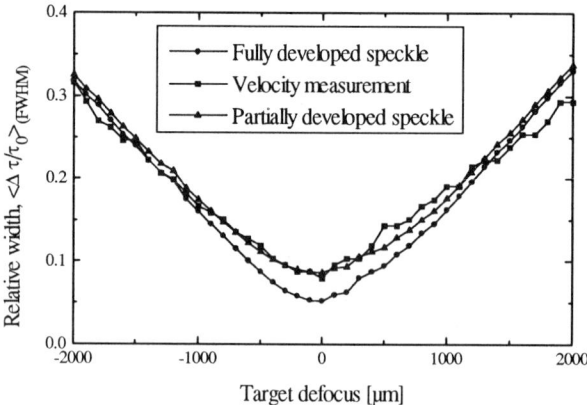

Fig. 7 Relative width measurements on silicon.

Thus, by inserting a value for $r_s$ for each defocussed spot, the relative width of the cross-covariance could be calculated. This is indicated by the curve with circular spots. The width of the cross-covariance was also measured for each value of the defocus using an analogue correlator (Hewlett Packard HP 3721A). This is also plotted on the same graph using square markers. The results given by Eq. 3 (circular markers) have been derived assuming that the surface exhibits fully developed speckle. However, it is seen that the surface does not behave as an ideal surface exhibiting fully developed speckle. A correction factor was obtained from the experimental results by calculating the offset in the width from the ideal at zero defocus. This offset was then applied to all point marked with a circular marker and replotted on the graph. This curve is shown with triangular markers and can be seen to closely resemble the experimentally obtained measurements for the relative width.

From this result, it was realised that the nature of the surface can affect the accuracy of velocity determination. More importantly, however, it also indicated that existing theories assuming surfaces exhibiting fully developed speckle are proving inadequate to model realistic surfaces, which exhibit *partially developed* speckle. With the aid of further investigations, it was found that the offset in the correlation width was closely related to the

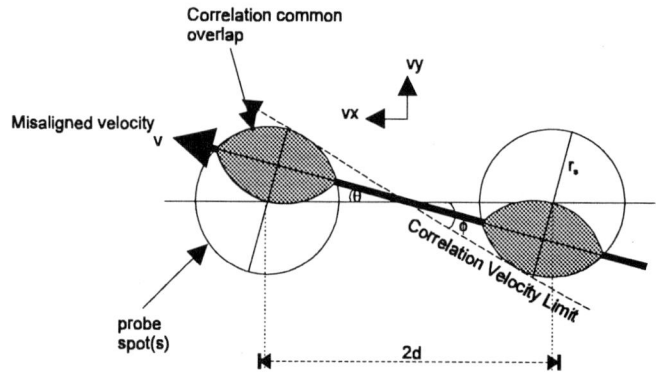

Fig. 8 Effect of velocity misalignment.

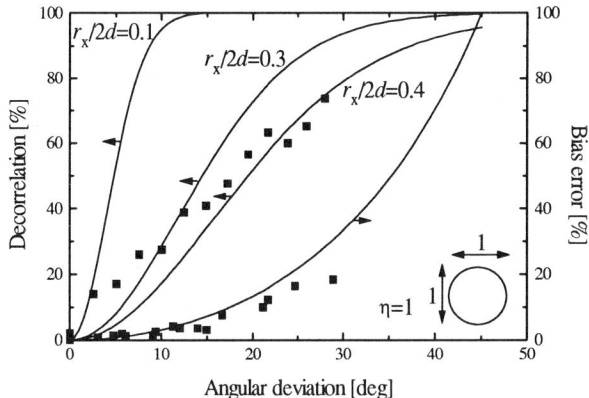

Fig. 9 Relationship between velocity error and decorrelation with misalignment (circular spots).

scale of the larger scale roughness encountered in many surfaces. This prompted an in depth theoretical analysis into the experimental findings and which has been published elsewhere[6]. The results obtained in this analysis have shown that the accuracy is dependent on the type of surface measured, and not only on the sensor system itself.

## 4.2. Accuracy due to misalignment

Fig. 8 shows a schematic of the two laser spots with a misaligned velocity. In an ideally aligned system, the velocity vector of the moving surface will pass through the centre of both spots, allowing the full area of the spot to be correlated. However, if the velocity is misaligned, then only a partial area of the two spots have a common correlation area. This is illustrated by the shaded segments in the figure. There are two consequences of misalignment. Firstly, due to the reduced correlation area, the correlation peak reduces in amplitude or decorrelates. Secondly, but more important, it can be seen by geometric considerations that the distance between the two correlation (shaded) areas are now longer than the spot centre-to-centre distance of $2d$. Thus, the maximum correlation will occur at a later time lag than under an aligned system or, in other words, the correlation peak will move to a greater time lag. As the system electronics will have $2d$ as the spot spacing, the result will be an erroneous velocity calculation.

Fig. 9 shows a graph showing the extent of velocity estimation error as a function of misalignment for circular spots. One can see that even for a 10 degree misalignment, decorrelation approaches 100%, even though the velocity error is around 5% for $r/2d=0.1$. The situation is better for larger spot sizes (0.3 and 0.4), which can tolerate a greater degree of correlation for increasing misalignment. However, the velocity error then becomes greater. As a means to circumvent this problem, Fig. 10 shows the same graph but with elliptical spot. It can be clearly seen that for a given amount of misalignment, the decorrelation and velocity error is very much reduced when compared with circular spot probes. In this case, specially designed diffractive optics would be required to provide the elliptical spots,

## 4.3. Influence of detector cross talk

As described earlier, each scattered spot is imaged by the optical system into its own detector. Thus, correlation takes place assuming that each detector is only detecting an image of its matching spot probe. However, there may be instances where the surface may

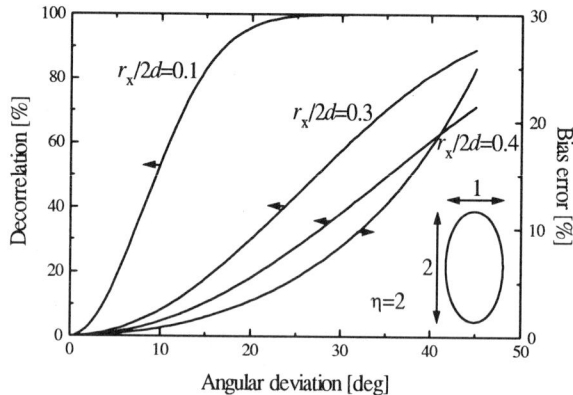

Fig. 10 Relationship between velocity error and decorrelation with misalignment (elliptical spots).

scatter light from one spot into *both* detectors. Thus, a partial *auto-correlation* will take place. Fig. 11 shows theoretical (solid curves) and experimental (dashed curves) correlations that have been obtained from the bulk testbed. One can see that for increased levels of crosstalk, the auto-correlation peak (i.e. the peak at zero time lag) becomes more pronounced. Again, the two effects observed are a decorrelated peak and a slight shift of the correlation peak to a shorter time-lag. Fig. 12 shows the velocity error as a function of cross talk. It seems that cross talk is an insensitive parameter, as levels of 50% cross talk (which is almost impossible to obtain even in a badly aligned system) will only produce, at maximum, 5% velocity error.

## 5. MINIATURISATION DESIGNS

With the wealth of experimental and theoretical results obtained from the testbed system, a miniaturised design was proposed. The essential elements of the system consisted of: 1) Twin laser devices; 2) Diffractive optical system; and, 3) Detector devices.

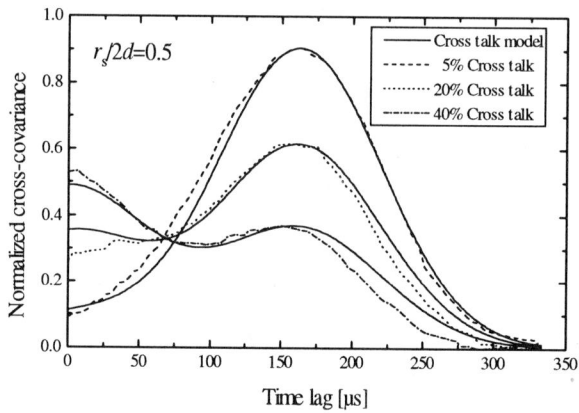

Fig. 11 Correlation graphs showing influence of cross talk.

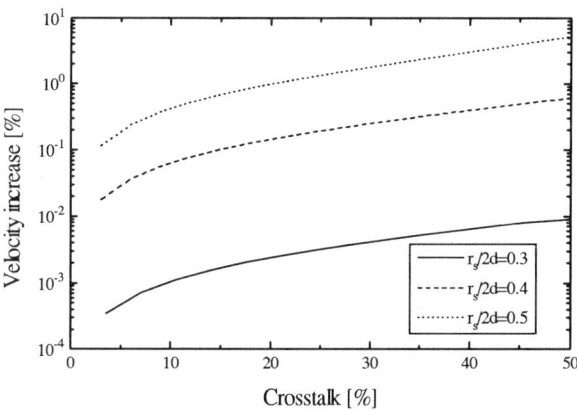

Fig. 12 Velocity error due to cross talk.

The laser devices were required to have a spot size:spot spacing ratio of 1:10. This is closely linked to the optical elements, which are required to focus the spots at a prescribed distance on the moving target. The detector devices are also required to have a prescribed spacing, dependent on the optical system. Fig. 13 shows a diagram for a sensor design. The lasers are a VCSEL array housed in a 20-pin lead DIP package. The tests made on the bulk system had shown the viability of VCSELs as light sources for the sensor. The Table I gives the specifications of the components.

Table I. Specifications of the components of the miniaturised sensor.

| Lens plate vertical tolerance | ±50 micron |
|---|---|
| Detector placement | Vertical: 60 micron<br>Horizontal:x ±20 micron<br>y ±100 micron |
| Spot size/spacing | 70/750 micron |
| Detector | 1.5 x 1.5 mm   (gap 25 micron) |
| Measuring velocity range | 0.3 - 44 m/s |
| Bounce (vertical) tolerance | 9mm |
| Acceptance angle | $12^0$ |
| Maximum velocity inaccuracy in acceptance | +2% |

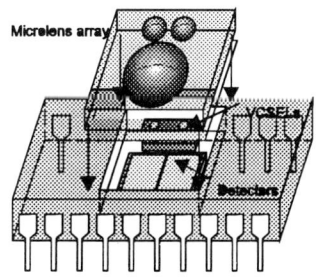

Fig. 13 Miniaturised sensor concept.

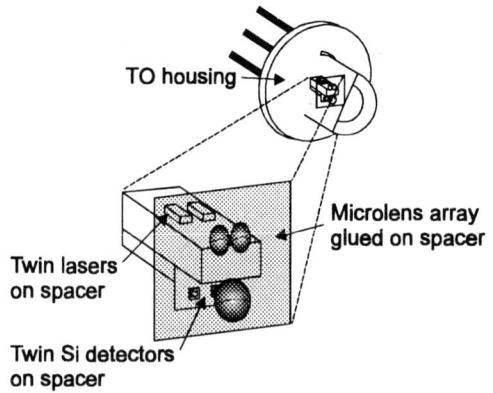

Fig. 14 Future sensor concept.

Active alignment would be required for the sensor, especially for the alignment of the diffractive optics. However, it has been designed so that all optics are housed on one lens "chip", in order to keep the alignment steps as few as possible. The lenses would be of hybrid (refractive/diffractive) type and be embossed with glass as a substrate. The detectors would be housed in the DIP housing in an offset position to the lasers. These would be mounted and fixed using a conductive adhesive, within the tolerances prescribed in the Table I.

The sensor depicted in Fig. 13 is a prototype design and will illustrate the viability of using miniaturised optoelectronic components for sensor technology. Fig. 14 shows a conceptual diagram showing a sensor more amenable to mass-production. The sensor is housed in a TO-cap, with edge-emitting lasers and detectors housed on a platform. The lens chip will then be mounted on the edge of the platform. A thorough design study of this sensor will take place in a time frame of 1-2 years.

## 6. CONCLUSIONS

In this Chapter, an illustration has been given of the complex investigation that was required to design a miniaturised sensor system. As a point of interest, it has been immensely difficult to obtain the individual components for the sensor. More often than not, the design was dictated by off-the-shelf components which were directed towards the telecommunications market. Due to the huge inertia of this industry, component manufacturers are reluctant to change designs to suit sensor applications. Thus, sensor designs necessarily have to comply with telecommunications-type devices. Custom designed components have the adverse affect of increasing the cost of the sensor to astronomical levels. Nevertheless, the paper has shown that sensors can be designed using optoelectronic components to provide a non-contact, robust and versatile sensor for measuring velocity.

## REFERENCES

1  D.H. Thompson, A tracer-particle fluid velocity meter incorporating a laser, *J. Phys.E:Sci. Instr.* 1 pp:929-932 (1968)
2  H.T. Yura and S.G. Hanson, Laser time-of-flight velocimetry: analytical solution to the optical system based on ABCD matrices, *J.Opt.Soc.Am.A* 10, pp:1918-1924 (1993)

3   J.W.Goodman, Statistical properties of laser speckle patterns, in *Laser speckle and related phenomena*, J.C.Dainty, ed., Springer-Verlag, Berlin (1975)
4   S.G. Hanson, L.R. Lindvold and L. Lading, A surface velocimeter based on a holographic optical element and semiconductor components, *Meas.Sci.Technol.* 7 pp: 69-78 (1996)
5   The VCSEL devices were kindly given by the Paul Scherrer Institute, Zûrich, Switzerland
6   B.Rose, H.Imam, S.G.Hanson and H.T.Yura, Effects of target structure on the performance of laser time-of-flight velocimeter systems, *Appl.Opt.* Vol. 36. no. 2 pp: 562-568 (1997)

# A ROBOTIC MICROSYSTEM FOR COLON VISUALISATION AND SAMPLING

L. Lencioni, P. Dario, M.C. Carrozza, B. Magnani, and S. D'Attanasio

MiTech Lab, Scuola Superiore Sant'Anna
Via Carducci 40, 56127 Pisa, Italy

## 1. INTRODUCTION

This Chapter describes the development of a new microrobotic system for colonoscopy. The project aims to investigate and exploit an innovative concept for endoscopy in the lower gastrointestinal tract. Colonoscopy is important since cancer of the colon and of the rectum is the second most malignant tumor in industrialised countries. Today endoscopic procedures subject patients to significantly less trauma than conventional surgery, but with present instrumentation colonoscopy is in many cases an unpleasant and painful procedure. The microrobot we have fabricated is able to propel itself in the colon is a semi-autonomous manner by inchworm locomotion obtained by means of pneumatic actuators controlled by shape memory alloy (SMA) microvalves. In this Chapter, the architecture of the microrobot, the propelling principle, the design and the fabrication of the SMA microvalves are illustrated.

Endoscopy belongs to the increasingly strategic field of MIT (Minimally Invasive Therapy). MIT indicates all those operations (diagnostic and surgical) whose main goal is to treat those organs - and only those - which are actually affected by pathological processes, while preserving as much as possible the integrity of healthy organs and tissues. The reduction of the access trauma due to the surgical operation involves important and evident advantages both for the patient and for the health care system, in terms of shortening the postoperative course and hospitalisation. A drawback is that the surgeon meets with some severe difficulties during most procedures, such as the reduction of visual and tactile feedback and some loss of dexterity, as documented in the literature (see Ref. 1). The challenge for research in medical technology for MIT is clear: to develop new and more powerful tools to restore to the surgeon the same capabilities he/she has in ordinary surgery.[2]

MIT is particularly important and widely used for diagnosis and surgery in the lower gastrointestinal tract (colonoscopy). In fact an increasing number of interventions are performed by colonoscopy. This procedure requires the insertion of an endoscope from the rectum to the colon, in order to visualise pathological conditions and to take samples of suspect tissue for biopsy.

At the moment, colonoscopy is performed by means of quite rigid endoscopic systems whose distal part can be bent up to 180 degrees by cables running along the endoscopic systems and controlled manually by the endoscopist through external knobs. Although the flexibility of the tip allows the endoscope to follow the tortuous path of the colon, the insertion of the endoscopic system requires the doctor to exert forces and rotations on the portion of the shaft outside the patient, thus causing discomfort to the patient due to pressure on and stretching of the intestine walls and adjacent tissues and organs. An additional cause of discomfort for the patient is the need for insufflating air into the colon, in order to open the normally collapsed colon lumen and to allow the insertion of the endoscopic system.

Removing as much as possible of the length of the endoscopy pipe, and eliminating or reducing the need for insufflating air, would substantially reduce discomfort and pain for the patient and possibly facilitate the procedure for the endoscopist. The objective of the work described in this paper is integrating all of the important functions of the endoscope in a miniature "head" capable of self propulsion, thus eliminating the stiff part of present endoscopes. In particular, our effort is focused towards the development of an innovative endoscopy system for the lower gastrointestinal tract.

The microrobot exploits microfabrication technologies and micromechatronics concepts (i.e., the integration of precision mechanisms, sensors, actuators, signal preprocessing electronics and embedded control).

The availability of a miniature robotic platform with self-propelling capabilities could provide a tremendous contribution to the development of a virtually unlimited range of new applications of MIT in many different regions in the human body.[3-5]

The proposed system for colonoscopy is shown in Fig. 1. The miniature robotic platform ("mothership") is connected to the outside by means of a thin and flexible "tail" pipe incorporating a few lumina. The most critical aspect of the miniature robot is the propulsion mechanism, which must be effective, easily controllable, safe and harmless. The

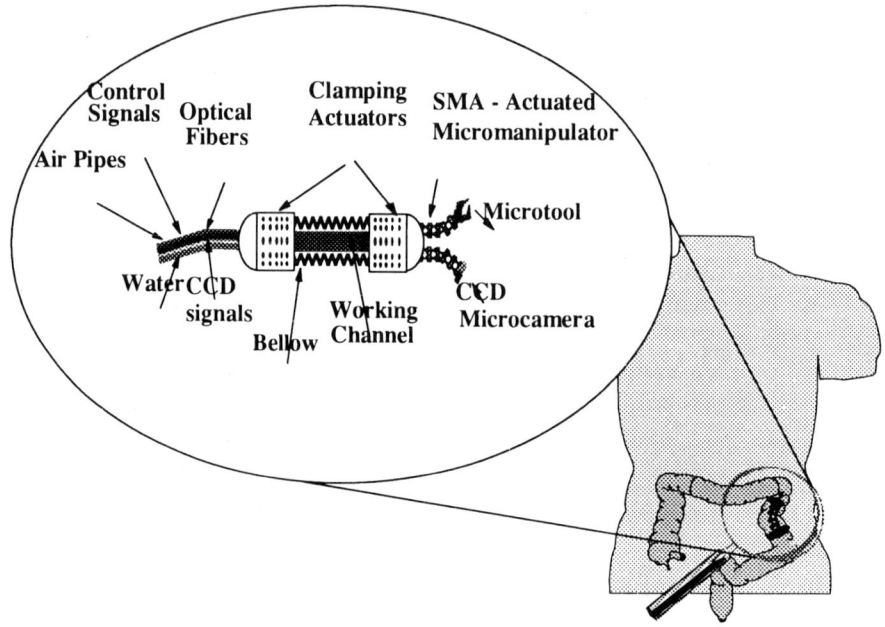

Fig. 1 The miniature robotic system.

inchworm propulsion is based on two "clamping" microactuators that allow the mothership to "adhere" to the colon wall, and on one extension (and contraction) microactuator that generates the "longitudinal" motion of the mothership. In addition to such longitudinal motion, the robot mothership could have two additional "bending" degrees of freedom, actuated by two separate microactuators, and intended to provide additional "dexterity" for navigation in the colon. All the actuators are controlled by a miniature pneumatic distributor based on new, custom designed microvalves.

A prototype of the mothership has been fabricated and some *in vitro* experimental tests have been performed in natural rubber tubes and pieces of fresh pig colon. These tests confirm that the inchworm locomotion principle is suitable for propelling the microrobot in the colon efficiently without significant damage of the tissue of the colon wall.

## 2. SYSTEM ARCHITECTURE

The miniature robotic endoscope offers the endoscopist the same basic functions as existing systems, that is, imaging and intervention capabilities. Fig. 2 shows the block diagram of the microrobotic architecture.

A high quality imaging system based on precision optics and a miniature CCD camera will be incorporated in the mothership. Moreover, virtually all of the existing surgical tools currently used during colonoscopy (and even new ones) could be inserted through a channel contained within the flexible tail. Tail channels are provided also for insufflating air, sampling fluids and flushing water.

In addition, the mothership will be equipped with two microarms to provide additional functions for diagnosis and intervention. One of the microarms will be used to steer the imaging microsystem towards different features of interest. Micromechatronics technologies allow incorporation of a virtually unlimited number of additional miniaturised sensing and intervention devices at the mothership and at the microarms. Specific "modules" (already existing, adapted, or purposely developed) will be mounted on the mothership and on the microarms, such as ultrasonic transducers, new teleoperated microtools for diagnosis and for surgical operation, optical microsystems for non-contact diagnosis and recognition of pathological tissue (such as instruments for spectroscopy, or for laser-induced fluorescence),

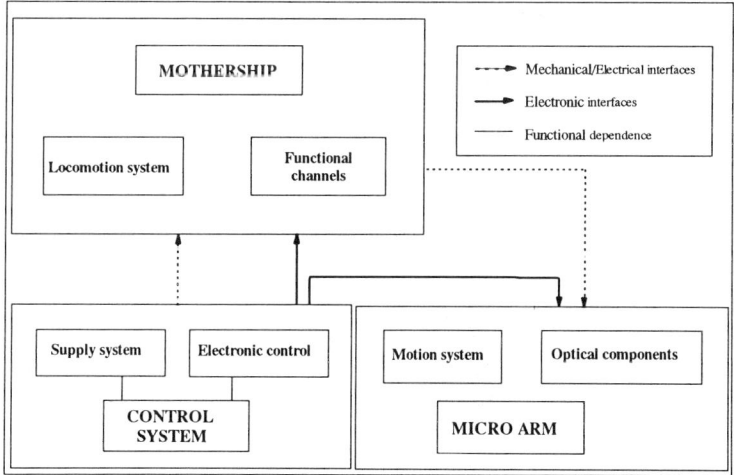

Fig. 2 Architecture of the microrobot system.

physical and chemical sensors (e.g. biosensors) able to perform *in situ* chemical analysis of colon tissue and of physiological fluids, and others.

The propulsion system, based on the inchworm principle, is integrated in the mothership body and is composed of three modules. Two modules have the primary role of providing traction to the microrobot by appropriately clamping the walls of the intestine, and are located at the two ends of the device. A third module, whose role is to extend the microrobot, is located between the two clamping modules. The sequence of propulsion steps is illustrated in Fig. 3, where the clamped module is indicated in grey.

In phase 1, the rear clamp is switched on, the front clamp is switched off, and the extensor module is in the contracted state. In phase 2 the extensor is in the extended state. In phase 3 the front clamp is also switched on. Subsequently, in phase 4 the rear clamp is switched off and in phase 5 the extensor is retracted. In phase 6 the rear clamp is switched off and in phase 7 the front clamp is switched off again. At this point the mechanism has returned to its original state, but it has moved forward by one stride length. The cycle can be repeated for continuous forward movement, or it can be reversed for rearward motion.

## 3. MICROFLUIDIC ACTUATORS

The three pneumatic actuators in the mothership (two for clamping and one for extension) are controlled by SMA-actuated micropneumatic valves.[6]

The valve is composed of a steel shutter that occludes the air conduit by pressing on a seal, and by an SMA spring which actuates the shutter. These components are enclosed in a plastic frame for electrical insulation.

At present the external dimensions of the valve body are 6 mm diameter and 19 mm length; however, since the diameter of the inner cylinder is only 2.4 mm, further miniaturisation is possible. The SMA spring has an average diameter of 1.45 mm, whereas the SMA wire diameter is 0.29 mm. The spring has been designed to operate at a maximum working pressure of about 0.5 MPa. The microvalve has been tested at various working

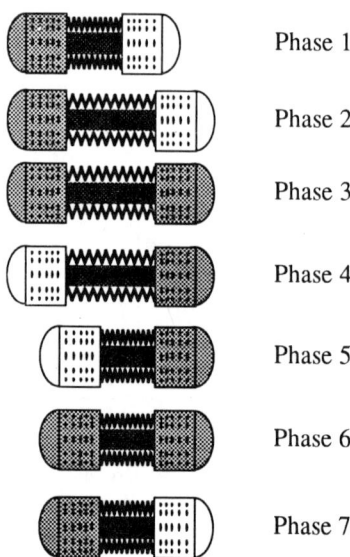

Fig. 3 Schematic diagram of the inchworm locomotion principle of the microrobot.

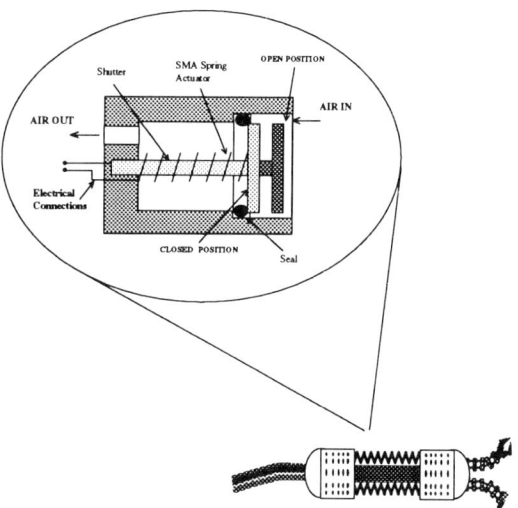

Fig. 4 Functional scheme of SMA valve.

pressures.

As shown in Fig. 4, the microvalve works as desired (that is, full communication is established between inlet and outlet) in a range of pressure useful for the functioning of the microrobot (0.3 to 0.5 MPa). This diagram is useful to define the electric current required to completely open the microvalve for different pressures.

Experimentally, fatigue failure occurred after more than 22,000 cycles for an inlet pressure of 0.4 MPa, at a frequency of 600 mHz.

## 4. CONCLUSIONS AND FUTURE WORK

In this Chapter the motivations for the design of a microrobot for inspection and intervention in the colon have been illustrated. The system under development will be strongly innovative and will introduce a totally new concept for endoscopy in general and colonoscopy in particular. The new concept (the physical separation between the operator and the active part of the instrument) has important implications: not only could colonoscopy evolve toward a more affordable and effective procedure, but virtually all of the existing areas of endoscopy and of MIT (and even areas thus far not investigated) could be affected.

Many scientific and technological issues must be addressed and solved, however, such as: a) effective and safe propulsion of the mothership; b) the control of locomotion and the intervention functions (including remote micromanipulation); c) the possibility of incorporating on board different instrumented modules for diagnosis, therapy and intervention; and d) multimodal feedback (visual, force, tactile) to the operator. Most of these aspects are presently being investigated in the authors' MiTech Laboratory at Scuola Superiore Sant'Anna in Pisa, Italy.

ACKNOWLEDGEMENTS. This work has been supported by the SAMA (Shape memory Alloy microactuators for Medical Applications) project sponsored by the BRITE - EURAM program of the Commission of European Communities. The authors would like to thank Mr. Carlo Filippeschi for his valuable technical collaboration.

**REFERENCES**

1. Cuschieri A., Ergonomics of minimal access surgery, *Surgery,* vol. 11, pp. 526-8 (1993)
2. Ikuta K., Nokata M., Aritomi S., Iper redundant active endoscope for minimum invasive surgery, in: *Proc. First Int. Symp. on Medical Robotics and Computer Assisted Surgery*, Pittsburg, PA (1994)
3. Schurr M.O., Buess G., Interdisciplinary technology development for minimally invasive therapy, in: *Mst News*, No. 13, pp.2-4 (1995)
4. Reynaerts D., Peirs J., Van Brussel H., Design of a shape memory alloy actuated gastrointestinal intervention system, in: *Proc. of MHS '95*, Nagoya, Japan (1995)
5. Dario P., Carrozza M.C., Allotta B., Guglielmelli E., Micromechatronics in medicine, *IEEE/ASME Transaction on Mechatronics*, vol. 1, no. 2, pp. 137-148 (1996)
6. Kunad H.K. and Muglitz J. Micromanipulators for minimal invasive surgery, *Proc. 1 st IARP Workshop on micro Robots and Systems,* Karlsruhe, Germany (1993)

# INDEX

Abbe number, 32
ABCD Law, 36
Aberrations, 27, 32, 35, 169, 195, 249, 389
Accessible modes, 55
ACES, 341, 342
Acoustic tunable optical filter, 328, 329, 333, 334, 335
Acousto-optical tunable filter, 328, 330, 332, 334, 335, 337
Actuation mechanisms, 211
Alignment, 121, 123, 133, 195, 200, 206, 215, 218, 220, 221, 319, 354, 367, 399, 408
All-optical beam steering, 68
All-optical signal processing, 65, 361
All-optical switching, 71, 74
Alloying, 251
Analysis of microobjects, 342
Anamorphic microlens array, 174
Anamorphic microlenses, 169, 170
AOTF, 328, 330, 332, 334, 335, 337
Artificial dielectrics, 122, 129, 130, 132
Astigmatic point imaging, 248
ATOF, 328, 329, 333, 334, 335
Automated interferometric methods, 342

Backplane interconnection, 249
Basis functions, 51, 52, 53
Batch processing fabrication, 190
Beam
    coupling, 239, 250
    dividers, 5
    formers, 252
    shaping, 27, 29, 33, 147, 158, 169, 251, 257, 277
    size controller device, 44

Bi-directional optical transceiver, 367
Binarization, 227, 234
Binary CGH, 223, 224, 225, 230, 232, 233, 234, 235, 252
Binary optics, 23, 130, 269
Binary profile, 25, 261
Blazed reflection grating, 367, 369
BOR, 266, 269, 276
BORC, 269, 276
Border rainbow, 266, 269, 276
Bragg
    gratings, 241, 293, 299, 300, 308, 309, 338
    law, 242
Broadband transmission, 169, 176

CAD, 133, 137, 138, 152
CAIBE, 135, 137, 138
CCD detectors, 341
Cell oriented coding, 226
Center high-mounted stop light, 274
Central peak, 227, 232, 233
Cerenkov
    effect, 359
    radiation, 357, 358, 362
    second harmonic generation, 357
    radiation, 358, 359
$CF_4$ plasma, 253
CGH, 133, 135, 137, 138, 158, 225, 251
Chemical etching, 115, 135, 199, 303
Chemically interactive materials, 371, 376
Chirped grating, 246
CHMSL, 274
Chromatic aberration, 32, 152, 249, 259, 266
CIM, 371, 374, 375

417

CMOS, 194, 195, 197, 204, 373
CO₂ laser, 114, 115, 118, 122, 251, 252, 253, 254
COC, 226, 230, 231, 234
Coding, 140, 141, 142, 224, 226, 227, 228, 232
Coherent light, 3, 19, 91, 341
Colon
    sampling, 411
    visualization, 411
Colonoscopy, 411, 412, 413, 415
Comb electrode, 214
    function, 14
Composite materials, 306, 327, 351
Computer generated holograms, 133, 138, 223, 224, 236, 237
Computerized interference microscopy, 344
Continuous surface profile, 141, 155
Couplers, 115, 117, 118, 239, 246, 249, 250, 260, 276, 293, 361, 362
Coupling device, 57, 281, 284, 288
Coupling elements, 239, 365
Crossed deposition, 170, 172, 174, 175, 176, 177
Crossed linear interaction zones, 169
Cross-sectional amplitude distribution, 244
Cylindrical microlenses, 170, 173, 174, 175, 176, 177
Cylindrical parabolic wavefront, 246

Dammann grating, 9, 277
Decay factor, 244, 245
Diagram of wave vectors, 243
Diamond like carbon films, 161
Dielectric grating, 55, 131
Dielectric strength, 213
Diffraction efficiency, 20, 24, 26, 27, 121, 123, 138, 139, 141, 142, 143, 144, 146, 147, 195, 241, 245, 246, 252, 254, 262
Diffractive lenses, 24, 26, 30, 31, 32, 33, 131, 133, 140, 147
Diffractive micro-optics, 147
Diffractive optical elements, 3, 23, 24, 119, 130, 131, 132, 133, 138, 139, 147, 159, 160, 169, 202, 239, 260, 276, 385, 387
Diffractive optics, 3, 4, 5, 6, 18, 20, 23, 28, 119, 122, 129, 134, 269, 276, 279, 381, 385, 405, 408
Dirac delta function, 14

Direct absorption-spectroscopy, 323
Dispersion, 23, 32, 59, 60, 61, 138, 139, 249, 266, 267, 293, 308, 329, 357, 362
    properties, 249
Displacement measurements, 315
DLC, 161, 163, 164, 165, 166
DMD, 197, 198, 205
DOE plane, 28, 29, 30
DOE, 3, 5, 8, 14, 19, 20, 24, 25, 26, 27, 28, 29, 30, 119, 121, 124, 127, 128, 130, 133, 135, 137, 138, 140, 141, 146, 147, 149, 150, 157, 159, 195, 196, 204, 251, 252, 260, 262, 264, 385
Dry etching, 113, 133, 196, 253, 254, 273
Dynamic focusing mirrors, 219
Dynamic light scattering, 311, 325

E-beam lithography, 133, 134, 151, 156
EBL, 149, 150, 152, 153, 158, 159
EFIE, 50, 52, 54
Elastic forces, 211
Electric field integral equation, 50, 52
Electric green function, 50
Electron beam lithography, 132, 159, 160, 181, 236
Electronic holography, 341, 343, 352
Electronic nose, 371, 375, 376, 377, 379, 380
Electronic packaging, 342, 347
Electrostatic actuation, 199, 214, 215
Embossing, 120, 125, 196 245, 271, 387, 389
Encoding, 131, 138, 158, 196, 225, 276, 277
Envelope phase functions, 172
EOH, 341, 343, 344, 348, 349, 352
Equivalence theorem, 49, 50, 51
Etching, 26, 104, 107, 112, 113, 115, 125, 126, 127, 128, 129, 130, 131, 133, 134, 135, 150, 155, 159, 161, 166, 167, 172, 173, 195, 196, 198, 209, 210, 218, 240, 252, 253, 303, 385
Extremely long focal lengths, 176

Fabrication
    aspects, 139
    methods, 26, 139, 170
    of microstructures, 365
    technologies, 119, 120, 125, 139, 147

tolerances, 33, 139, 140, 146, 147
Fast Fourier transform, 29
FBG, 293, 294, 296, 298, 301, 302, 305, 306, 308, 327, 328, 330, 332, 333, 337
FEM analysis, 347
FEM-EOH, 349
FFT, 29, 30, 264
Fiber
    optics, 169, 196, 260, 261, 306, 308, 309, 311, 313, 315, 318, 323, 325, 338, 386
    optic sensor, 308, 311, 313, 318, 338
Field
    rays, 36, 42
    measurements of methane, 324
Fill factor, 169, 261
Flexible substrates, 169, 172
Fluid-suspended particles, 317
Focal length, 11, 31, 32, 33, 140, 152, 158, 164, 169, 171, 174, 175, 176, 181, 196, 198, 225, 227, 228, 229, 250, 268, 269, 282, 283, 284, 286, 287, 288
Focusing grating, 248, 249, 250 362
Form birefringence, 18
Fourier
    coefficients, 6, 7, 8, 9, 10, 21, 22
    holograms, 226
    series, 6
    transform, 18, 29, 60, 225, 226, 228, 231, 264
Free space, 87, 189, 190, 202, 223, 224, 225, 229, 237, 239, 276
Free space optical interconnection, 223, 224, 225, 229, 237
Fresnel
    encoded lens, 148, 225, 227, 228, 234, 237
    hologram, 225, 227, 228, 233, 234, 235, 236
    losses, 26, 146
    numbers, 172
    transforms, 237
Friction, 163, 201, 211, 212
Full-field optical methods, 341

Gas absorption-measurement, 323
Gaussian spatial profile, 251
Gegenbauer polynomials, 53
Generators, 49, 50, 131, 276, 325, 373
Geometrical fill factors, 169, 170

GI, 312, 344, 352
GIRO grating, 54, 55, 56
Global optimization, 29
Graded index, 324
Grating
    equation, 241, 242, 246, 247, 248, 249
    interferometry, 341, 347, 349, 352
    vector, 28, 242, 243
GRIN, 24, 35, 36, 37, 45, 311, 312, 313, 315, 318, 319, 321, 323, 324, 325
GRIN-rod, 311, 314, 324
GSM, 47, 52, 53, 54
Guided second harmonic radiation, 360

Hardening, 251
HEBS material, 137
Hexagonal microlens arrays, 174
HFIE, 49, 50, 51, 53, 54
H-field integral equation, 49
High fill factor, 174, 176, 177, 195
Hole
    array masks, 170, 171
    mask method, 170, 172, 174
Hologram, 58, 61, 158, 224, 225, 226, 227, 228, 230, 231, 232, 234, 250, 264
Holographic equation, 247, 248
Holographic optical elements, 58, 61, 147, 224
Holographic techniques, 62, 114, 245
Hot embossing, 125, 139, 145
HS, 35, 36, 37, 38, 39, 40, 41, 42, 44, 45
Hybrid elements, 24, 139
Hybrid experimental/numerical methods, 341

IBE, 113
IFT, 29
Inchworm locomotion, 411, 413, 414
Inclined sinusoidal curve, 245
Industrial applications, 337
Injection moulding, 120, 125, 126, 139, 145
Ink-jet printer, 211, 216
Integral equation, 49, 51, 55
Integrated optic sensors, 353, 354, 356, 361
Integrated optic waveguides, 355
Integrated optics, 35, 45, 47, 104, 115, 117, 118, 131, 189, 190, 201, 203, 220, 250, 354, 362, 381, 382, 383, 384, 385

Integrated sensor, 218, 220
Interconnections, 62, 220, 223, 224, 225, 236, 250, 276, 347, 373, 381
Interconnector testing, 349
Intrinsic loss of an empty probe, 321
IO, 103, 105, 107, 110, 114, 115
Ion-beam etching, 240, 245, 396
IR, 121, 163, 179, 205, 369

Lagrange invariant, 32
Laser
    beam focusing, 164
        shaper, 260, 273
    diode tuning, 218
    manipulation, 398
    writing, 114, 118, 122, 123, 131, 139, 140, 142, 143, 145, 147, 259
LED, 193, 194, 204, 309
Legendre polynomials, 53, 93
LIGA, 199, 219, 365, 368, 370
    process, 365

Light
    deflector, 41, 45
    focuser, 38
    scattering, 317, 319, 397
    shifter, 41, 45
$LiNbO_3$ waveguides, 356, 357, 359, 361
Local approach
    to experimental mechanics, 342
    to material engineering, 342
LTV, 399, 400, 401, 402, 403

Magnetic field integral equation, 50
Magnetic Green function, 50
Masks, 114, 122, 123, 125, 127, 133, 134, 137, 169, 170, 171, 172, 174, 176, 204, 237, 252, 273, 385, 299
Mass fabrication, 365
MEMOS, 190, 196, 200, 201, 202
MEMS, 197, 202, 205, 206, 220, 341, 342, 352
Meridional plane, 247, 248
Method of moments, 51, 53, 55, 56
Micro total analysis systems, 371, 372, 373, 380, 387, 388
Microdrilling, 251
Microfabrication technologies, 103, 149, 412
Microlasers, 92, 193, 203
Micromachining, 200, 205, 206, 210, 215, 216, 218, 220, 253, 371, 398

Micromeasurements of displacement and strain, 342
Micromechanics, 203, 205, 209, 211, 212, 214, 216, 217, 219, 352, 395, 396
Micro-optical bench, 219
Micro-optical components, 33, 176
Micro-optics, 131, 147, 204, 206, 219, 387
Microparticles, 83, 88, 89, 91, 99, 395, 396, 398
Microprofilometer, 253
Microreliefs, 23
Microrobotic system, 411
Microspectrometer, 366, 368, 369, 370
Microsystems, 91, 161, 190, 193, 198, 199, 202, 203, 204, 219, 353, 367, 386, 396
Microtechnologies, 200
Miniaturised sensors, 399
Minimally invasive therapy, 416
MIT, 147, 160, 411, 412, 415
Modal impedance, 49
Modal transmission line, 47
Modal voltage, 48
Mode matching, 55
Modelling, 153, 154, 159, 280, 360, 379
Monitoring of polymerization processes, 319
Monodisperse particulate, 319, 325
Morphology dependent resonances, 84, 91, 94, 96
Multidimensional interconnects, 74
Multilayers, 169, 170, 174, 176, 180, 183, 184
Multilevel
    microreliefs, 23
    profile, 25

Nd:YAG laser, 67, 289, 358, 394, 395, 398
Nearly-autofocussing probe, 323
NIR, 323, 369
Nonlinear dynamics in laser, 92
Nonlinear effects, 66, 193, 361
Normal mode, 47, 48

Optical fibers, 105, 106, 200, 279, 281, 284, 288, 303, 311, 318, 319, 323, 325, 338, 386, 396
Optical interconnections, 196, 223, 225, 236

Optical microsystems, 190, 191, 196, 197, 200, 202, 204, 381, 386, 413
Optical power, 31, 195, 295, 328
Optical sensors, 353, 373, 380, 399
Optical spectrum analyser, 299
Optical tweezers, 391, 394, 395, 396, 397, 398
Optical waveguides, 42, 65, 115, 118, 181, 333, 354, 396
Optrodes, 320, 321
OSA, 33, 147, 177, 276, 328, 337
Out-of-plane displacements, 343, 349

Parabolic profile, 170
Partially coherent light, 19, 20
Partially reflecting anamorphic microlenses, 174, 175
Particle density, 319
Phase function, 6, 21, 24, 25, 26, 28, 29, 30, 32, 139, 140, 141, 158, 174, 257, 260, 262, 265, 269
Photoluminescent materials, 185
Photoresist mask, 240, 253
Piezoelectric actuation, 217
Pitch, 172, 173, 174, 175, 297, 299, 302, 312
Planar optical elements, 139
Planar optics, 202, 239
Planar structures, 246
Planar waveguide, 35, 41, 44, 45, 65, 66, 74, 105, 106, 110, 117, 250
Planetary rotation, 170, 172
Plasma, 105, 107, 113, 127, 132, 153, 162, 167, 172, 176, 253
  etching, 113, 127, 172, 176
PLD, 105, 161, 162, 163, 164, 167
PMMA, 114, 120, 122, 125, 134, 153, 157, 367, 369, 387
Pneumatic actuators, 411, 414
POC, 226, 228, 229, 232, 234
Point
  oriented coding, 226
  spread function, 29, 137, 138
Polychromatic sources, 259, 276
Polycrystalline materials, 346, 347
Polymethylmetacrylate, 114, 120, 122, 125, 134, 153, 157, 367, 369
Position dependent focal lengths, 172
Pressure measurements, 315
Primary imaging, 247
Propagation
  factor, 13, 243
  vectors, 241, 242
Proximity sensors, 315, 316, 317
PSF, 29
Pulsed laser deposition, 105, 161, 162, 163, 164, 167

QMB, 371, 376, 380
Quadratic gradient constant, 312
Quartz micro balance, 371, 380

Raman-Nath, 240
Rayleigh-Sommerfeld formula, 4
Ray-tracing, 27, 28, 30, 36, 141, 142, 262, 263
Rectangular grating profile, 245
Refractive index modulation, 245
Refractive lenses, 24, 32
Refractive microlenses, 387, 389
Replication, 3, 125, 133, 137, 138, 139, 145, 147, 156, 176, 196, 211, 389
RIBE, 113
RIE, 113, 123, 124, 127, 128, 130, 135, 158, 159
Ronchi grating, 14

Sagittal plane, 248
SAW, 328, 329, 330, 331, 333, 371, 372, 373, 374, 380
Scalar diffraction theory, 24, 141, 147, 252, 269
Scanners, 27, 198
Second harmonic generation, 66, 354, 357, 358, 360, 361, 362, 398
Secondary imaging, 247
Second-order nonlinear coefficients, 357
Second-order nonlinear waveguides, 357, 361
Self organizing map, 379
Sensor concept, 407, 408
Sensors, 194, 195, 199, 204, 211, 219, 293, 300, 302, 303, 305, 308, 313, 317, 324, 327, 330, 336, 352, 359, 361, 379, 381, 385, 399, 408, 412, 414
Shading masks, 170
Shallow profiles, 240
Shape memory alloy, 217, 411, 416
SHG, 66, 67, 70, 71, 72, 74
Simulated annealing, 10, 29, 30, 33, 260
Sinusoidal modulation, 240, 383
Slit mask arrays, 170, 176

SMA, 217, 411, 414, 415
Smoke detection, 319
Solitary wave interactions, 65
SOM, 379, 380
Spatial light modulators, 205, 223, 237
Spherical aberration, 32, 219, 284
Stop-position of a fast spindle, 315
Strain sensors, 303, 305, 308, 338
Stray light, 139, 144, 147
Stress relaxation, 307, 347
Substrate, 30, 62, 104, 106, 109, 113, 120, 134, 137, 142, 150, 153, 157, 162, 167, 170, 172, 176, 181, 183, 193, 197, 210, 217, 219, 239, 243, 245, 252, 254, 273, 355, 369, 374, 395, 408
Subwavelength structures, 17, 129
Superstrate, 243, 246, 247
Surface
    corrugation, 245
    relief, 17, 47, 48, 55, 119, 122, 126, 129, 130, 133, 139, 140, 141, 142, 145, 149, 250, 252
    roughness measurements, 344
    tension, 212, 218, 220

Talbot
    distance, 11, 13
    effect, 11, 13, 14, 21
TAS, 371, 372, 373, 380, 387, 388
Temperature
    measurements, 303
    sensors, 300, 356
Test functions, 51, 53
Tetrafluoropropyl methacrylate, 367
Texturing, 163, 251
TFPMA, 367
Thin-film
    microlens arrays, 171
    microstructures, 169
Three layer polymer waveguide, 369
Top-hat profile, 251
Total analysis systems, 371, 372, 373, 380, 387, 388
Total reflection, 44, 395
Tweezers, 391, 393, 395, 397
Twin image, 227, 229, 231, 232, 233

UV, 99, 115, 122, 123, 126, 127, 134, 150, 162, 166, 176, 179, 193, 272, 273, 294, 297, 309, 327, 395
    casting, 145

Vacuum deposition, 170, 172, 176
VCSEL, 61, 190, 191, 193, 194, 203, 407, 409
Vibration-and-temperature monitoring, 317
VLSI technique, 195, 342
Walk-off, 67, 68, 69, 70, 72, 74
Wave sensor array, 371
Waveguide, 37, 52, 55, 72, 74, 104, 106, 114, 117, 201, 203, 207, 218, 224, 246, 250, 333, 353, 383
    characterisation, 357, 358
    design, 357
    modelling, 361
Wavelenght division multiplexing, 203, 293, 308
WDM, 203, 293, 308
Wet etching, 113, 125, 126, 127, 153, 159, 216

XeCl laser, 166

Zero-order
    color, 259
    correction, 259
ZOC, 266, 268
ZOCC, 269, 276
Zone plates, 225, 250, 259, 268, 269, 270, 276
ZP fabrication, 152, 153